Medicinal Chemistry

Second Edition

Medicinal Chemistry

Second Edition

Gareth Thomas

University of Portsmouth

John Wiley & Sons, Ltd

Copyright © 2007   John Wiley & Sons Ltd, The Atrium, Southern Gate, Chichester,
West Sussex PO19 8SQ, England

Telephone (+ 44) 1243 779777

Email (for orders and customer service enquiries): cs-books@wiley.co.uk
Visit our Home Page on www.wileyeurope.com or www.wiley.com

All Rights Reserved. No part of this publication may be reproduced, stored in a retrieval system or transmitted in any form
or by any means, electronic, mechanical, photocopying, recording, scanning or otherwise, except under the terms of the
Copyright, Designs and Patents Act 1988 or under the terms of a licence issued by the Copyright Licensing Agency Ltd, 90
Tottenham Court Road, London W1T 4LP, UK, without the permission in writing of the Publisher. Requests to the
Publisher should be addressed to the Permissions Department, John Wiley & Sons Ltd, The Atrium, Southern Gate,
Chichester, West Sussex PO19 8SQ, England, or emailed to permreq@wiley.co.uk, or faxed to (+ 44) 1243 770620.

Designations used by companies to distinguish their products are often claimed as trademarks. All brand names and
product names used in this book are trade names, service marks, trademarks or registered trademarks of their respective
owners. The Publisher is not associated with any product or vendor mentioned in this book.

This publication is designed to provide accurate and authoritative information in regard to the subject matter covered.
It is sold on the understanding that the Publisher is not engaged in rendering professional services. If professional advice or
other expert assistance is required, the services of a competent professional should be sought.

Other Wiley Editorial Offices

John Wiley & Sons Inc., 111 River Street, Hoboken, NJ 07030, USA

Jossey-Bass, 989 Market Street, San Francisco, CA 94103-1741, USA

Wiley-VCH Verlag GmbH, Boschstr. 12, D-69469 Weinheim, Germany

John Wiley & Sons Australia Ltd, 33 Park Road, Milton, Queensland 4064, Australia

John Wiley & Sons (Asia) Pte Ltd, 2 Clementi Loop #02-01, Jin Xing Distripark, Singapore 129809

John Wiley & Sons Canada Ltd, 6045 Freemont Blvd, Mississauga, Ontario, L5R 4J3

Wiley also publishes its books in a variety of electronic formats. Some content that appears in print may not be
available in electronic books.

Anniversary Logo Design: Richard J. Pacifico

Library of Congress Cataloging-in-Publication Data
Thomas, Gareth, Dr.
  Medicinal chemistry : an introduction / Gareth Thomas. – 2nd ed.
       p. ; cm.
  Includes bibliographical references and index.
  ISBN 978-0-470-02597-0 (cloth : alk. paper) – ISBN 978-0-470-02598-7 (pbk. : alk. paper)
1. Pharmaceutical chemistry. I. Title.
  [DNLM: 1. Chemistry, Pharmaceutical. 2. Drug Design. 3. Drug Evaluation. 4. Pharmacokinetics. QV 744
T4567m 2007]
  RS403.T447 2007
  615'.19–dc22

                                                                                            2007026412

British Library Cataloguing in Publication Data

A catalogue record for this book is available from the British Library

ISBN 978-0-470-02597-0 (HB)
      978-0-470-02598-7 (PB)

Typeset in 10.5/13pt Times Roman by Thomson Digital
Printed and bound in Great Britain by Antony Rowe Ltd., Chippenham., Wiltshire
This book is printed on acid-free paper responsibly manufactured from sustainable forestry
in which at least two trees are planted for each one used for paper production.

Contents

Preface to the First Edition

This book is written for second, and subsequent, year undergraduates studying for degrees in medicinal chemistry, pharmaceutical chemistry, pharmacy, pharmacology and other related degrees. It assumes that the reader has a knowledge of chemistry at level one of a university life sciences degree. The text discusses the chemical principles used for drug discovery and design with relevant physiology and biology introduced as required. Readers do not need any previous knowledge of biological subjects.

Chapter 1 is intended to give an overview of the subject and also includes some topics of peripheral interest to medicinal chemists that are not discussed further in the text. Chapter 2 discusses the approaches used to discover and design drugs. The remaining chapters cover the major areas that have a direct bearing on the discovery and design of drugs. These chapters are arranged, as far as is possible, in a logical succession.

The approach to medicinal chemistry is kept as simple as possible. Each chapter has a summary of its contents in which the key words are printed in bold type. The text is also supported by a set of questions at the end of each chapter. Answers, sometimes in the form of references to sections of the book, are listed separately. A list of recommended further reading, classified according to subject, is also included.

Gareth Thomas

Preface to the Second Edition

This book is written for second and subsequent year undergraduates studying for degrees in medicinal chemistry, pharmaceutical chemistry, pharmacy, pharmacology and other related degrees. It assumes that the reader has a knowledge of chemistry at Level 1 of a university life science degree. The text discusses the chemical principles used for drug discovery and design with relevant physiology and biology introduced as required. Readers do not need any previous knowledge of biological subjects.

The second edition of *Medicinal Chemistry, an Introduction* has a new layout that I hope presents the subject in a more logical form. The main changes are that Chapter 2 has been rewritten as three separate chapters, namely, structure–activity and quantitative structure relationships, computer-aided drug design and combinatorial chemistry. Two new chapters entitled Drugs from Natural Sources and Drug Development and Production have been added. The text has been simplified and extended where appropriate with a number of case histories, new examples and topics. Among the new topics are a discussion of monoclonal antibodies and photodynamic drugs. The inclusion of the new chapters and new material has necessitated a reduction in the biological and chemical introductions to some topics and the omission of some material included in the first edition. Furthermore, the reader should be aware that there are many more drugs and targets than those discussed in this text.

Chapter 1 introduces and gives an overview of medicinal chemistry. This is followed by chapters that discuss the principal methods used in drug design and the isolation of drugs from natural sources. Chapters 7–14 are concerned with a discussion of more specialised aspects of medicinal chemistry. The final two chapters outline drug and analogue synthesis, development and production. Appropriate chapters have an outline introduction to the relevant biology. Each chapter is supported by a set of questions. Answers to these questions, sometimes in the form of references to sections and figures in the book, are listed separately. An updated list of further reading, classified according to subject, is also included.

Gareth Thomas

Acknowledgements

I wish to thank all my colleagues and students, past and present, whose help enabled this second edition of my book to be written. In particular I would like to rethank all those who helped me with the first edition. I would like particularly to thank the following for their help with the second edition: Dr L. Banting; Dr J. Brown for once again acting as my living pharmacology dictionary; Dr P. Cox for his advice on molecular modelling; Dr J. Gray for proofreading the sections on monoclonal antibodies; Dr P. Howard for bringing me up to date with advances in combinatorial chemistry and allowing me to use his lecture notes; Dr Tim Mason, Mr A. Barrow and Dr D. Brimage; Dr A. Sautreau for proofreading and correcting Chapter 6; Robin Usher and his colleagues at Mobile Library Link One for their help in obtaining research papers; Dr. G. White; and Professor D. Thurston for his support. My thanks are also due to Dr J. Fetzer of Tecan Deutschland GmbH, Crailsheim, Germany for the pictures of the equipment used in high-throughput screening. I also wish to acknowledge that the main source of the historical information given in the text is *Drug Discovery, a History*, by W. Sneader, published by John Wiley and Sons Ltd.

I would like to offer a very special thanks to the dedicated NHS medical teams who have treated my myeloma over the past years. Without their excellent care I would not have been here to have written this book. I would particularly like to thank Dr R. Corser, Dr T. Cranfield and the other doctors of the Haematology Department at the Queen Alexandra Hospital, Portsmouth, the nurses and ancillary staff of Ward D16, Queen Alexandra Hospital, Portsmouth, Dr K. Orchard, Dr C. Ottensmier and their respective staff at Southampton General Hospital and the nurses and ancillary staff of Wards C3 and C6 at Southampton General Hospital.

Finally, I would like to thank my wife for the cover design for the first Edition and the sketches included in this text. Her support through the years has been an essential contribution to my completing the text.

Abbreviations

A	Adenine
Abe	Abequose
AC	Adenylate cyclase
ACE	Angiotensin-converting enzymes
ACh	Acetyl choline
ADAPT	Antibody-directed abzyme prodrug therapy
ADEPT	Antibody-directed enzyme prodrug therapy
ADME	Absorption, distribution, metabolism and elimination
ADR	Adverse drug reaction
AIDS	Acquired immuno deficiency syndrome
Ala	Alanine
AMP	Adenosine monophosphate
Arg	Arginine
Asp	Aspartate
ATP	Adenosine triphosphate
AUC	Area under the curve
AZT	Zidovudine
BAL	British anti-Lewisite
BESOD	Bovine erythrocyte superoxide dismutase
C	Cytosine
CaM	Calmodulin
cAMP	Cyclic adenosine monophosphate
Cbz	N-(Benzyloxycarbonyloxy)succinamide
Cl	Clearance
CNS	Central nervous system
CoA	Coenzyme A
CoMFA	Comparative molecular field analysis
CYP-450	Cytochrome P-450 family
Cys	Cysteine
C_x	Concentration of x
dATP	Deoxyadenosine triphosphate
d.e.	Diastereoisomeric excess
DHF	Dihydrofolic acid
DHFR	Dihydrofolate reductase
DMPK	Drug metabolism and pharmacokinetics
DNA	Deoxyribonucleic acid
dTMP	Deoxythymidylate-5'-monophosphate
dUMP	Deoxyuridylate-5'-monophosphate
EC	Enzyme Commission
EDRF	Endothelium-derived relaxing factor

EDTA	Ethylenediaminotetraacetic acid
e.e.	Enantiomeric excess
ELF	Effluent load factor
EMEA	European Medicines Evaluation Agency
EPC	European Patent Convention
EPO	European Patent Office
E_s	Taft steric parameter
F	Bioavailability
FAD	Flavin adenine dinucleotide
FDA	Food and Drug Administration (USA)
FdUMP	5-Fluoro-2′-deoxyuridyline monophosphate
FGI	Functional group interconversion
FH_4	Tetrahydrofolate
FMO	Flavin monooxygenases
Fmoc	9-Fluorenylmethoxychloroformyl group
FUdRP	5-Fluoro-2′-deoxyuridylic acid
G	Guanine
GABA	γ-Aminobutyric acid
GC	Guanylyl cyclase
GDEPT	Gene-directed enzyme prodrug therapy
GDP	Guanosine diphosphate
GI	Gastrointestinal
Gln	Glutamine
Glu	Glutamatic acid
Gly	Glycine
5′-GMP	Guanosine 5′-monophosphate
GSH	Glutathione
GTP	Guanosine triphosphate
HAMA	Human anti-mouse antibodies
Hb	Haemoglobin
HbS	Sickle cell haemoglobin
His	Histidine
HIV	Human immunodeficiency disease
hnRNA	Heterogeneous nuclear RNA
HTS	High-throughput screening
IDDM	Insulin-dependent diabetes mellitus
Ig	Immunoglobins
Ile	Isoleucine
IP_3	Inositol-1,4,5-triphosphate
IV	Intravenous
IM	Intramuscular
KDO	2-Keto-3-deoxyoctanoate
k_x	Reaction rate constant for reaction x
LDA	Lithium diisopropylamide

LDH	Lactose dehydrogenase
Leu	Leucine
Lys	Lysine
MA(A)	Marketing authorisation (application)
Mab	Monoclonal antibody
mACh	Muscarinic cholinergic receptor
MAO	Monoamine oxidase
MCA	Medicines Control Agency
MESNA	2-Mercaptoethanesulphonate
Met	Methionine
MO	Molecular orbital
Moz	4-Methoxybenzyloxychloroformyl group
MR	Molar refractivity
mRNA	Messenger RNA
nACh	Nicotinic cholinergic receptor
NAD$^+$	Nicotinamide adenine dinucleotide (oxidised form)
NADH	Nicotinamide adenine dinucleotide (reduced form)
NADP$^+$	Nicotinamide dinucleotide phosphate (oxidised form)
NADPH	Nicotinamide dinucleotide phosphate (reduced form)
NAG	β-N-Acetylglucosamine
NAM	β-N-Acetylmuramic acid
NCI	National Cancer Institute (USA)
NOS	Nitric oxide synthase
P-450	Cytochrome P-450 oxidase
PABA	p-Aminobenzoic acid
PCT	Patent Cooperation Treaty
PDT	Photodynamic therapy
PEG	Polyethyene glycol
PG	Prostaglandin
Phe	Phenylalanine
PO	Per os (by mouth)
pre-mRNA	Premessenger RNA
Pro	Proline
ptRNA	Primary transcript RNA
QSAR	Quantitative structure–activity relationship
Q_x	Rate of blood flow for x
RMM	Relative molecular mass
RNA	Ribonucleic acid
S	Svedberg units
SAM	S-Adenosylmethionine
SAR	Structure–activity relationship
Ser	Serine
SIN-1	3-Morpholino-sydnomine
T	Thymine

TdRP	Deoxythymidylic acid
THF	Tetrahydrofolic acid
Thr	Threonine
tRNA	Transfer RNA
Tyr	Tyrosine
U	Uracil
UDP	Uridine diphosphate
UDPGA	Uridine diphosphate glucuronic acid
UdRP	Deoxyuridylic acid
Val	Valine
V_d	Volume of distribution
WHO	World Health Organization

1

An introduction to drugs, their action and discovery

1.1 Introduction

The primary objective of medicinal chemistry is the design and discovery of new compounds that are suitable for use as drugs. This process involves a *team of workers* from a wide range of disciplines such as chemistry, biology, biochemistry, pharmacology, mathematics, medicine and computing, amongst others.

The discovery or design of a new drug not only requires a discovery or design process but also the synthesis of the drug, a method of administration, the development of tests and procedures to establish how it operates in the body and a safety assessment. Drug discovery may also require fundamental research into the biological and chemical nature of the diseased state. These and other aspects of drug design and discovery require input from specialists in many other fields and so medicinal chemists need to have an outline knowledge of the relevant aspects of these fields.

1.2 What are drugs and why do we need new ones?

Drugs are strictly defined as chemical substances that are used to prevent or cure diseases in humans, animals and plants. The *activity* of a drug is its pharmaceutical effect on the subject, for example, analgesic or β-blocker, whereas its *potency* is the quantitative nature of that effect. Unfortunately the term drug is also used by the media and the general public to describe the substances taken for their psychotic rather than medicinal effects. However, this does not mean that these substances cannot be used as drugs. Heroin, for example, is a very effective painkiller and is used as such in the form of diamorphine in terminal cancer cases.

Medicinal Chemistry, Second Edition Gareth Thomas
© 2007 John Wiley & Sons, Ltd

Heroin

Drugs act by interfering with biological processes, so no drug is completely safe. *All* drugs, including those non-prescription drugs such as aspirin and paracetamol (Fig. 1.1) that are commonly available over the counter, act as poisons if taken in excess. For example, overdoses of paracetamol can causes coma and death. Furthermore, in addition to their beneficial effects most drugs have non-beneficial biological effects. Aspirin, which is commonly used to alleviate headaches, can also cause gastric irritation and occult bleeding in some people The non-beneficial effects of some drugs, such as cocaine and heroin, are so undesirable that the use of these drugs has to be strictly controlled by legislation. These unwanted effects are commonly referred to as *side effects*. However, side effects are not always non-beneficial; the term also includes biological effects that are beneficial to the patient. For example, the antihistamine promethazine is licenced for the treatment of hayfever but also induces drowsiness, which may aid sleep.

Aspirin

Paracetamol

Figure 1.1   Aspirin and paracetamol

Drug resistance or tolerance (*tachyphylaxis*) occurs when a drug is no longer effective in controlling a medical condition. It arises in people for a variety of reasons. For example, the effectiveness of barbiturates often decreases with repeated use because the body develops mixed function oxidases in the liver that metabolise the drug, which reduces its effectiveness. The development of an enzyme that metabolises the drug is a relatively common reason for drug resistance. Another general reason for drug resistance is the *downregulation* of receptors (see section 8.6.1). Downregulation occurs when repeated stimulation of areceptor results in the receptor being broken down. This results in the drug being less effective because there are fewer receptors available for it to act on. However, downregulating has been utilised therapeutically in a number of cases. The continuous use of gonadotrophin releasing factor, for example, causes gonadotrophin receptors that control the menstrual cycle to be downregulated. This is why gonadotrophin-like drugs are used as contraceptives. Drug resistance may also be due to the appearance of a significantly high proportion of drug-resistant strains of microorganisms. These strains arise naturally and can rapidly multiply and become the currently predominant strain of that microorganism. Antimalarial drugs are proving less effective because of an increase in the proportion of drug-resistant strains of the malaria parasite.

New drugs are constantly required to combat drug resistance even though it can be minimised by the correct use of medicines by patients. They are also required for improving the treatment of existing diseases, the treatment of newly identified diseases and the production of safer drugs by the reduction or removal of adverse side effects.

1.3   Drug discovery and design: a historical outline

Since ancient times the peoples of the world have had a wide range of natural products that they use for medicinal purposes. These products, obtained from animal, vegetable and mineral sources, were sometimes very effective. However, many of the products were very toxic and it is interesting to note that the Greeks used the same word *pharmakon* for both poisons and medicinal products. Information about these ancient remedies was not readily available to users until the invention of the printing press in the fifteenth century. This led to the widespread publication and circulation of Herbals and Pharmacopoeias, which resulted in a rapid increase in the use, and misuse, of herbal and other remedies. Misuse of tartar emetic (antimony potassium tartrate) was the reason for its use being banned by the Paris parliament in 1566, probably the first recorded ban of its type. The usage of such remedies reached its height in the seventeenth century. However, improved communications between practitioners in the eighteenth and nineteenth centuries resulted in the progressive removal of preparations that were either ineffective or too toxic from Herbals and Pharmacopoeias. It also led to a more rational development of new drugs.

The early nineteenth century saw the extraction of pure substances from plant material. These substances were of consistent quality but only a few of the compounds isolated proved to be satisfactory as therapeutic agents. The majority were found to be too toxic although many, such as morphine and cocaine for example, were extensively prescribed by physicians.

The search to find less toxic medicines than those based on natural sources resulted in the introduction of synthetic substances as drugs in the late nineteenth century and their widespread use in the twentieth century. This development was based on the structures of known pharmacologically active compounds, now referred to as *leads*. The approach adopted by most nineteenth century workers was to synthesise structures related to that of the lead and test these compounds for the required activity. These lead-related compounds are now referred to as *analogues*.

The first rational development of synthetic drugs was carried out by Paul Ehrlich and Sacachiro Hata who produced arsphenamine in 1910 by combining synthesis with reliable biological screening and evaluation procedures. Ehrlich, at the beginning of the nineteenth century, had recognised that both the beneficial and toxic properties of a drug were important to its evaluation. He realised that the more effective drugs showed a greater selectivity for the target microorganism than its host. Consequently, to compare the effectiveness of different compounds, he expressed a drug's selectivity and hence its effectiveness in terms of its chemotherapeutic index, which he defined as:

$$\text{Chemotherapeutic index} = \frac{\text{Minimum curative dose}}{\text{Maximum tolerated dose}} \qquad (1.1)$$

At the start of the nineteenth century Ehrlich was looking for a safer antiprotozoal agent with which to treat syphilis than the then currently used atoxyl. He and Hata tested and catalogued in terms of his therapeutic index over 600 structurally related arsenic compounds. This led to their discovery in 1909 that arsphenamine (Salvarsan) could cure mice infected with syphilis. This drug was found to be effective in humans but had to be used with extreme care as it was very toxic. However, it was used up to the mid- 1940s when it was replaced by penicillin.

Atoxyl   Arsphenamine (Salvarsan)

Ehrlich's method of approach is still one of the basic techniques used to design and evaluate new drugs in medicinal chemistry. However, his chemotherapeutic index has been updated to take into account the variability of individuals and is now defined as its reciprocal, the therapeutic index or ratio:

$$\text{Therapeutic index} = \frac{\text{LD}_{50}}{\text{ED}_{50}} \qquad (1.2)$$

where LD_{50} is the lethal dose required to kill 50 per cent of the test animals and ED_{50} is the dose producing an effective therapeutic response in 50 per cent of the test animals. In theory, the larger a drug's therapeutic index, the greater is its margin of safety. However, because of the nature of the data used in their derivation, therapeutic index values can only be used as a limited guide to the relative usefulness of different compounds.

The term *structure–activity relationship* (*SAR*) is now used to describe Ehrlich's approach to drug discovery, which consisted of synthesising and testing a series of structurally related compounds (see Chapter 3). Although attempts to quantitatively relate chemical structure to biological action were first initiated in the nineteenth century, it was not until the 1960s that Hansch and Fujita devised a method that successfully incorporated quantitative measurements into structure–activity relationship determinations (see section 3.4.4). The technique is referred to as *QSAR* (*quantitative structure–activity relationship*).

QSAR methods have subsequently been expanded by a number of other workers. One of the most successful uses of QSAR has been in the development in the 1970s of the antiulcer agents cimetidine and ranitidine. Both SAR and QSAR are important parts of the foundations of medicinal chemistry.

Cimetidine   Ranitidine

At the same time as Ehrlich was investigating the use of arsenical drugs to treat syphilis, John Langley formulated his theory of *receptive substances*. In 1905 Langley proposed that so-called receptive substances in the body could accept either a stimulating compound, which would cause a biological response, or a non-stimulating compound, which would prevent a biological response. These ideas have been developed by subsequent workers and the theory of *receptors* has become one of the fundamental concepts of medicinal chemistry. *Receptor sites* (see Chapter 8) usually take the form of pockets, grooves or other cavities in the surface of certain proteins and glycoproteins in the living organism. They should not be confused with active sites (see section 9.3), which are the regions of enzymes where metabolic chemical reactions occur. It is now accepted that the binding of a chemical agent, referred to as a *ligand* (see section 8.1), to a receptor sets in motion a series of biochemical events that result in a biological or physiological effect. Furthermore, a drug is most effective when its structure or a significant part of its structure, both as regards molecular shape and electron distribution (*stereoelectronic structure*), is complementary with the stereoelectronic structure of the receptor responsible for the desired biological action. Since most drugs are able to assume a number of different conformations, the conformation adopted when the drug binds to the receptor is known as its *active conformation*.

The section of the structure of a ligand that binds to a receptor is known as its *pharmacophore*. The sections of the structure of a ligand that comprise a pharmacophore may or may not be some distance apart in that structure. They do not have to be adjacent to one another. For example, the quaternary nitrogens that are believed to form the pharmacophore of the neuromuscular blocking agent tubocrarine are separated in the molecule by a distance of 115.3 nm.

Tubocrarine

The concept of receptors also gives a reason for side effects and a rational approach to ways of eliminating their worst effects. It is now believed that side effects can arise when the drug binds to either the receptor responsible for the desired biological response or to different receptors.

The mid- to late twentieth century has seen an explosion of our understanding of the chemistry of disease states, biological structures and processes. This increase in knowledge has given medicinal chemists a clearer picture of how drugs are distributed through the body, transported across membranes, their mode of operation and metabolism. This knowledge has enabled medicinal chemists to place groups that influence its absorption, stability in a bio-system, distribution, metabolism and excretion into the molecular structure of a drug. For example, the *in situ* stability of a drug and hence its potency could be increased by rationally modifying the molecular structure of the drug. Esters and N-substituted amides, for example, have structures with similar shapes and electron distributions (Fig. 1.2a) but N-substituted amides hydrolyse more slowly than esters. Consequently, the replacement of an ester group by an N-substituted amide group *may* increase the stability of the drug without changing the nature of its activity. This *could possibly* lead to an increase in either the potency or time of duration of activity of a drug by improving its chances of reaching its site of action. However, changing a group or introducing a group may change the nature of the activity of the compound. For example, the change of the ester group in procaine to an amide (procainamide) changes the activity from a local anaesthetic to an antiarrhythmic (Fig. 1.2b).

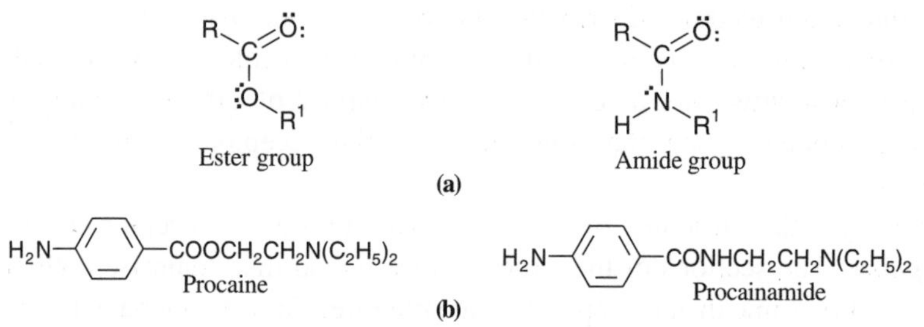

Figure 1.2 **(a)** The similar shapes and outline electronic structures (stereoelectronic structures) of amide and ester groups. **(b)** Procaine and procainamide

Drugs normally have to cross non-polar lipid membrane barriers (see sections 7.2 and 7.3) in order to reach their site of action. As the polar nature of the drug increases it usually becomes more difficult for the compound to cross these barriers. In many cases drugs whose structures contain charged groups will not readily pass through membranes. Consequently, charged structures can be used to restrict the distribution of a drug. For example, quaternary ammonium salts, which are permanently charged, can be used as an alternative to an amine in a structure in order to restrict the passage of a drug across a membrane. The structure of the anticholinesterase neostigmine, developed from physostigmine, contains a quaternary ammonium group that gives the molecule a permanent charge. This stops the molecule from crossing the blood–brain barrier, which prevents unwanted CNS activity. However, its analogue miotine can form the free base. As a result, it is able to cross lipid membranes and causes unwanted CNS side effects.

Serendipity has always played a large part in the discovery of drugs. For example, the development of penicillin by Florey and Chain was only possible because Alexander Fleming noted the inhibition of *staphylococcus* by *Penicillium notatum*. In spite of our increased knowledge base, it is still necessary to pick the correct starting point for an investigation if a successful outcome is to be achieved and luck still plays a part in selecting that point. This state of affairs will not change and undoubtedly luck will also lead to new discoveries in the future. However, modern techniques such as *computerised molecular modelling* (see Chapter 4) and *combinatorial chemistry* (see Chapter 5) introduced in the 1970s and 1990s, respectively, are likely to reduce the number of intuitive discoveries.

Two of the factors necessary for drug action are that the drug fits and binds to the target. *Molecular modelling* allows the researcher to predict the three-dimensional shapes of molecules and target. It enables workers to check whether the shape of a potential lead is complementary to the shape of its target. It also allows one to calculate the *binding energy* liberated when a molecule binds to its target (see section 4.6). Molecular modelling has reduced the need to synthesise every analogue of a lead compound. It is also often used retrospectively to confirm the information derived from other sources. Combinatorial chemistry originated in the field of peptide chemistry but has now been expanded to cover other areas. It is a group of related techniques for the simultaneous production of large numbers of compounds, known as *libraries*, for biological testing. Consequently, it is used for structure–activity studies and to discover new lead compounds. The procedures may be automated.

1.3.1 The general stages in modern-day drug discovery and design

At the beginning of the nineteenth century drug discovery and design was largely carried out by individuals and was a matter of luck rather than structured investigation. Over the last century, a large increase in our general scientific knowledge means that today drug discovery requires considerable *teamwork*, the members of the team being specialists in various fields, such as medicine, biochemistry, chemistry, computerised molecular modelling, pharmaceutics, pharmacology, microbiology, toxicology, physiology and pathology. The approach is now more structured but a successful outcome still depends on a certain degree of luck.

The modern approach to drug discovery/design depends on the objectives of the project. These objectives can range from changing the pharmacokinetics of an existing drug to discovering a completely new compound. Once the objectives of the project have been decided the team will select an appropriate starting point and decide how they wish to proceed. For example, if the objective is to modify the pharmacokinetics of an existing drug the starting point is usually that the drug and design team has to decide what structural modifications need to be investigated in order to achieve the desired modifications. Alternatively, if the objective is to find a new drug for a specific disease the starting point may be a knowledge of the biochemistry of the disease and/or the microorganism responsible for that disease (Fig. 1.3). This may require basic research into the biochemistry of the disease

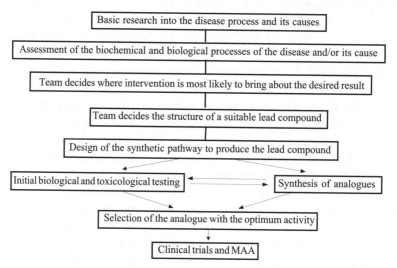

Figure 1.3   The general steps in the discovery of a new drug for a specific disease state

causing process before initiating the drug design investigation. The information obtained is used by the team to decide where intervention would be most likely to bring about the desired result. Once the point of intervention has been selected the team has to decide on the structure of a compound, referred to as a *lead compound*, that could possibly bring about the required change. A number of candidates are usually considered but the expense of producing drugs dictates that the team has to choose only one or two of these compounds to act as the lead compound. The final selection depends on the experience of the research team.

Lead compounds are obtained from a variety of sources that range from extracting compounds from natural sources (see Chapter 6), synthesis using combinatorial chemistry (see Chapter 5), searching data bases and compound collections (see section 1.5.6) for suitable candidates and ethnopharmacological sources (see section 1.5.1). However, whatever the objective and starting point, all investigations start with the selection of a suitable bioassay(s) (see section 6.2), which will indicate whether the compound is likely to be active against the diseased state and also if possible the potency of active compounds. These assays are often referred to as *screening* programmes. They may also be carried out at different stages in drug discovery in order to track active compounds. Once an active lead has been found, it is synthesised and its activity determined. SAR studies (see Chapter 3) are then carried out by synthesising and testing compounds, referred to as *analogues*, that are structurally related to the lead in order to find the structure with the optimum activity. These studies may make use of QSAR (see section 3.4) and computational chemistry (see Chapter 4) to help discover the nature of this optimum structure for activity. This analogue would, if economically viable, be developed and ultimately, if it met the MAA regulations, placed in clinical use (see Chapter 16).

1.4   Leads and analogues: some desirable properties

1.4.1   Bioavailability

The activity of a drug is related to its *bioavailability*, which is defined as the fraction of the dose of a drug that is found in general circulation (see section 11.5). Consequently, for a compound to be suitable as a lead it must be bioavailable. In order to assess a compound's bioavailability it must be either available off the shelf or be synthesised. Synthesis of a compound could result in the synthesis of an inactive compound, which could be expensive both in time and money. In order to avoid unnecessary work and expense in synthesising inactive molecules, Lipinski *et al.* proposed a set of four rules that would predict whether a molecule was likely to be orally bioavailable. These rules may be summarised as having:

- a molecular mass less than 500;

- a calculated value of log P less than 5;

- less than ten hydrogen bond acceptor groups (e.g. -O- and -N-, etc.);

- less than five hydrogen bond donor groups (e.g. NH and OH, etc.).

where P is the calculated partition coefficient for the octanol/water system (see section 2.12.1). Any compound that fails to comply with two or more of the rules is unlikely to be bioavailable, that is, it is unlikely to be active. Lipinski's rules are based on multiples of five and so are often referred to as the *rule of fives*. Other researchers have developed similar methods to assess the bioavailability of molecules prior to their synthesis. However, it should be realised that Lipinski's and other similar rules are only guidelines.

1.4.2 Solubility

Solubility is discussed in more detail in Chapter 2. However, a requirement for compounds that are potential drug candidates is that they are soluble to some extent in both lipids and water. Compounds that readily dissolve in lipid solvents are referred to as *lipophilic* or *hydrophobic* compounds. Their structures often contain large numbers of non-polar groups, such as benzene rings and ether and ester functional groups. Compounds that do not readily dissolve in lipids but readily dissolve in water are known as *hydrophilic* or *lipophobic* compounds. Their structures contain polar groups such as acid, amine and hydroxy functional groups. The balance between the polar and non-polar groups in a molecule defines its lipophilic character: compounds with a high degree of *lipophilic character* will have a good lipid solubility but a poor water solubility; conversely, compounds with a low degree of *lipophilic character* will tend to be poorly soluble in lipids but have a good solubility in water. It is desirable that leads and analogues have a balance between their water solubility and their lipophilicity. Most drugs are administered either as aqueous or solid preparations and so need to be water soluble in order to be transported through the body to its site of action. Consequently, poor water solubility can hinder or even prevent the development of a good lead or analogue. For example, one of the factors that hindered the development of the anticancer drug taxol was its poor water solubility. This made it difficult to obtain a formulation for administrationby intravenous infusion, the normal route for anticancer drugs (see section 6.8). However, careful design of the form in which the drug is administered (the *dosage form*) can in many instances overcome this lack of water solubility (see sections 2.13 and 6.8). Drugs also require a degree of lipid solubility in order to pass through membranes (see section 7.3.3). However, if it has too high a degree of lipophilicity it may become trapped in a membrane and so become ineffective. The lipophilicity of a compound is often represented by the partition coefficient of that compound in a defined solvent system (see section 2.12.1).

1.4.3 Structure

The nature of the structures of leads and analogues will determine their ability to bind to receptors and other target sites. Binding is the formation, either temporary or permanent, of chemical bonds between the drug or analogue with the receptor (see sections 2.2 and 8.2). Their nature will influence the operation of a receptor. For example, the binding of most drugs or analogues takes the form of an equilibrium (see section 8.6.1) in which the drug or analogue forms weak, electrostatic bonds, such as hydrogen bonds and van der Waals' forces, with the receptor. Ultimately the drug or analogue is removed from the vicinity of the receptor by natural processes and this causes the biological processes due to the receptor's activity to stop. For example, it is thought that the local anaesthetic benzocaine (see section 7.4.3) acts in this manner. However, some drugs and analogues act by forming strong covalent bonds with the receptor and either prevent it operating or increase its duration of operation. For example, melphalan, which is used to treat cancer, owes its action to the strong covalent bonds it forms with DNA (see section 10.13.4).

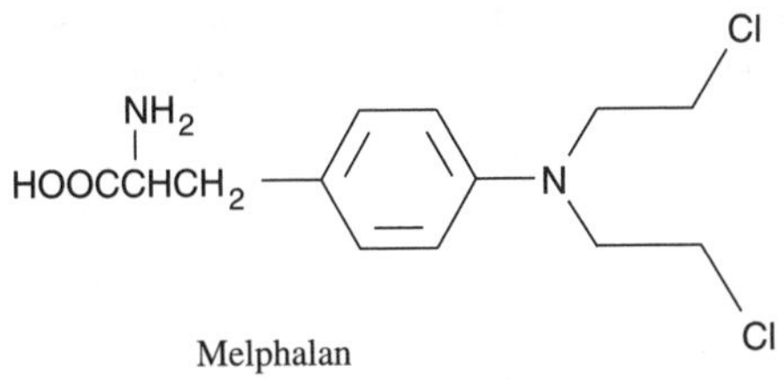

Melphalan

A major consideration in the selection of leads and analogues is their stereochemistry. It is now recognised that the biological activities of the individual enantiomers and their racemates may be very different (see section 2.3 and Table 1.1). Consequently, it is necessary to pharmacologically evaluate individual enantiomers as well as any racemates. However, it is often difficult to obtain specific enantiomers in a pure state (see section 15.3). Both of these considerations make the production of optically active compounds expensive and so medicinal chemists often prefer to synthesise lead compounds that are not optically active. However, this is not always possible and a number of strategies exist to produce compounds with specific stereochemical centres (see sections 6.5 and 15.3).

1.4.4 Stability

Drug stability can be broadly divided into two main areas: stability after administration and shelf-life.

Stability after administration

A drug will only be effective if, after administration, it is stable enough to reach its target site in sufficient concentration (see section 1.6) to bring about the desired effect. However, as soon as a drug is administered the body starts to remove it by metabolism (see section 1.7.1 and Chapter 12). Consequently, for a drug to be effective it must be stable long enough after administration for sufficient quantities of it to reach its target site. In other words, it must not be metabolised *too* quickly in the circulatory system. Three strategies are commonly used for improving a drug's *in situ* stability, namely:

* modifying its structure;

* administering the drug as a more stable prodrug (see section 12.9.4);

* using a suitable dosage form (see section 1.6).

The main method of increasing drug stability in the biological system is to prepare a more stable analogue with the same pharmacological activity. For example, pilocarpine,

Table 1.1 Variations in the biological activities of stereoisomers

First stereoisomer	Second stereoisomer	Example
Active	Activity of same type and potency	The R and S isomers of the antimalarial chloroquine have equal potencies 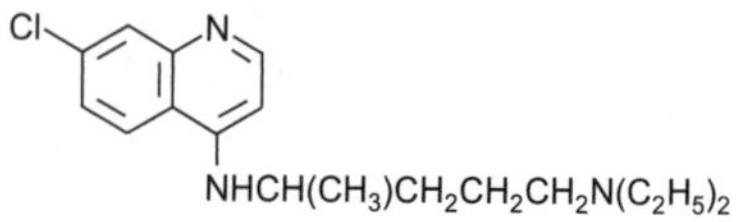
Active	Activity of same type	The E isomer of diethylstilbestrol, an oestrogen but weaker, is only 7% as active as the Z isomer 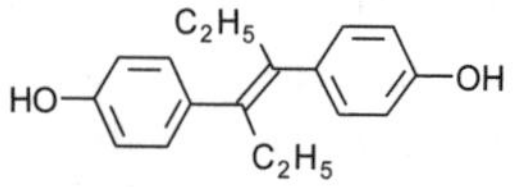
Active	Activity of a different type	S-Ketamine is an anaesthetic whereas R-Ketamine has little anaesthetic action but is a psychotic agent 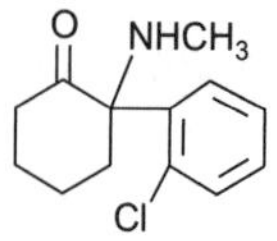
Active	No activity	S-α-Methyldopa is a hypertensive drug but the R isomer is inactive 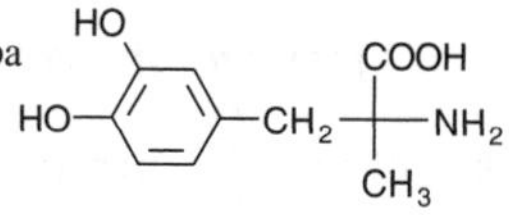
Active	Active but different side effects	Thalidomide: the S isomer is a sedative and has teratogenic side effects; the R isomer is also a sedative but has no teratogenic activity 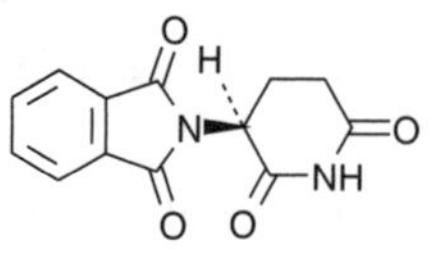

which is used to control glaucoma, rapidly loses it activity because the lactone ring readily opens under physiological conditions. Consequently, the lowering of intraocular pressure by pilocarpine lasts for about three hours, necessitating administration of 3–6 doses a day. However, the replacement of C-2 of pilocarpine by a nitrogen yields an isosteric carbamate that has the same potency as pilocarpine but is more stable. Although this analogue was discovered in 1989 it has not been accepted for clinical use.

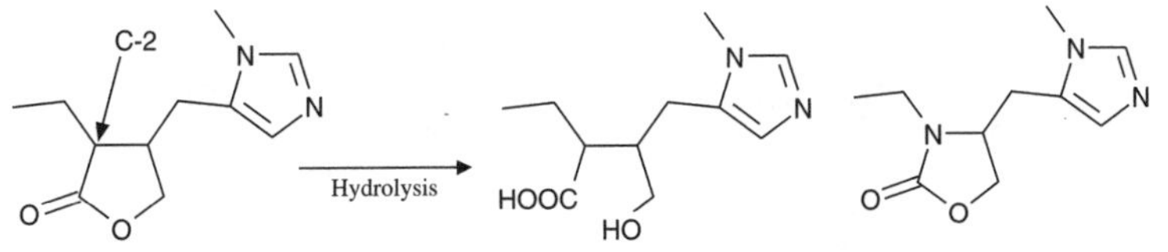

Pilocarpine   Carbamate analogue

The *in situ* stability of a drug may also be improved by forming a complex with a suitable reagent. For example, complexing with hydroxypropyl-β-cyclodextrin is used to improve both the stability and solubility of thalidomide, which is used to inhibit rejection of bone marrow transplants in the treatment of leukaemia. The half-life of a dilute solution of the drug is increased from 2.1 to 4.1 hours while its aqueous solubility increases from 50 to 1700 μg ml.$^{-1}$

Cyclodextrins are bottomless flower-pot-shaped cylindrical oligosaccharides consisting of about 6–8 glucose units. The exterior of the 'flower-pot' is hydrophilic in character whilst the interior has a hydrophobic nature. Cyclodextrins are able to form inclusion complexes in which part of the guest molecule is held within the flower-pot structure (Fig. 1.4). The hydrophobic nature of the interior of the cyclodextrin structure probably means that hydrophobic interaction plays a large part in the formation and stability of the complex. Furthermore, it has been found that the stability of a drug *in situ* is often improved when the active site of a drug lies within the cylinder and decreased when it lies outside the cylinder. In addition, it has been noted that the formation of these complexes may improve the water solubility, bioavailability and pharmacological action and reduce the side effects of some drugs. However, a high concentration of cyclodextrins in the blood stream can cause nephrotoxicity.

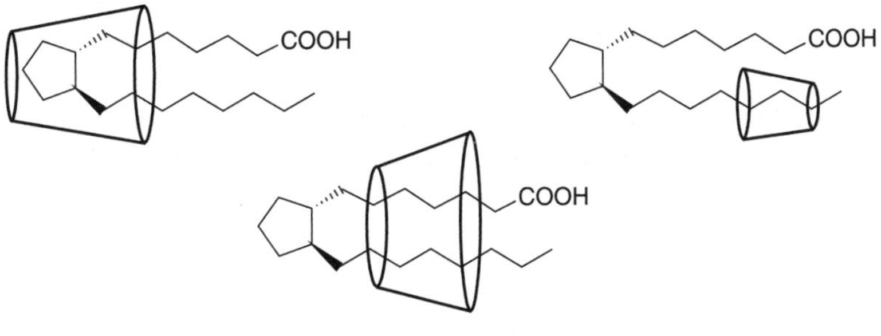

Figure 1.4 Schematic representations of the types of inclusion complexes formed by cyclodextrins and prostaglandins. The type of complex formed is dependent on the cavity size

Prodrug formation can also be used to improve drug stability. For example, cyclophosphamide, which is used to treat a number of carcinomas and lymphomas, is metabolised in the liver to the corresponding phosphoramidate mustard, the active form of the drug.

Cyclophosphamide

Phosphoramidate mustard

The highly acidic gastric fluids can cause extensive hydrolysis of a drug in the gastrointestinal tract (GI tract). This will result in poor bioavailability. However, drug stability in the gastrointestinal tract can be improved by the use of enteric coatings, which dissolve only when the drug reaches the small intestine.

In many cases, but not all (see sections 2.2 and 8.2), once a drug has carried out its function it needs to be removed from the body. This occurs by metabolism and excretion and so a potential drug should not be too stable that it is not metabolised. Furthermore, the drug should not accumulate in the body but be excreted. These aspects of drug stability should be investigated in the preclinical and clinical investigations prior to the drug's release onto the market.

Shelf-life

Shelf-life is the time taken for a drug's pharmacological activity to decline to an unacceptable level. This level depends on the individual drug and so there is no universal specification. However, 10 per cent decomposition is often taken as an acceptable limit provided that the decomposition products are not toxic.

Shelf-life deterioration occurs through microbial degradation and adverse chemical interactions and reactions. Microbial deterioration can be avoided by preparing the dosage form in the appropriate manner and storage under sterile conditions. It can also be reduced by the use of antimicrobial excipients. Adverse chemical interactions between the components of a dosage form can also be avoided by the use of suitable excipients. Decomposition by chemical reaction is usually brought about by heat, light, atmospheric oxidation, hydrolysis by atmospheric moisture and racemisation. These may be minimised by correct storage with the use of refrigerators, light-proof containers, air-tight lids and the appropriate excipients.

1.5   Sources of leads and drugs

Originally drugs and leads were derived from natural sources. These natural sources are still important sources of lead compounds and new drugs, however the majority of lead compounds are now discovered in the laboratory using a variety of sources, such as local folk remedies (ethnopharmacology), investigations into the biochemistry of the pathology of disease states and high-throughput screening of compound collections (see Chapter 5), databases and other literature sources of organic compounds.

1.5.1 Ethnopharmaceutical sources

The screening of local folk remedies (*ethnopharmacology*) has been a fruitful source of lead compounds and many important therapeutic agents. For example, the antimalarial quinine from cinchona bark, the cardiac stimulants from foxgloves (Fig. 1.5) and the antidepressant reserpine isolated from *Rauwolfia serpentina.*

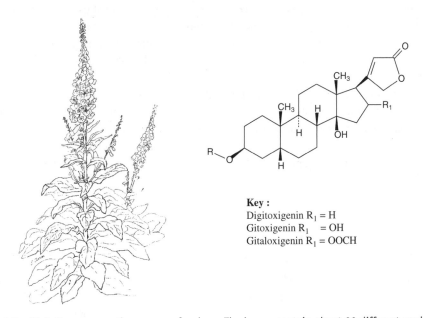

Key :
Digitoxigenin R_1 = H
Gitoxigenin R_1 = OH
Gitaloxigenin R_1 = OOCH

Figure 1.5 *Digitalis purpurea*, the common foxglove. The leaves contain about 30 different cardioactive compounds. The major components of this mixture are glycosides, with aglycones of digitoxigenin, gitoxigenin and gitaloxigenin. Two series of compounds are known, those where R, the carbohydrate residue (glycone) of the glycoside, is either a tetrasaccharide or a trisaccharide chain. Many of the compounds isolated were formed by drying of the leaves prior to extraction. Digitoxin, a trisaccharide derivative of digitoxigenin, is the only compound to be used clinically to treat congestive heart failure and cardiac arrhythmias

1.5.2 Plant sources

In medicinal chemistry, 'plant' includes trees, bushes, grasses, etc., as well as what one normally associates with the term plant. All parts of a plant, from roots to seed heads and flowers, can act as the source of a lead. However, the collecting of plant samples must be carried out with due consideration of its environmental impact. In order to be able to repeat the results of a collection and if necessary cultivate the plant to ensure supplies of the compounds produced by the plant, it is essential that a full botanical record of the plant is made if it does not already exist. This record should contain a description and pictures of the plant and any related species, where it was found (GPS coordinates) and its growing conditions. A detailed record of the collection of the samples taken must also be kept since the chemical constitution of a plant can vary with the seasons, the method used for its collection, its harvest site storage and method of preparation for onward transportation to the investigating laboratory. If the plant material is to be shipped to a distant destination it must be protected from decomposition by exposure to inappropriate environmental conditions, such as a damp atmosphere or contamination by insects, fungi and micro-organisms.. The drying of so-called *green samples* for storage and shipment can give rise to chemical constituent changes because of enzyme action occurring during the drying process. Consequently, extraction of the undried green sample is often preferred, especially as chemical changes due to enzyme action is minimised when the green sample is extracted with aqueous ethanol.

Plant samples are normally extracted and put through screening programmes (see Chapter 6). Once screening shows that a material contains an active compound the problem becomes one of extraction, purification and assessment of the pharmacological activity. However, the isolation of a pure compound of therapeutic value can cause ecological problems. The anticancer agent Taxol (Fig. 1.6), for example, was isolated from the bark of

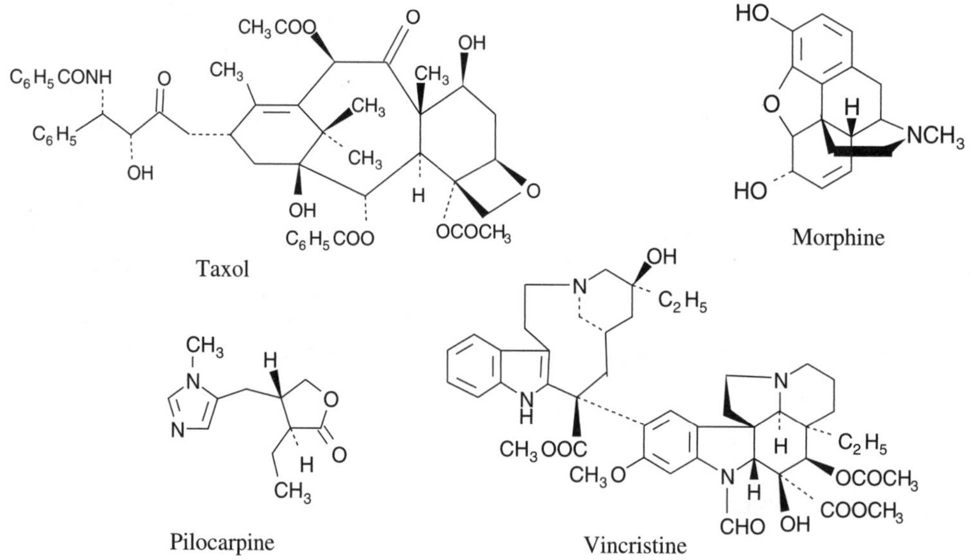

Figure 1.6 Examples of some of the drugs in clinical use obtained from plants. Taxol and vincristine are anticancer agents isolated from *Taxus breifolia* and *Vinca rosea* Linn, respectively. Pilocarpine is used to treat glaucoma and is obtained from *Pilocarpus jaborandi* Holmes *Rutaceae*. Morphine, which is used as an analgesic, is isolated from the opium poppy

the Pacific Yew tree (see section 6.8). Its isolation from this source requires the destruction of this slow-growing tree. Consequently, the production of large quantities of Taxol from the Pacific Yew could result in the wholesale destruction of the tree, a state of affairs that is ecologically unacceptable.

A different approach to identifying useful sources is that used by Hostettmann and Marston, who deduced that owing to the climate African plants must be resistant to constant fungal attack because they contain biologically active constituents. This line of reasoning led them to discover a variety of active compounds (Fig. 1.7).

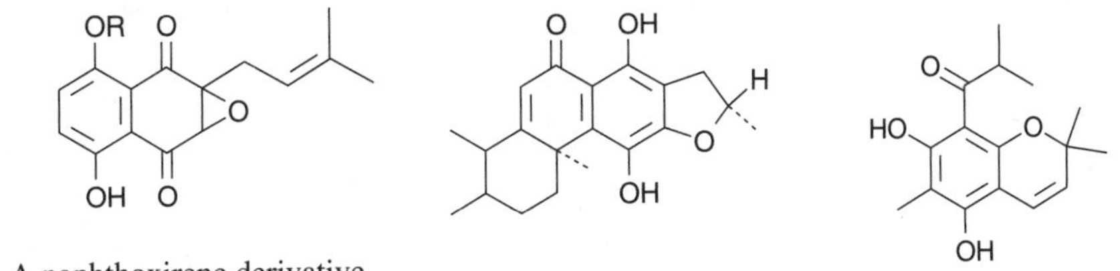

A naphthoxirene derivative
Key: R = β-D-glucopyaranosyl   Uncinatone   A chromene

Figure 1.7 Examples of the antifungal compounds discovered by Hostettmann and Marston

A number of the drugs in clinical use today have been obtained from plant extracts (see Fig. 1.6). Consequently, it is vitally important that plant, shrub and tree sources of the world are protected from further erosion as there is no doubt that they will yield further useful therapeutic agents in the future.

1.5.3 Marine sources

Prior to the mid-twentieth century little use was made of marine products in either folk or ordinary medicine. In the last 40 years these sources have yielded a multitude of active compounds and drugs (Fig. 1.8) with potential medical use. These compounds exhibit a range of biological activities and are an important source of new lead compounds and drugs. However, care must be taken so that exploitation of a drug does not endanger its marine sources, such as marine microorganisms, fungi, shellfish, sponges, plants and sea snakes. Marine microorganisms and fungi may be grown in fermentation tanks on a commercial scale. Microbial fermentation is a batch process, the required compound being extracted from the mature organisms by methods based on those outlined in Chapter 6. As well as drugs, microbial fermentation is also used to produce a wide range of chemicals for use in industry.

Marine sources also yield the most toxic compounds known to man. Some of these toxins, such as tetrodotoxin and saxitoxin (Fig. 1.8), are used as tools in neurochemical research work, investigating the molecular nature of action potentials and Na$^+$ channels

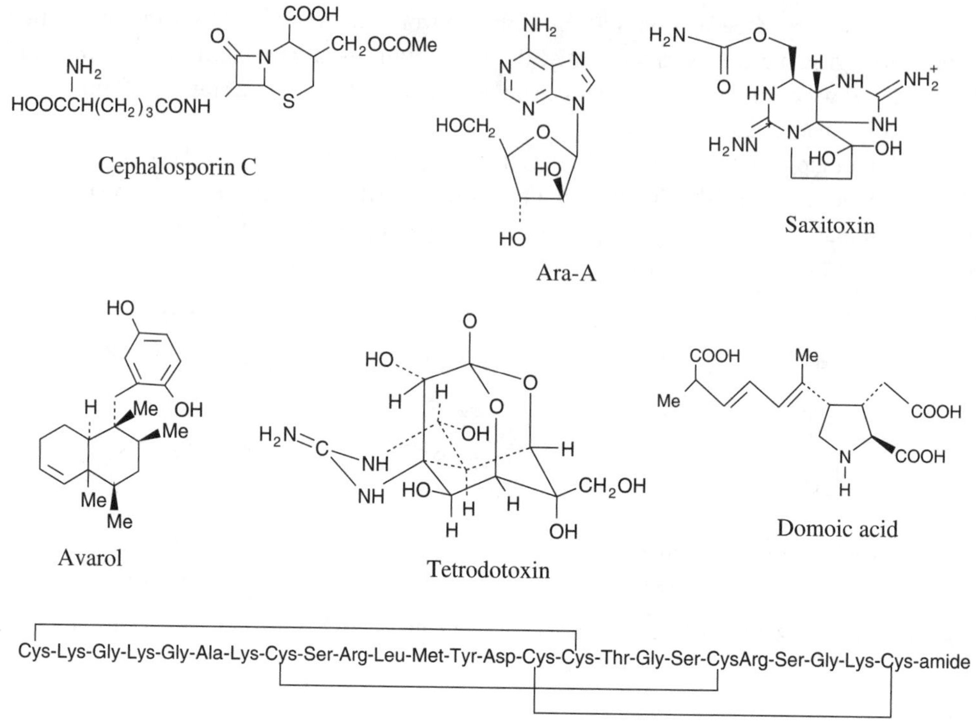

Figure 1.8 Examples of active compounds isolated from marine sources (Me represents a methyl group). Avarol is reported to be an immunodeficiency virus inhibitor. It is extracted from the sponge *Disidea avara*. The antibiotic cephalosporin C was isolated from the fungus *Acremonium chrysogenium (Cephalosporin acremonium)*. It was the lead for a wide range of active compounds, a number of which are used as drugs (see section 7.5.2). Domoic acid, which has anthelmintic properties, is obtained from *Chondria armata*. Tetrodotoxin and saxitoxin exhibit local anaesthetic activity but are highly toxic to humans. Tetrodotoxin is found in fish of the order *Tetraodontiformis* and saxitoxin is isolated from some marine dinoflagellates. Ara-A is an FDA - approved antiviral isolated from the sponge *Tethya crypta*. Ziconotide is the active ingredient of Prialt, which is used to treat chronic pain. It is an analogue of the ω-conopeptide MVIIA, which occurs in the marine snail *Conus magnus*.

(see sections 7.2.2 and 7.4.3). Although tetrodotoxin and saxitoxin are structurally different they are both believed to block the external opening of these channels.

1.5.4 Microorganisms

The inhibitory effect of microorganisms was observed as long ago as 1877 by Louis Pasteur, who showed that microbes could inhibit the growth of anthrax bacilli in urine. Later in 1920 Fleming demonstrated that *Penicillin notatum* inhibited *staphylococcus*

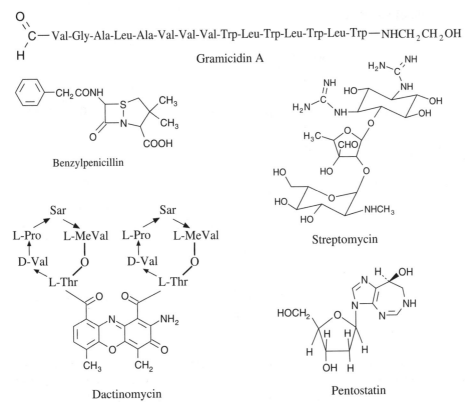

Figure 1.9 Examples of drugs produced by microbial fermentation. Gramicidin A, benzylpenicillin (penicillin G) and streptomycin are antibiotics isolated from *Bacillus brevis*, *Penicillin notatum* and *Streptomyces griseus*, respectively. The anticancer agents dactinomycin and pentostatin are obtained from *Streptomyces parvulus* and *Streptomyces antibioticus*, respectively

cultures, which resulted in the isolation of penicillin (Fig. 1.9) by Chain and Florey in 1940. In 1941 Dubos isolated a pharmacologically active protein extract from *Bacillus brevis* that was shown to contain the antibiotic gramicidin. This was concurrent with Waksman who postulated that soil bacteria should produce antibiotics as the soil contains few pathogenic bacteria from animal excreta. His work on soil samples eventually led Schatz *et al.* to the discovery and isolation in 1944 of the antibiotic streptomycin from the actinomycete *Streptomyces griseus*. This discovery triggered the current worldwide search for drugs produced by microorganisms. To date several thousand active compounds have been discovered from this source, for example the antibiotic chloramphenicol (*Streptomyces venezuelae*), the immunosuppressant cyclosporin A (*Tolypocladium inflatum* Gams) and antifungal griseofulvin (*Penicillium griseofulvum*) (Fig. 1.9). An important advantage of using microorganisms as a source is that, unlike many of the marine and plant sources, they are easily collected, transported and grown in fermentation tanks for use in industry.

1.5.5   Animal sources

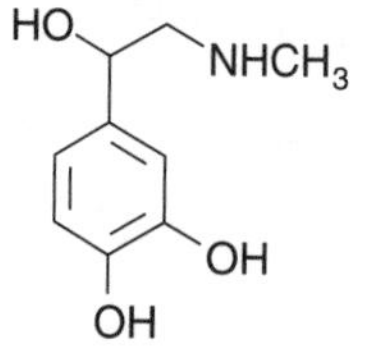

Adrenaline

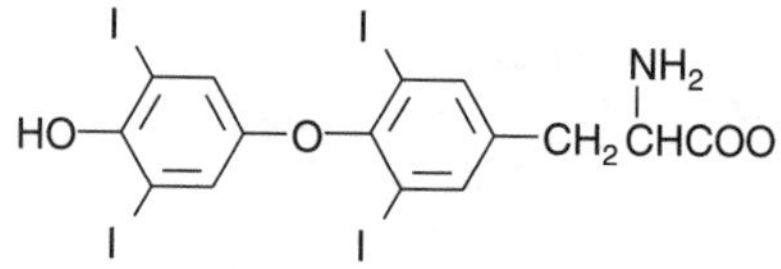

Thyroxine

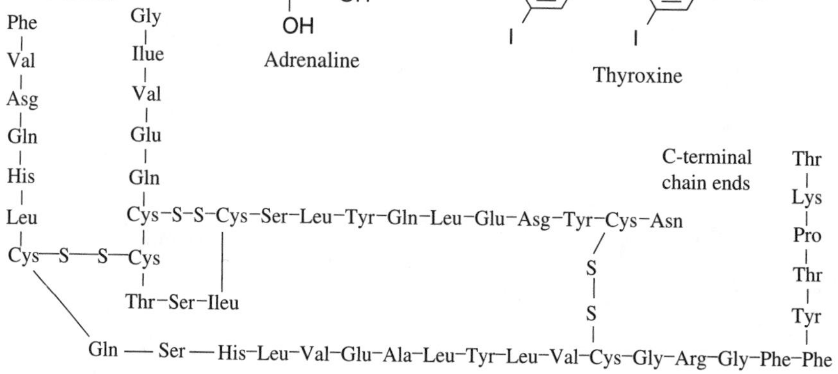

Insulin

Animal-derived products have been used since ancient times. However, it was not until the late nineteenth century that thyroid and adrenal medullary extracts were used to treat patients. Investigation of adrenal medullary extracts resulted in the isolation of the pure hormone adrenaline (epinephrine) in 1901. However, it was not until 1914 that pure thyroxine was isolated. This was followed in 1921 by the isolation of insulin from pancreatic extracts by Banting and Best. This enabled insulin to be produced commercially from bovine and porcine sources. Some insulin is still produced from these sources. However, in the later part of the twentieth century insulin was produced from bacteria using recombinant genetic engineering (see section 10.15.2). Animal sources are still used for hormone research but are seldom used to commercially produce drugs.

1.5.6   Compound collections, data bases and synthesis

All large pharmaceutical companies maintain extensive collections of compounds known as *libraries*. Smaller libraries are held by certain universities. An important approach to lead discovery is to put the members of these libraries through an appropriate *high-throughput screening* (HTS) (see section 5.6). Screening of large numbers of compounds can be very expensive so pharmaceutical companies tend to test groups of compounds that are selected using criteria specified by the company. These criteria may consist of similar chemical structures, chemical and physical properties, classes of compound and the structure of the target.

Pharmaceutical companies also maintain databases of compounds and their properties where known. Leads are found by searching these data bases for compounds that meet the companies' criteria. These compounds, known as *hits*, are synthesised, if necessary, before being tested for biological activity by HTS.

The pharmaceutical industry makes extensive use of combinatorial chemistry (see Chapter 5) to synthesise compounds for testing in high-throughput screens. Molecular modelling techniques (see Chapter 4) may be used to support this selection by matching the structures of potential leads to either the structures of compounds with similar activities or the target domain. The latter requires a detailed knowledge of both the three-dimensional structures of the ligand and target site.

1.5.7 The pathology of the diseased state

An important approach to lead compound selection is to use the biochemistry of the pathology of the target disease. The team select a point in a critical pathway in the biochemistry where intervention may lead to the desired result. This enables the medicinal chemist to either suggest possible lead compounds or to carry out a comprehensive literature and database search to identify compounds found in the organism (*endogenous compounds*) and compounds that are not found in the organism (*exogenous compounds* or *xenobiotics*) that may be biologically active at the intervention site. Once the team have decided what compounds might be active, the compounds are synthesised so that their pharmaceutical action may be evaluated.

1.5.8 Market forces and 'me-too drugs'

The cost of introducing a new drug to the market is extremely high and continues to escalate. One has to be very sure that a new drug is going to be profitable before it is placed on the market. Consequently, the board of directors' decision to market a drug or not depends largely on information supplied by the accountancy department rather than ethical and medical considerations. One way of cutting costs is for companies to produce drugs with similar activities and molecular structures to those of their competitors. These drugs are known as the 'me-too drugs'. They serve a useful purpose in that they give the practitioner a choice of medication with similar modes of action. This choice is useful in a number of situations, for example when a patient suffers an adverse reaction to a prescribed drug or on the rare occasion that a drug is withdrawn from the market.

1.6 Methods and routes of administration: the pharmaceutical phase

The form in which a medicine is administered is known as its *dosage form.* Dosage forms can be subdivided according to their physical nature into liquid, semisolid and solid formulations. Liquid formulations include solutions, suspensions, and emulsions. Creams, ointments and gels are normally regarded as semisolid formulations, whilst tablets, capsules and moulded products such as suppositories and pessaries are classified as solid formulations. These dosage forms normally consist of the active constituent and other ingredients (*excipients*). Excipients can have a number of functions, such as fillers (bulk providing agent), lubricants, binders, preservatives and antioxidants. A change in the nature of the excipients can significantly affect the release of the active ingredient from the dosage form. For example, the anticonvulsant phenytoin was found to be rapidly absorbed when lactose is used as a filler. This resulted in patients receiving toxic doses. In contrast, when calcium sulphate was used as a filler, the rate of absorption was so slow that the patient did not receive a therapeutic dose.

Phenytoin

Changes in the preparation of the active principle, such as the use of a different solvent for purification, can affect the bioavailability of a drug (see section 11.5) and consequently its effectiveness. This indicates the importance of having all-inclusive quality control procedures for drugs, especially when they reach the manufacturing stage.

The design of dosage forms lies in the field of the pharmaceutical technologist but it should also be considered by the medicinal chemist when developing a drug from a lead compound. It is no use having a wonder drug if it cannot be packaged in a form that makes it biologically available as well as acceptable to the patient. Furthermore, the use of an incorrect dosage form can render the medicine ineffective and potentially dangerous.

Drugs are usually administered topically or systemically. The routes are classified as being either *parenteral* or *enteral* (Fig. 1.10). Parenteral routes are those which avoid the gastrointestinal tract (GI tract), the most usual method being intramuscular injection (IM). However, other parental routes are intravenous injection (IV), subcutaneous injection (SC) and transdermal delivery systems. Nasal sprays and inhalers are also parenteral routes. The enteral route is where drugs are absorbed from the alimentary canal (given orally, PO), rectal and sublingual routes. The route selected for the administration of a drug will depend on the chemical stability of the drug, both when it is across a membrane (*absorption*) and in transit to the site of action (*distribution*). It will also be influenced by the age and physical and mental abilities of the patients using that drug. For example, age-related metabolic changes often result in elderly patients requiring lower dosages of the drug to achieve the desired clinical result. Schizophrenics and patients with conditions that require constant medication are particularly at risk of either overdosing or underdosing. In these cases a slow-release intramuscular injection, which need only be given once in every two to four

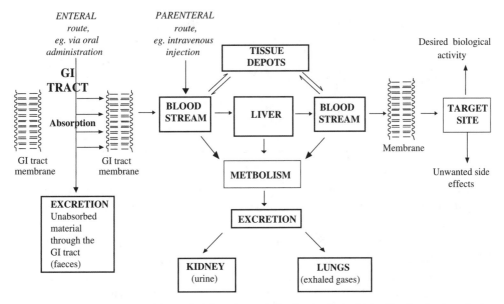

Figure 1.10   The main routes of drug administration and distribution in the body. The distribution of a drug is also modified by metabolism, which can occur at any point in the system

weeks rather than a daily dose, may be the most effective use of the medicine. Consequently, at an appropriately early stage in its development, the design of a drug should also take into account the nature of its target groups. It is a waste of time and resources if it is found that a drug that is successful in the laboratory cannot be administered in a convenient manner to the patient.

Once the drug enters the blood stream it is distributed around the body and so a proportion of the drug is either lost by excretion, metabolism to other products or is bound to biological sites other than its target site. As a result, the dose administered is inevitably higher than that which would be needed if all the drug reached the appropriate site of biological action. The dose of a drug administered to a patient is the amount that is required to reach and maintain the concentration necessary to produce a favourable response at the site of biological action. Too high a dose usually causes unacceptable side effects, whilst too low a dose results in a failure of the therapy. The limits between which the drug is an effective therapeutic agent is known as its *therapeutic window* (Fig. 1.11). The amount of a drug the plasma can contain, coupled with *elimination* processes (see section 11.4) that irreversibly remove the drug from its site of action, results in the drug concentration reaching a so-called *plateau* value. Too high a dose will give a plateau above the therapeutic window and toxic side effects. Too low a dose will result in the plateau below the therapeutic window and ineffective treatment.

The dose of a drug and how it is administered is called the *drug regimen*. Drug regimens may vary from a single dose taken to relieve a headache, regular daily doses taken to counteract the effects of epilepsy and diabetes, to continuous intravenous infusions for

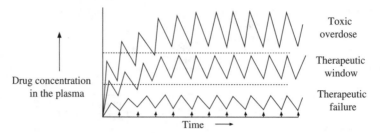

Figure 1.11   A simulation of a therapeutic window for a drug, given in fixed doses at fixed time intervals (↑)

seriously ill patients. Regimens are designed to maintain the concentration of the drug within the therapeutic window at the site of action for the period of time that is required for therapeutic success.

The design of an effective dosage regimen requires not just a knowledge of a drug's biological effects but also its *pharmacokinetic* properties, that is, its rate of absorption, distribution, metabolism and elimination from the body. It is possible for a drug to be ineffective because of the use of an incorrect dosage regimen. When quinacrine was introduced as a substitute for quinine in the 1940s it was found to be ineffective at low dose levels or too toxic at the high dose levels needed to combat malaria. Quinacrine was only used successfully after its pharmacokinetic properties were studied. It was found to have a slow elimination rate and so in order to maintain a safe therapeutic dose it was necessary to use large initial doses but only small subsequent maintenance doses to keep the concentration within its therapeutic window. This dosage regimen reduced the toxicity to an acceptable level.

Quinacrine

1.7   Introduction to drug action

The action of a drug is believed to be due to the interaction of that drug with enzymes, receptors and other molecules found in the body. When one or more drug molecules bind to the target endogenous and exogenous molecules, they cause a change in or inhibit the biological activity of these molecules. The effectiveness of a drug in bringing about these changes usually depends on the stability of the drug–target complex, whereas the medical success of the drug intervention usually depends on whether enough drug molecules bind to sufficient target molecules to have a marked effect on the course of the disease state.

The degree of drug activity is directly related to the concentration of the drug in the aqueous medium in contact with the target molecule. The factors effecting this concentration in a biological system can be classified into the *pharmacokinetic phase* and the *pharmacodynamic phase* of drug action. The pharmacokinetic phase concerns the study of the parameters that control the journey of the drug from its point of administration to its point of action. The pharmacodynamic phase concerns the chemical nature of the relationship between the drug and its target, in other words, the effect of the drug on the body.

1.7.1 The pharmacokinetic phase (ADME)

The pharmacokinetic phase of drug action includes the *Absorption, Distribution, Metabolism* and *Excretion* (*ADME*) of the drug. Many of the factors that influence drug action apply to all aspects of the pharmacokinetic phase. Solubility (see Chapter 2), for example, is an important factor in the absorption, distribution and elimination of a drug. Furthermore, the rate of drug dissolution (see section 11.5.1) controls its activity when that drug is administered as a solid or suspension by enteral routes (see section 1.6)

Absorption

Absorption isusually defined as *the passage of the drug from its site of administration into the general circulatory system after enteral administration.* The use of the term does not apply to parenteral administration discussions. The most common enteral route is by oral administration. Drugs administered in this way take ab out 24 hours to pass through the gastrointestinal tract (GI tract). Individual transit times for the stomach and small intestine are about 20 minutes and 6 hours, respectively. Compounds may be absorbed throughout the length of the GI tract but some areas will suit a drug better than others.

The absorption of drugs through membranes and tissue barriers (see Chapter 7) can occur by a number of different mechanisms (see section 7.3). However, in general, neutral molecules are more readily absorbed through membranes than charged species. For example, ionisation of orally administered aspirin is suppressed in the stomach by acids produced by the parietal cells in the stomach lining. As a result, it is absorbed in this uncharged form through the stomach lining into the blood stream in significant quantities.

$$\text{COOH} \quad \text{OCOCH}_3 \quad + H_2O \quad \rightleftharpoons \quad \text{COO}^- \quad \text{OCOCH}_3 \quad + \quad H^+$$

Aspirin

The main structural properties of a drug governing its good absorption from the GI tract are its aqueous solubility (see Chapter 2) and the balance between its polar (hydrophilic) and non-polar (hydrophobic) groups (see section 1.4.2). If the drug's water solubility is too low it will pass through the GI tract without a significant amount being absorbed. Drugs that are too polar will tend to be absorbed by paracellular diffusion, which is only readily available in the small intestine and is usually slower than transcellular diffusion undergone by less polar and non-polar compounds (see section 11.5.2). Drugs that are absorbed by transcellular diffusion are usually absorbed along the whole length of the GI tract. If the drug is too non-polar (lipophilic) it will be absorbed into and remain within the lipid interior of the membranes of the cells forming the membrane. The Lipinski rule of fives (see section 1.4.1) is useful for assessing whether a compound is likely to be absorbed from the GI tract. However, this rule does have its limitations and the results of its use should only be used as a guide and not taken as being absolute.

The degree of absorption can also be related to the surface area of the region of tissue over which the absorption is occurring and the time the drug spends in contact with that region. For example, it can be shown by calculation using the Henderson–Hasselbalch equation (see section 2.11) that aspirin will be almost fully ionised in the small intestine. Consequently, aspirin should not be readily absorbed in this region of the GI tract. However, the very large surface area of the small intestine (300 m^2) together with the time spent in this region (~ 6 hours) results in aspirin being absorbed in significant quantities in this region of the GI tract. Examples of *some* of the other factors that can effect the degree of absorption of a drug are:

- the pH of the medium from which absorption occurs (see section 2.11);

- the drug's partition coefficient (see section 3.7.2);

- the drug's dosage form (see section 1.6);

- the drug's particle size (see section 11.5.1); and

- for orally administered drugs, in either solid or emulsion form, their rate of dissolution (see section 11.5.1), amongst others.

It should be noted that the form of the drug that is absorbed is not necessarily the form that is responsible for its action. Benzocaine, for example, is absorbed as its neutral molecule but acts in its charged form.

$$H_2N-\langle \rangle-COOCH_2CH_3 \rightleftharpoons H_3\overset{+}{N}-\langle \rangle-COOCH_2CH_3$$

Inactive form transported through membranes   Active form

Benzocaine

Distribution

Distribution is the transport of the drug from its initial point of administration or absorption to its site of action. The main route is through the circulation of the blood although some distribution does occur via the lymphatic system. Once the drug is absorbed it is rapidly distributed throughout all the areas of the body reached by the blood. This means that the chemical and physical properties of blood will have a considerable effect on the concentration of the drug reaching its target site.

Drugs are transported in the blood stream as either a solution of drug molecules or bound to the serum proteins, usually albumins. The binding of drugs to the serum proteins is usually reversible.

$$Drug \rightleftharpoons Drug-Serum\ Protein\ complex$$

Drug molecules bound to serum proteins have no pharmacological effect until they are released from those proteins. Consequently, this equilibrium can be an important factor in controlling a drug's pharmacological activity (see section 11.4.1). However, it is possible for one drug to displace another from a protein if it forms a more stable complex, that is, has a stronger affinity for that protein. This aspect of protein binding can be of considerable importance when designing drug regimens involving more than one drug. For instance, the displacement of antidiabetic agents by aspirin can trigger hypoglycaemic shock and so aspirin should not be used by patients taking these drugs. Protein binding also allows drugs with poor water solubility to reach their target site. The drug–protein complex acts as a depot maintaining the drug in sufficient concentration at the target site to bring about a response. However, a low plasma protein concentration can also affect the distribution of a drug in some diseases such as rheumatoid arthritis as the 'reduced transport system' is unable to deliver a sufficient concentration of the drug to its target site. Protein binding can also increase the duration of action if the drug–protein complex is too large to be excreted through the kidney by glomerular filtration.

Major factors that influence distribution are the solubility and stability of drugs in the biological environment of the blood. Sparingly water-soluble compounds may be deposited in the blood vessels, leading to restriction in blood flow. This deposition may be influenced by the common ion effect (see section 2.4.1). Drug stability is of particular importance in that serum proteins can act as enzymes that catalyse the breakdown of the drug. Decompositions such as these can result in a higher dose of the drug being needed in order to achieve the desired pharmacological effect. This increased dose increases the risk of toxic side effects in the patient. However, the active form of some drugs is produced by the decomposition of the administered form of the drug. Drugs that function in this manner are known as *prodrugs* (see section 12.9). The first to be discovered, in 1935, was the bactericide prontosil. Prontosil itself is not active but is metabolised *in situ* to the antibacterial sulphanilamide. Its discovery paved the way to the devolvement of a wide range of antibacterial sulphonamide (sulfa) drugs. These were the only effective antibiotics available until the general introduction of penicillin in the late 1940s.

Prontosil   Sulphanilamide

The distribution pattern of a drug through the tissues forming the blood vessels will depend largely on the nature of the tissue (see section 7.2.9) and on the drug's lipid solubility. For example, in general, the pH of the tissues ($\sim$ pH 7.0) forming blood vessels is less basic than that of the plasma ($\sim$ pH 7.4). Acidic drugs such as aspirin, which ionise in aqueous solution, exist largely in the form of their anions in the slightly basic plasma. Since uncharged molecules are transferred more readily than charged ions these acidic anions tend to remain in the plasma and not move out of the plasma into the tissues surrounding the blood vessel. Consequently, acids have a tendency to stay in the plasma rather than pass into the surrounding tissue. Conversely, a significant quantity of a basic drug tends to exist as neutral molecules in the plasma. As a result, bases are more likely to pass into the tissues surrounding the plasma. Furthermore, once the base has passed into the tissue the charged form of the base is likely to predominate and so the drug will tend to remain in the tissue. This means that the base is effectively removed from the plasma, which disturbs its equilibrium in the plasma to favour the formation of the free base, which results in further absorption of the base into the tissue (Fig. 1.12). As a result, basic drugs, unlike acidic drugs, are likely to be more widely distributed in tissues.

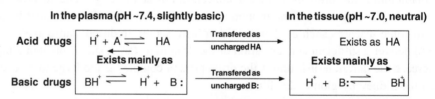

Figure 1.12 The species involved in the transfer of acidic and basic drugs from the plasma to the surrounding tissues

A drug's lipophilicity (see sections 1.4.2 and 3.7.2) will also influence its distribution. Highly lipophilic drugs can readily enter and accumulate in the fatty deposits of humans. These fatty deposits, which form up to 15 per cent in the body weight of normal individuals and 50 per cent in obese persons, can act as pharmacologically inert depots for drugs, which could terminate their action. For example, the concentration of the ultra-short-acting anaesthetic thiopental rapidly falls after administration to an ineffective level because it accumulates in the fatty tissue deposits of the body. It is slowly released from these deposits in concentrations that are too low to cause a pharmacological response.

The distribution of drugs to the brain entails having to cross the *blood–brain barrier (BBB)* (see section 7.2.9). This barrier protects the brain from both exogenous and

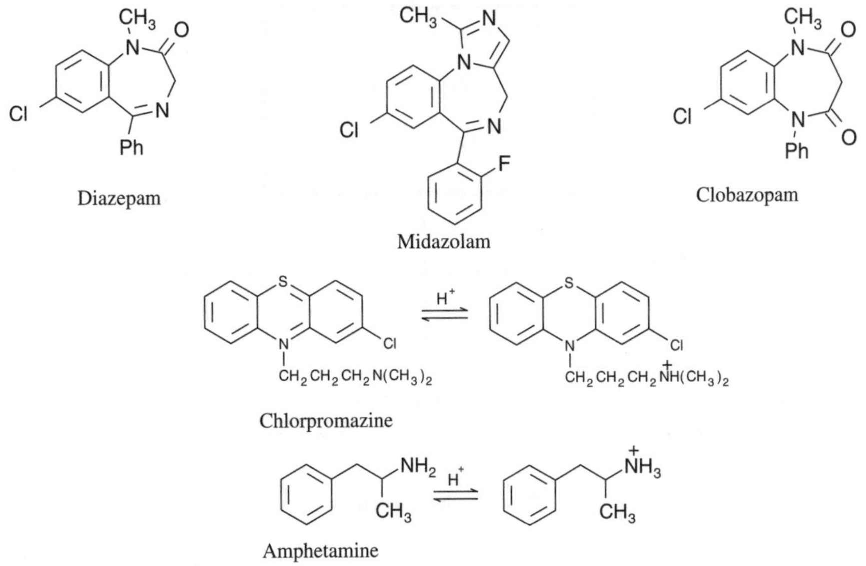

Figure 1.13 The structures of some of the drugs that are able to cross the blood–brain barrier

endogenous compounds. The extent to which lipophilic drugs are able to cross this barrier varies: highly lipophilic drugs, such as diazepam, midazolam and clobazopam (Fig. 1.13), are rapidly absorbed while less lipophilic drugs are absorbed more slowly. Polar drugs are either unable to cross the BBB or only do so to a very limited extent. For example, calculations using the Henderson–Hasselbalch equation (see section 2.11) show that at blood pH, 99.6 per cent of amphetamine and 98.4 per cent of chlorpromazine exist in their charged forms (Fig. 1.13). However, these polar drugs are still sufficiently lipid soluble to cross the BBB. Some polar drugs may cross by an active transport mechanism (see section 7.3.5). Other polar endogenous compounds such as amino acids, sugars, nucleosides and small ions (Na^+, Li^+, Ca^{2+} and K^+) are also able to cross the BBB.

Metabolism

Drug metabolism (see Chapter 12) is the biotransformation of the drug into other compounds (*metabolites*) that are usually more water soluble than their parent drug and are usually excreted in the urine. It usually involves more than one route and results in the formation of a succession of metabolites (Fig. 1.14). These biotransformations occur mainly in the liver but they can also occur in blood and other organs such as the brain, lungs and kidneys. Drugs that are administered orally usually pass through the liver before reaching the general circulatory system. Consequently, some of the drug will be metabolised before it reaches the systemic circulation. This loss is generally referred to as either the *first-pass effect* or *first-pass metabolism* (see section 11.4.1). Further metabolic losses will also be encountered before the drug reaches its target site, which means it is

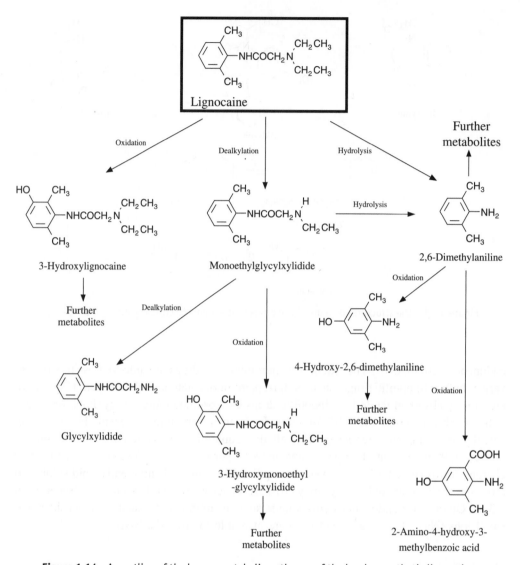

Figure 1.14   An outline of the known metabolic pathways of the local anaesthetic lignocaine

important to administer a dose large enough for sufficient of the drug to reach its target site.

Drug metabolism may produce metabolites that are pharmacologically inert, have the same or different action to the parent drug or are toxic (see section 12.2). Exceptions are prodrugs (see sections 1.8.4 and 12.9) where metabolism is responsible for producing an active drug, for example the non-steroidal anti-inflammatory agent sulindac is metabolised to the active sulphide (Fig. 1.15). In addition, the metabolic products of a drug may be used as leads for the development of a new drug.

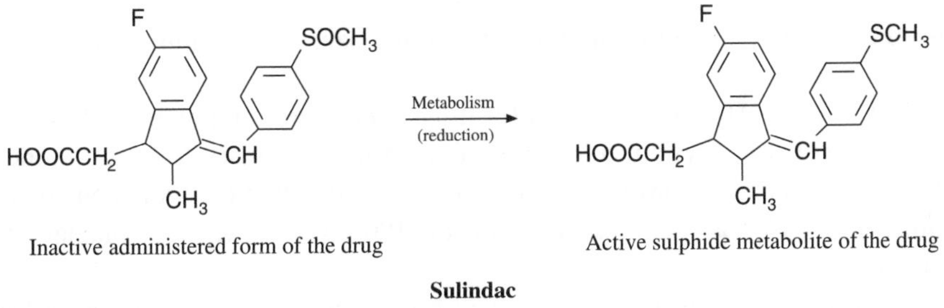

Inactive administered form of the drug

Active sulphide metabolite of the drug

Sulindac

Figure 1.15 An outline of the metabolic pathway for the formation of the active form of sulindac

Excretion

Excretion is the process by which unwanted substances are removed from the body. The main excretion route for drugs and their metabolites is through the kidney in solution in the urine. However, a significant number of drugs and their metabolic products are also excreted via the bowel in the faeces. Other forms of drug excretion, such as exhalation, sweating and breast feeding, are not usually significant except in specific circumstances. Pregnant women and nursing mothers are recommended to avoid taking drugs because of the possibility of biological damage to the foetus and neonate. For example, the use of thalidomide by pregnant mothers in the 1960s resulted in the formation of drug-induced malformed foetuses (*teratogenesis*). It has been estimated that the use of thalidomide led to the birth of 10,000 severely malformed children.

In the kidneys drugs are excreted by either *glomerular filtration* or *tubular secretion*. However, some of the species lost by these processes are reabsorbed by a recycling process known as *tubular reabsorption*. In the kidney, the glomeruli act as a filter allowing the passage of water, small molecules and ions but preventing the passage of large molecules and cells. Consequently, glomerular filtration excrets small unbound drug molecules but not the larger drug–protein complexes. Tubular secretion on the other hand is an active transfer process (see section 7.3.5) and so both bound and unbound drug molecules can be excreted. However, both of these excretion systems have a limited capacity and not all the drug may be eliminated. In addition, renal disease can considerably increase or decrease the rate of drug excretion by the kidney.

Tubular reabsorption is a process normally employed in returning compounds such as water, amino acids, salts and glucose that are important to the well-being of the body from the urine to the circulatory system, but it will also return drug molecules. The mechanism of reabsorption is mainly passive diffusion (see section 7.3.3), but active transport (see section 7.3.5) is also involved, especially for glucose and lithium ions. The reabsorption of acidic and basic compounds is dependent on the pH of the urine. For example, making the urine alkaline in cases of poisoning by acidic drugs, such as aspirin, will cause these drugs to form ionic salts, which will result in a significantly lower tubular reabsorption since the passage of the charged form of a drug across a lipid membrane is more difficult than the passage of the uncharged form of that drug. Similarly, in cases of poisoning by basic drugs such as amphetamines, acidification of the urine can, for a similar reason, reduce reabsorption.

Control of urinary pH is also required for drugs whose concentration reaches a level in the urine that results in crystallisation (crystalluria) in the urinary tract and kidney with subsequent tissue damage. For example, it is recommended that the urine is maintained at an alkaline pH and has a minimum flow of 190 ml h^{-1} when sulphonamides are administered.

Excretion also occurs via the intestines and bowel through *biliary clearance* from the liver. The liver is linked to the intestine by the bile duct and some compounds are excreted by this route. However, very large molecules are metabolised to smaller compounds before being excreted. However, a fraction of some of the excreted drugs are reabsorbed through the *enterohepatic cycle*. This reabsorption can be reduced by the use of suitable substances in the dosage form, for example the ion exchange resin cholestyramine is used to reduce cholesterol levels by preventing its reabsorption.

Lead optimisation and ADME

A drug must reach its site of action in sufficient quantity to be effective. One of the tasks of the medicinal chemist is to take an active compound and modify the structure to achieve the desired ADME properties. However, having satisfactory ADME properties is not the only requirement for a new drug. A drug candidate must also be:

- potentially effective in treating a patient;

- free of existing patents;

- produced in sufficient quantities;

- capable of being dispensed in a dosage form acceptable to the patient;

- must not be too toxic for use;

- must not exhibit teratogenicity or mutagenicity;

- and commercial development must be cost effective.

Failure to comply with these additional aspects of drug discovery and design will mean that work on the candidate is discontinued before the project proceeds past its preliminary stages.

1.7.2   The pharmacodynamic phase

Pharmacodynamics is concerned with the result of the interaction of drug and body at its site of action, that is, what the drug does to the body. It is now known that a drug is most effective when its shape and electron distribution, that is, its *stereoelectronic structure*, is complementary with the stereoelectronic structure of the target site.

The role of the medicinal chemist is to design and synthesise a drug structure that has the maximum beneficial effects with a minimum of toxic side effects. This design has to take into account the stereoelectronic characteristics of the target site and also such factors as the drug's stability *in situ*, its polarity and its relative solubilities in aqueous media and lipids. The stereochemistry of the drug is particularly important as stereoisomers often have different biological effects which range from inactive to highly toxic (see section 1.4.3 and Table 1.1).

Drugs act at their target site by either inhibiting or stimulating a biological process with, hopefully, beneficial results to the patient. To bring about these changes the drug must bind to the target site, that is, its potency will depend on its ability to bind to that site. This binding is either reversible or permanent. In the former case, the bonding is due to weak electrostatic bonds such as hydrogen bond and van der Waals' forces. The binding takes the form of a dynamic equilibrium with the drug molecules repeatedly binding to and being released from their target site (see section 8.6). Consequently, in this instance, a drug's duration of action will depend on how long it remains at the target site. Permanent binding usually requires the formation of strong covalent bonds between the drug and its target. In this case, the duration of action will depend on the strength of the bond. However, in both cases, the drug structure must contain appropriate functional groups in positions that correspond to the appropriate structures in the target site.

1.8 Classification of drugs

Drugs are classified in different ways depending on where and how the drugs are being used. The methods of interest to medicinal chemists are chemical structure and pharmacological action, which includes the site of action and target system. However, it is emphasised that other classifications, such as the nature of the illness, are used both in medicinal chemistry and other fields depending on what use is to be made of the information. In all cases, it is important to bear in mind that most drugs have more than one effect on the body and so a drug may be listed in several different categories within a classification scheme.

1.8.1 Chemical structure

Drugs are grouped according to the structure of their carbon skeletons or chemical classifications, for example steroids, penicillins and peptides. Unfortunately in medicinal chemistry this classification has the disadvantage that members of the same group often exhibit different types of pharmaceutical activity. Steroids, for example, have widely differing activities: testosterone is a sex hormone, spironolactone is a diuretic and fusidic acid is an antibacterial agent (Fig. 1.16).

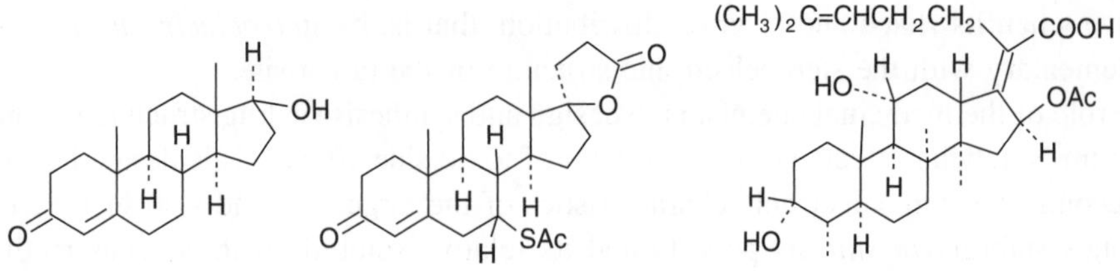

Testosterone Spirolactone Fusidic acid

Figure 1.16 Examples of the diversity of action of compounds belonging to the same class (Ac≡acetyl)

Classification by means of chemical structure is useful to medicinal chemists who are concerned with synthesis and structure–activity relationships.

1.8.2 Pharmacological action

This classification lists drugs according to the nature of their pharmacodynamic behaviour, for example diuretics, hypnotics, respiratory stimulants and vasodilators. This classification is particularly useful for doctors looking for an alternative drug treatment for a patient.

1.8.3 Physiological classification

The World Health Organization (WHO) has developed a classification based on the body system on which the drug acts. This classification specifies seventeen sites of drug action. However, a more practical method but less detailed system often used by medicinal chemists is based on four classifications, namely:

1. *Agents acting on the central nervous system (CNS).* The central nervous system consists of the brain and spinal cord. Drugs acting on the CNS are the *psychotropic* drugs that effect mood and the *neurological* drugs required for physiological nervous disorders such as epilepsy and pain.

2. *Pharmacodynamic agents.* These are drugs that act on the body, interfering with the normal bodily functions. They include drugs such as vasodilators, respiratory stimulants and antiallergy agents.

3. *Chemotherapeutic agents.* Originally these were drugs such as antibiotics and fungicides that destroyed the microorganisms that were the cause of a disease in an unwitting host. However, the classification has also now become synonymous with the drugs used to control cancer.

4. *Miscellaneous agents.* This class contains drugs that do not fit into the other three categories, for example hormones and drugs acting on endocrine functions.

1.8.4 Prodrugs

Prodrugs are compounds that are pharmacologically inert but converted by enzyme or chemical action to the active form of the drug at or near their target site. For example, levodopa, used to treat Parkinson's syndrome, is the prodrug for the neurotransmitter dopamine. Dopamine is too polar to cross the blood–brain barrier but there is a transport system for amino acids such as levodopa. Once the prodrug enters the brain it is decarboxylated to the active drug dopamine (Fig. 1.17).

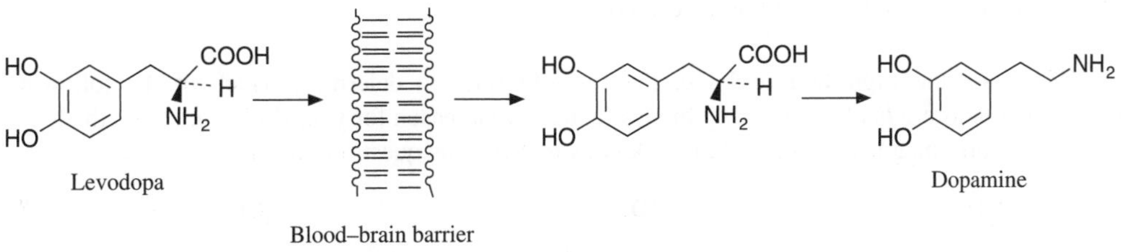

Figure 1.17   A schematic representation of the formation of dopamine from levodopa

1.9   Questions

1 Predict, giving a reason for the prediction, the *most likely* general effect of the stated structural change on either the *in situ* stability or the pharmacological action of the stated drug.

  (a) The introduction of *ortho* ethyl groups in dimethylaminoethyl 4-aminobenzoate.

  (b) The replacement of the amino group in the CNS stimulant amphetamine ($PhCH_2CH(NH_2)CH_3$) by a trimethylammonium chloride group.

  (c) The replacement of the ester group in the local anaesthetic ethyl 4-aminobenzoate (benzocaine).

2 State the general factors that need to be considered when designing a drug.

3 Explain the meaning of the terms: (a) lead compound, (b) dosage form, (c) enteral administration of drugs, (d) drug regimen, (e) prodrug, (f) pharmacophore and (g) excipient.

4 Define the meaning of the terms pharmacokinetic phase and pharmacodynamic phase in the context of drug action. List the main general factors that affect these phases.

5   The drug amphetamine ($PhCH_2CH(NH_2)CH_3$) binds to the protein albumin in the blood stream. Predict how a reduction in pH would be expected to influence this binding? Albumin is negatively charged at pH 7.4 and electrically neutral at pH 5.0.

6   Discuss the general effects that stereoisomers could have on the activity of a drug. Draw the *R* and *S* isomers of the anaesthetic ketamine. Indicate which of the structures you have drawn is mainly responsible for its anaesthetic activity.

7   Suggest strategies for improving the stability of compound A in the gastrointestinal tract. What could be the general effect of these strategies on the pharmaceutical action of compound A?

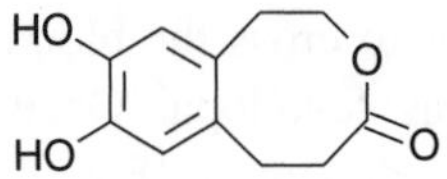

8   Explain the meaning of the term receptor.

9   What are the Lipinski rules? Use the Lipinski rules to determine which of the following compounds are *likely* to be orally bioavailable. Give a reason for your decision. (*Note*: the log *P* values are imaginary but should be taken as real for this question only!)

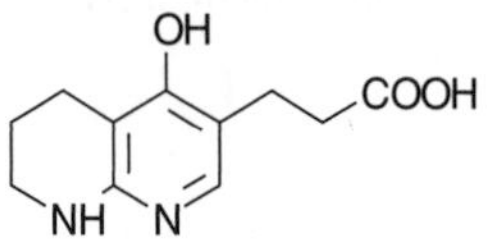

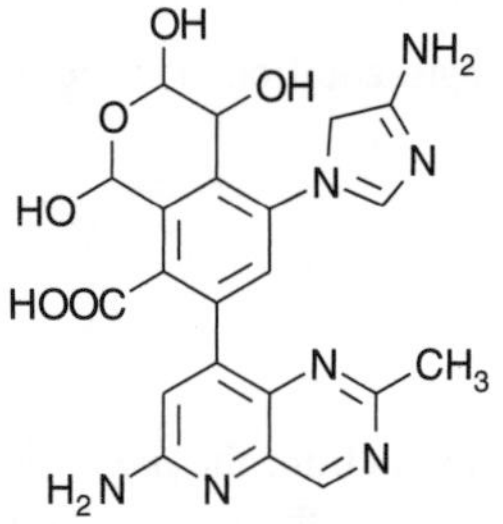

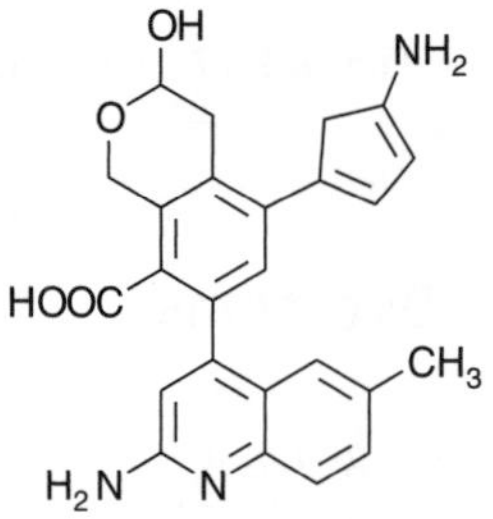

10   List the desirable requirements for a lead.

2
Drug structure and solubility

2.1 Introduction

The chemical structure and solubility of a drug have a significant influence on its activity (see sections 1.4.2 and 1.4.3). This chapter discusses aspects of these properties of compounds in the context of drug discovery and design.

2.2 Structure

Drugs act by binding to their target domains. This binding is only possible if their stereoelectronic structures are complementary to those of their target domains. Compounds are believed to bind to their target site by either weak electrostatic bonds such as hydrogen bonds and van der Waals' forces or stronger covalent bonds. The action of compounds that use weak bond to bind to their target domain is believed to be due to these bonds being repeatedly broken and reformed (see section 8.6.1). Action that is due to strong covalent bond formation is believed to be based on the compound reacting with a key compound in the biological pathway of the diseased state to form a stable compound that is inactive in that biological pathway. For example, many anticancer drugs act by forming strong covalent bonds with DNA (see section 10.13.4). However, regardless of the type of bond formed to the target, the bonds can only be formed if the compound can approach close enough to its target. Consequently, a potential drug candidate must have a chemical structure and a shape that are compatible with those of its target domain.

The overall shape of the structure of a molecule is an important consideration when designing an analogue. Some structural features impose a considerable degree of rigidity into a structure whilst others make the structure more flexible. Other structures give rise to stereoisomers, which can exhibit different potencies, types of activity and unwanted side effects (see Table 1.1). This means that it is necessary to pharmacologically evaluate individual

Medicinal Chemistry, Second Edition   Gareth Thomas
© 2007 John Wiley & Sons, Ltd

stereoisomers and racemates. Consequently, the stereochemistry of compounds must be taken into account when selecting them for use as leads and analogues. Molecular modelling (see Chapter 4) is frequently used to evaluate the fit and binding characteristics of a lead to its target before starting out on a drug development programme. It may also be used at different stages of the development to assess how the analogues of the drug are binding to the target site. However, the extent to which one can exploit this technique will depend on our knowledge of the structure and biochemistry of the target biological system.

2.3   Stereochemistry and drug design

It is now well established that the shape of a molecule is normally one of the most important factors that affect drug activity. Consequently, the overall shape of the structure of a molecule is an important consideration when designing an analogue. Some structural features impose a considerable degree of rigidity into a structure whilst others make the structure more flexible. Other structures give rise to stereoisomers which can exhibit different potencies, types of activity and unwanted side effects (see section 1.7.2 and Table 1.1). Furthermore, it has been shown that it is possible for some enantiomers to racemise under physiological conditions. For example, thalidomide developed in the 1950s as a sleeping pill was originally marketed as its racemate. This drug was found after use to be teratogenic, causing foetal abnormalities. Further research work using mice showed that it was the S-enantiomer that had the teratogenic properties while the R-enantiomer was a sedative with no teratogenic properties. However, it was also found using rabbits that both enantiomers racemise under physiological conditions. Thalidomide is now used in the treatment of multiple myeloma.

S-Thalidomide (sedative)   R-Thalidomide (teratogenic)

This and other similar are examples make it necessary to pharmacologically evaluate both the individual enantiomers and racemates of a drug candidate. Consequently, the stereochemistry of compounds must be taken into account when selecting them for use as leads and analogues. However, the extent to which one can exploit these structural features will depend on our knowledge of the structure and biochemistry of the target biological system.

2.3.1   Structurally rigid groups

Groups that are structurally rigid are unsaturated groups of all types and saturated ring systems (Fig. 2.1). The former includes esters and amides as well as aliphatic conjugated systems and aromatic and heteroaromatic ring systems. The binding of these rigid structures

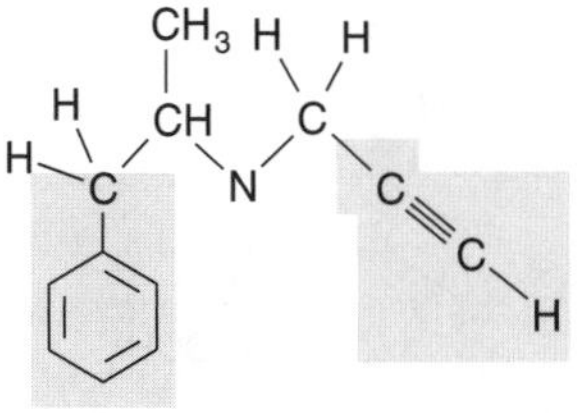

Selegiline (MAO inhibitor)

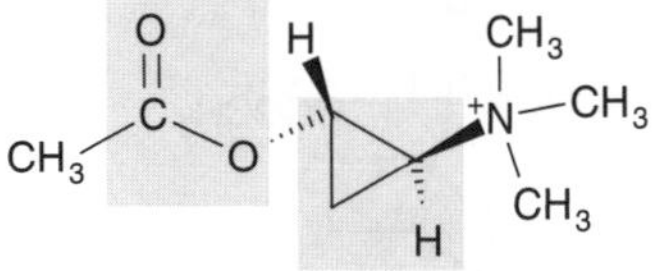

1-Ethoxycarbonyl-2-trimethylaminocyclopropane (acetylcholine mimic)

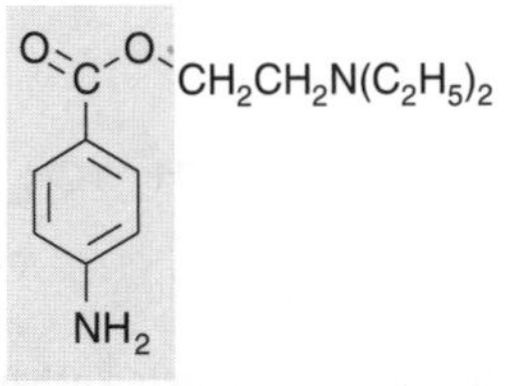

Procaine (local anaesthetic)

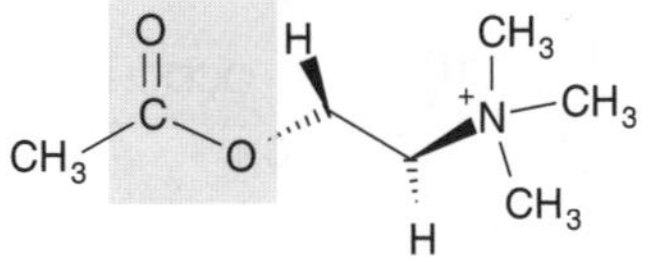

Acetylcholine

Figure 2.1 Examples of structural groups that impose a rigid shape on sections of a molecule. The shaded areas represent the rigid sections of the molecule

to a target site can give information about the shape of that site as well as the nature of the interaction between the site and the ligand. Rigid structures may also be used to determine the conformation assumed by a ligand when it binds to its target site (see section 2.3.2). Furthermore, the fact that the structure is rigid means it may be replaced by alternative rigid structures of a similar size and shape to form analogues that may have different binding characteristics and possibly, as a result, a different activity or potency.

2.3.2 Conformation

Early work in the 1950s and early 1960s by Schueler and Archer suggested that the flexibility of the structures of both ligands and receptors accounted for the same ligand being able to bind to different subtypes of a receptor (see section 8.3). Archer also concluded that a ligand appeared to assume different conformations when it bound to the different subtypes of a receptor. For example, acetylcholine exhibits both muscarinic and nicotinic activity. Archer *et al.* suggested that the muscarinic activity was due to the *anti* or staggered conformation whilst the nicotinic activity was due to the *syn* or eclipsed form (Fig.2.2). These workers based this suggestion on their observation that the *anti* conformation of 2-tropanyl ethanoate methiodide preferentially binds to muscarinic receptors whilst the *syn* conformation binds preferentially to nicotinic receptors. The structures of both of these compounds contain an acetyl choline residue locked in the appropriate conformation by the ring structure. This and subsequent investigations led to the conclusion that the development of analogues with restricted or rigid conformations could result in the selective binding of drugs to target sites, which could result in very active drugs with reduced unwanted side effects.

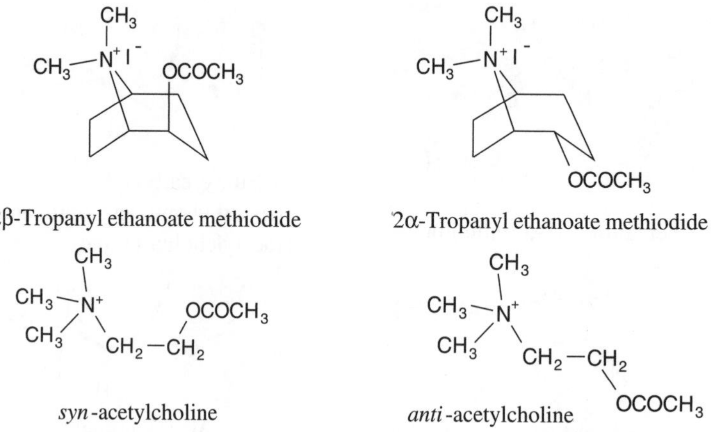

2β-Tropanyl ethanoate methiodide

2α-Tropanyl ethanoate methiodide

syn-acetylcholine

anti-acetylcholine

Figure 2.2   The syn and anti conformers of acetylcholine and 2-tropanyl ethanoate methiodide

The main methods of introducing conformational restrictions are by using bulky substituents, unsaturated structures or small ring systems. Small ring systems are usually the most popular choice (Fig. 2.3). In all cases the structures used must be chosen with care because there will always be the possibility that steric hindrance will prevent the binding of the analogue to the target. A further limitation is knowing which bond to restrict. Even in simple molecules numerous eclipsed, staggered and gauche conformations are possible (Fig.2.4). However, if sufficient information is available molecular modelling (see Chapter 4) may be used to overcome these challenges.

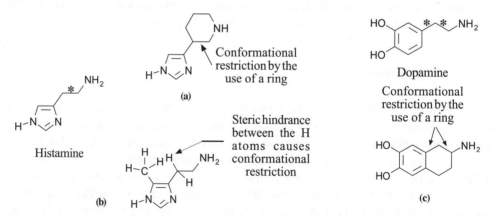

Figure 2.3   Examples of the use of conformational restrictions to produce analogues of histamine and dopamine. Bonds marked * can exhibit free rotation and form numerous conformers. **(a)** Conformational restriction of the bond marked * in histamine by the use of a ring structure. **(b)** Conformational restriction of the bond marked * in histamine by the use of steric hindrance. **(c)** Conformational restriction of the bonds marked * in dopamine by the use of a ring structure

The biological data obtained using restricted conformation analogues can be of use in determining the most bioactive conformation of the ligand. If the analogue exhibits either the same or a greater degree of activity as the lead compound it may be concluded that the

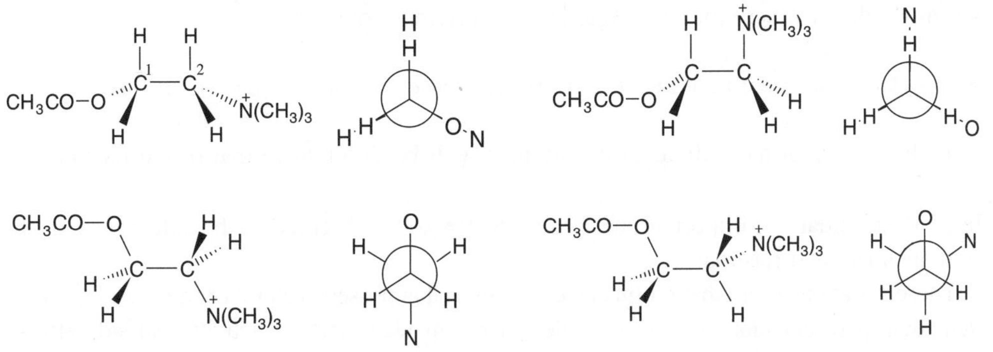

Figure 2.4 Examples of some of the major conformations of the C1–C2 bond of acetylcholine. Other conformations occur about the C–N and C–O single bonds

analogue has the correct conformation for binding to that site. However, if the analogue exhibits no activity the result could be due to either steric hindrance between the restricting group and the target or the analogue having an incorrect conformation. In this case molecular modelling may be of some assistance in assessing whether a conformation would fit the intended target site (see sections 4.5 and 4.9).

It has also been observed that a degree of flexibility in a drug often improves the action of that drug. This is logical when one remembers that a drug has to bind to its target to initiate its action. Consequently, a conformationally constrained drug may not interact strongly with its target site due to a poor fit to that site. A more flexible structure may be able to adjust to give a better fit to its target site. Furthermore, a flexible drug may be able to reach the target site more easily than a more ridged drug.

2.3.3 Configuration

Configurational centres impose a rigid shape on sections of the molecule in which they occur. However their presence gives rise to geometric and optical isomerism. Since these stereoisomers have different shapes and properties they will often behave differently in biological systems. Biologically active stereoisomers, for example, will often exhibit differences in their potencies and/or activities (see Table 1.1). These pharmacological variations are particularly likely when a stereochemical centre is located in a critical position in the structure of the molecule. The consequence of these differences is that it is necessary to make and test separately the individual stereoisomers and the racemate of a drug. An exception *may* be made when a stereoisomer of a potential drug candidate is extracted from a natural source (see Chapter 6).

The stereochemistry of a drug will have an influence on both the pharmacodynamic and pharmacokinetic properties of a compound. In the pharmacodynamic phase stereoisomers and, where applicable, their racemates, usually exhibit the following general properties:

• almost identical activities and potencies;

- almost identical activities but significantly different potencies;

- completely different activities (in an extreme case one isomer may be inactive);

- the behaviour of the individual enantiomers will be different to that of the racemate.

In all these situations important differences may also be observed in the side effects of the stereoisomers and racemates.

    These variations mean that promising compounds that possess a chiral centre must have their individual pure enantiomers and any relevant racemates tested for activity and side effects. Consequently, medicinal chemists usually try to avoid synthesising analogues that contain chiral centres because of the difficulty of obtaining pure enantiomers and the increased cost of their testing and development. However, it may not be possible or necessary to test all the enantiomers of a compound when it is isolated from a natural source (see Chapter 6).

The influence of configuration on ADME

Absorption   The stereoselective nature of some of the processes occurring in ADME may result in stereoisomers exhibiting different pharmacokinetic properties. In absorption the rates of absorption of the individual pure enantiomers of drugs that are absorbed by active transport (see section 7.3.5) may be different. For example, (−)norgestrel is absorbed at twice the rate of (+)norgestrel through buccal and vaginal membranes whilst L-dopa is more rapidly absorbed than its enantiomer D-dopa. However, absorption of enantiomers by passive diffusion (see section 7.3.3) is not normally stereoselective so the rates of absorption of enantiomers are usually identical. Furthermore racemates may be absorbed at a different rate to their pure individual enantiomers.

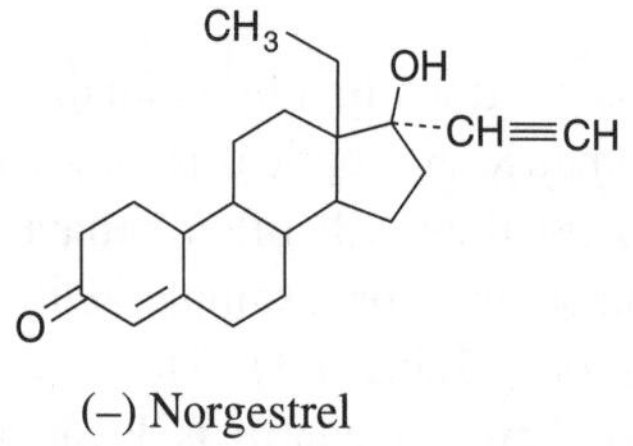

(−) Norgestrel 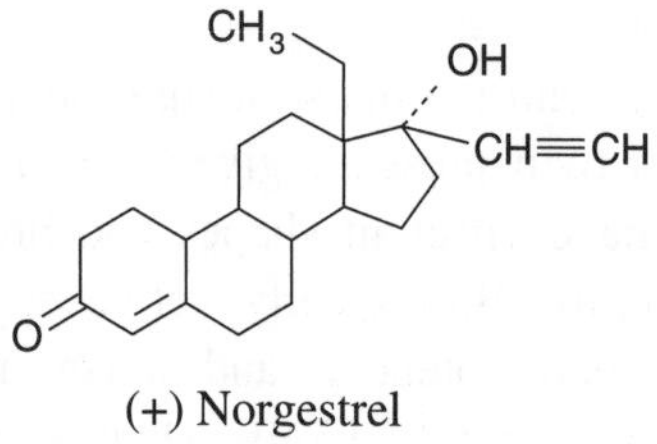   (+) Norgestrel

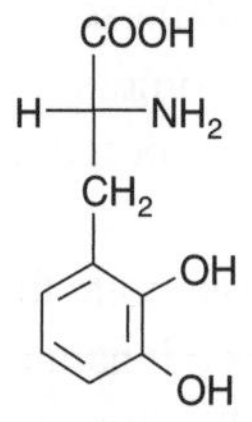

D-Dopa 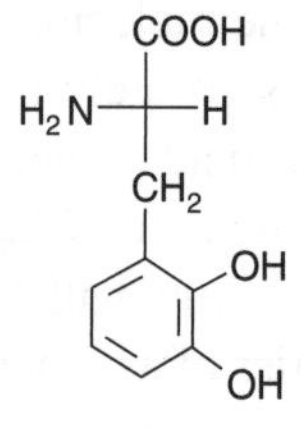   L-Dopa

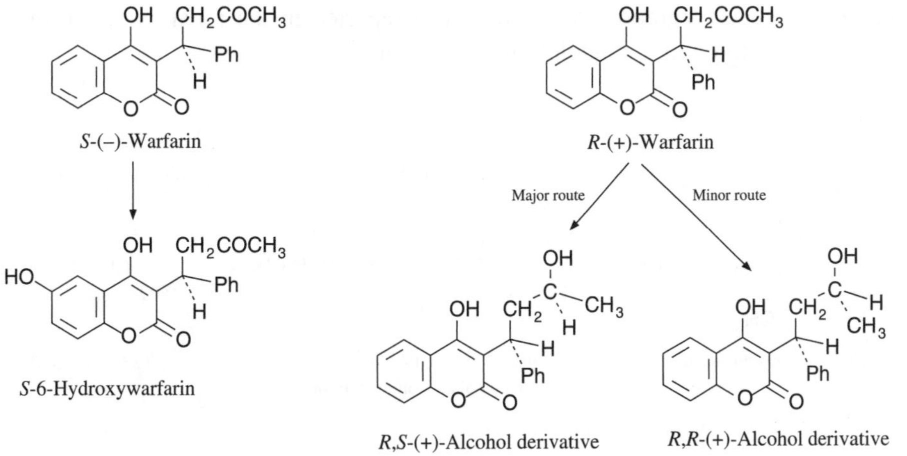

Figure 2.5 The different routes used in the metabolism of warfarin

Distribution Stereoselectivity appears to have little influence on the transport of stereo-isomers through the circulatory system. However, it has been shown that some enantiomers preferentially bind to a specific plasma protein. In humans, for example, *R*-propanolol preferentially binds to human albumin while the *S* isomer prefers α-acid glycoprotein. In contrast, stereoselectivity may, depending on the mechanism of transfer, influence the movement of drugs across the membranes separating one body compartment from another.

Metabolism Many metabolic processes are stereoselective. Consequently, stereoisomers that are metabolised by these processes may exhibit different behaviour patterns. For example, they might, like the enantiomers of warfarin, be metabolised by different routes to form different metabolites (Fig.2.5). Alternatively, their *in vivo* stabilities may be different. For example, the plasma half-life of *S*-indacrinone is 2–5 hours but the value for the *R* isomer is 10–12 hours. These and other differences between stereoisomers are believed to be often due to the stereospecific nature of the actions of the enzymes involved in metabolic processes.

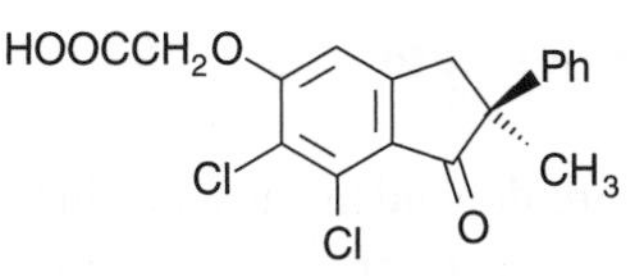

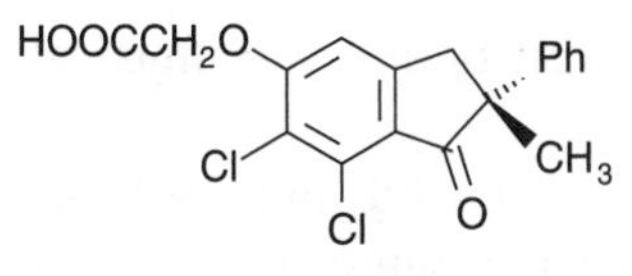

Excretion   A modest stereoselectivity has been reported in the renal excretion of a few drugs, such as chloroquine, pindolol and terbutaline.

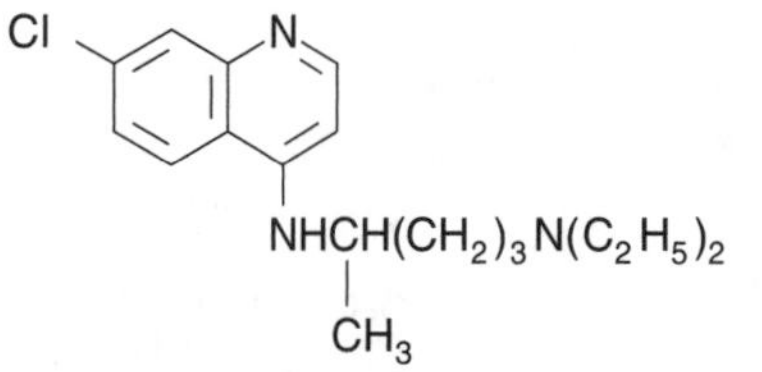

Chloroquine (antimalarial)

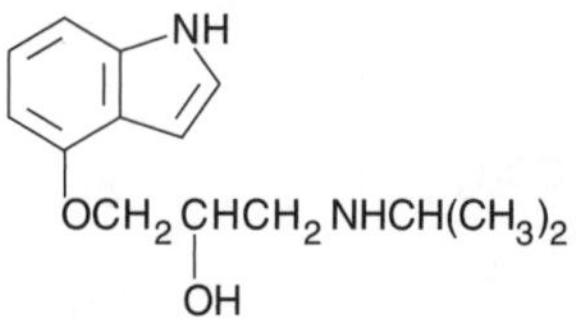

Pindolol (antihypertensive, antianginal and antiarrhythmic)

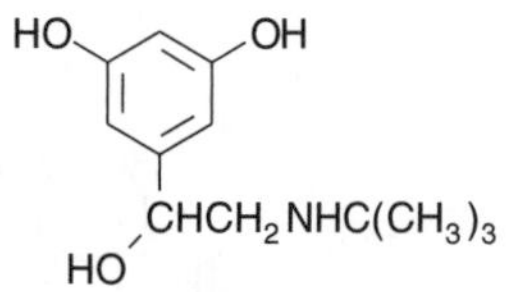

Terbuline (antifungal)

2.4   Solubility

The solubility of a drug both in water and lipids is an important factor in its effectiveness as a therapeutic agent and in the design of its dosage form For example, the absorption of drugs from the GI tract into the circulatory system by passive diffusion (see section 7.3) depends on them being water soluble. Moreover, the passage of drugs through other membranes will also depend on them having the correct balance of water and lipid solubilities (see sections 1.4.2 and 1.7.1). A drug's distribution through the circulatory system, and hence its action, will also depend to some extent on it having a reasonable water solubility. In addition, to be effective, most drugs have to be administered in dosage forms that are water-soluble.

A drug's solubility depends on both the chemical structure of the compound (see section 2.8) and the nature of the solvent (see section 2.6). Where a compound can exist in different polymorphic forms its solubility will also depend on its polymorphic form.

Several methods of predicting the solubility of a compound in a solvent have been proposed but none of these methods is accurate and comprehensive enough for general use. Consequently, the solubility of a drug is always determined by experiment. Since most drugs are administered at room temperature (25°C) and the body's temperature is 37°C the solubility of drugs is usually measured and recorded at these temperatures. However, the correlation between the activities of a series of drugs with similar structures and their solubilities in water is usually poor. This indicates that there are other factors playing important roles in controlling drug activity.

2.4.1   Solubility and the physical nature of the solute

The solubility of solids in all solvents is temperature dependent, usually increasing with increase in temperature. However, there are some exceptions. In medicinal chemistry the solubilities at room temperature (for drugs administered in solution) and body temperatures are of primary importance. The solubility of sparingly soluble substances that ionise in water can be recorded in the usual solubility units (cm^3 per 100 cm^3[% v/v], grams per 100 g of solvent [% w/w], g^3 per 100 cm^3[% w/v], molarity and molality). It may also be recorded in terms of its so-called *solubility product* (K_{sp}). Solubility product is the equilibrium constant for a heterogeneous system at constant temperature, comprising of a saturated solution of a sparingly soluble salt C_xA_y in contact with undissolved solid salt. It is defined as:

$$K_{sp} = [C^+]^x[A^-]^y \qquad (2.1)$$

K_{sp} is a measure of the limit of solubility of the salt, the larger its value the more soluble the salt.

The solubility of a solute that ionises in solution will be depressed by the presence of an ion from a different source. For example, the presence of A^- ions from an ionic compound BA that produces A^- ions in solution will depress the ionisation of an ionic compound CA and hence its solubility. This phenomenon is known as the *common ion effect*. For example, the hydrogen ions produced in the stomach will reduce the ionisation of all sparingly water-soluble acidic drugs in the stomach, which can improve their absorption into the blood stream through the stomach wall by increasing the concentration of unionised drug molecules in solution (see section 1.7.1). Uncharged molecules are normally transported more easily through biological membranes than charged molecules (see section 7.3.3). However, it must be realised that the degree of ionisation is not the only factor that could affect the absorption of a drug.

The solubility of liquids in solvents usually increases with temperature. However, the situation is complicated by the fact that some solute–solvent systems can exist as two or more immiscible phases at certain combinations of temperature and composition.

The solubility of a gas in a liquid depends on the temperature and pressure of the gas, its structure (see section 2.8) and the nature of the solvent (see section 2.9). As the temperature rises the solubility of almost all gases decreases. For example, air and carbon dioxide-free water can be prepared by boiling the water to remove these gases. At constant temperature the solubility (C_g) of a gas that does not react with the liquid is directly proportional to its partial pressure (P_g). This relationship is expressed mathematically by Henry's Law:

$$C_g = K_gP_g \qquad (2.2)$$

where K_g is a constant at constant temperature. The value of K_g is a characteristic property of the gas.

Henry's Law applies separately to each of the components of a mixture of gases and not the mixture as a whole. It is obeyed by many gases over a wide range of pressures. However, gases that react with the liquid usually show wide deviations from Henry's Law. For example, at constant temperature sparingly soluble gases such as hydrogen, nitrogen and oxygen dissolve in water according to Henry's Law but very soluble gases such as ammonia and hydrogen chloride that react with water do not obey

Henry's Law. Sparingly soluble gases that slightly react with water show small deviations from Henry's Law.

A consequence of Henry's Law is that the concentration of a gas dissolved in a biological fluid is often expressed in terms of its partial pressure as against the more conventional mass-based units (see section 2.4). For example, in healthy people the partial pressure of oxygen (pO_2) in arterial blood is about 100 mmHg whilst in venous blood it is about 40 mmHg.

The increase in the solubility of a gas with increase in pressure is the basis of hyperbaric medicine. In some situations the cells of the body cannot obtain sufficient oxygen even when the patient breaths in pure oxygen. For example, in carbon monoxide poisoning the carbon monoxide binds to the haemoglobin, which prevents it taking up oxygen in the lungs. As a result, the tissues are starved of oxygen and the patient could die. To alleviate this condition pure oxygen is administered under pressure to the patient. The increase in pressure results in oxygen directly dissolving in the plasma, which keeps the tissues sufficiently supplied for recovery to occur. However, pure oxygen is also highly toxic and must only be administered under strictly controlled conditions.

2.5   Solutions

A solution consists of particles, molecules or ions usually of the order of 0.1–1 nm in size, dispersed in a solvent. The small size of the solute means it cannot be detected by the naked eye and so solutions have a uniform appearance. As a solute particle moves through the solvent it is usually surrounded by a region of solvent molecules that move with it through the solution. This phenomenon is called *solvation* or, if it involves water molecules, *hydration*. The nature of this interaction between the solute particles and the solvated solvent molecules is not fully understood. However, it is believed that solvated molecules are bound to the solute by a variety of weak attractive forces such as hydrogen bonding, van der Waals' forces and dipole–dipole interactions (Fig. 2.6). The solvated solvent molecules are believed to stabilise the solution by preventing the solute particles coagulating into large enough particles to be precipitated. It follows that the stronger the solvation, the more stable the solution and the better the solubility of the solute.

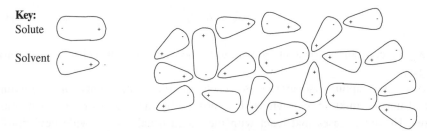

Figure 2.6   A representation of the dipole–dipole electrostatic attractions between molecules of the same and different types. Dipole–dipole attractions occur where the positive and negative ends of the dipoles are in close proximity

Solutes are generally classified as either *polar* or *non-polar*. *Polar solutes* have permanent dipoles and so there are strong electrostatic attractive forces between their particles and the polar water molecules. As a result, polar compounds will, where they are able, form stable aqueous solutions. On the other hand, *non-polar solutes* either have no dipole or one that is considerably smaller than those found in polar solutes. The attractive forces between non-polar molecules and water are likely to be weak and so non-polar molecules are usually less soluble in water than polar compounds. However, non-polar compounds are normally more soluble in non-aqueous solvents such as hexane and lipids than polar compounds. Little is known about the structure of the solutions formed when a solute dissolves in a liquid lipid. The stability of these solutions is believed to be due to hydrophobic interactions, hydrogen bonding and other dipole–dipole attractive forces between the solute and the lipid molecules. In practice, a rough guide to the solubility of a compound in a solvent may be predicted using the rule of thumb: *like dissolves like*.

A number of the physical properties of solutions vary with changes in physical conditions such as temperature and pressure. Solutions whose properties are *linearly* proportional to the concentration of the solute are referred to as *ideal solutions*. Conversely, solutions whose properties are not linearly proportional to the concentration of the solute are known as *non-ideal* solutions. This non-linear behaviour is due to the influence of intermolecular interactions in the solution. In general, the more concentrated a solution, the more likely it will exhibit non-ideal behaviour. However, apparent non-ideal behaviour will also be exhibited by solutes that dissociate or associate in solution if the degrees of association and dissociation are not taken into account.

2.6 The importance of water solubility

A drug's solubility and behaviour in water is particularly important since the cells in our bodies normally contain about 65 per cent water. In living matter water acts as an inert solvent, a dispersing medium for colloidal solutions and a nucleophilic reagent in numerous biological reactions. Furthermore, hydrogen bonding and hydrophobic interactions in water influence the conformations of biological macromolecules, which in turn affect their biological behaviour. Water solubility also makes drug toxicity testing, bioavailability evaluation and clinical application easier. As a result, it is necessary to assess the water solubility of drug candidates and, if required, design a reasonable degree of water solubility into their structures in an early point in their development.

Drugs administered orally as a solid or in suspension have to dissolve in the aqueous gastric fluid (*dissolution*, see section 11.5.1) before they can be absorbed and transported via the systemic circulation to their site of action. The rate and extent of dissolution of a drug is a major factor in controlling the absorption of that drug. This is because the concentration of the drug (see section 7.3.3) in the fluid in the gut lumen is one of the main factors governing the transfer of the drug through the membranes of the gastrointestinal tract (GI tract). The rate of dissolution depends on the surface area of the solid, which is dependent on both the physical nature of the dosage form of the drug and the chemical structure of the drug. However, the extent of dissolution depends only on the drug's solubility, which depends on the chemical structure of the drug. The dosage form is a formulation problem that is normally beyond the remit of medicinal chemist but the design of the structure of lead compounds with regard to solubility is within the realm of the medicinal chemist.

Once the drug has entered the circulatory system, either by absorption or by direct administration, its water solubility will influence its ease of transport to the body compartments available to that drug. Drugs that are sparingly soluble in water may be deposited on route to their site of action, which can clog up blood vessels and damage organs. For example, many sulphonamides, such as sulphamethoxazole, tend to crystallise in the kidney, which may result in serious liver and kidney damage. Water solubility also affects the ease of drug transport through cell membranes found throughout the general circulatory system.

Although a reasonable degree of water solubility is normally regarded as an essential requirement for a potential drug it is possible to utilise poor water solubility in drug action and therapy. For example, pyrantel embonate, which is used to treat pinworm and hookworm infestations of the GI tract, is insoluble in water. This poor water solubility coupled with the polar nature of the salt means that the drug is poorly absorbed from the gut and so the greater part of the dose is retained in the gastrointestinal tract, the drug's site of action. The low water solubility of a drug can also be used to produce drug depots, chewable dosage forms and mask bitter tasting drugs because taste depends on the substance forming an aqueous solution.

Pyrantel embonate
(Embonate)

Sulphamethoxazole (antibacterial)

The reactivity of water will also affect the stability of the drug in transit. Hydrolysis by water is one of the main routes for the metabolism of drugs containing ester, amide and other hydrolysable groups. For example, one of the metabolic pathways of the local anaesthetic lignocaine is hydrolysis to the amine.

Lignocaine

2,6-Dimethylaniline

The production of compounds with the required degree of water solubility early in the development of a new drug can considerably reduce the overall cost of development since it does not cause any delays in the later stages of development. For example, if it was found to be necessary to produce a water-soluble analogue of a lead compound at a later stage in the development the new analogue would have to be put through the same comprehensive testing procedure as the lead compound. This would necessitate the repeating of expensive toxicity and bioavailability trials, which could result in an expensive delay in the trials programme and possibly production. The importance of water solubility in drug action means that one of the medicinal chemist's development targets for a new drug is s to develop analogues that have the required degree of water solubility.

2.7 Solubility and the structure of the solute

The structure of a compound will influence its solubility in water and lipids. Its water solubility will depend on the number and nature of the polar groups in its structure as well as the size and nature of the compound's carbon–hydrogen skeleton. In general, the higher the ratio of polar groups to the total number of carbon atoms in the structure, the more water soluble the compound. Polar groups that ionise in water will usually result in a higher water solubility than those that do not ionise. However, aromatic compounds do tend to be less soluble in water than the corresponding non-aromatic compounds. Using these observations it is possible to compare, in a very general manner, the relative water solubilities of compounds with similar carbon skeletons.

The lipid solubility of a compound depends on the nature and number of non-polar groups in its structure. In general, the greater the number of non-polar groups in the structure of a compound, the greater that compound's lipid solubility. Consequently, the lipid solubility of analogues may be improved by replacing polar groups by significantly less polar groups or non-polar groups. Potential drug candidates whose structures contain both polar and non-polar groups will exhibit a degree of solubility in both water and lipids.

The water solubility of a lead compound can be improved by three general methods: salt formation (see section 2.8); by incorporating water solubilising groups into its structure (see section 2.9), especially those that can hydrogen bond with water; and the use of special dosage forms (see section 2.10). In salt formation, the activity of the drug is normally unchanged although its potency may be different. However, when new structural groups are incorporated into the structure of a drug the activity of the drug could be changed. Consequently, it will be necessary to carry out a full trials programme on the new analogue. Both of these modifications can be a costly process if they have to be carried out at a late stage in drug development. The use of specialised dosage forms does not usually need extensive additions to the trials programme but these formulation methods are only suitable for use with some drugs. However, in some circumstances it should be noted that poor water solubility is a desirable property for a drug (see section 2.6).

2.8   Salt formation

Salt formation usually improves the water solubility of acidic and basic drugs because the salts of these drugs dissociate in water to produce hydrated ions:

$$\text{Salt} \rightleftharpoons \text{Cation} + \text{Anion}$$

Hydrogen and hydroxide ions can disturb this equilibrium if they combine with the appropriate cation or anion to form less soluble acids or bases. Consequently, the pH of the biological fluid may affect the solubility of a drug and, as a result, its activity. In general, increasing the hydrophilic nature of the salt should increase its water solubility. However, there are numerous exceptions to this generalisation and each salt should be treated on its merits. Acidic drugs are usually converted to their metallic or amino salts whilst the salts of organic acids are normally used for basic drugs (Table 2.1).

Table 2.1 Examples of the acids and bases used to form the salts of drugs

Anions and anion sources	Cations and cation sources
Ethanoic acid $\sim$ ethanoate (CH_3COO^-)	Sodium $\sim$ sodium ion (Na^+)
Citric acid $\sim$ citrate (*see* Fig. 2.15)	Calcium $\sim$ calcium ion (Ca^{2+})
Lactic acid $\sim$ lactate (*see* 2.15)	Zinc $\sim$ zinc ion (Zn^{2+})
Tartaric acid $\sim$ tartrate (*see* Fig. 2.15)	Diethanolamine $\sim R_2NH_2^+$ (*see* Fig. 2.15)
Hydrochloric acid $\sim$ chloride (Cl^-)	*N*-Methylglucamine $\sim RNH_2CH_3$ (*see* Fig. 2.15)
Sulphuric acid $\sim$ sulphate (SO_4^{2-})	2-Aminoethanol ($HN_2CH_2CH_2OH$) $\sim RNH_3^+$
Sulphuric acid $\sim$ hydrogen sulphate (HSO_4^-)	

The degree of water solubility of a salt will depend on the structure of the acid or base used to form the salt. For example, acids and bases whose structures contain water solubilising groups will form salts with a higher water solubility than compounds that do not contain these groups. (Fig. 2.7). However, if a drug is too water soluble it will not dissolve in lipids and so will not usually be readily transported through lipid membranes. This normally results in either its activity being reduced or the time for its onset of action being increased. It should also be noted that the presence of a high concentration of chloride ions in the stomach will reduce the solubility of sparingly soluble chloride salts because of the common ion effect (see section 2.4.1).

Water-insoluble salts are often less active than the water-soluble salts since it is more difficult for them to reach their site of action (see sections 1.4.2 and 1.7.1). However, in some cases this insolubility can be utilised in delivering the drug to its site of action. For example, pyrantel embonate (see section 2.6), which is used to treat pinworm and hookworm infestations of the GI tract, is insoluble in water and so is not removed by absorption from the GI tract, its site of action.

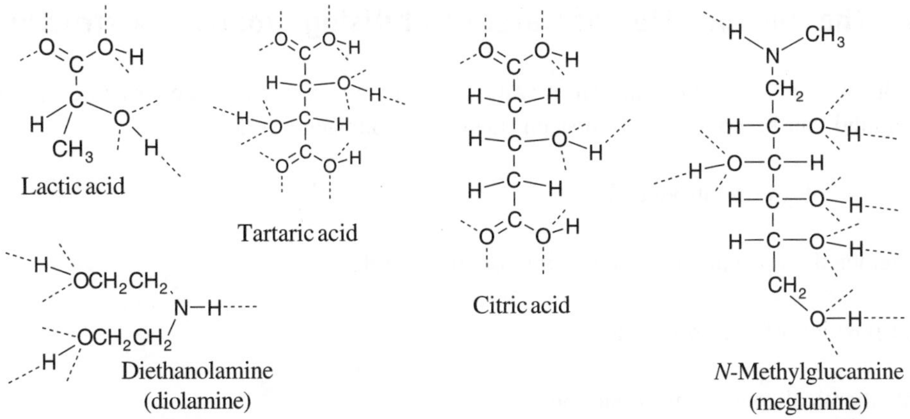

Figure 2.7 Examples of the structures of acids and bases whose structures contain water solubilising groups. The possible positions of hydrogen bonds are shown by the dashed lines; lone pairs are omitted for clarity. At room temperature it is highly unlikely that all the possible hydrogen bonds will be formed. *Note*: hydrogen bonds are not shown for the acidic protons of the acids as these protons are donated to the base on salt formation. Similarly no hydrogen bonds are shown for the lone pairs of the amino groups because these lone pairs accept a proton in salt formation

Some water-insoluble salts dissociate in the small intestine to liberate the component acid and base. This property has been utilised in drug delivery, for example erythromycin stearate dissociates in the small intestine to liberate the antibiotic erythromycin, which is absorbed as the free base. Salts with a low water solubility can also be used as a drug depot. For example, penicillin G procaine has a solubility of about 0.5 g in 100 g of water. When this salt is administered as a suspension by intramuscular injection it acts as a depot by slowly releasing penicillin. Salt formation is also used to change the taste of drugs to make them more palatable to the patient. For example, the antipsychotic chlorpromazine hydrochloride is water soluble but has a very bitter taste that is unacceptable to some patients. However, the water-insoluble embonate salt is almost tasteless and so is a useful alternative as it can be administered orally in the form of a suspension.

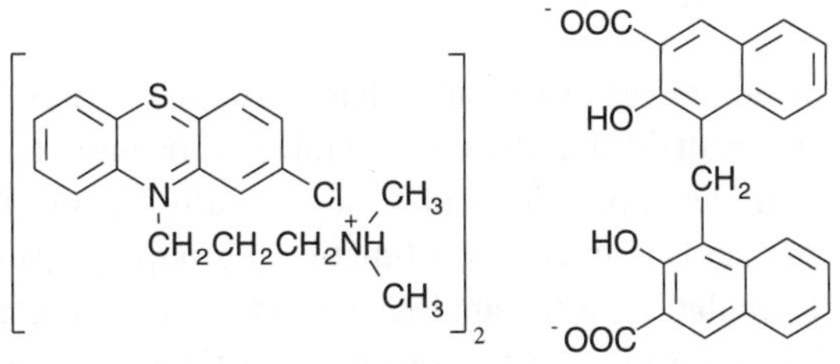

Chlorpromazine embonate

2.9   The incorporation of water solubilising groups in a structure

The discussion of the introduction of water solubilising groups into the structure of a lead compound can be conveniently broken down into four general areas:

- the type of group introduced;

- whether the introduction is reversible or irreversible;

- the position of incorporation;

- the chemical route of introduction.

2.9.1   The type of group

The incorporation of polar groups into the structure of a compound will normally result in the formation of an analogue with a better water solubility than its parent lead compound. Polar groups that either ionise or are capable of relatively strong intermolecular forces of attraction with water will usually result in analogues with an increased water solubility. For example, the incorporation of strongly polar alcohol, amine, amide, carboxylic acid, sulphonic acid and phosphorus oxyacid groups, which form relatively stable hydrates with water, would be expected to result in analogues that are more water soluble than those formed by the introduction of the less polar ether, aldehyde and ketonic functional groups. The introduction of acidic and basic groups is particularly useful since these groups can be used to form salts (see section 2.8), which would give a wider range of dosage forms for the final product. However, the formation of zwitterions by the introduction of either an acid group to a structure containing a base or a base group into a structure containing an acid group can reduce water solubility. Introduction of weakly polar groups such as carboxylic acid esters, aryl halides and alkyl halides will not significantly improve water solubility and can result in enhanced lipid solubility. As well as individual functional groups, multifunctional group structures such as glucose residues can also be introduced. In all cases the degree of solubility obtained by the incorporation cannot be accurately predicted since it also depends on other factors. Consequently, the type of group introduced is generally selected on the basis of previous experience.

The incorporation of acidic residues into a lead structure is less likely to change the type of activity but can result in the analogue exhibiting haemolytic properties. Furthermore, the introduction of an aromatic acid group usually results in anti-inflammatory activity whilst carboxylic acids with an alpha functional group may act as chelating agents. It also means that the formulation of the analogue as its salt is restricted to mainly metallic cations. This can result in a surfeit of these ions in the patient, which could be detrimental. In addition, the introduction of an acid group into drugs whose structures contain basic groups would result in zwitterion formation in solution with a possible reduction in solubility.

Basic water solubilising groups have a tendency to change the mode of action since bases often interfere with neurotransmitters and biological processes involving amines. However, their incorporation does mean that the analogue can be formulated as a wide variety of acid salts. The introduction of a basic group into drugs whose structures contain acid groups could result in zwitterion formation in solution with a possible reduction in solubility. Non-ionisable groups do not have the disadvantages of acidic and basic groups.

2.9.2   Reversible and irreversible groups

The type of group selected also depends on the degree of permanency required. Groups that are bound directly to the carbon skeleton of the lead by less reactive C–C, C–O and C–N bonds are likely to be irreversibly attached to the lead structure. Groups that are linked to the lead by ester, amide, phosphate, sulphate and glycosidic links are more likely to be metabolised from the resulting analogue to reform the parent lead as the analogue is transferred from its point of administration to its site of action. Compounds with this type of solubilising group are acting as prodrugs (see section 12.8) and so their activity is more likely to be the same as the parent lead compound. However, the rate of loss of the solubilising group will depend on the nature of the transfer route and this could affect the activity of the drug.

2.9.3   The position of the water solubilising group

The position of the new water solubilising group will depend on the reactivity of the lead compound and the position of its pharmacophore. Initially, the former requires a general appraisal of the chemistry of the functional groups found in the lead compound. For example, if the lead structure contains aromatic ring systems it may undergo electrophilic substitution in these ring systems whilst aldehyde groups are susceptible to oxidation reduction, nucleophilic addition and condensation. This general reactivity must be taken into account when selecting a method to introduce a water solubilising group.

In order to preserve the type of activity exhibited by the lead compound, the water solubilising group should be attached to a part of the structure that is not involved in the drug–receptor interaction. Consequently, the route used to introduce a new water solubilising group and its position in the lead structure will depend on the relative reactivities of the pharmacophore and the rest of the molecule. The reagents used to introduce the new water solubilising group should be chosen on the basis that they do not react with, or in close proximity to, the pharmacophore. This will reduce the possibility of the new group affecting the relevant drug–receptor interactions.

2.9.4   Methods of introduction

Water solubilising groups are best introduced at the beginning of a drug synthesis although they may be introduced at any stage. Introduction at the beginning avoids the problem of a later introduction changing the type and/or nature of the drug–receptor interaction. A wide variety of routes may be used to introduce a water solubilising group, the one selected depending on the type of group being introduced and the chemical nature of the target structure. Many of these routes require the use of protecting agents to prevent unwanted reactions of either the water solubilising group or the lead structure.

Carboxylic acid groups by alkylation

Carboxylic acid groups can be introduced by alkylation of alcohols, phenols and amines with suitably substituted acid derivatives: O-alkylation may be achieved by a Williamson's synthesis using both hydroxy- and halo-substituted acid derivatives, whereas only halo-substituted derivatives are used for N-alkylation (Fig. 2.8).

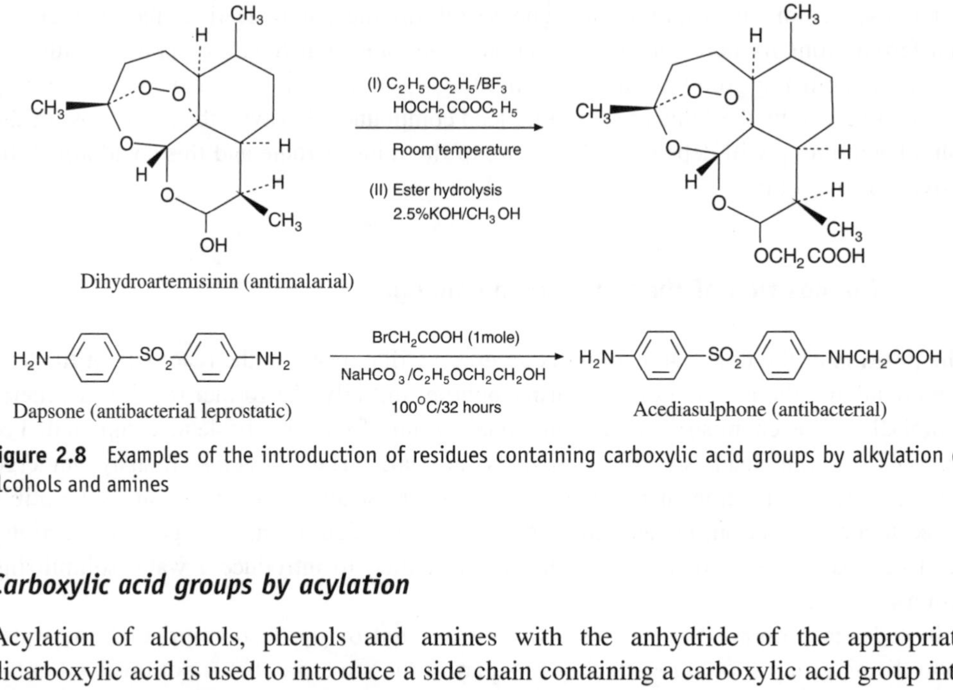

Figure 2.8 Examples of the introduction of residues containing carboxylic acid groups by alkylation of alcohols and amines

Carboxylic acid groups by acylation

Acylation of alcohols, phenols and amines with the anhydride of the appropriate dicarboxylic acid is used to introduce a side chain containing a carboxylic acid group into the lead structure. For example, succinic anhydride is used to produce chloramphenicol sodium succinate and succinyl sulphathiazole. The resulting esters and amides can be formulated as their metallic or amine salts. However, since the esters are liable to hydrolysis in aqueous solution the stability of the resulting analogue in aqueous solution must be assessed. For example, chloramphenicol sodium succinate is so unstable in aqueous solution that it is supplied as a lyophilised powder that is only dissolved in water when it is required for use. This solution must be used within 48 hours.

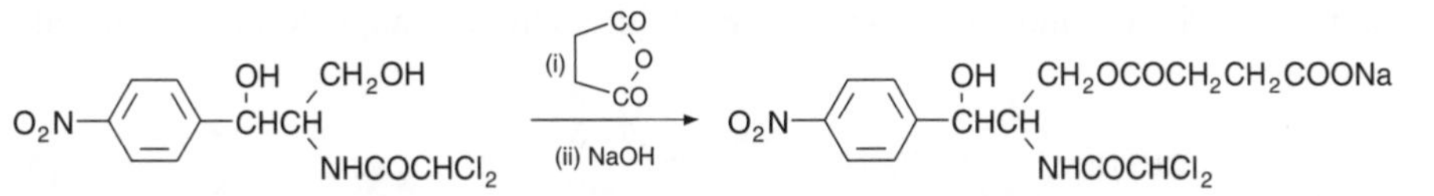

Chloramphenicol (antibacterial)   Chloramphenicol sodium succinate (antibacterial)

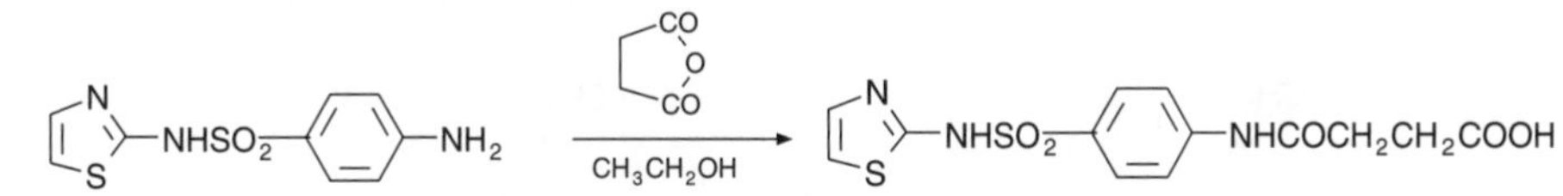

Sulphathiazole (antibacterial)   Succinyl sulphathiazole (antibacterial)

Phosphate groups

A number of phosphoric acid halide derivatives have been successfully used to attach phosphate groups to hydroxy groups in drug structures. The hydroxy groups of the acid halide must normally be protected by a suitable protecting group (Fig. 2.9). These protecting groups are removed in the final stage of the synthesis to reveal the water solubilising phosphate ester. The resulting phosphate esters tend to be more stable in aqueous solution than carboxylic acid esters.

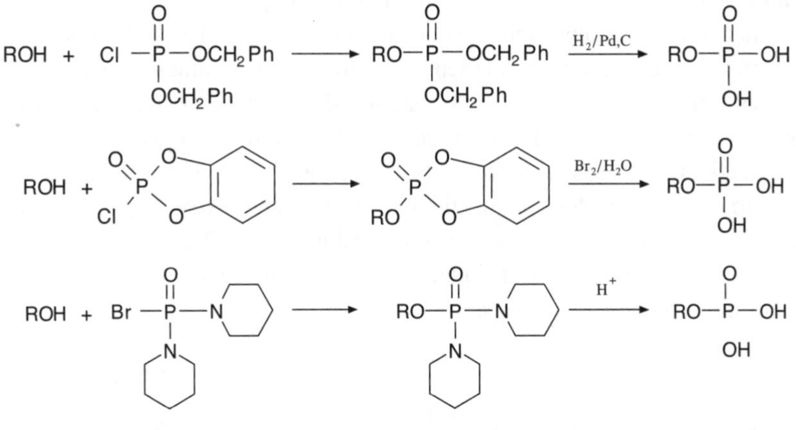

Figure 2.9 Some general methods of phosphorylating alcohols

Sulphonic acid groups

Sulphonic acid groups can be incorporated into the structures of lead compounds by direct sulphonation with concentrated sulphuric acid.

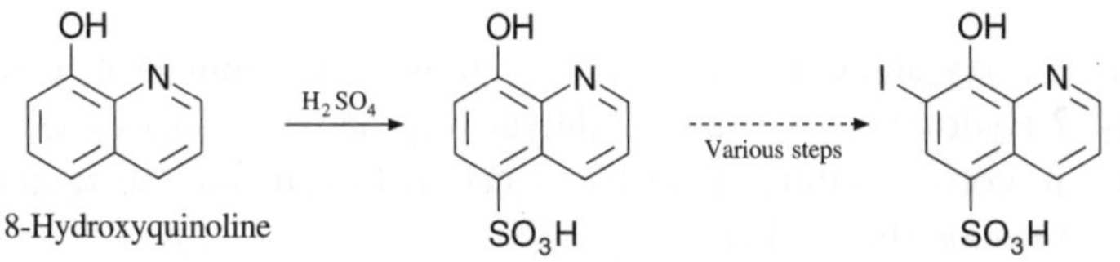

8-Hydroxyquinoline   8-Hydroxy-7-iodo-5-quinolinesulphonic acid (topical antiseptic)

Alternative routes are the addition of sodium bisulphite to conjugated C=C bonds and the reaction of primary and secondary amines with sodium bisulphate and methanal.

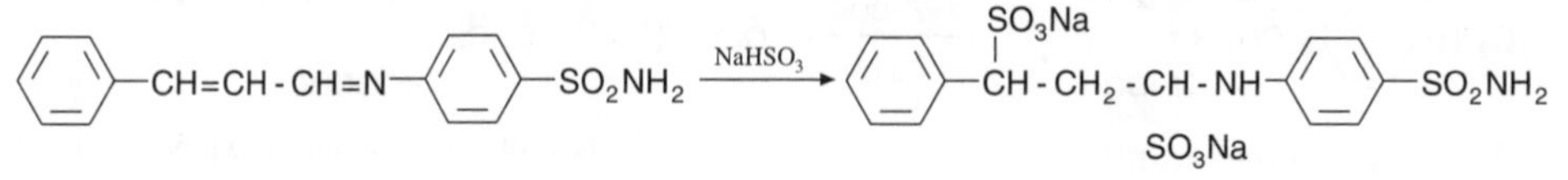

N^4-Cinnamylidenesulphanilamide

Noprylsulphamide (antibacterial)

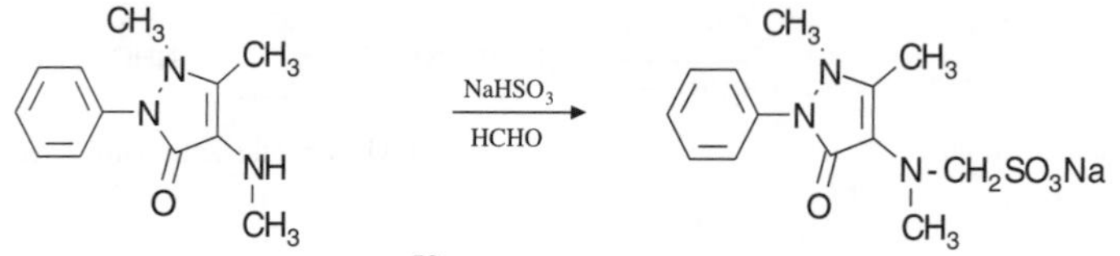

Noraminopyrine (analgesic, antipyretic)

59

Dipyrone (analgesic, antipyretic)

Incorporation of basic groups

Water solubilising groups containing basic groups can be incorporated into a lead structure by the alkylation and acylation of alcohols, phenols and amines. Alkylation is achieved by the use of an alkyl halide whose structure contains a basic group whilst acylation usually involves the use of acid halides and anhydrides. In both alkylation and acylation the basic groups in the precursor structure must be either unreactive or protected by suitable groups (Fig.2.10).

Amide derivatives are usually more stable than esters. Esters are often rapidly hydrolysed in serum, the reaction being catalysed by serum esterases. In these cases the analogue effectively acts as a prodrug (see section 12.8). The introduction of water solubilising amino-acid residues using standard peptide chemistry preparative methods has also been used successfully to introduce basic residues (Fig. 2.11).

The Mannich reaction is also frequently used to introduce basic groups into both aromatic and non-aromatic structures.

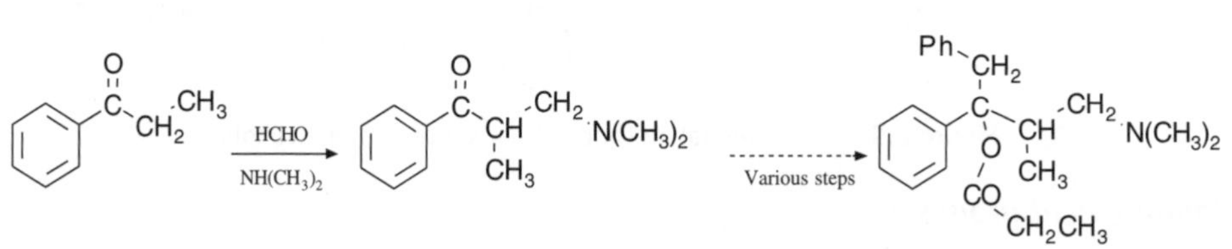

4-Nitrobenzoyl chloride

Dextropropoxyphene (narcotic analgesic)

Polyhydroxy and ether residues

The introduction of polyhydroxy and ether chains has been used in a number of cases to improve water solubility. 2-Hydroxyethoxy and 2,3-dihydroxypropoxy residues have been introduced by reaction of the corresponding monochlorinated hydrin and the use of suitable epoxides, amongst other methods (Fig. 2.12).

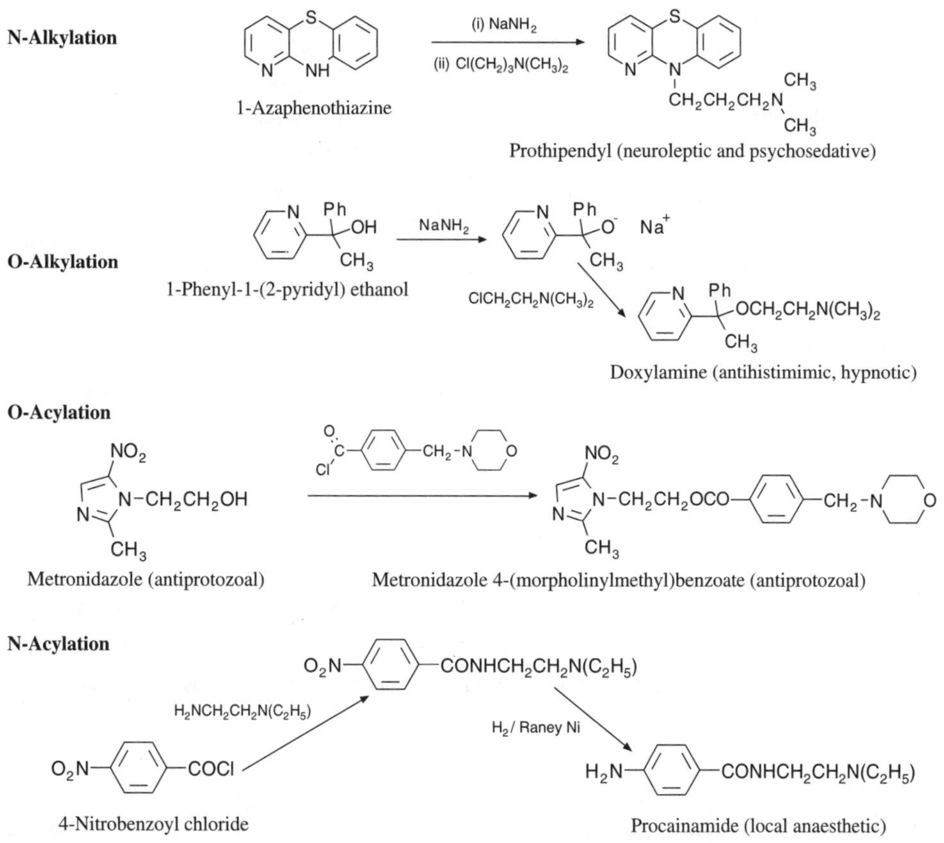

Figure 2.10 Examples of the introduction of basic groups by alkylation and acylation. These basic groups are used to form the more water-soluble salts of the drug, in which form it is usually administered

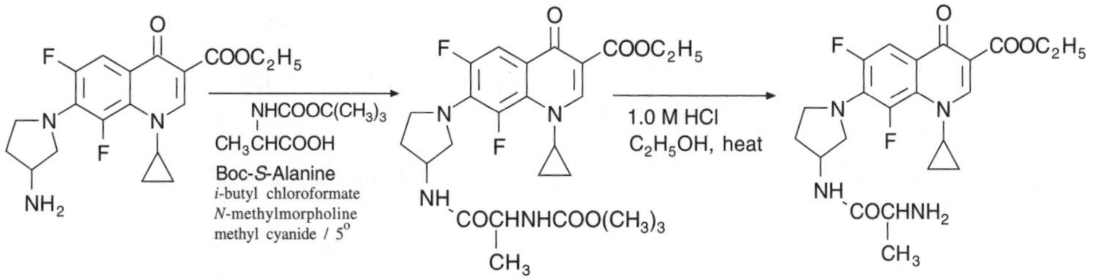

Figure 2.11 The introduction of amino acid residues using Boc(t-butyloxycarbonyl)as a protecting group. The second step is the removal of the protecting group

Sugar residues as water solubilising groups are rarely incorporated by O-glycosidic links but they have been attached to a number of drugs through N-glycosidic links involving the nitrogen atoms of amines and hydrazide groups (Fig. 2.13), as well as N-acylation of amino sugars by suitable acid halides.

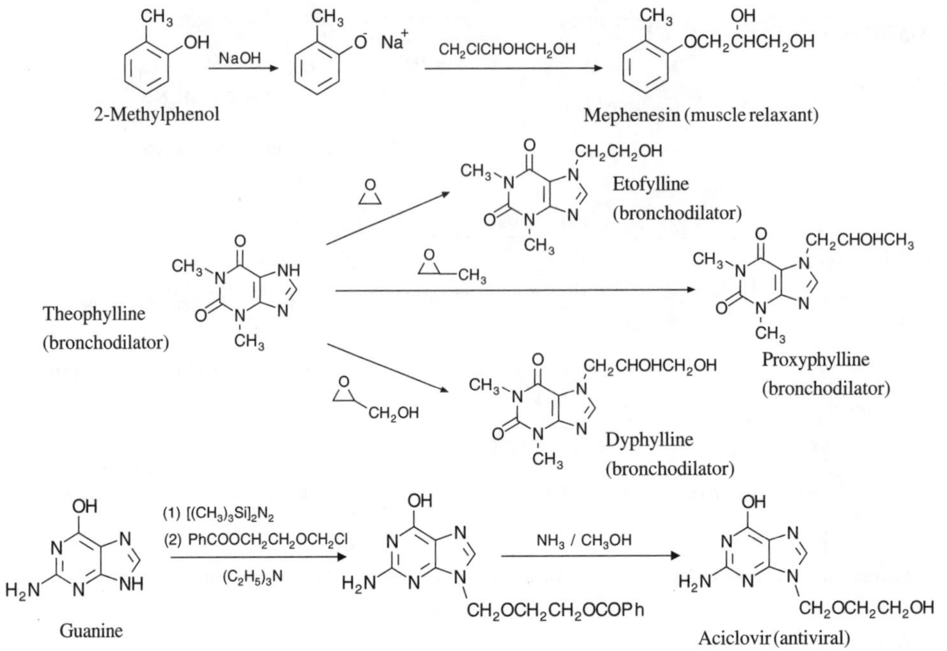

Figure 2.12   Examples of the methods of introducing polyhydroxy and ether groups into a lead structure

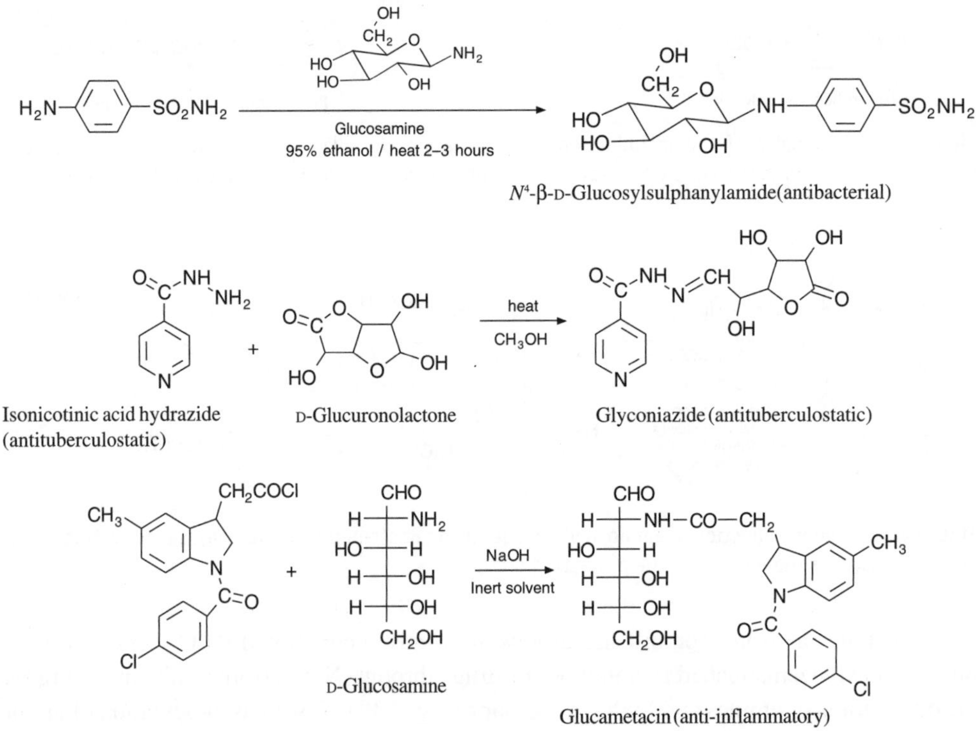

Figure 2.13   Examples of the use of sugar residues to increase water solubilities of drugs

2.9.5   Improving lipid solubility

The commonest way to improve lipid solubility is to either introduce non-polar groups into the structure or replace polar groups by less polar groups. Methyl, fluoro and chloro groups (see sections 3.4.1 and 3.4.2) are commonly used for this purpose. These introductions may be made using appropriate organic chemical reactions.

2.10   Formulation methods of improving water solubility

The delivery of water-insoluble or sparingly soluble drugs to their site of action can be improved by the formulation of the dosage form. Techniques include the use of cosolvents, colloidal particles, surfactants (see section 2.13), micelles (see section 2.13.2), liposomes (see section 2.13.3) and complexing with water-soluble compounds such as the cyclodextrins (see section 1.4.4)

2.10.1   Cosolvents

The addition of a water-soluble solvent (cosolvent) can improve the water solubility of a sparingly soluble compound. Cosolvents for pharmaceutical use should have minimal toxic effects and not affect the stability of the drug. Consequently, the concentration of the cosolvent used must be within the acceptable degree of toxicity associated with that cosolvent. In pharmaceutical preparations, the choice of solvents is usually limited to alcohols such as ethanol, propan-2-ol, 1,2-dihydroxypropane, glycerol and sorbitol and some low molecular mass polyethylene glycols, however other solvents are sometimes used. Mixtures of cosolvents are also used to achieve the required degree of solubility. For example, paracetamol is formulated as an elixir in an aqueous sucrose solution by the use of a mixture of ethanol and 1,2-dihydroxypropane.

2.10.2   Colloidal solutions

Sparingly water-soluble drugs and potential drugs can be dispersed in an aqueous medium as colloidal sized particles with diameters of 1–1000 nm. Colloidal solutions are known generally as *sols*. Water-based colloids are often referred to as *hydrosols*. Hydrosols offer a potential method for the formulation of drugs that are sparingly soluble in water. For example, hydrosols of the poorly water-soluble drugs cyclosporin and isradipine have been prepared and found to be stable for five or more days. Hydrosols stored as spray-dried powders have been successfully reconstituted as liquid sols after several years of storage under cool dry conditions. Hydrosols used for parenteral delivery systems normally contain colloidal particles with a diameter of less than 200 nm because colloidal particles below this size will not block small capillaries. They may be prepared by dissolving a high concentration of the drug in an organic solvent that is miscible with water. This concentrated solution is rapidly mixed with an aqueous solution containing a suitable stabiliser. These stabilisers may be either an electrolytic salt or a polymer. Both of these types of stabiliser act by being adsorbed onto the surface of the colloidal particles. In the case of electrolytic stabilisers the colloidal particles gain an electrostatic charge, which repels other colloidal particles and prevents them coagulating into particles large enough to precipitate. Polymer stabilisers prevent coagulation by steric hindrance. The degree of stabilisation will depend on the nature of both the hydrosol and the stabiliser. However, all hydrosols will slowly crystallise and because of this are normally stored as either freeze-dried or spray-dried powders.

Colloidal solutions can also be prepared by mechanically grinding larger particles into colloidal sized particles. This process is not very efficient and in most cases only about 5 per cent of the material is converted into a sol. It also produces particles with very different sizes.

Sparingly water-soluble drugs may also be delivered using so-called *nanoparticles*. These are colloidal particles that are either solid (nanospheres) or hollow (nanocapsules). The drug is either adsorbed on the exterior or trapped in the interior of both types of nanoparticle. To date they have been used only for experimental and clinical testing.

Nanoparticles are formed by a variety of methods using a wide range of materials, for example polysaccharides, proteins, polyacrylates and polyamides. The wide range of precursors means that nanoparticles may be produced with properties that can be used to test specific aspects of drug activity.

2.10.3   Emulsions

Emulsions are systems in which one liquid is dispersed as fine droplets with diameters of 0.1–100 μm in a second liquid. Dosage forms usually consist of either oil-in-water (o/w) where the dispersed medium is the oil or water-in-oil (w/o) where water is the dispersed phase. More complex systems where either water droplets are encased in oil drops dispersed in water (w/o/w) or the reverse where oil droplets are encased in water droplets dispersed in an oil medium (o/w/o) are also known. These emulsion systems are intrinsically unstable and so the formation of a stable emulsion normally requires additional components known as emulsifying agents. These are surfactants (see section 2.13) such as glyceryl monostearate and polyoxyethylene sorbitan monooleate (Tween 80) that dissolve in both water and oils. Emulsions are frequently made using mixtures of emulsifying agents as this has been found to give more stable mixtures than those prepared using a single emulsifying agent.

All types of emulsion can be used as a delivery vehicle for liquid drugs that are not very soluble in water. However, this form of drug delivery is not widely used even though there is some evidence that administering some drugs as emulsions improves their bioavailability. For example, griseofulvin orally administered to rats is more readily absorbed from a corn oil-in-water emulsion than other types of orally administered dosage forms (Fig. 2.14).

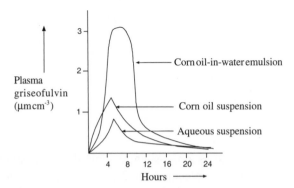

Figure 2.14 The change in plasma concentrations of griseofulvin after administration using different dosage forms. Reproduced with permission from P. J. Carrigan and T. R. Bates, © Wiley-Liss, Inc, a subsidiary of John Wiley & Sons, Ltd, 1973 *J. Pharm. Sci.*, **62**, 1476 (1973).

2.11 The effect of pH on the solubility of acidic and basic drugs

Aqueous biological fluids are complex systems that contain a variety of different solutes. These species will have an effect on each other's relative solubilities and the solubilities of drugs or xenobiotics introduced into the fluid. They also control the pH of the fluid, which is of considerable importance to the bioavailability of acidic and basic drugs. An acidic or basic pH will either enhance or reduce the ionisation of these drugs with subsequent changes in their solubility and absorption through membranes. The degree of ionisation of weak monobasic *acidic* drugs at different pH values may be calculated using the Henderson–Hasselbalch equation:

$$pK_a = pH + \log \frac{[\text{Non-ionised form}]}{[\text{Ionised form}]} \tag{2.3}$$

Example 2.1. *The pK_a of aspirin, a weak acid, is 3.5. Calculate the degree of ionisation of aspirin in the (a) stomach and (b) intestine if the pH of the contents of the stomach is 1 and the pH of the contents of the intestine is 6.*

Substitute the values of pH and pK_a into equation (2.3) for each of the two problems:

(a) $3.5 = 1 + \log \dfrac{[unionised\ form]}{[ionised\ form]}$

therefore:

$\log \dfrac{[unionised\ form]}{[ionised\ form]} = 3.5 - 1 = 2.5$

(b) $3.5 = 6 + \log \dfrac{[unionised\ form]}{[ionised\ form]}$

therefore:

$\log \dfrac{[unionised\ form]}{[ionised\ form]} = 3.5 - 6 = -2.5$

and

$$\frac{[unionised\ form]}{[ionised\ form]} = antilog\ 2.5$$

$$= 316.23$$

and

$$\frac{[unionised\ form]}{[ionised\ form]} = antilog - 2.5$$

$$= \frac{1}{antilog\ 2.5}$$

$$= \frac{1}{316.23}$$

These figures show that aspirin is only slightly ionised in the stomach (one ionised molecule for every 316 unionised molecules) but is almost completely ionised (316 ionised molecules for every one unionised molecule) in the intestine. As a result, under the pH conditions specified in Example 2.1 aspirin will be more readily absorbed in the stomach than in the intestine since drugs are more easily transferred through a membrane in their unionised form. The low degree of ionisation of aspirin in the stomach accounts for the relative ease of its absorption from the stomach even though aspirin is almost insoluble in water at 37°C. However, the degree of ionisation is not the only factor influencing a drug's absorption and although it may offer a good explanation for the behaviour of one drug it does not explain the rate and ease of absorption of all drugs. Other factors may be more significant. For example, the weakly acidic drugs thiopentone, secobarbitone and barbitone (barbital) have pK_a values of 7.6, 7.9 and 7.8, respectively. Consequently, their degrees of ionisation in water are almost the same but their rates of absorption from the stomach are very different. This indicates that other factors besides ionisation influence the transport of these drugs through the stomach membrane. This observation is supported by the fact that the partition coefficients (see section 3.7.2) for the chloroform/water system of these drugs are considerably different, namely:

thiopentone > 100, secobarbitone > 23 and barbitone > 0.7

The degree of ionisation of weak monoacidic *basic* drugs at different pH values may also be calculated in a similar manner using the Henderson–Hasselbalch equation for the monoacidic bases:

$$pK_a = pH + \log \frac{[Ionised\ form]}{[Non\text{-}ionised\ form]} \tag{2.4}$$

The presence of soluble acids and bases in biological fluids can result in the formation of the corresponding salts of acidic and basic drugs, which usually have different solubilities to those of the parent drugs. Salt formation is frequently used to enhance the water solubility of a sparingly soluble drug (see section 2.8). However, salt formation does not always result in an increase in solubility.

The solubility of compounds such as peptides and proteins, which contain both acidic and basic groups, is complicated by internal salt formation. Peptides and proteins have their lowest solubilities near their isoelectric points (Fig. 2.15) where the solution contains the

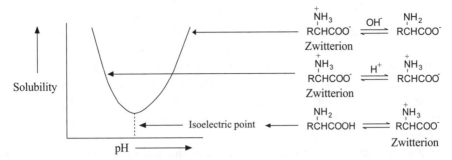

Figure 2.15 A typical solubility curve for a peptide. R represents the rest of the structure of the peptide or protein. It should be realised that the acidic and basic groups do not have to be adjacent to each other in a peptide or protein and that these compounds usually have a number of free acid and base groups

internal salt (zwitterion). On the lower pH value side of the isoelectric point a peptide or protein will form the corresponding cation, whilst on the higher pH value side they will form the corresponding anion. Both the cationic and anionic forms of the peptide will exhibit a higher solubility than its zwitterion. These variations of solubility will affect the therapeutic effectiveness of the peptide and so will influence the design of their dosage forms.

Biological fluids, such as plasma, contain dissolved electrolytes and non-electrolytes. The presence of an electrolyte in an aqueous solution will usually increase the solubilities of any other electrolytes in solution provided that the electrolytes do not have any ions in common (see section 2.4.1). However, the presence of electrolytes reduces the solubility of non-electrolytes in aqueous solution. The cations and anions from the electrolyte form stronger bonds with the water molecules and form hydrates more readily with water than the non-electrolyte molecules. As a result, the ions displace the non-electrolyte molecules from their weaker hydrates with a subsequent reduction in the solubility of the non-electrolyte.

$$\text{Non-electrolyte hydrate} + \text{Ion} \rightarrow \text{Ion hydrate} + \text{Non-electrolyte}$$

If sufficient electrolyte is added to an aqueous solution the non-electrolyte is precipitated out of solution. The process is referred to as *salting out*. Peptides and proteins are especially sensitive to salting out and the technique is often used to isolate these compounds from solution. Consequently, the solubility of peptide drugs in electrolytes is a factor that needs to be considered in the design of the dosage forms used in trials and with patients. The presence of dissolved non-electrolytes reduces the solubility of electrolytes. This is because the presence of the non-electrolyte reduces the dielectric constant of the system, which reduces the degree of ionisation of the electrolyte and results in a corresponding decrease in the solubility of the electrolyte.

2.12 Partition

The transport of a drug to its site of action normally involves the drug having to pass through a large number of lipid membranes. Consequently, the relative solubilities of a drug in aqueous medium or lipid is of considerable importance in the transport of that drug to its site of action, especially at the aqueous medium/lipid interface. *Partition coefficients* are a measure of the way a compound distributes itself between two immiscible solvents and so attempts have been made to correlate the activities of drugs with their lipid/water partition coefficients. These correlations have been used with some degree of success to predict the activities of compounds that could be potential drugs. However, the results are only valid in situations where solubility and transport by diffusion through a membrane are the main factors controlling drug action.

It is not easy to measure partition coefficients *in situ* and so, instead, the less accurate organic solvent/aqueous solution model systems are used. The partition coefficients of these model systems are usually calculated assuming ideal solutions, using equation (2.5).

$$\text{Partition coefficient}(P) = \frac{[\text{Drug in the organic phase}]}{[\text{Drug in an aqueous phase}]} \tag{2.5}$$

The partition coefficient P is a constant for the specified system provided that the temperature is kept constant and ideal dilute solutions are used. However, a five degree change in temperature does not normally cause a significant change in the value of the partition coefficient. Since the charged form of a drug is not easily transferred through a membrane, a more useful form of equation (2.5) for biological investigations is:

$$P = \frac{[\text{Non-ionised drug in the organic phase}]}{[\text{Non-ionised drug in an aqueous phase}]} \tag{2.6}$$

The values of partition coefficients are normally measured at either 25°C or 37°C and for a particular solute will be different for different solvent systems. *n*-Octanol is the most commonly used organic phase in pharmacological investigations but other organic solvents such as butanol, chloroform and olive oil are also used. The aqueous phase is either water or a phosphate buffer at pH 7.4, the pH of blood. A high P value for the partition coefficient indicates that the compound will readily diffuse into lipid membranes and fatty tissue. The compound is said to be *hydrophobic* (*water hating*) and have a high *hydrophobicity*. A low P value indicates that the compound is reluctant to enter lipid material and prefers to stay in the more polar aqueous medium. Compounds of this type are said to be *hydrophilic* (*water loving*) and have a low hydrophobicity. Hydrophobicity can have a significant effect on biological activity. For example, compounds with relatively high hydrophobicity (relatively high P value) will easily enter a membrane but will be reluctant to leave and so will not be readily transported through the membrane if diffusion is the only transport mechanism for the drug. This means that the drug could fail to reach its site of action in sufficient quantity to be effective (see section 11.1) unless hydrophobicity is the dominant factor in its action. Conversely, analogues with relatively low hydrophobicities will not easily diffuse into a membrane and so could al so fail to reach their site of action in effective quantities.

If hydrophobicity is the most important factor in drug action, an increase in hydrophobicity usually results in an increase in action. For example, general anaesthetics are believed to act by dissolving in cell membranes. The octanol/water partition coefficients of the anaesthetics diethyl ether, chloroform and halothane are 0.98, 1.97 and 2.3, respectively, which indicates that halothane would be the most soluble of these in lipid membranes. This corresponds to their relative activities, with halothane being the most potent. Consequently, the hydrophobicity of a compound in terms of its partition coefficient is one of the factors considered in the QSAR approach to drug design (see Section 3.7).

2.12.1 Practical determination of partition coefficients

The traditional method is to mutually saturate the two liquids involved in the determination by shaking them together for a set period of time. At the end of this time the solute compound is added and the shaking is continued in a constant temperature bath until the system has reached equilibrium. The concentration of the solute is determined by a method appropriate for the solute and a value for the partition calculated using equations (2.5) and/ or (2.6). This method of determining partition coefficients usually takes several hours and so is not always feasible when a significant amount of the compound decomposes within this time frame. A number of alternative methods are available, such as using high-pressure liquid chromatograpy (see Section 6.7.2). This method is based on partitioning the compound between a solid non-polar support such as octadecyl silica or a non-polar support with a coating of octanol or other-non polar oil and a mobile phase containing some water. The partition coefficient may be calculated using:

$$Log\ P = m\ Log(\text{Retention time}) + c \qquad (2.7)$$

where m and c are constants. Since equation 2.7 is of he form $y = mx + c$, it represents a straight line. Consequently, the values of m and c may be obtained by mesuring the retention times of a group of related compounds whose partition coefficients are known and analysing the results using statistical methods such as regression analysis (see section 3.7.1). The accuracy of the method will depend on the compounds chosen to calculate m and c. This and similar methods, have other limitations and so many partition constants are now calculated using a number of different theoretical methods (see section 2.12.2).

Partition coefficient values are most frequently calculated using an octanol/aqueous buffer model. Alternative systems may be used depending on the nature of the investigation. For example, the most commonly used buffer is pH 7.4, however a buffer with a pH of 6.5 may be used if the investigation for gastrointestinal absorption. Furthermore, the blood–brain barrier has been modelled using alkane/water systems. However, the poor solubility of many compounds in alkanes is a severe limitation on the use of this model system.

2.12.2   Theoretical determination of partition coefficients

The practical measurement of P values is not always as easy as equation (2.6) would suggest. Consequently, a number of theoretical methods have been developed to calculate P values. The most popular of these methods is based on measuring the P values of a large number of compounds and using statistical analysis methods to relate differences in their structures to differences in their P values. This approach, initially developed by Rekker, has been extended by Hansch and co-workers by the use of computer programs. The CLOGP program developed by Hansch and co-workers divides the structure of a molecule into suitable fragments, extracts the corresponding numerical values from its data base and calculates the P value of the compound.

The partition coefficients of a compound in an organic/water phase system may be calculated from the P value of the same compound in a different organic/water phase system. Experimental work has shown that for ideal solutions the two values are related by the expression:

$$Log \ P' = a \ Log \ P + b \tag{2.8}$$

where P' is the partition coefficient in the new organic solvent/water system, P is the partition coefficient in the octanol/water system and a and b are constants characteristic of the new organic solvent/water system. The values of a and b are properties of the system and in ideal solutions are independent of the nature of the solute (Table 2.2). Consequently, the values of a and b for a particular solvent system may be calculated by measuring the

Table 2.2   The values of a and b for various organic/water phase systems relative to the octanol/water system where $a = 1$ and $b = 0$

Solvent/water system relative to the octanol/water system	a	b
Butanol	0.70	0.38
Chloroform	1.13	− 1.34
Cyclohexane	0.75	0.87
Diethyl ether	1.13	− 0.17
Heptane	1.06	− 2.85
Oleyl alcohol	0.99	− 0.58

values of P and P' for a compound and assigning standard values for a and b in one of the solvent systems. For example, when $a = 1$ and $b = 0$ for the water/octanol system, then $a = 1.13$ and $b = -0.17$ for the diethyl ether/water system.

2.13   Surfactants and amphiphiles

Amphiphiles are compounds that have a region in their molecules that likes a solvent, that is, dissolves in it, and also a region that dislikes the same solvent, that is, is insoluble in that

Table 2.3 Examples of surfactants. Reproduced from G. Thomas, *Chemistry for Pharmacy and the Life Sciences*, 1996, by permission of Prentice Hall, a Pearson Education Company

Compound	Structural formula Hydrophobic end ... Hydrophilic end
Cationic surfactants	
Sodium stearate	$CH_3(CH_2)_{16} \ldots COO^- \, Na^+$
Anionic surfactant	
Dodecylpyridinium chloride	$C_{12}H_{25} \ldots C_5H_5N^+ \, Cl^-$
Dodecylamine hydrochloride	$CH_3(CH_2)_{11} \ldots NH_3 \, Cl^-$
Ampholytic surfactants	
Dodecyl betaine	$C_{12}H_{25}N(CH_3)_2CH_2 \ldots COO^-$
Non-ionic surfactants	
Heptaoxyethylene monohexyldecyl ether	$CH_3(CH_2)_{15} \ldots (OCH_2CH_2)_7OH$
Polyoxyethylene sorbitan monolaurate	The polyoxyethylene ethers of the lauric acid esters of sorbitan

solvent. *Surfactants* are compounds that lower the surface tension of water (Table 2.3). Their structures contain both strong hydrophilic (polar group) and strong hydrophobic groups (non-polar group) and so they dissolve in both polar and non-polar solvents. They are classified as cationic, anionic, ampholytic and non-ionic surfactants depending on the nature of their hydrophilic groups. Cationic surfactants have a positively charged hydrophilic group while anionic surfactants have a negatively charged hydrophilic group. Ampholytic surfactants have electrically neutral structures that contain both positive and negative charges. They are zwitterions. Non-ionic surfactants do not form ions in solution. The terms amphiphile and surfactant are frequently used interchangeably. This will be the case in this text, which will only consider their properties in aqueous solution.

Surfactants are frequently used to prepare aqueous solutions of compounds that are insoluble or sparingly soluble in water. The hydrophobic part of the structure of the surfactant binds to the compound and the strong water affinity of the hydrophilic part of the surfactant effectively pulls the compound into solution in the water. This behaviour is also the basis of detergent action. The surfactant molecules (detergent) bind to the dirt particles and the strong water affinity of their polar groups pulls the dirt particles into suspension in water. This solubilising effect of surfactants is also of considerable importance in the design of dosage forms.

In biological systems surfactants dissolve in both the aqueous medium and lipid membranes and so tend to accumulate at the interfaces between these phases. This property is the reason for the antiseptic and disinfectant action of some non-ionic and quaternary ammonium surfactants. Surfactants such as cetylpyridinium chloride and octoxynol-9 (Fig. 2.16) partially dissolve in the lipid membrane of the target cell. This lowers the surface tension of the cell membrane, which results in lysis and the death of the cell (see section 7.2.5).

Octoxynol-9 and nonoxynol-9 are different from other surfactant antiseptics in that they do not dissolve in the cell membranes of pathogens. They dissolve in the cell membranes of

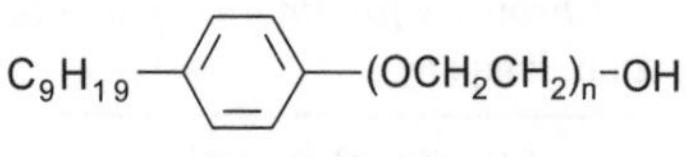

Nonoxynol-9 (spermaticide)

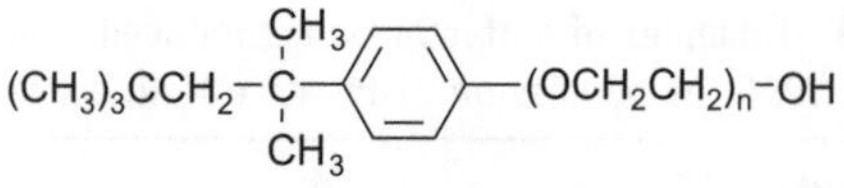

Octoxynol-9 (spermaticide)

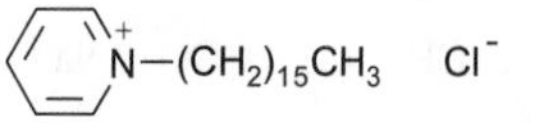

Cetylpyridinium chloride (disinfectant)

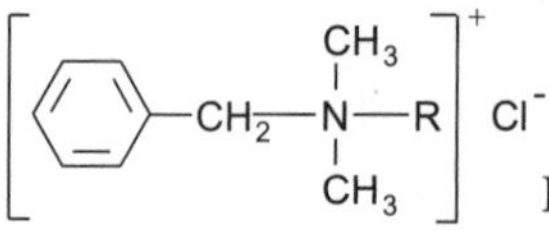

Benzalkonium chloride (antiseptic)

Figure 2.16 The structures of some biologically active surfactants. A number of nonoynols and octoxynols are known, their structures containing different numbers of ethylene oxide units. In each case the number after the name indicates the average number of ethylene oxide units (n) per molecule in the prepared sample. Benzalkonium chloride is a mixture of compounds where R ranges from C_8 to C_{18}

spermatozoa. This immobilises the sperm which allows these compounds to be used as spermicides in birth control.

Naturally occurring surfactants are involved in a number of bodily functions. For example, bile salts, which are produced in the livers, play an essential part in the digestion of lipids in the intestine. Surfactants produced in the membranes of the aveoli prevent the accumulation of water and mucus in the lungs. Furthermore, a number of drugs with structures that contain a suitable balance between hydrophobic and hydrophilic groups have also been reported to exhibit surfactant properties (Fig. 2.17).

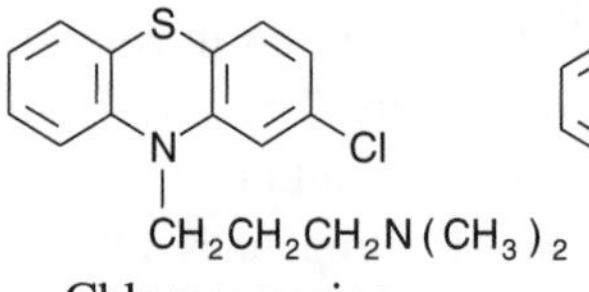

Chlorpromazine (antipsychotic)

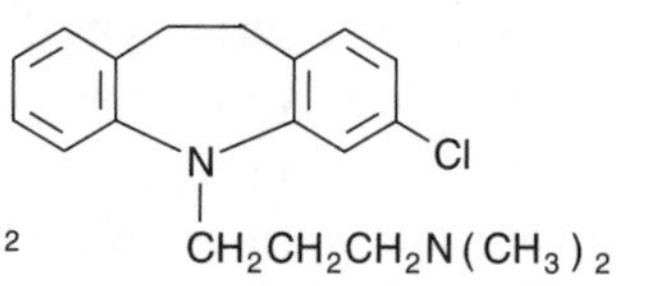

Imipramine (antipsychotic)

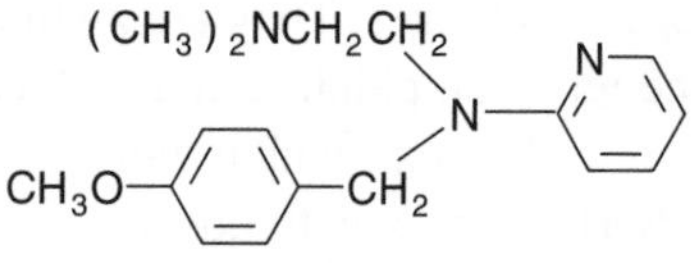

Pyrilamine (antihistaminic)

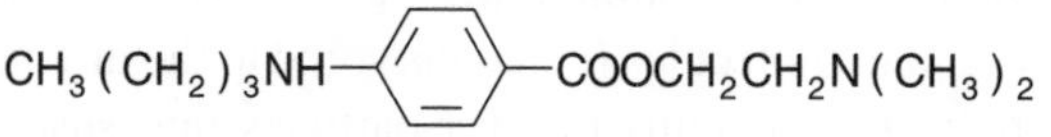

Tetracaine (local anaesthetic)

Figure 2.17 The structures of some drugs that have been reported to exhibit surfactant properties as well their normal pharmacological activity

As the concentration of a surfactant dissolved in water increases, the system changes from a true solution to a colloidal solution. The surfactant molecules find it more energetically favourable to form colloidal aggregates known as *micelles* in which the hydrophobic sections of the molecules effectively form a separate organic phase with the hydrophilic part of the molecule in the aqueous medium (Fig. 2.18). The concentration at which micelles begin to form is known as the *critical micelle concentration* (*cmc*). It is temperature dependent and is usually measured at 25°C. For example, the cmc for sodium dodecyl sulphate is about

0.08 mol dm^{-1} at 25°C. At concentrations just above the cmc the micelles tend to be spherical in shape. As the concentration increases the micelle changes from a sphere to cylindrical, laminar and other forms (Fig. 2.18). Micelles may also be formed when the temperature of a constant concentration of a surfactant is varied. The temperature at which micelle formation occurs is known as the *critical micelle temperature* (*cmt*).

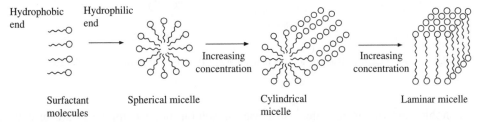

Figure 2.18 Micelle structure and its variation with concentration. Reproduced from G. Thomas, *Chemistry for Pharmacy and the Life Sciences*, 1996, by permission of Prentice Hall, a Pearson Education Company

Many of the physical properties of surfactant solutions, such as surface tension, electrical conductivity and osmotic pressure exhibit an abrupt change at the cmc point (Fig. 2.19). Consequently, changes in these physical measurements can be used to indicate the onset of micelle formation and also determine the cmc value for a surfactant. However, the actual value will depend on the physical property used to determine the cmc point. The most widely method is based on surface tension measurements.

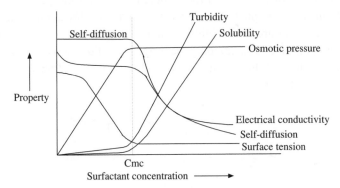

Figure 2.19 General examples of the changes in the physical properties of surfactants at the cmc point. Reproduced with permission from I. W. Hamley, *Introduction to Soft Matter Polymers, Colloids, Amphiphiles and Liquid Crystals*, fig. 4.5, © John Wiley and Sons Ltd, 2000

2.13.1 Drug solubilisation

Incorporation into suitable micelles can be used to solubilise water-insoluble and sparingly soluble drugs. The way in which a drug is incorporated into the micelle depends on the structure of the drug (Fig. 2.20a). Non-polar compounds tend to accumulate in the

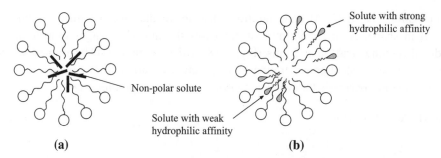

Figure 2.20   Solubilisation of solutes in a micelle. (**a**) Non-polar solutes. (**b**) Polar solutes

hydrophobic core of the micelle whilst water-insoluble polar compounds are orientated with their polar groups towards the surface of the micelle (Fig. 2.20b). The position of the polar compound in the micelle will depend on the relative affinities of the polar group of the solute molecule for the aqueous medium and the non-polar sections of the molecule for the hydrophobic core of the micelle. A relatively strong affinity for the aqueous medium will result in the polar group of the solute being near or on the surface of the micelle whilst a weak affinity for the aqueous medium will result in the polar group being located further into the interior of the micelle. In all cases, the solute molecules are held in the micelle by intermolecular forces of attraction such as hydrogen bonds and hydrophobic bonding.

Incorporation into micelles is used to deliver water-insoluble and sparingly soluble compounds to their site of action both for clinical use and testing purposes. For example, parenterally administered amphotericin B used to treat life threatening fungal infections. The salts of amphotericin are only slightly soluble in water and so it is administered as a colloidal dispersion of micelles containing the drug and sodium deoxycholate. It is believed to act by forming a channel through the fungal cell wall (see section 7.4.1), which allows the cell contents to leak out.

Amphotericin B

The use of micelles as a delivery vehicle has a number of disadvantages. The drug's absorption and hence its activity is dependent on it being released from the micelle. In addition, the surfactant molecules forming the micelles frequently irritate mucous membranes and many are haemolytically active. Furthermore, ionic surfactants can react with anionic and cationic drug substances and so non-ionic surfactants are more widely used in the preparation of dosage forms. Cationic surfactants are, however, used as preservatives. Incorporation of a drug into a micelle can also result in a more rapid decomposition of some compounds because their molecules are in close proximity to each other in the micelle. However, this enhanced reactivity is utilised in synthetic chemistry to promote reactions.

Incorporation into micelles has also been shown to reduce the rates of hydrolysis and oxidation of susceptible drugs. All types of surfactant have been shown to improve the stability of these drugs, the degree of protection usually increasing the further the drug penetrates into the core of the micelle. For example, benzocaine has been shown to be more stable to alkaline hydrolysis than homatropine in the presence of non-ionic surfactants. This is probably because benzocaine is less polar than homatropine and its molecules are located deeper in the core of the micelle.

Benzocaine

Homatropine

Micelles are also involved in the digestion of triglycerides by mammals. These fats are first emulsified by mechanical action in the upper part of the small intestine (the duodenum). This is followed by hydrolysis of the triglycerides to 2-monoglycerides and fatty acids, the reaction being catalysed by pancreatic enzymes. The fatty acids and 2-monoglyerides combine with bile salts released from the gall bladder to form mixed micelles (see section 2.13.2). These mixed micelles are transported to the wall of the intestine where the bile salts, fatty acid and 2-monoglycerides are released The fatty acids and the 2-monoglycerides are absorbed into the cells of the duodenal lining and are reconstituted into triglycerides, which are incorporated with other lipids and proteins into lipoprotein complexes called *chylomicrons*. These chylomicrons are delivered to the lymph and from there to the circulatory system. Meanwhile the more polar bile salts pass further down the intestine before being absorbed into the circulatory system and recycled.

2.13.2 Mixed micelles as drug delivery systems

Mixed micelles are formed by mixtures of surfactants. Suitable selection of the surfactants results in a mixed micelle that has low haemolytic and membrane irritant actions. For example, diazepam has been solubilised and stabilised by mixed micelles produced from lecithin and sodium cholate. Sodium cholate is very haemolytically active but this action is considerably reduced by the presence of the lecithin, which has no haemolytic activity. It appears that the mixing of haemolytically active and haemolytically non-active surfactants either reduces or stops this form of biological activity.

2.13.3   Vesicles and liposomes

Vesicles are aggregates formed from spherical bilayers of amphiphiles. *Liposomes* are vesicles formed from lipids. For example, phospholipids can act as surfactants. They will, when dispersed in water, spontaneously organise themselves into liposomes provided that the temperature is below the so-called chain melting temperature of the lipid. This is the temperature at which the lipid changes from a solid to a liquid crystal. In their simplest form, liposomes consist of a roughly spherical bilayer of phospholipid molecules surrounding an interior core of water (Fig. 2.21). The polar heads of the exterior and interior lipid molecules are orientated towards the exterior and interior water molecules. More complex liposomes have structures that consist of a number of concentric bilayer shells.

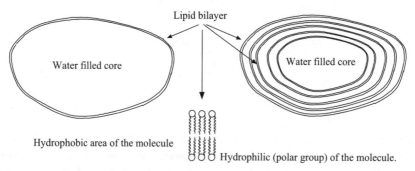

Figure 2.21   The structures of liposomes

Liposomes are used as drug delivery systems for a wide variety of agents. The drug is incorporated into the interior of the liposome, hydrophilic drugs usually occupying the aqueous core whilst hydrophobic drugs are normally found in the lipid double layers. The liposomes used are stable at normal body temperatures but break down at higher temperatures. Consequently, in theory, the drug is protected from degradation and safely transported through the biological system until the liposome reaches an area of infection where the temperature is higher. Here the liposome decomposes, releasing the drug, hopefully at the desired site of action. In spite of this specific action only a few drugs, such as the antimycotic amphotericin B and the anticancer drugs doxorubicin and daunorubicin, have been successfully delivered to their sites of action in this manner.

   Liposomes have a double layer structure similar to biological membranes. Consequently, they have been used as model membranes to study the diffusion of substances into and out of the liposome with a view to correlating these observations to the *in situ* behaviour of drugs.

2.14   Questions

1 What is the significance of the stereoelectronic structure of a drug as far as drug design is concerned?

2 Outline why the stereochemistry of a compound is important in drug design.

3 Outline by means of suitable examples what the significance of (a) structurally rigid groups, (b) conformations and (c) configuration has on the design of new drugs.

4 Why is water solubility an important factor in the action of drugs?

5 Define the meaning of the terms: (a) ideal solution and (b) polar solute.

6 The solubility (% w/w) of oxygen in water at a pressure of 1 atmosphere and a temperature of 20°C is 4.25 mg. Calculate the solubility of oxygen in water when the water is in contact with air saturated with water vapour at a pressure of 1 atmosphere and a temperature of 20°C if the partial pressure of oxygen in water saturated air is 160 mmHg.

7 List the structural features that would indicate whether a compound is likely to be reasonably water-soluble. Illustrate the answer by reference to suitable examples.

8 Outline three general methods by which the water solubility of a compound could be improved without affecting its type of biological action.

9 Suggest, by means of chemical equations, one route for the introduction of each of the following residues into the structure of 4-hydroxybenzenesulphonamide.

(a) an acid residue,

(b) a basic residue,

(c) a neutral polyhydroxy residue.

10 Calculate the degree of ionisation of codeine, pK_a 8.2, in a solution with a pH of 2. Predict how the degree of ionisation could affect the ease of absorption of codeine in (a) the stomach and (b) the intestine when the pH of the stomach fluid is 2.0 and the pH of the intestinal fluid is 6.

11 What general structural features are characteristic of surfactants.

12 Suggest the most appropriate organic/aqueous medium for use in determining P values in the following cases.

(a) CNS activity.

(b) GI tract absorption.

(c) Buccal absorption

13 Explain how micelles and liposomes could be used as drug delivery vehicles.

3

Structure–activity and quantitative structure relationships

3.1 Introduction

The *structure–activity relationship (SAR)* approach to drug discovery is based on the observation that compounds with similar structures to a pharmacologically active drug are often themselves biologically active. This activity may be either similar to that of the original compound but differ in potency and unwanted side effects or completely different to that exhibited by the original compound. These structurally related activities are commonly referred to as structure–activity relationships. A study of the structure–activity relationships of a lead compound and its analogues may be used to determine the parts of the structure of the lead compound that are responsible for both its beneficial biological activity, that is, its *pharmacophore* (see section 1.3), and also its unwanted side effects. This information may be used to develop a new drug that has increased activity by selecting the structure with the optimum activity, a different activity from an existing drug and fewer unwanted side effects.

The *quantitative structure–activity relationship (QSAR)* is an attempt, based on the SAR approach, to remove the element of luck from drug discovery. It uses physicochemical properties (*parameters*) to represent drug properties that are believed to have a major influence on drug action. Parameters must be properties that are capable of being represented by a numerical value. These values are used to produce a general equation relating drug activity with the parameters. This equation enables medicinal chemists to predict the activity of analogues and, as a result, determine which analogue is most likely to produce the desired clinical response. Its use takes some of the guess work out of deciding which analogues of a lead to synthesise. This has the knock-on effect of reducing cost, a major consideration in all commercial companies.

Medicinal Chemistry, Second Edition   Gareth Thomas
© 2007 John Wiley & Sons, Ltd

3.2   Structure–activity relationship (SAR)

Structure–activity relationship studies are usually carried out by making minor changes to the structure of a lead to produce analogues and assessing the effect that these structural changes have on biological activity (see section 3.6 for a case study). The investigation of numerous lead compounds and their analogues has made it possible to make some broad generalisations about the biological effects of specific types of structural changes. These changes may be conveniently classified as changing:

- the size and shape of the carbon skeleton (see section 3.3.1);

- the nature and degree of substitution (see section 3.3.2);

- the stereochemistry of the lead (see section 2.3).

The selection of the changes required to produce analogues of a particular lead is made by considering the activities of compounds with similar structures and also the possible chemistry and biochemistry of the intended analogue. It is believed that structural changes that result in analogues with increased lipid character may either exhibit increased activity because of better membrane penetration (Fig. 3.1a; $n = 3$–6) or reduced activity because of a reduction in their water solubility (Fig. 3.1b). However, whatever the change, its effect on

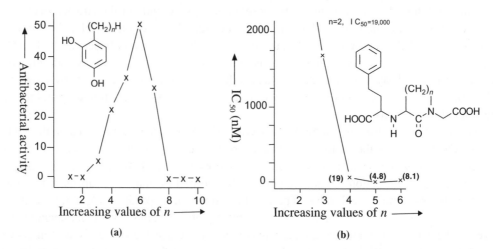

Figure 3.1 Examples of the variation of response curves with increasing numbers of inserted methylene groups. **(a)** A study by Dohme *et al.* on the variation of antibacterial activity of 4-alkyl substituted resorcinols. **(b)** Inhibition of ACE by enalaprilat analogues (Thorsett). The values in parentheses are the IC_{50} values for that analogue

water solubility, transport through membranes, receptor binding and metabolism and other pharmacokinetic properties of the analogue should be considered as far as is possible before embarking on what could be an expensive synthesis. Furthermore, changing the structure of the lead compound could result in an analogue that is too big to fit its intended target site. Computer-assisted molecular modelling (see Chapter 4) can alleviate this problem provided that the structure of the target is known or can be simulated with some degree of accuracy. However, it is emphasised that although it is possible to predict the effect of structural changes there will be numerous exceptions to the predictions and so all analogues must be synthesised and tested.

3.3 Changing size and shape

The shapes and sizes of molecules can be modified in a variety of ways, such as:

- changing the number of methylene groups in chains and rings;

- increasing or decreasing the degree of unsaturation;

- introducing or removing a ring system.

3.3.1 Changing the number of methylene groups in chains and rings

Increasing the number of methylene groups in a chain or ring increases the size and the *lipophilicity* (see section 1.4.1) of the compound. It is believed that any increase in activity with increase in the number of methylene groups is probably due to an increase in the lipid solubility of the analogue, which gives a better membrane penetration (Fig.3.1a). Conversely, a decrease in activity with an increase in the number of methylene groups is attributed to a reduction in the water solubility of the analogues (Fig. 3.1b). This reduction in water solubility can result in the poor distribution of the analogue in the aqueous media as well as the trapping of the analogue in biological membranes (see section 2.12). A further problem with large increases in the numbers of inserted methylene groups in chain structures is micelle formation (see section 2.13.1). Micelle formation produces large aggregates that, because of their shape, cannot bind to active sites and receptors.

Introducing chain branching, different sized rings and the substitution of chains for rings, and vice versa, may also have an effect on the potency and type of activity of analogues. For example, the replacement of the sulphur atom of the antipsychotic chlorpromazine by -CH_2-CH_2- produces the anti depressant clomipramine.

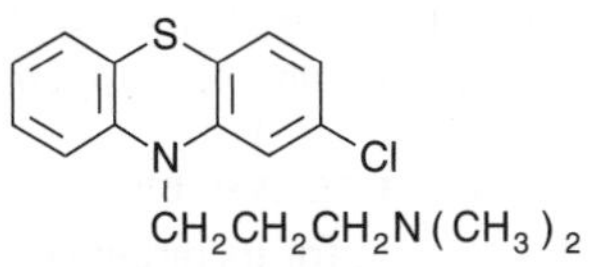

Chlorpromazine

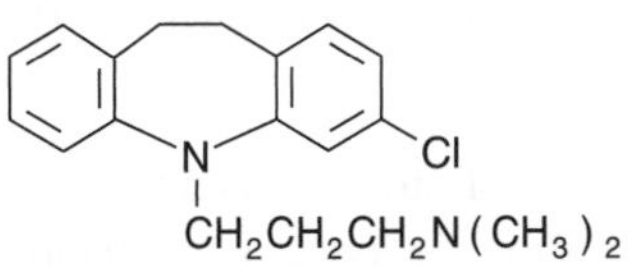

Clomipramine

3.3.2 Changing the degree of unsaturation

The removal of double bonds increases the degree of flexibility of the molecule, which may make it easier for the analogue to fit into active and receptor sites by taking up a more suitable conformation. However, an increase in flexibility could also result in a change orloss of activity.

The introduction of a double bond increases the rigidity of the structure. It may also introduce the complication of E and Z isomers, which could have quite different activities (see Table 1.1). The analogues produced by the introduction of unsaturated structures into a lead compound may exhibit different degrees of potency or different types of activities. For example, the potency of prednisone is about 30 times greater than that of its parent compound cortisol, which does not have a 1–2 C=C bond. The replacement of the S atom of the antipsychotic phenothiazine drugs by a -CH=CH- group gives the antidepressant dibenzazepine drugs, such as protriptyline.

Cortisol   Prednisone   Phenothiazine drugs   Protriptyline (Vivactil)

The introduction of a C=C group will often give analogues that are more sensitive to metabolic oxidation. This may or may not be a desirable feature for the new drug. Furthermore, the reactivity of the C=C frequently causes the analogue to be more toxic than the lead.

3.3.3 Introduction or removal of a ring system

The introduction of a ring system changes the shape and increases the overall size of the analogue. The effect of these changes on the potency and activity of the analogue is not generally predictable. However, the increase in size can be useful in filling a hydrophobic pocket in a target site, which might strengthen the binding of the drug to the target. For example, it has been postulated that the increased inhibitory activity of the cyclopentyl analogue (rolipram) of 3-(3,4-dimethyloxyphenyl)-butyrolactam towards cAMP phosphodiesterase is due to the cyclopentyl group filling a hydrophobic pocket in the active site of this enzyme.

3-(3,4-Dimethoxyphenyl)-butyrolactam   Rolipram, an antidepressant, is ten times more
                                              active than 3-(3,4-dimethoxyphenyl)-butyrolactam.

The incorporation of smaller, as against larger, alicyclic ring systems into a lead structure reduces the possibility of producing an analogue that is too big for its target site. It also reduces the possibility of complications caused by the existence of conformers. However, the selection of the system for a particular analogue may depend on the objective of the alteration. For example, the antidepressant tranylcypromine is more stable than its analogue 1-amino-2-phenylethene.

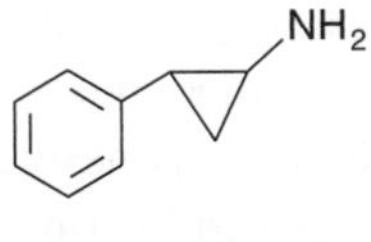

Tranylcypromine 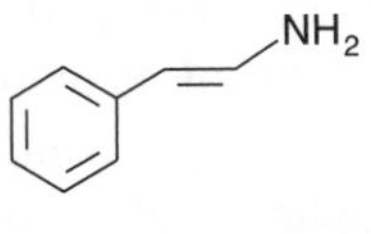   1-Amino-2-phenylethene

The insertion of aromatic systems into the structure of the lead will introduce rigidity into the structure as well as increase the size of the analogue. The latter means that small aromatic systems such as benzene and five-membered heterocyclic systems are often preferred to larger systems. However, the π electrons of aromatic systems may or may not improve the binding of the analogue to its target site. Furthermore, heterocyclic aromatic systems will also introduce extra functional groups into the structure, which could also affect the potency and activity of the analogue. For example, the replacement of the *N*-dimethyl group of chlorpromazine by an *N*-methylpiperazine group produces an analogue (prochlorperazine) with increased antiemetic potency but reduced neuroleptic activity. It has been suggested that this change in activity could be due to the presence of the extra tertiary amine group.

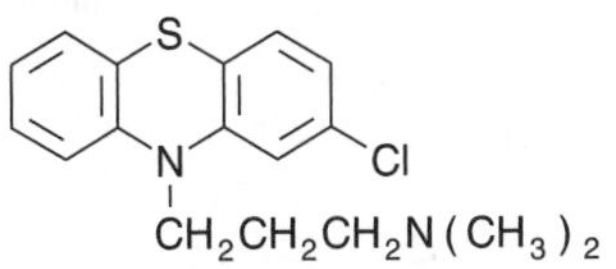

Chlorpromazine 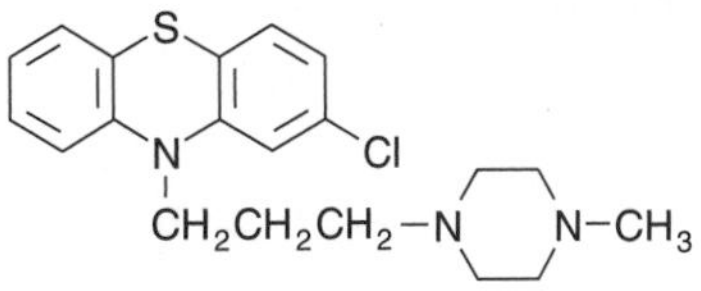   Prochlorperazine

The incorporation of ring systems, especially larger systems, into the structure of a lead can be used to produce analogues that are resistant to enzymic attack by sterically hindering the access of the enzyme to the relevant functional group. For example, the resistance of diphenicillin to β-lactamase is believed to be due to the diphenyl group preventing the enzyme from reaching the β-lactam. It is interesting to note that 2-phenylbenzylpenicillin is not resistant to β-lactamase attack. In this case, it appears that the diphenyl group is too far away from the β-lactam ring to hinder the attack of the β-lactamase.

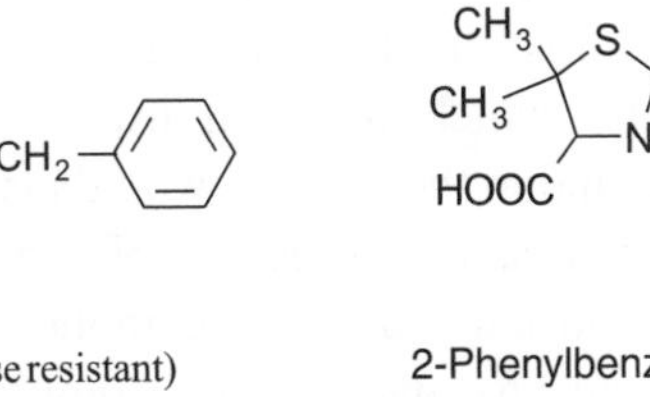

Diphenicillin (β-lactamase resistant)

Benzylpenicillin (not β-lactamase resistant)   2-Phenylbenzylpenicillim (not β-lactamase resistant)

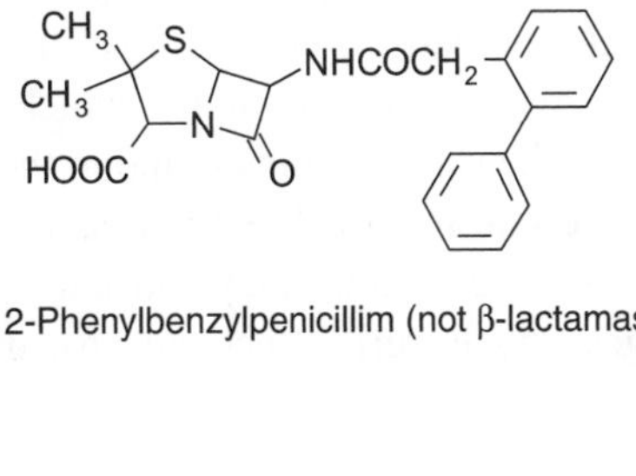

Many of the potent pharmacologically active naturally occurring compounds, such as the alkaloids morphine and curare, have such complex structures that it would not be economic to synthesise them on a large scale. Furthermore they also tend to exhibit unwanted side effects. However, the structures of many of these compounds contain several ring systems. In these cases, one approach to designing analogues of these compounds centres around determining the pharmacophore and removing any surplus ring structures. It is hoped that this will also result in the loss of any unwanted side effects. The classic example illustrating this type of approach is the development of drugs from morphine (Fig. 3.2).

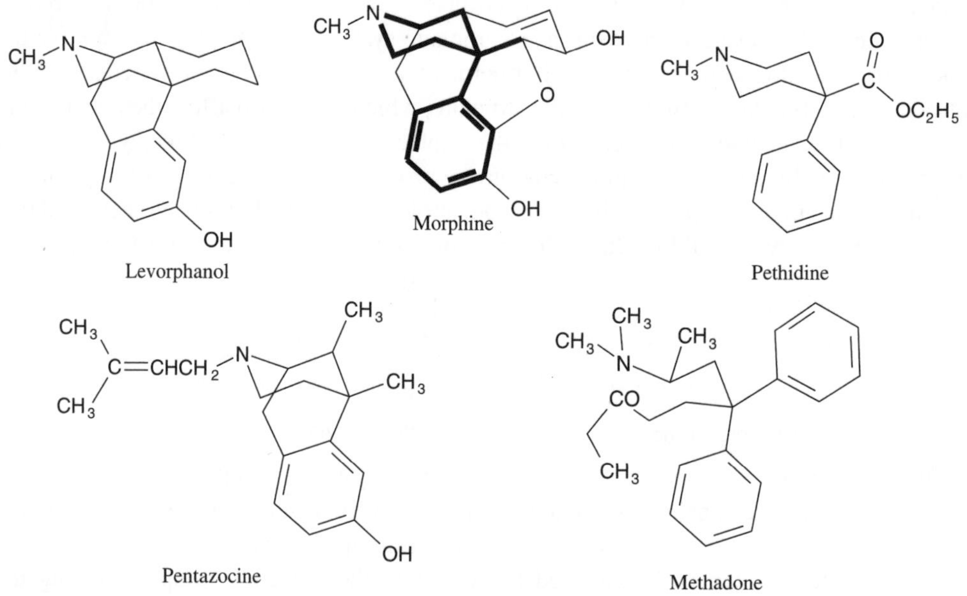

Figure 3.2   The pharmacophore of morphine was found to be the structure represented by the heavy type bonds. Pruning and modification of the remaining structure of morphine resulted in the development of (**a**) the more potent but still highly addictive levorphanol, (**b**) the very much less potent pethidine, (**c**) the less potent and less addictive pentazocine and (**d**) the equally potent but much less addictive methadone, amongst other drugs

3.4   Introduction of new substituents

The formation of analogues by the introduction of new substituents into the structure of a lead may result in an analogue with significantly different chemical and hence different pharmacokinetic properties. For example, the introduction of a new substituent may cause significant changes in lipophilicity, which affect transport of the analogue through membranes and the various fluids found in the body. It would also change the shape, which could result in conformational restrictions that affect the binding to target site. In addition the presence of a new group may introduce a new metabolic pathway for the analogue. These changes will in turn affect the pharmacodynamic properties of the analogue. For example, they could result in an analogue with either increased or decreased potency, duration of action, metabolic stability and unwanted side effects. The choice of substituent will depend on the properties that the development team decide to enhance in an attempt to meet their objectives. Each substituent will impart its own characteristic properties to the analogue. However, it is possible to generalise about the effect of introducing a new substituent group into a structure but there will be numerous exceptions to the predictions.

3.4.1 Methyl groups

The introduction of methyl groups usually increases the lipophilicity of the compound and reduces its water solubility as shown by an increase in the value of the partition coefficient (Table 3.1). It should improve the ease of absorption of the analogue into a biological membrane but will make its release from biological membranes into aqueous media more difficult (see section 2.13). The introduction of a methyl group may also improve the binding of a ligand to its receptor by filling a pocket on the target site.

Table 3.1 The change in the partition coefficients (P) of some common compounds when methyl groups are introduced into their structures. The greater the value of P the more lipid soluble the compound. Benzene and toluene values were measured using an n-octanol/water system whilst the remaining values were measured using an olive oil/water system

Compound	Structure	P	Analogue	Structure	P
Benzene		135	Toluene		490
Acetamide	CH_3CONH_2	83	Proprionamide	$CH_3CH_2CONH_2$	360
Urea	NH_2CONH_2	15	N-Methylurea	$CH_3NHCONH_2$	44

The incorporation of a methyl group can impose steric restrictions on the structure of an analogue. For example, the *ortho*-methyl analogue of diphenhydramine exhibits no antihistamine activity. Harmes *et al.* suggest this to be due to the *ortho*-methyl group restricting rotation about the C–O bond of the side chain. This prevents the molecule adopting the conformation necessary for antihistamine activity. It is interesting to note that the *para*-methyl analogue is 3.7 times more active than diphenhydramine (Fig. 3.3).

The incorporation of a methyl group can have one of three general effects on the rate of metabolism of an analogue. It can result in either (1) an increased rate of metabolism due to oxidation of the methyl group, (2) an increase in the rate of metabolism due to demethylation by the transfer of the methyl group to another compound or (3) a reduction in the rate of metabolism of the analogue.

1. A methyl group bound to an aromatic ring or a structure may be metabolised to a carboxylic acid, which can be more easily excreted. For example, the antidiabetic tolbutamide is

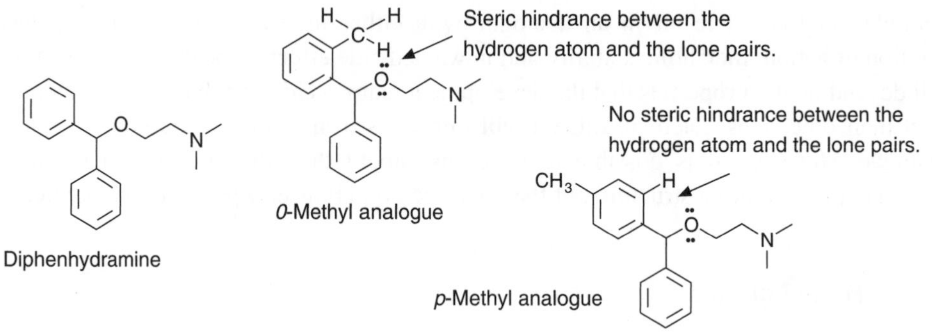

Figure 3.3 Steric hindrance of a methyl analogue of diphenylhydramine, which is believed to be responsible for the analogues, lack of activity

metabolised to its less toxic benzoic acid derivative. The introduction of a reactive C–CH_3 group offers a detoxification route for lead compounds that are too toxic to be of use.

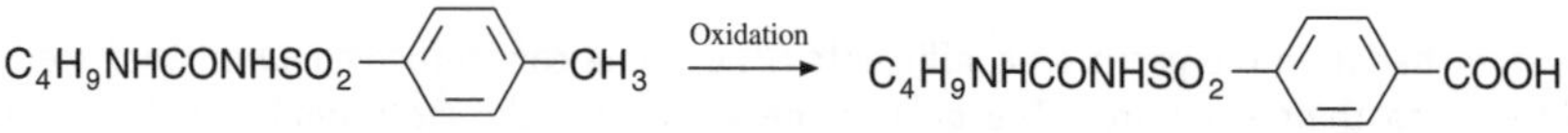

Tolbutamide Less toxic metabolite

2. Demethylation is more likely to occur when the methyl group is attached to positively charged nitrogen and sulphur atoms, although it is possible for any methyl group attached to a nitrogen, oxygen or sulphur atom to act in this manner (see Table 12.1). A number of methyl transfers have been associated with carcinogenic action.

3. Methyl groups can reduce the rate of metabolism of a compound by masking a metabolically active group, thereby giving the analogue a slower rate of metabolism than the lead. For example, the action of the agricultural fungicide nabam is due to it being metabolised to the active diisothiocyanate. N-Methylation of nabam yields an analogue that is in active because it cannot be metabolised to the active diisothiocyanate.

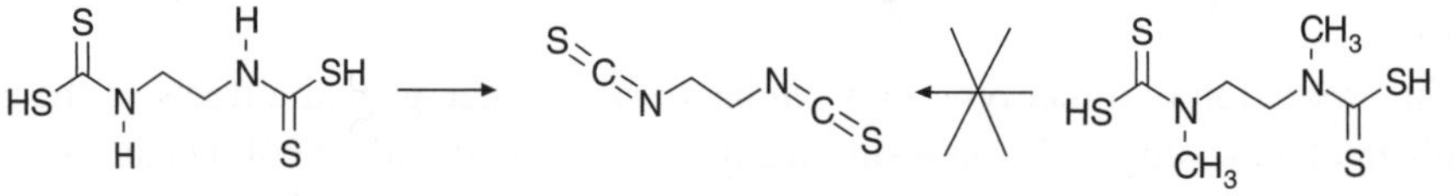

Nabam 1,2-Diisothiocyanoethane Dimethyl analogue of nabam

Methylation can also reduce the unwanted side effects of a drug. For example, mono-and di-*ortho*-methylation with respect to the phenolic hydroxy group of paracetamol produces analogues with reduced hepatotoxicity. It is believed that this reduction is due to the methyl groups preventing metabolic hydroxylation of these *ortho* positions.

Paracetamol   o,o'-Dimethyl analogue of paracetamol

Larger alkyl groups will have similar effects. However, as the size of the group increases the lipophilicity will reach a point where it reduces the water solubility to an impractical level. Consequently, most substitutions are restricted to methyl and ethyl groups.

3.4.2 Halogen groups

The incorporation of halogen atoms into a lead results in analogues that are more lipophilic and so less water soluble. Consequently, halogen atoms are used to improve the penetration of lipid membranes. However, there is an undesirable tendency for halogenated drugs to accumulate in lipid tissue.

The chemical reactivity of halogen atoms depends on both their point of attachment to the lead and the nature of the halogen. Aromatic halogen groups are far less reactive than aliphatic halogen groups, which can exhibit considerable chemical reactivity. The strongest and least reactive of the aliphatic carbon–halogen bonds is the C–F bond. It is usually less chemically reactive than aliphatic C–H bonds. The other aliphatic C–halogen bonds are weaker and so more reactive, their reactivity increasing as one moves down the periodic table. They are normally more chemically reactive than aliphatic C–H bonds. Consequently, the most popular halogen substitutions are the less reactive aromatic fluorine and chlorine groups. However, the presence of electron withdrawing ring substituents may increase their reactivity to unacceptable levels. Trifluorocarbon groups (-CF$_3$) are sometimes used to replace chlorine as these groups are of a similar size. These substitutions avoid introducing a very reactive centre and hence a possible site for unwanted side reactions into the analogue. For example, the introduction of the more reactive bromo group can cause the drug to act as an alkylating agent.

The changes in potency caused by the introduction of a halogen or halogen containing group will, as with substitution by other substituents, depend on the position of the substitution. For example, the antihypertensive clonidine with its o,o'-chloro substitution is more potent than its p,m-dichloro analogue. It is believed that the bulky o-chlorine groups impose a conformational restriction on the structure of clonidine that probably accounts for its increased activity.

Clonidine ED$_{50}$ 0.01 mg kg^{-1} ED$_{50}$ 3.00 mg kg^{-1}

3.4.3   Hydroxy groups

The introduction of hydroxy groups into the structure of a lead will normally produce analogues with an increased hydrophilic nature and a lower lipid solubility. It also provides a new centre for hydrogen bonding, which could influence the binding of the analogue to its target site. For example, the *ortho*-hydroxylated minaprine analogue binds more effectively to M_1-muscarinic receptors than many of its non-hydroxylated analogues.

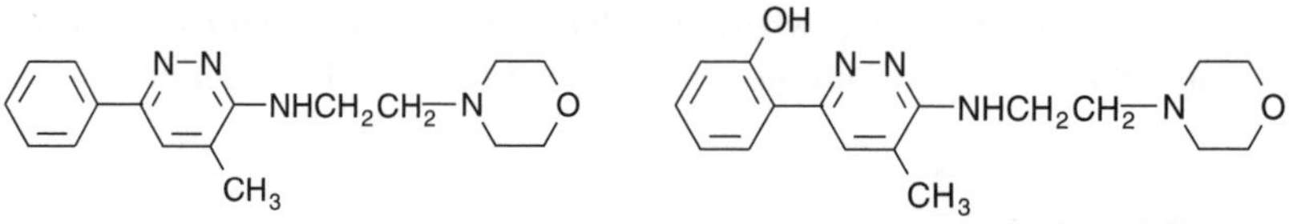

Minaprine   The *ortho*-hydroxylated analogue

The introduction of a hydroxy group also introduces a centre that, in the case of phenolic groups, could act as a bacterioside, whilst alcohols have narcotic properties. However, the presence of hydroxy groups opens a new metabolic pathway (see Tables 12.1 and 12.2) that can either act as a detoxification route or prevent the drug reaching its target.

3.4.4   Basic groups

The basic groups usually found in drugs are amines, including some ring nitrogen atoms, amidines and guanidines. All these basic groups can form salts in biological media. Consequently, incorporation of these basic groups into the structure of a lead will produce analogues that have a lower lipophilicity but an increased water solubility (see section 2.9). This means that the more basic an analogue, the more likely it will form salts and the less likely it will be transported through a lipid membrane

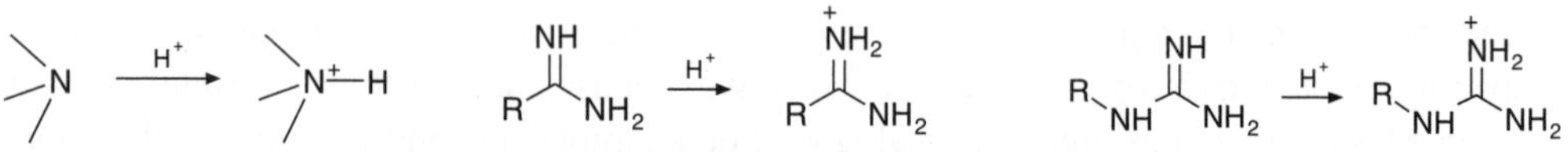

All types of amine   Amidines   Guanidines

The introduction of basic groups may increase the binding of an analogue to its target by hydrogen bonding between that target and the basic group (Fig. 3.4a). However, a number

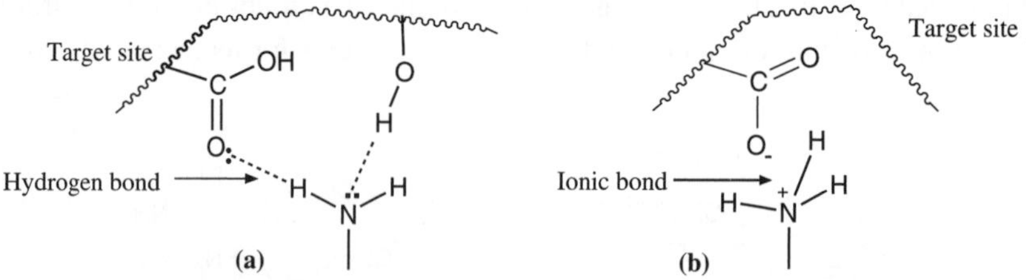

Figure 3.4 Possible sites of (a) hydrogen bonding between a target site and amino groups and (b) ionic bonding between amine salts and a target site

of drugs with basic groups owe their activity to salt formation and the enhanced binding that occurs due to the ionic bonding between the drug and the target (Fig 3.4b). For example, it is believed that many local anaesthetics are transported to their site of action in the form of their free bases but are converted to their salts, which bind to the appropriate receptor sites.

The incorporation of aromatic amines into the structure of a lead is usually avoided since aromatic amines are often very toxic and are often carcinogenic.

3.4.5   Carboxylic and sulphonic acid groups

The introduction of acid groups into the structure of a lead usually results in analogues with an increased water but reduced lipid solubility. This increase in water solubility may be subsequently enhanced by *in vivo* salt formation. In general the introduction of carboxylic and sulphonic acid groups into a lead produces analogues that can be more readily eliminated (see Table 12.2).

The introduction of carboxylic acid groups into small lead molecules may produce analogues that have a very different type of activity or are inactive. For example, the introduction of a carboxylic acid group into phenol results in the activity of the compound changing from being a toxic antiseptic to the less toxic anti-inflammatory salicylic acid. Similarly, the incorporation of a carboxylic acid group into the sympathomimetic phenylethylamine gives phenylalanine, which has no sympathomimetic activity. However, the introduction of carboxylic acid groups appears to have less effect on the activity of large molecules.

Phenol   Salicylic acid   Phenylethylamine   Phenylalanine

Sulphonic acid groups do not usually have any effect on the biological activity but will increase the rate of excretion of an analogue.

3.4.6   Thiols, sulphides and other sulphur groups

Thiol and sulphide groups are not usually introduced into leads in SAR studies because they are readily metabolised by oxidation (see Table 12.1). However, thiols are sometimes introduced into a lead structure when improved metal chelation is the objective of the SAR study. For example, the antihypertensive captopril was developed from the weakly active carboxyacylprolines by replacement of their terminal carboxylic acid group with a thiol group, which is a far better group for forming complexes with metals than carboxylic acids (see section 9.12.2).

The introduction of thiourea and thioamide groups is usually avoided since these groups may produce goitre, a swelling on the neck due to enlargement of the thyroid gland.

3.5 Changing the existing substituents of a lead

Analogues can also be formed by replacing an existing substituent in the structure of a lead by a new substituent group. The choice of group will depend on the objectives of the design team. It is often made using the concept of *isosteres*. Isosteres are groups that exhibit some similarities in their chemical and/or physical properties. As a result, they can exhibit similar pharmacokinetic and pharmacodynamic properties. In other words, the replacement of a substituent by its isostere is more likely to result in the formation of on analogue with the same type of activity as the lead than the totally random selection of an alternative substituent. However, luck still plays a part and an isosteric analogue may have a totally different type of activity from its lead.

Classical isosteres were originally defined by Erlenmeyer as being atoms, ions and molecules which had identical outer shells of electrons. This definition has now been broadened to include groups that produce compounds that can sometimes have similar biological activities (Table 3.2). These groups are frequently referred to as *bioisosteres* in orderto distinguish them from classical isosteres.

A large number of drugs have been discovered by isosteric and bioisosteric interchanges. For example, the replacement of the 6-hydroxy group of hypoxanthine by a thiol group gave the anti tumour drug 6-mercaptopurine whilst the replacement of hydrogen in the 5-position

Table 3.2 Examples of bioisosteres. Each horizontal row represents a group of structures that are isosteric

Classical Isosteres	Bioisosteres
$-CH_3$, $-NH_2$, -OH, -F, -Cl. -Cl, -SH $-PH_2$ -Br, Isopropyl $-CH_2-$, -NH-, -O-, -S- $-COCH_2R$, -CONHR, -COOR, -COSR -HC=, -N= **In rings:** -CH=CH-, -S- -O-, -S-, $-CH_2-$, -NH- -CH=, -N-	

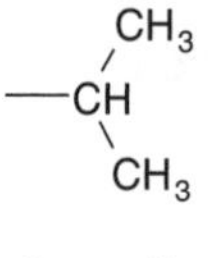

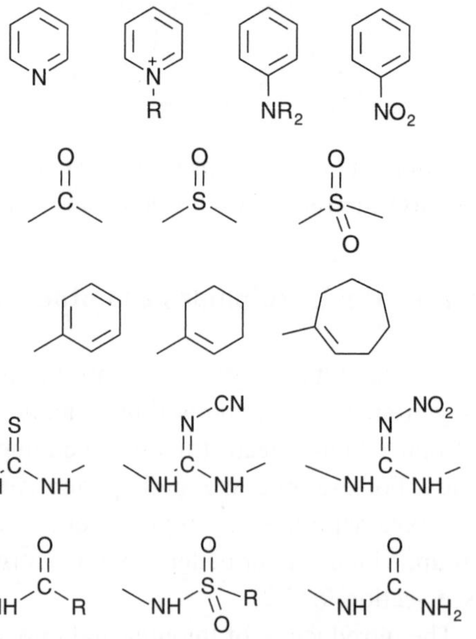

of uracil by fluorine resulted in fluorouracil, which is also an anti tumour agent. However, not all isosteric changes yield compounds with the same type of activity: the replacement of the -S- of the neuroleptic phenothiazine drugs by either -CH=CH- or -CH$_2$CH$_2$- produces the dibenzazepines, which exhibit antidepressant activity (Fig. 3.5).

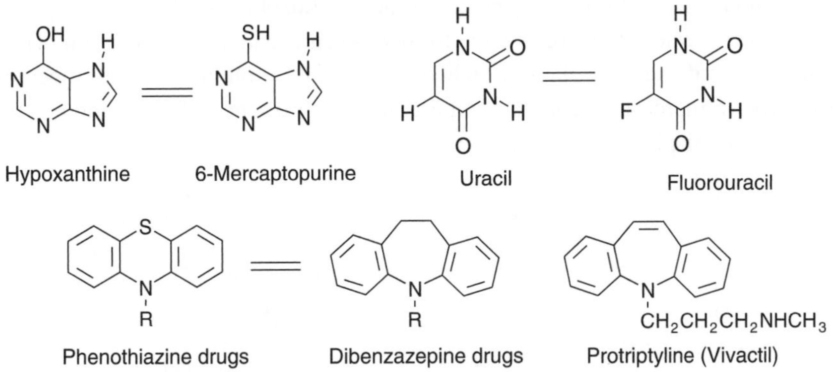

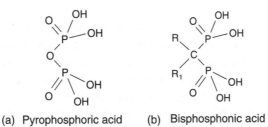

Hypoxanthine   6-Mercaptopurine   Uracil   Fluorouracil

Phenothiazine drugs   Dibenzazepine drugs   Protriptyline (Vivactil)

Figure 3.5 Examples of drugs discovered by isosteric replacement

3.6 Case study: a SAR investigation to discover potent geminal bisphosphonates

Bisphosphonates, or diphosphonates as they were originally known, are derivatives of bisphosphonic acid (Fig. 3.6), which is an analogue of the naturally occurring pyrophosphoric acid. They were synthesised in the nineteenth century for use as antiscalers and anticorrosive agents. However, it was not until the late 1960s that they were recognised as potential sources of drugs. Bisphosphonates are now used to treat diseases such as osteoporosis, Paget's disease and myeloma, which involve a loss of calcium and other minerals from the bone.

(a) Pyrophosphoric acid   (b) Bisphosphonic acid

Figure 3.6 The structures of (a) pyrophosphoric and (b) bisphosphonic acids

Bisphosphonates were first used clinically in the 1970s and 1980s to inhibit mineral loss from bones (bone reabsorption). However, the early drugs, such as clodronic and etidronic acids (Fig. 3.7a), exhibited only moderate potency. The second generation of bisphosphonates (Fig.3.7b) are much more effective. A study of this second generation of compounds indicated that the activity of bisphosphonates was probably due to two structural features. The first is that the phosphonate groups have a high affinity for bone minerals, which is increased by the presence of a hydroxy group on carbon 1. This increased affinity for bone minerals results in the bisphosphonates being incorporated into the bone structure, where their resistance to metabolism reduces reabsorption. The second is that the nature of the side chain attached to the bisphosphonate group appeared to be responsible for an increase in potency, in particular the presence of an amino group and its distance from the bisphosphonate groups.

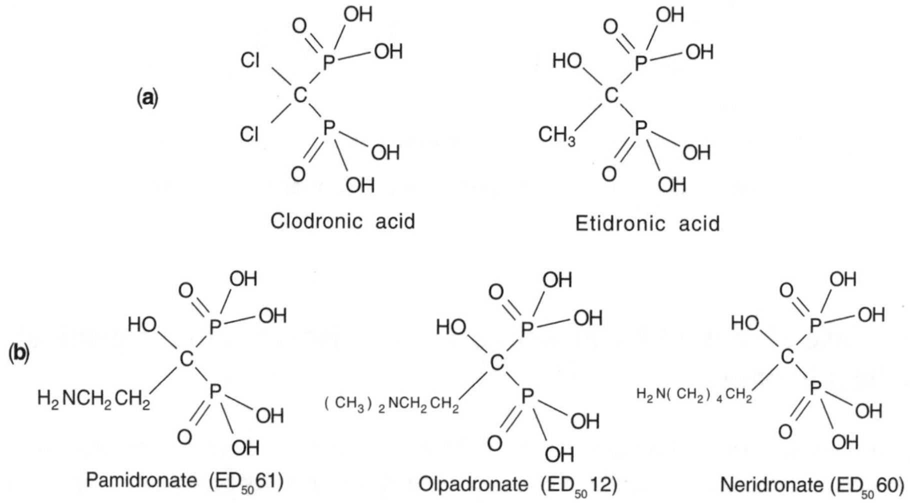

Figure 3.7 Examples of (**a**) first- and (**b**) second-generation bisphosphonates. Note: bisphonates are often referred to by their salt names. The ED_{50} is recorded in $\mu g\ kg^{-1}$. It is the dose that reduces hypercalchaemia induced in thyroparathyroidetomised rats by 50 per cent

The potency of pamidronate (Fig. 3.7) coupled with its wide therapeutic window and patient tolerance resulted in it being selected in 1986 as a lead in a SAR study to improve on its potency and therapeutic window. This study was carried out by L. L. Widler *et al.* at Novartis Pharma Research, Arthritis and Bone Metabolism Therapeutic Area, Basel, Switzerland. They used methods based on the general schemes in Figure 3.8 to synthesise over 80 structurally related geminal bisphosphonate and bisphosphinate analogues of pamidronate. The information gained from a study of the second-generation bisphosphonates (Fig. 3.7b) meant that the research team concentrated on producing compounds in which R (Fig. 3.6b) is a hydroxy group and R_1 (Fig. 3.6b) is a side chain containing at least one nitrogen atom normally separated from the bisphosphonate group by a methylene chain of 1, 2 or 3 carbon atoms. Examples of some of the compounds synthesised and tested in the SAR study are

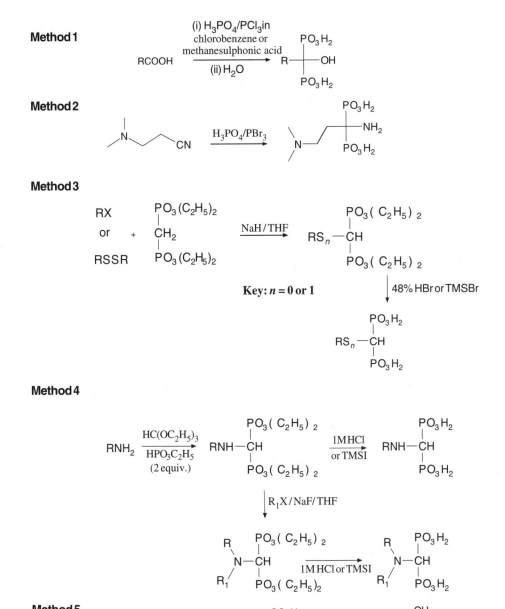

Figure 3.8 General methods for the synthesis of bisphosphonates (methods 1–4) and bisphosphinates (method 5). TMSI is trimethylsilyl iodide

listed in Table 3.3. They give the reader an idea of both the nature and range of compounds examined in the study. All the compounds synthesised were tested for antiresorptive action using a test based on thyroparathyroidectomised (TPTX) rats. Hypercalcaemia was induced in these rats using 1,25-dihydroxyvitamin D. Assessment was carried out *in vivo* and the results were expressed as an ED_{50}, which is the dose in $\mu g \ kg^{-1}$ that reduces the hypercalcaemia by 50 per cent. The results of the testing showed that the most potent compound was zoledronic acid with an ED_{50} of 0.07 $\mu g \ kg^{-1}$.

Zoledronice acid (Zometa, ED_{50} 0.07 $\mu g/kg^{-1}$)

Zoledronic acid has undergone preclinical and clinical trials that have shown that low doses are a safe and effective therapy for bone deminerilisation diseases. A Phase II trial using a 4 mg annual intravenous dose has indicated that it may be effective for treating postmenopausal osteoporosis. Furthermore, a large Phase III trial has also indicated that it was significantly better than pamidronate in the treatment for tumour-induced hypercalcaemia. A rare unwanted side effect is osteonecrosis.

3.7 Quantitative structure–activity relationship (QSAR)

The success of the SAR approach to drug design depends not only on the knowledge and experience of the design team but also a great deal of luck. QSAR is an attempt to remove this element of luck from drug design by establishing a mathematical relationship in the form of an equation between biological activity and measurable physicochemical parameters. These parameters are used to *represent* properties such as lipophilicity, shape and electron distribution, which are believed to have a major influence on the drug's activity. They are normally defined so that they are in the form of numbers derived from practical data that are thought to be related to the property that the parameter represents. This makes it possible either to measure or to calculate these parameters for a group of compounds and relate their values to the biological activity of these compounds by means of mathematical equations using statistical methods such as regression analysis (see section 3.7.1). These equations may be used by the medicinal chemist to make a more informed choice as to which analogues to prepare. For example, it is often possible to use statistical data from other compounds to calculate the theoretical value of a specific parameter for an as yet unsynthesised compound. By substituting this value in the appropriate equation relating activity to that parameter' it is possible to calculate the theoretical activity of this unknown compound. Alternatively, the equation could be used to determine the value x of the parameter y that would give optimum activity. As a result, only analogues that have values of the parameter y in the region of x need be synthesised.

Table 3.3 A selection of the results of the SAR study to discover a bisphosphonate drug more potent than pamidronate. The ED_{50} is recorded in $\mu g\ kg^{-1}$. It is the dose that reduces hypercalcaemia induced in TPTX rats by 50. The compounds with similar types of structure are grouped vertically under the relevant general structural formulae. Second-generation compounds are named; all others have been given a reference number. The candidates for synthesis were selected according to the information available from both the team's initial results and the work of other groups of workers at the time. Adapted with permission from L. L. Widler *et al.*, *J. Med. Chem.*, **45**, 3721–3738, © 2002, American Chemical Society

General structural formulae

Compound	R	R_1	R_2	ED_{50}
1 Pamidronate	H	H	H	61
2	H	H	CH_3	3.4
3 Olapadronate	CH_3	CH_3	H	12
4	CH_3	CH_3	CH_3	18
5 Ibandronate	C_5H_{11}	CH_3	H	1.2
6	C_5H_{11}	CH_3	CH_3	65

Compound	Ring	R	n	ED_{50}
7		H	2	10
8		H	3	25
9		H	5	250
10		Ph	2	70
11		4-Cl-Ph	2	3.5
12		H	2	5.6
13		Ph	2	~11
14		Ph	3	100
15		Ph	5	>300
16		3-F-Ph	2	30

General structural formula

Compound	Ring	ED_{50}
21		50
22		250
23		~2500
24		>3000

Compound	Ring	R	n	ED_{50}
17			2	25
18			2	>300
19		Me	2	~400
20		Ph	2	>10000

General structural formulae

Table 3.3 (*continued*)

Compound	R	ED_{50}	Compound	Ring	R	R_1	R_2	n	ED_{50}
25		>2000	29		H	H	H	1	0.07
			30		H	H	H	2	45
26		800	31		Me	H	H	1	3
			32		H	Me	Me	1	1.5
27		40	33						>300
28		7	34						600

The main properties of a drug that appear to influence its activity are its, lipophilicity, the electronic effects within the molecule and the size and shape of the molecule (steric effects). Lipophilicity is a measure of a drug's solubility in lipid membranes. This is usually an important factor in determining how easily a drug passes through lipid membranes (see section 7.3). It is used as a measure of the ease of distribution of a drug to its target site. The electronic effects of the groups within the molecule will affect its electron distribution, which in turn has a direct bearing on how easily and permanently the molecule binds to its target molecule. Drug size and shape will determine whether the drug molecule is able to get close enough to its target site in order to bind to that site. The parameters commonly used to represent these properties are partition coefficients and lipophilic substitution constants for lipophilicity (see section 3.7.2), Hammett σ constants for electronic effects (see section 3.7.3) and Taft E_s steric constants for steric effects (see section 3.7.4). Consequently, this text will be largely restricted to a discussion of the use of these constants. However, the other parameters mentioned in this and other texts are normally used in a similar fashion.

QSAR-derived equations take the general form:

$$\text{Biological activity} = \text{Function \{parameter(s)\}} \qquad (3.1)$$

in which the activity is normally expressed as $\log[1/(\text{concentration term})]$ where the concentration term is usually C, the minimum concentration required to cause a defined biological response. The precise details of the function are obtained by the use of statistical mathematical methods such as regression analysis (see section 3.7.1) using an appropriate computer program. The accuracy of equations obtained in this manner is normally assessed as part of the computer package used in their generation. Where there is a poor correlation between the values of a specific parameter and the drug's activity' other parameters must be playing a more important part in the drug's action and so they must also be incorporated into the QSAR equation.

QSAR studies are normally carried out on groups of related compounds. However, QSAR studies on structurally diverse sets of compounds are becoming more common. In both instances it is important to consider as wide a range of parameters as possible.

3.7.1 Regression analysis

Regression analysis is a group of mathematical methods used to obtain mathematical equations relating different sets of data that have been obtained from experimental work or calculated using theoretical considerations. The data are fed into a suitable computer program, which, on execution, produces an equation that represents the line that is the best fit for those data. For example, suppose that an investigation indicated that the relationship between the activity and the partition coefficients of a number of related compounds appeared to be linear (Fig. 3.9). These data could be represented mathematically in the form of the straight line equation $y = mx + c$. Regression analysis would calculate the values of m and c that gave the line of best fit to the data.

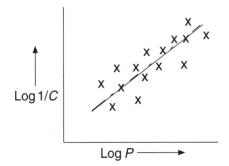

Figure 3.9 A hypothetical plot of the activity (log 1/C) of a series of compounds against the logarithm of their partition coefficients (log P)

Regression equations do not indicate the accuracy and spread of the data. Consequently, they are normally accompanied by additional data, which as a minimum requirement should include the number of observations used (n), the standard deviation of the observations(s) and the regression coefficient(r).

The value of the regression coefficient (0–1) is a measure of how closely the data match the equation. A value of $r = 1$ indicates a perfect match. In medicinal chemistry r values greater than 0.9 are usually regarded as representing an acceptable degree of accuracy provided that they are obtained using a reasonable number of results with a suitable standard deviation.

The value of $100r^2$ is a measure of the percentage of the data that can be satisfactorily explained by the regression analysis. For example, a value of $r = 0.90$ indicates that 81 per cent of the results can be satisfactorily explained by regression analysis using the parameters specified. It indicates that 19 per cent of the data are not satisfactorily explained by these parameters and so indicates that the use of an additional parameter(s) might give a more acceptable account of the results. Suppose, for example, regression analysis using an extra parameter gave a regression coefficient of 0.98. This shows that 96.04 per cent of the data are now satisfactorily accounted for by the chosen parameters.

When one is dealing with a linear relationship the analysis is usually carried out using the method of least squares. Other mathematical methods are also used for non-linear relationships.

3.7.2 The lipophilic parameters

Two parameters are commonly used to relate drug absorption and distribution with biological activity, namely, the partition coefficient (P) and the lipophilic substituent constant (π). The former parameter refers to the whole molecule whilst the latter is related to substituent groups.

Partition coefficients (P)

A drug has to pass through a number of biological membranes in order to reach its site of action. Consequently, partition coefficients (see section 2.12) were the obvious parameter to use as a measure of the movement of the drug through these membranes. The nature of the relationship obtained depends on the range of P values for the compounds used. If this range is small the results may, by the use of regression analysis (see section 3.7.1), be expressed as a straight line equation having the general form:

$$\log(1/C) = k_1 \log\ P + k_2 \tag{3.2}$$

where k_1 and k_2 are constants. This equation indicates a linear relationship between the activity of the drug and its partition coefficient. A number of examples of this type of correlation are known (Table 3.4).

Over larger ranges of P values the graph of log (1/C) against log P often has a parabolic form (Fig. 3.10) with a maximum value (log $P°$). The existence of this maximum value implies that there is an optimum balance between aqueous and lipid solubility for maximum biological activity. Below $P°$ the drug will be reluctant to enter the membrane whilst above $P°$ the drug will be reluctant to leave the membrane. Log $P°$ represents the optimum partition coefficient for biological activity. This means that analogues with partition coefficients near this optimum value are likely to be the most active and worth further investigation. Hansch et al. showed that many of these parabolic relationships could be represented reasonably accurately by equations of the form:

$$\log(1/C) = -k_1 (\log P)^2 + k_2 \log P + k_3 \tag{3.3}$$

Table 3.4 Examples of linear relationships between log($1/C$) against log P. Equations (2) and (3) are adapted from C. Hansch, Drug design I (1971). Ed. E. J. Ariens and reproduced by permission of Academic Press and C. Hansch. $r=$ regression coefficient, $n=$ number of compounds tested and $s=$ standard deviation

---

(1) Toxicity of alcohols to red spiders:
$$\log(1/C) = 0.69 \log P + 0.16 \qquad r = 0.979, \; n = 14, \; s = 0.087$$

(2) The binding of miscellaneous neutral molecules to bovine serum:
$$\log(1/C) = 0.75 \log P + 2.30 \qquad r = 0.96, \; n = 42, \; s = 0.159$$

(3) The binding of miscellaneous molecules to haemoglobin:
$$\log(1/C) = 0.71 \log P + 1.51 \qquad r = 0.95, \; n = 17, \; s = 0.16$$

(4) Inhibition of phenols on the conversion of P-450 to P-420 cytochromes:
$$\log(1/C) = 0.57 \log P + 0.36 \qquad r = 0.979, \; n = 13, \; s = 0.132$$

The values of the constants k_1, k_2 and k_3 in equation (3.3) are normally determined by either regression analysis (see section 3.7.1) or other statistical methods. For example, a study of the inducement of hypnosis in mice by a series of barbiturates showed that the correlation could be expressed by the equation:

$$\log(1/C) = -0.44(\log P)^2 + 1.58 \log P + 1.93 \, (r = 0.969) \tag{3.4}$$

This equation has a maximum log $P°$ at about 2.0. Hansch *et al.* showed that a range of non-specific hypnotic drugs with widely different types of structure were found to have log P values around 2. This implies that it is the solubility of these different drugs in the membrane rather than their structures that is the major factor in controlling their activity. On the basis of these and other partition studies, Hansch suggested in the mid-1960s that any organic compound with a log P value of approximately 2 would, provided it was not rapidly metabolised or eliminated, have some hypnotic properties and would be rapidly transported into the CNS. Subsequent practical evidence gives some support to this assertion. The fact that the thiobarbiturates have log P values of about 3.1 suggests that

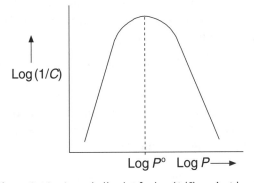

Figure 3.10 A parabolic plot for log ($1/C$) against log P

these drugs probably have a different site of action from those of the barbiturates. The larger value also suggests that a more lipophilic receptor is involved.

Monitoring the change in log P has been used in drug design. For example, compound I (log $P = 2.57$) is a cardiotonic agent. However, its use resulted in unwanted CNS side effects in some patients. Replacement of the methoxy group by the approximately same sized but more hydrophilic methyl sulphone residue gave the cardiotonic drug sulmazole with a value of 1.17 for log P.

Compound I (log $P = 2.57$)   Sulmazole (log $P = 1.17$)

The accuracy of the correlation of drug activity with partition coefficient will also depend on the solvent system used as a model to measure the partition coefficient values. The n-octanol/water system is frequently chosen because it has the most extensive data base. However, more accurate results may be obtained if the organic phase is matched to the area of biological activity being studied. For example, n-octanol usually gives the most consistent results for drugs absorbed in the GI tract whilst less polar solvents such as olive oil frequently give more consistent correlations for drugs crossing the blood–brain barrier (see sections 1.7.1 and 7.2.9). More polar solvents such as chloroform give more consistent values for buccal absorption (soft tissues in the mouth). However, when correlating P values with potency it should be borne in mind that the partition coefficient of a drug is usually only one of a number of parameters influencing its activity. Consequently, in cases where there is a poor correlation between the partition coefficient and the drug's activity, other parameters must be playing a more important part in the action of the drug.

Lipophilic substituent constants (π)

Lipophilic substituent constants are also known as hydrophobic substituent constants. They represent the contribution that a group makes to the partition coefficient and were defined by Hansch and co-workers by the equation:

$$n = \log P_x - \log P_H \qquad (3.5)$$

where P_H and P_x are the partition coefficients of the standard compound and its monosubstituted derivative, respectively. For example, the value of π for the chloro group of chlorobenzene could be calculated from the partition coefficients for benzene and chlorobenzene in the octanol/water system:

$$\pi_{Cl} = \log P_{(C_6H_6Cl)} - \log P_{(C_6H_6)} \qquad (3.6)$$

and substituting the appropriate π values:

$$\pi_{Cl} = 2.84 - 2.13 = 0.71$$

The values of π will vary depending on the solvent system used to determine the partition coefficients. However, most π values are determined using the *n*-octanol/water system. A positive π value indicates that a substituent has a higher lipophilicity than hydrogen and so will probably increase the concentration of the compound in the *n*-octanol layer and by inference its concentration in the lipid material of biological systems. Conversely, a negative π value shows that the substituent has a lower lipophilicity than hydrogen and so probably increases the concentration of the compound in the aqueous media of biological systems.

The π value for a specific substituent will vary with its structural environment (Table 3.5). Consequently, average values or the values relevant to the type of structures being investigated may be used in determining activity relationships. Where several substituents are present the value of π for the compound is the sum of the π values of each of the separate substituents, namely:

$$\pi = \pi(\text{substituent 1}) + \pi(\text{substituent 2}) + \ldots\ldots\ldots \pi(\text{substituent } n) \qquad (3.7)$$

Table 3.5 Examples of the variations of π values with chemical structure

Substituent X	Aliphatic systems R-X	$\langle\text{C}_6\text{H}_5\rangle$–X	O_2N–$\langle\text{C}_6\text{H}_4\rangle$–X	HO–$\langle\text{C}_6\text{H}_4\rangle$–X
-H	0.00	0.00	0.00	0.00
-CH$_3$	0.50	0.56	0.52	0.49
-F	-0.17	0.14		0.31
-Cl	0.39	0.71	0.54	0.93
-OH	-1.16	-0.67	0.11	-0.87
-NH$_2$		-1.23	-0.46	-1.63
-NO$_2$		-0.28	-0.39	0.50
-OCH$_3$	0.47	-0.02	0.18	-0.12

Lipophilic substituent constants can be used as an alternative to the partition coefficient when dealing with a series of analogues in which only the substituents are different. This usage is based on the assumption that the lipophilic effect of the unchanged part of the structure is similar for each of the analogues. Consequently, the π values of substituents indicate the significance of the contribution of that substituent to the lipophilicity of the molecule. Furthermore, biological activity–π relationships that have high regression constants and low standard deviations demonstrate that the substituents are important in determining the lipophilic character of the drug.

Lipophilic substituent constants can also be used to calculate theoretical partition coefficients for whole molecules using equation (3.5). For example, the calculated partition coefficient for 1,3-dimethylbenzene would be given by:

$$\pi = \log P_{1,3 \text{ dimethylbenzene}} - \log P_{\text{benzene}} \tag{3.8}$$

However, the value of π for compounds containing more than one substituent is the sum of the π values for each of the substituent groups (equation 3.7) and so using the π value given for the methyl group in methylbenzene given in Table 3.5 the value of π in equation (3.8) for the two methyl groups of 1,3-dimethylbenzene is:

$$\pi = 0.56 + 0.56 = 1.12$$

Therefore, as $\log P_{\text{benzene}}$ for benzene is 2.13, substituting in equation (3.8):

$$1.12 = \log P_{1,3-\text{dimethylbenzene}} - 2.13$$

and

$$\log P_{1,3-\text{dimethylbenzene}} = 3.25$$

This is in good agreement with the experimental value of 3.20 for $\log P_{1,3-\text{dimethylbenzene}}$. These calculated values are sometimes referred to as $C \log P$ values in order to distinguish them from experimentally determined P values. They are often in good agreement with the experimentally determined values provided that the substituents are not sterically crowded. However, poorer agreements are often found when substituents are located close to each other in the molecule. In addition, strong electron interactions between substituent groups can result in inaccurate theoretical P values.

Distribution coefficients

Partition coefficient values do not take into account the fact that many compounds ionise in aqueous solution. The extent of this ionisation will have a significant effect on the absorption and distribution of these drugs (see sections 1.7.1 and 2.8). Consequently, in Hansch (see section 3.7.4) and other forms of mathematical analyses of drug behaviour the lipophilicity of ionisable compoundsis often represented by the value of their distribution coefficients (D), which is defined as the ratio of the concentrations of the unionised and the ionised compound between an organic solvent and an aqueous medium. For example the distribution coeficient of the acid HA is given by:

$$D = \frac{[\text{HA}_{(\text{organic})}]}{[\text{H}^+_{(\text{aqueous})}] + [\text{A}^-_{(\text{aqueous})}]} \tag{3.9}$$

Since the ionisation of acids and bases depends on the pH of the aqueous medium it can be shown that for acids:

$$\text{Log}(P/D - 1) = \text{pH} - pK_a \tag{3.10}$$

and for bases BH dissociating to $B + H^+$:

$$Log(P/D - 1) = pK_a - pH \qquad (3.11)$$

These equations allow the calcuation of the effective lipophilicity of a compound at any pH provided that the pK_a and the value of P are known for the same solvent system. Distribution coefficients are normally used as log D values, a large number of which are available from databases (see sections 3.7.4 and 4.10).

3.7.3   Electronic parameters

The distribution of the electrons in a drug molecule will have a considerable influence on the distribution and activity of a drug. In order to reach its target a drug normally has to pass through a number of biological membranes (see section 7. 3). As a general rule, non-polar and polar drugs in their unionised form are usually more readily transported through membranes than polar drugs and drugs in their ionised forms (see section 7.3.3). Furthermore, once the drug reaches its target site the distribution of electrons in its structure will control the type of bonds it forms with that target (see section 8.2), which in turn affects its biological activity. In other words, the electron distribution in a drug molecule will have an effect on how strongly that drug binds to its target site, which in turn affects its activity.

   One of the attempts to quantify the electronic effects of groups on the physicochemical properties of compounds was made by Hammett in 1940. He introduced his *substitution constant* (σ) in order to predict the values of equilibrium and rate constants. This constant is now extensively used as a parameter for the electronic effects of the structure of a drug on its activity.

The Hammett constant (σ)

The distribution of electrons within a molecule depends on the nature of the electron withdrawing and donating groups found in that structure. For example, benzoic acid is weakly ionised in water:

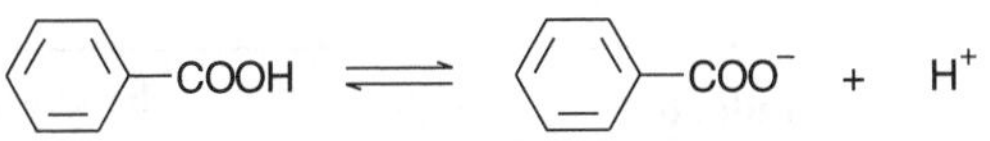

Substitution of a ring hydrogen by an electron withdrawing substituent (X), such as a nitro group, will weaken the O–H bond of the carboxyl group and stabilise the carboxylate anion (Fig. 3.11). This will move the equilibrium to the right which means that the substituted compound is a stronger acid than benzoic acid ($K_X > K$). It also means that at equilibrium more of the nitrobenzoic acid will exist as anions, which could make its transfer through membranes more difficult than that of the weaker lessionised benzoic acid. Conversely, the introduction of an electron donor substituent (X) such as a methyl group into the ring strengthens the acidic O–H group and reduces the stability of the carboxylate anion. This

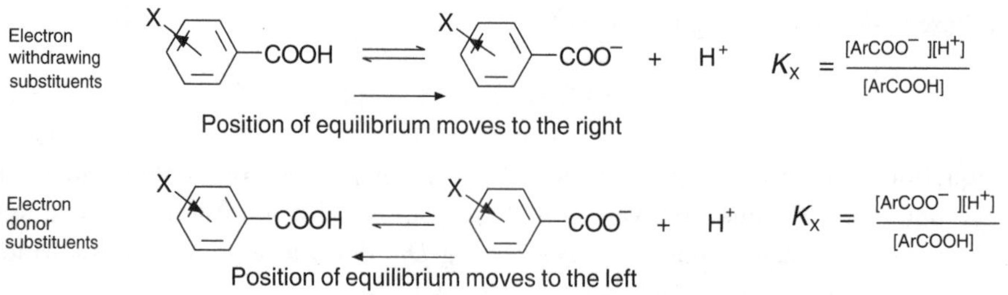

Figure 3.11 The effect of electron withdrawing and donor groups on the position of equilibrium of substituted benzoic acids

moves the equilibrium to the left, which means that the compound is a weaker acid than benzoic acid ($K > K_X$). This in turn means that it has fewer anions in solution at equilibrium than benzoic acid and so could pass through membranes more easily than benzoic acid. In addition to the effect that changes in the electron distribution have on transfer through membranes, they will also have an effect on the binding of these acids to a target site. These observations show that it is possible to use equilibrium constants *to compare* the electron distributions of *structurally similar groups of compounds.*

Hammett used equilibrium constants to study the relationship between the structure of aromatic acids and acid strength. In the course of this study he calculated constants, which are now known as *Hammett substituent constants* (σ_X) or simply *Hammett constants,* for a variety of ring substituents (X) of benzoic acid, using this acid as the comparative reference standard (Table 3.6). These constants are related to the type and extent of the electron distribution in these aromatic acids and as a result are now used as electron distribution parameters in QSAR studies.

Table 3.6 Examples of the different electronic substitution constants used in QSAR studies. Inductive substituent constants (σ_1) represent the contribution the inductive effect makes to Hammett constants and can be used for aliphatic compounds. Taft substitution constants (σ^*) refer to aliphatic substituents and use the 2-methyl derivative of ethanoic acid (propanoic acid) as the reference point. The Swain–Lupton constants represent the contributions due to the inductive (F) and mesomeric or resonance (R) components of Hammett constants. Reproduced with permission from H.J. Smith and H. Williams, *An Introduction to the Principles of Drug Design and Action*, 3rd edn, table 5.3, 1998, Harwood Academic Publishers

| Substituent | Hammett constants | | Inductive constant | Taft constant | Swain–Lupton constants | |
	σ_m	σ_p	σ_1	σ^*	F	R
-H	0.00	0.00	0.00	0.49	0.00	0.00
-CH$_3$	− 0.07	− 0.17	− 0.05	0.00	0.04	− 0.13
-C$_2$H$_5$	− 0.07	− 0.15	− 0.05	− 0.10	− 0.05	− 0.10
-Ph	0.06	− 0.01	0.10	0.60	0.08	− 0.08
-OH	0.12	− 0.37	0.25	–	0.29	− 0.64
-Cl	0.37	0.23	0.47	–	0.41	− 0.15
-NO$_2$	0.71	0.78	–	–	0.67	0.16

Hammett constants (σ_X) are defined as:

$$\sigma_X = \log \frac{K_X}{K} \tag{3.12}$$

that is:

$$\sigma_X = \log K_X - \log K \tag{3.13}$$

and so, as $pK_a = -\log K_a$:

$$\sigma_X = pK - pK_X \tag{3.14}$$

A negative value for σ_X indicates that the substituent is acting as an electron donor group since $K \gg K_X$. Conversely, a positive value for σ_X shows that the substituent is acting as an electron withdrawing group as $K < K_X$. The value of σ_X varies with the position of the substituent in the molecule. Consequently, this position is usually indicated by the use of the subscripts *o*, *m* and *p*. Where a substituent has opposite signs depending on its position on the ring it means that in one case it is acting as an electron donor and in the other as an electron withdrawing group. This is possible because the Hammett constant includes both the inductive and mesomeric (resonance) contributions to the electron distribution. For example, the σ_m Hammett constant for the methoxy group of *m*-methoxybenzoic acid is 0.12 whilst for *p*-methoxybenzoic acid it is -0.27. In the former case the electronic distribution is dominated by the inductive (I or F) contribution whilst in the latter case it is controlled by the mesomeric (M) or resonance (R) effect (Fig. 3.12).

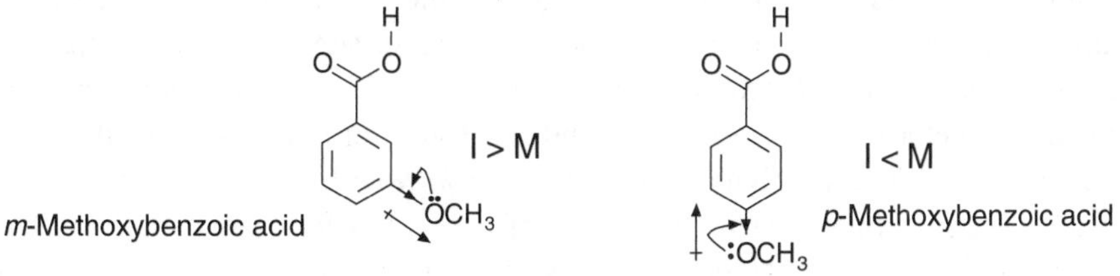

Figure 3.12 The inductive and mesomeric effects of *m*-methoxybenzoic and *p*-methoxybenzoic acids

Hammett postulated that the σ values calculated for the ring substituents of a series of benzoic acids could also be valid for those ring substituents in a different series of similar aromatic compounds. This relationship has been found to be in good agreement for the *meta* and *para* substituents of a wide variety of aromatic compounds but not for their *ortho* substituents. The latter is believed to be due to steric hindrance and other effects, such as intramolecular hydrogen bonding, playing a significant part in the ionisations of compounds with *ortho* substituents. Hammett substitution constants also suffer from the disadvantage that they only apply to substituents directly attached to a benzene ring. Consequently, a number of other electronic constants (Table 3.6) have been introduced and used in QSAR studies in a similar manner to the Hammett constants. However, Hammett substitution constants are probably still one of the most widely used electronic parameters for QSAR studies.

Attempts to relate biological activity to the values of Hammett substitution and similar constants have been largely unsuccessful since electron distribution is not the only factor involved (see section 3.7). However, a successful attempt to relate biological activity to structure using Hammett constants was the investigation by Fukata and Metcalf into the effectiveness of diethyl aryl phosphates for killing fruit flies. This investigation showed that the activity of these compounds is dependent only on electron distribution factors. Their results may be expressed by the relationship:

$$Log(1/C) = 2.282\sigma - 0.348 \tag{3.15}$$

Diethyl aryl phosphate insecticides

This equation shows that the greater the positive value for σ, the greater the biological activity of the analogue. This type of knowledge enables one to predict the activities of analogues and synthesise the most promising rather than spend a considerable amount of time synthesising and testing all the possible analogues.

3.7.4 Steric parameters

In order for a drug to bind effectively to its target site the dimensions of the pharmacophore of the drug must be complementary to those of the target site (see sections 8.2. and 9.3). The Taft steric parameter (E_s) was the first attempt to show the relationship between a measurable parameter related to the shape and size (bulk) of a drug and the dimensions of the target site and a drug's activity. This has been followed by Charton's steric parameter (v), Verloop's steric parameters and the molar refractivity (MR), amongst others. The most used of these additional parameters is probably the molar refractivity. However, in all cases the required parameter is calculated for a set of related analogues and correlated with their activity using a suitable statistical method such as regression analysis (see section 3.7.1). The results of individual investigations have shown varying degrees of success in relating the biological activity to the parameter. This is probably because little is known about the finer details of the three-dimensional structures of the target sites.

The Taft steric parameter (E_S)

Taft in 1956 used the relative rate constants of the acid-catalysed hydrolysis of α-substituted methyl ethanoates (Fig. 3.13) to define his steric parameter because it had been

$$CH_3COOCH_3 \qquad XCH_2COOCH_3 + H_2O \xrightarrow{H^+} XCH_2COOH + CH_3OH$$
$$\text{(a)} \qquad\qquad\qquad\qquad\qquad\qquad \text{(b)}$$

Figure 3.13 (a) Methyl ethanoate. (b) The hydrolysis of α-substituted methyl ethanoates

shown that the rates of these hydrolyses were almost entirely dependent on steric factors. He used methyl ethanoate as his standard and defined E_s as:

$$E_s = \log \frac{k_{(XCH_2COOCH_3)}}{k_{(CH_3COOCH_3)}} = \log k_{(XCH_2COOCH_3)} - \log k_{(CH_3COOCH_3)} \qquad (3.16)$$

where k is the rate constant of the appropriate hydrolysis and the value of $E_s = 0$ when X = H. It is assumed that the values for E_s (Table 3.7) obtained for a group using the hydrolysis data are applicable to other structures containing that group. The methyl-based E_s values can be converted to H-based values by adding -1.24 to the corresponding methyl-based values.

Table 3.7 Examples of the Taft steric parameter E_s

Group	E_s	Group	E_s	Group	E_s
H-	1.24	F-	0.78	CH_3O-	0.69
CH_3-	0.00	Cl-	0.27	CH_3S-	0.19
C_2H_5-	-0.07	F_3C-	-1.16	$PhCH_2$-	-0.38
$(CH_3)_2CH$-	-0.47	Cl_3C-	-2.06	PhOCH-	0.33

Taft steric parameters have been found to be useful in a number of investigations. For example, regression analysis has shown that the antihistamine effect of a number of related analogues of diphenhydramine (Fig. 3.14) was related to their biological response (BR) by equation (3.17), where E_s is the sum of the *ortho* and *meta* E_s values in the most highly substituted ring.

$$\log BR = 0.440E_s - 2.204 \quad (n = 30, \; s = 0.307 \; r = 0.886) \qquad (3.17)$$

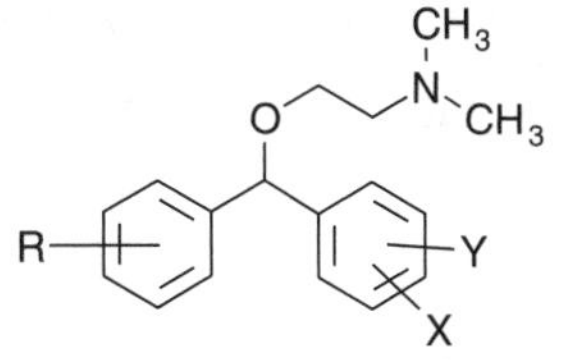

Figure 3.14 The general formula of diphenhydramine analogues

Regression analysis also showed that the biological response was related to the Hammett constant by the relationship:

$$\log BR = 2.814\sigma - 0.223 \quad (n = 30, \ s = 0.519, \ r = 0.629) \tag{3.18}$$

A comparison of the standard deviations (s) for equations (3.17) and (3.18) shows that the calculated values for the Hammett constants σ for each of the analogues are more scattered than the calculated values for the corresponding Taft E_s values. Furthermore, although both the r and s values for equation (3.17) are reasonable, those for equation (3.18) are unacceptable. This indicates that the antihistamine activity of these analogues appears to depend more on steric than electronic effects. This deduction is supported by the fact that using regression analysis to obtain a relationship involving both the Hammett and Taft constants does not lead to a significant increase in the r and s values (equation 3.16).

$$\log BR = 0.492E_s - 0.585\sigma - 2.445 \quad (n = 30, \ s = 0.301, \ r = 0.889) \tag{3.19}$$

Taft constants suffer from the disadvantage that they are determined by experiment. Consequently, the difficulties in obtaining the necessary experimental data have significantly limited the number of values recorded in the literature.

Molar refractivity (MR)

The molar refractivity is a measure of both the volume of a compound and how easily it is polarised. It is defined as:

$$MR = \frac{(n^2 - 1)M}{(n^2 + 2)\rho} \tag{3.20}$$

where n is the refractive index, M is the relative mass and ρ is the density of the compound. The M/ρ term is a measure of the molar volume whilst the refractive index term is a measure of the polarisability of the compound. Although MR is calculated for the whole molecule, it is an additive parameter and so the MR values for a molecule can be calculated by adding together the MR values for its component parts (Table 3.8).

Table 3.8 Examples of calculated Mr values. Reproduced by permission of John Wiley and sons, Ltd. from C. Hansch and A.J. Leo, *Substituent constants for correlation Analysis in Chemistry and Biology, 1979*

Group	MR	Group	MR	Group	MR
H-	1.03	F-	0.92	CH_3O-	7.87
CH_3-	5.65	Cl-	6.03	HO-	2.85
C_2H_5-	10.30	F_3C-	5.02	CH_3CONH-	14.93
$(CH_3)_2CH$-	14.96	O_2N-	7.63	CH_3CO-	11.18

The other steric parameters

A wide variety of other parameters have been used to relate the steric nature of a drug on activity. These can be broadly divided into those that apply to sections of the molecule and those that involve the whole molecule. The former include parameters such as van der Waals' radii, Charton's steric constants and the Verloop steric parameters. They have all been used in a similar manner to the Taft steric parameters to correlate biological activity to structure with varying degrees of success.

Hansch analysis

Hansch analysis is based on the attempts by earlier workers, notably Richardson (1867), Richet (1893), Meyer (1899), Overton (1901), Ferguson (1939) and Collander (1954), to relate drug activity to measurable chemical properties. Hansch and co-workers in the early 1960s proposed a multiparameter approach to the problem based on the lipophilicity of the drug and the electronic and steric influences of groups found in its structure. They realised that the biological activity of a compound is a function of its ability to reach and bind to its target site. Hansch proposed that drug action could be divided into two stages:

1. The transport of the drug to its site of action.

2. The binding of the drug to the target site.

He stated that the transport of the drug is like a 'random walk' from the point of administration to its site of action. During this 'walk' the drug has to pass through numerous membranes and so the ability of the drug to reach its target is dependent on its lipophilicity. Consequently, this ability could be expressed mathematically as a function of either the drug's partition coefficient P or the π value(s) of appropriate substituents (see sections 3.3 and 2.13). However, on reaching its target the binding of the drug to the target site depends on the shape, electron distribution and polarisability of the groups involved in the binding. A variety of parameters are now used to describe each of these aspects of drug activity, the most common ones being the Hammett electronic σ (see section 3.7.3) and Taft E_s constants (see section 3.7.4).

Hansch postulated that the biological activity of a drug could be related to all or some of these factors by simple mathematical relationships based on the general format:

$$\log 1/C = k_1(\text{partition parameter}) + k_2(\text{electronic parameter}) + k_3(\text{steric parameter}) + k_4$$

$$(3.21)$$

where C is the minimum concentration required to cause a specific biological response and k_1, k_2, k_3 and k_4 are numerical constants obtained by feeding the data into a suitable computer statistical package. For example, the general equations relating activity to all of these parameters often takes the general form:

$$\log 1/C = k_1 P - k_2 P^2 + k_3 \sigma + k_4 E_s + k_5 \tag{3.22}$$

where other parameters could be substituted for P, σ and E_s. For example, π may be used instead of P and MR for E_s. However, it is emphasised that these equations, which are collectively known as Hansch equations, do not always contain the main three types of parameter (Table 3.9). The numerical values of the constants in these equations are obtained by feeding the values of the parameters into a suitable computer package. These

Table 3.9 Examples of simple Hansch equations

Compound	Activity	Hansch equation
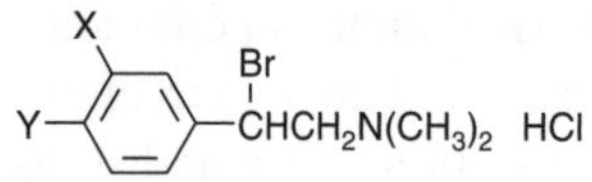	Antiadrenergic	$\log 1/C = 1.22\pi - 1.59\sigma + 7.89$ ($n = 22$; $s = 0.238$; $r = 0.918$)
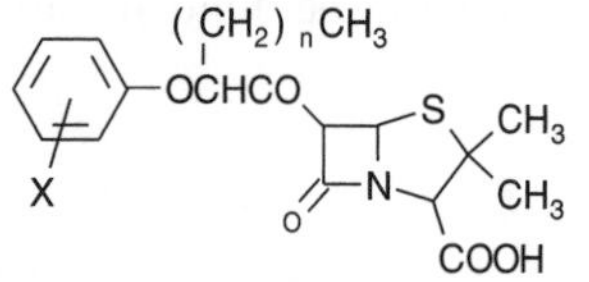	Antibiotic (*in vivo*)	$\log 1/C = -0.445\pi - 5.673$ ($n = 20$; $r = 0.909$)
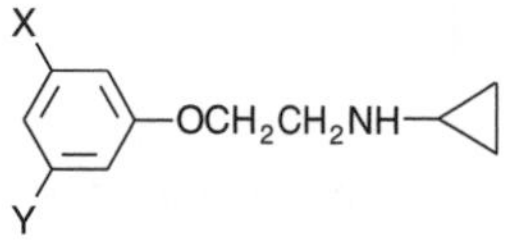	MAO inhibitor (humans)	$\log 1/C = 0.398\pi + 1.089\sigma + 1.03E_s + 4.541$ ($n = 9$; $r = 0.955$)
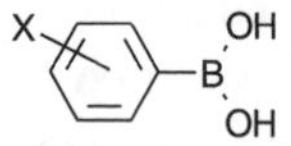	Concentration (C_b) in the brain after 15 min	$\log C_b = 0.765\pi - 0.540\pi^2 + 1.505$

parameter values are obtained either from the literature (eg. π, σ, E_s) or determined by experiment (eg. C, P, etc.).

Many QSAR investigations involve varying more than one ring substituent. In these cases the values of the same parameter for each substituent are expressed in the Hansch equation as either the sum of the individual parameters or are treated as independent individual parameters. For example, in the hypothetical case of a benzene ring with two substituents X and Y the Hammett constants could be expressed in the Hansch equation as either $k_1 \Sigma(\sigma_X + \sigma_Y)$ or $k_1 \sigma_X + k_2 \sigma_Y$. A comprehensive list of many of the parameters used in Hansch analysis may be found in a review by Tute in *Advances in Drug Research* 1971, **6**,1.

Hansch equations can be used to give information about the nature of the mechanism by which drugs act as well as predict the activity of as yet unsynthesised analogues. In the former case the value of the constant for a parameter gives an indication of the importance of the influence of that parameter in the mechanism by which the drug acts. Consider, for example, a series of analogues whose activity is related to the parameters π and σ by the hypothetical Hansch equation:

$$\log 1/C = 1.78\pi - 0.12\sigma + 1.674 \qquad (3.23)$$

The small value of the coefficient for σ relative to that of π in equation (3.23) shows that the electronic factor is not an important feature of the action of these drugs. Furthermore, the value of the regression coefficient (r) for the equation will indicate whether suitable parameters were used for its derivation. Values of r that are significantly lower than 0.9 indicate that either the parameter(s) used to derive the equation were unsuitable or that there is no relationship between the compounds used and their activity. This suggests that the mechanisms by which these compounds act may be very different and therefore unrelated.

Predictions of the activities of as yet unsynthesised analogues are useful in that they allow the medicinal chemist to make a more informed choice as to which analogues to synthesise. However, these predictions should only be made within the limits used to establish that relationship. For example, if a range of partition coefficients with values from 3 to 8 were used to obtain an activity relationship–partition coefficient equation then this equation should not be used to predict activities of compounds with partition coefficients of less than 3 and greater than 8.

Hansch equation activity predictions that are widely different from the observed values suggest that the activity of a compound is affected by factors that have not been included in the derivation of the Hansch equation. The discovery of this type of anomaly may give some insight into the mechanism by which the compound acts. For example, a study by Hansch on the activity of penicillins against a strain of *Staphylococcus aureus* in mice gave the *in vivo* relationship:

$$\log 1/C = -0.445\pi + 5.673 \quad (n = 20, \ s = 0.191, \ r = 0.909) \qquad (3.24)$$

This relationship predicts that a penicillin with branched side chain $PhOCH(CH_3)CONH-$ should be less active than a penicillin with the similar sized unbranched side chain $PhOCH_2CONH-$ (Fig. 3.15) because the branched side chain has a higher π value. However, Hansch found that the penicillin with the branched side chain was the more active. He suggested that this anomaly could be explained by the fact that branched side chains were more resistant to metabolism than straight side chains.

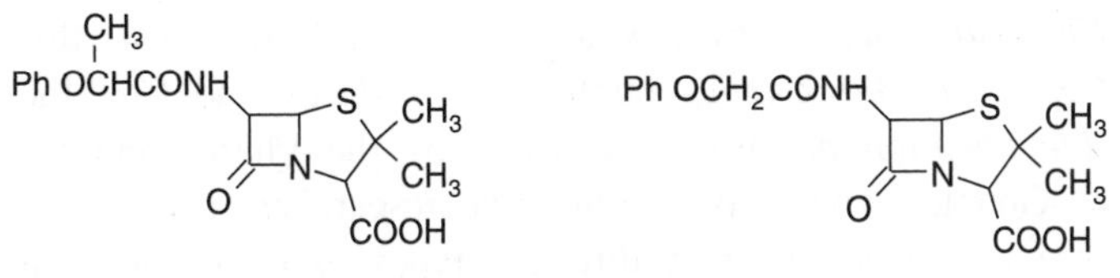

Figure 3.15 Penicillin analogues

The accuracy of Hansch equations and hence the success of QSAR investigations depends on using enough analogues, the accuracy of the data and the choice of parameters:

1. The greater the number of analogues used in a study, the greater the probability of deriving an accurate Hansch equation. A rough rule of thumb is that the minimum number of compounds used in the study should be not less than $5x$, where x is the number of parameters used to obtain the relationship. Where substituents are being varied it is also necessary to use as wide a variety of substituents in as large a range of differentpositions as possible.

2. The accuracy of the biological data used to establish a Hansch equation will affect the accuracy of the derived relationship because its value depends to some extent on the subject used for the measurement. Consequently, it is necessary to take a statistically viable number of measurements of any biological data, such as the activity of an individual compound, and use an average value in the derivation of the Hansch equation. Furthermore, extreme parameter values should not be used since they are likely to dominate the regression analysis and so give less accurate Hansch equations. Their presence suggests that the parameter is either not suitable or there is no possible correlation.

3. The choice of parameter is important because one parameter may give an equation with an acceptable regression constant whilst another may give an equation with an unacceptable regression constant. For example, it may be possible to successfully correlate the dipole moments of a series of analogues with biological activity but fail to obtain a correlation for the same series of analogues using Hammett constants as a parameter. Furthermore, some parameters like π and σ are interrelated and so their use in the same equation can lead to confusion in interpreting the equation.

Hansch equations are normally use to either determine and assess the factors controlling an analogue's activity or predict the structure with the optimum activity. In the latter case, the Hansch equation is used to determine the so-called *ideal parameter values* that would give the most active analogue and relate these values to substituents with values that either correspond to or are the best fit to these ideal values. These values may be found in tables. However, it is often easier to relate the *ideal values* from the Hansch equation to those of the best substituents for a structure using a more visual method such as a Craig plot (see below).

In order to carry out a Hansch analysis it is usually necessary to obtain the required parameter data from a suitable source, by calculation or by experiment. Sources are normally data books or computerised data bases. Three commonly used books are: *The Handbook of Chemistry and Physics*, which is often referred to as the *Rubber Book*, published by CRC; *the Merck index* published by Merck and Co., Inc.; and *Clarks' Isolation and Identification of Drugs* published by the Pharmaceutical Press. Search engines such as Google (www.google.com) give researchers access to many different data bases. These data bases contain different types of data, for example ECOTOX lists toxicological and some physicochemical data, the ChemExper chemical directory (www. chemexper.com) lists many melting points and boiling points and Biobyte's (www.biobyte.com) Thor Masterfile data base contains a large number of log P, log D and pK_a values and other data. Other data bases are listed in Table 4.2 (see section 4.10). Most of these data bases may be searched for the required data using the name of a substance, name of a structural fragment, chemical structure, structural fragments, type of activity, Chemical Abstracts Service (CAS) registry numbers and other parameters. A number of these searches are free but others require the payment of a fee. The values for the properties listed in data bases are obtained by calculation using a wide range of specialist computer programs (see section 4.10), by deductive methods (see section 2.12.2) and determined by experiment. However, because of experimental error some experimentally determined values are now regarded as being less accurate than the calculated values.

A serious disadvantage of Hansch analysis is that its parameters do not take into account the three-dimensional nature of biological systems. Three-dimensional QSAR (see section 4.9) is an attempt to remedy this aspect of these equations.

Craig plots

Craig plots are two-dimensional plots of one parameter against another (Fig. 3.16). The plot is divided into four sections corresponding to the positive and negative values of the parameters. They are used in conjunction with an already established Hansch equation for a series of related aromatic compounds, to select aromatic substituents that are likely to

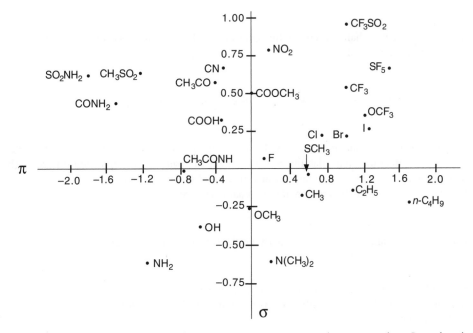

Figure 3.16 An example of a Craig plot of *para* Hammett constants σ against *para* π values. Reproduced with permission from M.E. Wolf (ed.), *Burgers Medicinal Chemistry*, 4th edn, Past 1, p. 343, 1980 John Wiley & Sons, Ltd.

produce the most active analogues. For example, suppose that a Hansch analysis carried out on a series of aromatic compounds yields the Hansch equation:

$$\log 1/C = 2.67\pi - 2.56\sigma + 3.92 \tag{3.25}$$

To obtain a high value for the activity ($1/C$) and as a result a low value for C, it is necessary to pick substituents with a positive π value and a negative σ value. In other words, if high activity analogues are required, the substituents should be chosen from the lower right-hand quadrant of the plot. However, it is emphasised that the use of a Craig plot does not guarantee that the resultant analogues will be more active than the lead because the parameters used may not be relevant to the mechanism by which the analogue acts.

3.8   Questions

1   State the full wording of the abbreviation 'SAR'. Describe the general way in which SAR is used to develop a drug. Illustrate the answer by reference to the changes in the activities of 4-alkylresorcinols caused by changes in the length of the 4-alkyl group.

2   (a) Explain why chlorine and fluorine are normally the preferred halogen substituents for a SAR investigation.

    (b) What alternative halogen containing group could be used in place of chlorine? Give one reason for the use of this group.

3   Explain the meaning of the terms: (a) bioisostere and (b) pharmacophore.

4   Suggest how the introduction of each of the following groups into the structure of a lead could be expected to affect the bioavailability of the resultant analogue: (a) a sulphonic acid group, (b) a methyl group and (c) a thiol group. Assume that these groups are introduced into the section of the lead, structure that does not contain its pharmacophore.

5   Outline the fundamental principle underlying the QSAR approach to drug design.

6   Lipophilicity, shape and electron distribution all have a major influence on drug activity. State the parameters that are commonly used as a measure of these properties in the QSAR approach to drug design.

7   (a) Describe the approach to drug design known as Hansch analysis.

(b) Phenols are antiseptics. Hansch analysis carried out on a series of phenols with the general structure A yielded the Hansch equation:

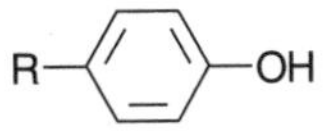

$$R-\bigcirc-OH$$

(A)

$$\text{Log } 1/C = 1.5\pi - 0.2\sigma + 2.3$$
$$(n = 23, s = 0.13, r = 0.87)$$

What are (i) the significance of the terms n, s and r, (ii) the relative significance of the lipophilicity and electronic distribution of a phenol of type A on its activity and (iii) the effect of replacing the R group of A by a more polar group.

8 (a) How are Craig plots used in Hansch analysis? (b) Use Figure 3.16 to predict the structures of the analogues of compound A in question 7 that would be likely to have a high antiseptic activity.

9 Table 3.3 lists the results of a SAR investigation to discover a bisphosphonate drug more potent than pamidronate. The ED_{50} of pamidronate is 61. Discuss the significance of the ED_{50} values of the following groups of compounds in the context of the SAR investigation:

(a) Compounds 1–6 inclusive.

(b) Compounds 7–23 inclusive.

10 The regression analysis equations relating the Taft and Hammette parameters, with the biological response (BR) for the analogues of a lead compound are:

(1) $\log \text{BR} = 2.972E_s - 0.164$ $\qquad\qquad$ $(n = 30, \ s = 0.307 \ r = 0.886)$

(2) $\log \text{BR} = 0.369\sigma - 2.513$ $\qquad\qquad$ $(n = 30, \ s = 0.519, \ r = 0.629)$

(3) $\log \text{BR} = 0.442E_s - 0.505\sigma - 2.445$ $\quad$ $(n = 30, \ s = 0.301, \ r = 0.927)$

Comment on the relative importance of electronic and steric effects in governing the activity of a drug.

4

Computer-aided drug design

4.1 Introduction

The development of powerful desk top and larger computers has led computer mathematicians to develop molecular modelling packages that can be used to solve sets of mathematical equations. This ability has enabled chemists to predict the structures and the values of the physical properties of known, unknown, stable and unstable molecular species. The prediction process is based on devising and solving mathematical equations related to the property required. These equations are obtained using a so-called 'model' (see section 4.1.1). The reliability of the mathematical methods used to obtain and solve the equations is well known and so in most cases it is possible to obtain a reliable estimate of the accuracy of the results. In some cases the calculated values are believed to be more accurate than the experimentally determined figures because of the higher degree of experimental error in the experimental work. Graphics packages have also been developed, which convert the data for the structure of a chemical species into a variety of easy to understand visual formats (Fig. 4.1). Consequently, in medicinal chemistry, it is now possible to visualise the three-dimensional shapes of exogenous and endogenous compounds, commonly referred to as *ligands* (see section 1.3), and their target sites. Furthermore, sophisticated computational chemistry packages also allow the medicinal chemist to evaluate the interactions between a compound and its target site before synthesising that compound (see section 4.5). This means that the medicinal chemist need only synthesise and test the most promising of the compounds, which considerably increases the chances of discovering a potent drug. It also significantly reduces the cost of development.

Molecular modelling is a complex subject and it is not possible to cover it in depth in this text. For workers wishing to use it as a tool in drug design it will be necessary to either ask a competent *computational chemist* to make the necessary calculations and graphic

Medicinal Chemistry, Second Edition   Gareth Thomas
© 2007 John Wiley & Sons, Ltd

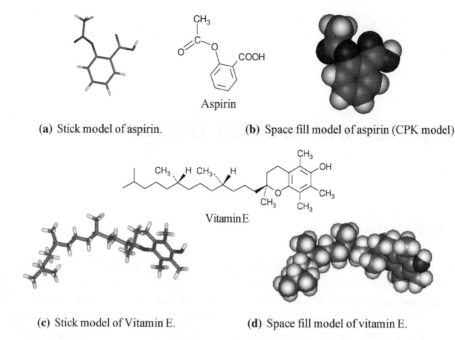

(a) Stick model of aspirin.   (b) Space fill model of aspirin (CPK model)

(c) Stick model of Vitamin E.   (d) Space fill model of vitamin E.

Figure 4.1 Examples of some of the formats used by graphics packages to display molecular models on computer screens

conversions or to treat the computer as a *black box* and use the relevant computer program according to its manufacturer's instructions. In both approaches to molecular modelling, it is essential that the drug designer has a basic understanding of the fundamental concepts of the methods used in order to avoid making incorrect deductions, as well as to appreciate the limitations of the methods.

4.1.1   Models

Models are used to visualise concepts in science. These models may take the form of flow diagrams or mathematical equations or a combination of both. They are often constructed by taking a property of a compound and determining the factors that have a significant influence on or have a major contribution to that property (see section 3.7). The effects of these factors may usually be expressed by a general equation of the form:

$$\text{Property} = \sum (\text{factors influencing the property}) + \text{a constant value (optional)} \quad (4.1)$$

In mathematical models each factor is represented either by using a logical mathematical relationship or a previously determined mathematical relationship for an analogous system.

These relationships are used to replace the appropriate factors in the equation and, as a result, build up an equation defining the property being studied. For example, suppose a system 1 has a property A that is influenced by three factors B, C and D. The general equation for A becomes:

$$A = B + C + D + \text{a constant} \tag{4.2}$$

Suppose $B = b$, $C = (c + x)/y$ and $D = d/z$, where b, c, x, y, d and z all have either experimentally measurable or theoretically calculable values, then:

$$A = b + (c + x)/y + d/z + \text{a constant} \tag{4.3}$$

This equation is used to predict values of A for different values of B, C and D. The data obtained may then be either used to determine other properties of system 1 or predict the value of A for different values of B, C and D.

4.1.2 Molecular modelling methods

The three-dimensional shapes of both a compound and its target site may be determined by X-ray crystallography or computational methods. The most common computational methods are based on either *molecular* or *quantum mechanics*. Both of these approaches produce equations for the total energy of the structure. In these equations the positions of the atoms in the structure are represented by either Cartesian or polar coordinates (Fig. 4.2). In the past, the initial values of these atomic coordinates were set by the modeller. However, it is now customary to construct models from existing structural fragments (see section 4.2.1). Modern computer programs will automatically set up the coordinates of the atoms in the first fragment from the program's data base. As additional fragments are added the computer automatically adjusts the coordinates of the atoms of these additional fragments to values that are relative to those of the first fragment since it is the relative positions of the atoms that are important as regards the energy of the structure and not the absolute positions of the atoms. Once the energy equation is established, the

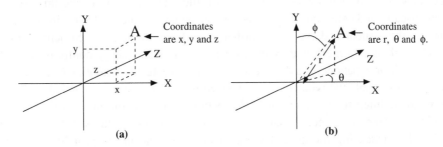

Figure 4.2 (a) Cartesian and (b) polar coordinates of the point A

computer computes the set of coordinates that correspond to a minimum total energy value for the system. These coordinates are converted into the required visual display by the graphics package (Fig. 4.1). However, although the calculations made by computers are always accurate, the calculated result should be checked for accuracy against experimental observations. In this respect it is essential that the approximations on which the calculations are based are understood. For example, most calculations are based on a frozen molecule at 0 K in a vacuum and so do not take into account that the structure is vibrating and also the influence of the medium in which the chemical species is found. Calculations taking these factors into account will undoubtedly give a more realistic picture of the structure.

Quantum mechanics calculations are more expensive to carry out because they require considerably more computing power and time than molecular mechanics calculations. Consequently, molecular mechanics is generally more useful for calculations involving large structures of interest to the medicinal chemist and so this chapter will concentrate on this method. To save time and expense, structures are often built up using information obtained from data bases, such as the Cambridge and Brookhaven data bases. Information from data bases may also be used to check the accuracy of the modelling technique. However, in all cases, the accuracy of the structures obtained will depend on the accuracy of the data used in their determination. Furthermore, it must be appreciated that the molecular models produced by computers are a caricature of reality that simply provide us with a useful picture for design and communication purposes. It is important to realise that we still do not know what molecules actually look like!

4.1.3   Computer graphics

In molecular modelling the data produced are converted into visual images on a computer screen by graphics packages. These images may be displayed as space fill, CPK (Corey–Pauling–Koltun), stick, ball and stick, mesh and ribbon (see Fig. 4.1 and Figs 4.3a, 4.3b and 4.3c). Ribbon representations are usually used to depict large molecules, such as nucleic acids and proteins. Each of these formats can, if required, use a colour code to represent the different elements, for example carbon atoms are usually green, oxygen is red and nitrogen is blue. However, most graphics packages will allow the user to change this code. The program usually indicates the three-dimensional nature of the molecule by making the colours of the structure darker, the further it is from the viewer. Structures may be displayed in their minimum energy conformations or other energy conformations. They may be shrunk or expanded to a desired size as well as rotated about either the x or y axis. These facilities enable the molecule to be viewed from different angles and also allows the structure to be fitted to its target site (see section 4.5). In addition, it is possible using molecular dynamics (see section 4.4) to show how the shape of the structure might vary with time by visualising the natural vibrations of the molecule (Fig. 4.3d) as a moving image on the screen. However, it is emphasised that both the stationary and moving images

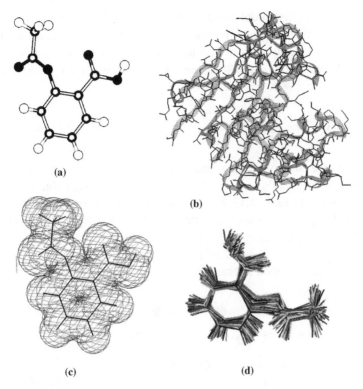

(a)

(b)

(c)   (d)

Figure 4.3 **(a)** Ball and stick representation of aspirin. **(b)** Ribbon representation of dihydrofolate reductase. **(c)** Mesh representation of aspirin. This representation simultaneously uses both the space fill and stick structures of the molecule. **(d)** Molecular dynamics representation of aspirin at 500 K. The relative movements of the atoms with time within the molecule are indicated by the use of multiple lines between the atoms

shown on the screen are useful caricatures and not pictures of the real structure of the molecule.

4.2 Molecular mechanics

Molecular mechanics is the more popular of the methods used to obtain molecular models as it is simpler to use and requires considerably less computing time to produce a model. The molecular mechanics method is based on the assumption that the relative positions of the nuclei of the atoms forming a structure are determined by the forces of attraction and repulsion operating in that structure. It assumes that the total *potential energy* (E_{Total}) of a molecule is given by the sum of all the energies of the attractive and repulsive forces between the atoms in the structure. These energies are calculated using a *mechanical model* in which these atoms are represented by balls whose mass is proportional to their relative atomic masses joined by mechanical springs corresponding to the covalent bonds in the structure. This model means that the laws and equations of classical physics can be used to determine the potential energies of the forces of attraction and repulsion operating between the atoms in a molecule. Using this model, E_{Total} may be expressed mathematically by equations, known as *force fields*. These equations normally take the general form:

$$E_{Total} = \Sigma E_{Stretching} + \Sigma E_{Bend} + \Sigma E_{Torsion} + \Sigma E_{vdW} + \Sigma E_{Coulombic} \qquad (4.4)$$

where $E_{Stretching}$ is the bond stretching energy (Fig. 4.4), E_{Bend} is the bond energy due to changes in bonding angle (Table 4.1), $E_{Torsion}$ is the bond energy due to changes in the conformation of a bond (Table 4.1), E_{vdW} is the total energy contribution due to van der Waals' forces and $E_{Coulombic}$ is the electrostatic attractive and repulsive forces operating in the molecule between atoms carrying a partial or full charge. Other energy terms, such as one for hydrogen bonding, may be added as required. Each of these energy terms includes expressions for all the specified interactions between all the atoms in the molecule.

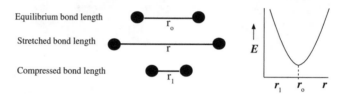

Figure 4.4 Bond stretching and compression related to the changes in the potential energy (*E*) of the system

The values of each of the energy terms in equation (4.4) are calculated by considering the mechanical or electrical nature of the structure that the energy term represents. For example, the $E_{Stretching}$ for a pair of atoms joined by a single covalent bond may be estimated by considering the bond to be a mechanical spring that obeys *Hooke's Law*. If *r* is the stretched length of the bond and r_0 is the ideal bond length, that is, the length the bond wants to be, then:

$$E_{Stretching} = \frac{1}{2}k(r - r_0)^2 \qquad (4.5)$$

where *k* is the force constant, which may be thought of as being a measure of the strength of the spring, in other words a measure of the strength of a bond. The larger the value of *k*, the stronger the bond. For example, C–C bonds have a smaller *k* value than C=C bonds, that is, C=C bonds are stronger than C–C bonds. In reality, more complex mathematical expressions, such as those given by the Morse function, would probably be used to describe bond stretching.

The value of $\Sigma E_{Stretching}$ in the force field equation (equation 4.4) for a structure is given by the sum of appropriate expressions for *E* for every pair of bonded atoms in the structure.

Table 4.1 Some of the expressions commonly used to calculate the energy terms given in equation (4.4)

Term	Expression	Model for:
E_{Bend}	$E_{Bend} = \frac{1}{2}k_\theta(\theta - \theta_o)^2$ where θ_o is the ideal bond angle, that is, the minimum energy positions of the three atoms	*Bond bending* 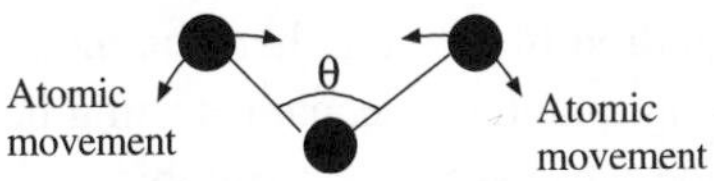
$E_{Torsion}$	$E_{Torsion} = \frac{1}{2}k_\phi[1 + \cos m(\phi + \phi_{offset})]$ where k_ϕ is the energy barrier to rotation about the torsion angle ϕ, m is the periodicity of the rotation and ϕ_{offset} is the ideal torsion angle relative (minimum energy positions for the atoms) to a staggered arrangement of the two atoms	*Rotation about a single bond* 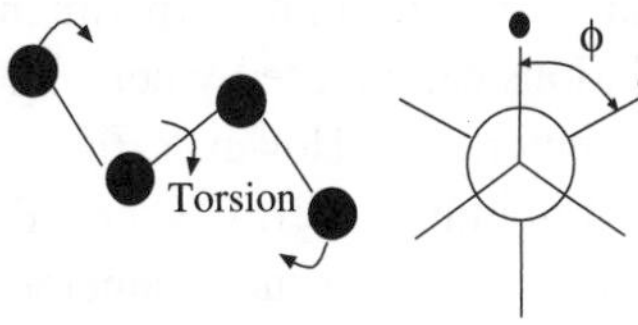
E_{vdW}	$E_{vdW} = \varepsilon\left[\left(\frac{r_{min}}{r}\right)^{12} - 2\left(\frac{r_{min}}{r}\right)^6\right]$ where r_{min} is the distance between two atoms i and j when the energy is at a minimum ε and r is the actual distance between the atoms. This equation is known as the Lennard–Jones 6–12 potential. The $(\;)^6$ term in this equation represents attractive forces whilst the $(\;)^{12}$ term represents short-range repulsive forces between the atoms	*Van der Waals' non-bonded interactions* 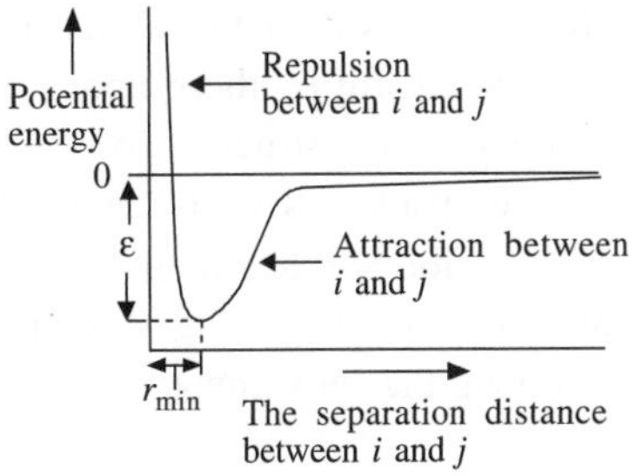
$E_{Coulombic}$	$E_{Coulombic} = \frac{q_j q_i}{D r_{ij}}$ where q_j and q_j are the point charges on atoms i and j, r_{ij} is the distance between the charges and D is the dielectric constant of the medium surrounding the charges	*Electrostatic coulombic interactions* 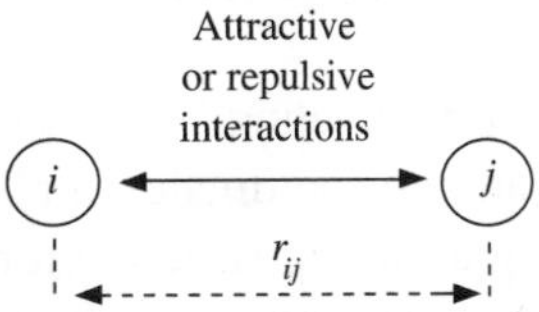

For example, using the Hook's Law model, for a molecule consisting of three atoms bonded a–b–c the expression would be:

$$\Sigma E_{Stretching} = E_{a\text{-}b} + E_{b\text{-}c} \tag{4.6}$$

that is, the expression for $\Sigma E_{Stretching}$ in the force field for the molecule would be:

$$\Sigma E_{Stretching} = \frac{1}{2}k_{(a\text{-}b)}\left(r_{(a\text{-}b)} - r_{o(a\text{-}b)}\right)^2 + \frac{1}{2}k_{(b\text{-}c)}\left(r_{(b\text{-}c)} - r_{o(b\text{-}c)}\right)^2 \tag{4.7}$$

The other energy terms in the force field equation for a structure are treated in a similar manner using expressions appropriate to the mechanical or electrical model on which the energy term is based (Table 4.1). These expressions may be the equations given in Table 4.1 but, depending on the nature of the system being modelled, other equations may be a more appropriate way of mathematically describing the mechanical or electrical model.

The values of the parameters r, r_o, k, etc. used in the expressions for the energy terms in equation (4.4) are either obtained/calculated from experimental observations or calculated using quantum mechanics using best fit methods. Experimental calculations are based on a wide variety of spectroscopic techniques, thermodynamic data measurements and crystal structure measurements for interatomic distances. Unfortunately, values are often difficult to obtain since accurate experimental data are not always available. Quantum mechanical calculations can be used when experimental information is not available but are expensive on computer time. However, this method does give better values for structures that are not in the minimum energy state. The best fit values are obtained by looking at related structures with known parameter values and using the values from the parts of these structures that most resemble the structure being modelled. Parameter values are also stored in the data bases of the molecular modelling computer programs.

Solving the force field of a molecule by means of a computer gives the coordinates (see section 4.1.1) of all the atoms in the molecule. These coordinates are displayed in an appropriate form by the graphics package (Figs 4.1, 4.3a, 4.3b and 4.3c). It is not usually necessary to construct a force field for a structure since a number of 'off the shelf' computer packages of force field equations (such as MM3, CVFF and AMBER), the programs for solving them and the graphics programs for displaying them are readily available. These packages will normally include molecular dynamics programs for visualising the movement of the atoms in a structure (see section 4.3).

4.2.1 Creating a molecular model using molecular mechanics

The first step in creating a molecular model using the molecular mechanics approach is to obtain the atomic coordinates of all the atoms in the structure. This is usually carried out by constructing an initial structure for the molecule, the method depending on the modelling program used. Three popular routes are:

1. Joining fragments obtained from the program's data base (see section 4.2.1.1).

2. Sketching a two-dimensional structure and using the program to turn it into a three-dimensional structure.

3. Converting an existing model of a similar compound into the initial structure by adding or subtracting fragments from the program's data base.

All these approaches produce an initial structure that contains the required atomic coordinates. These coordinate values are used by the computer to calculate an initial value of E_{Total} for the model. This initial energy value is minimised by the computer iteratively (consecutive repetitive calculations) changing the values of the atomic coordinates in the equation for the force field until a minimum energy value is obtained. The values of the atomic coordinates corresponding to this minimum energy value are used to visualise the model on the monitor screen in an appropriate format (Fig 4.1s, 4.3a, 4.3b and 4.3c).

Constructing a molecular model of paracetamol starting with fragments

The program used in this discription is INSIGHT II. The fragments used are found in the program's data base They are initally put together in a reasonably sensible manner to give a structure that does not allow for steric hinderence (Fig. 4.5). At this point it is necessary to check that the computer has selected atoms for the structure whose configurations correspond to the types of bonding required in the structure. In other words, if an atom is double bonded in the structure, the computer has selected a form of the atom that is double bonded. These checks are carried out by matching a code for the atoms on the screen against the code given in the manual for the program and replacing atoms where necessary.

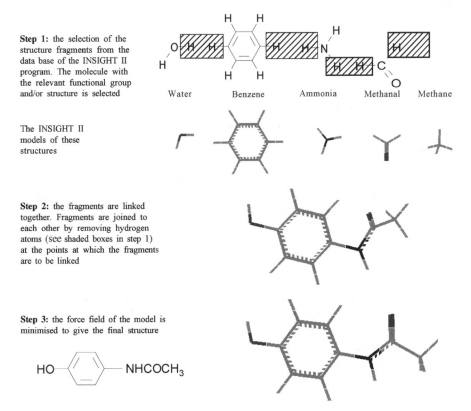

Step 1: the selection of the structure fragments from the data base of the INSIGHT II program. The molecule with the relevant functional group and/or structure is selected

Water   Benzene   Ammonia   Methanal   Methane

The INSIGHT II models of these structures

Step 2: the fragments are linked together. Fragments are joined to each other by removing hydrogen atoms (see shaded boxes in step 1) at the points at which the fragments are to be linked

Step 3: the force field of the model is minimised to give the final structure

HO—⟨benzene⟩—NHCOCH₃

Figure 4.5 An outline of the steps involved using INSIGHT II to produce a stick model of the structure of paracetamol

At this stage the structure displayed is not necessarily in its minimum potential energy conformation. However, the program can be instructed to iteratively change the atomic coordinates of the model to give a minimum value for E_{Total}. As a result of this change, the structure on the monitor screen assumes a conformation corresponding to this minimum energy state. This conformation may be presented in a number of formats depending on the requirements of the modeller (Figs 4.1, 4.3a, 4.3b and 4.3c).

The energy minimising procedure also automatically twists the molecule to allow for steric hindrance. However, the energy minimising process is not usually very sophisticated. It stops when the force field reaches the nearest *local* minimum energy value even though this value is not necessarily the lowest minimum energy value or *global* minimum energy value for the structure (Fig. 4.6). For example, butane has three minimum energy conformations, namely two skew conformers and one staggered conformation (Fig. 4.7). The skew conformers correspond to local minimum energy states while the staggered is the global minimum energy conformer. Consequently, it may be necessary to use a more sophisticated computer procedure, *molecular dynamics* (see section 4.3.1), to obtain the global minimum energy value and as a result the best model for the molecule. This final structure may be moved around the screen and expanded or reduced in size. It may also be rotated about the x or y axis to view different elevations of the molecule.

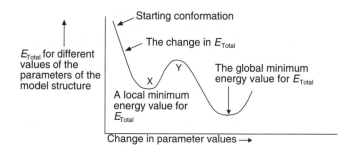

Figure 4.6   A representation of the change in the value of E_{Total}, demonstrating how the computation could stop at a local (X) rather than the true (global) minimum value. The use of molecular dynamics gives the structure kinetic energy, which allows it to overcome energy barriers, such as Y, to reach the global minimum energy structure of the molecule

The molecular mechanics method requires considerably less computing time than the quantum mechanical approach and may be used for large molecules containing more than a thousand atoms. This means that it may be used to model target sites as well as drug and analogue molecules. As well as being used to produce molecular models, it may also be used to provide information about the binding of molecules to receptors (see section 4.7) and the conformational changes (see section 4.3.1) that occur in the molecule. However, molecular mechanics is not so useful for computing properties such as electron density, which are related to the electron cloud. Furthermore, it is important to realise that the accuracy of the structure obtained will depend on the quality and appropriatness of the parameters used in the force field. Moreover, molecular mechanical calculations are

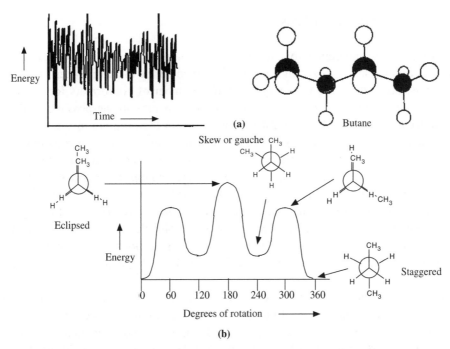

Figure 4.7 **(a)** Molecular dynamics trajectory for the rotation of the C_2–C_3 bond in butane at 600°C using the CAChe program. Moving the cursor along the energy trajectory causes the structure of butane on the right to assume the corresponding conformation. Reproduced from W. B. Smith, *Introduction to Theoretical Organic Chemistry and Molecular Modelling*, 1996, by permission of Wiley-VCH, Inc. **(b)** A plot of the change in energy with rotation about the C_2–C_3 bond in butane showing the corresponding conformations

normally based on isolated structures at 0 K and do not normally take into account the effect of the environment on the structure.

4.3 Molecular dynamics

Molecular mechanics calculations are made at 0 K, that is, on structures that are frozen in time and so do not show the natural motion of the atoms in those structures. Molecular dynamics programs allow the modeller to show the dynamic nature of molecules by simulating the natural motion of the atoms in a structure. This motion, which is time and temperature dependent, is modelled by including terms for the *kinetic energy* of the atoms in the structure in the force field by using equations based on Newton's laws of motion. The solution of these force field equations gives coordinates that show how the positions of the atoms in the structure vary with time. These variations are displayed on the monitor as a moving picture. The appearance of this picture will depend on the force field selected for the structure and the temperature and time interval used for integration of the Newtonian equations. Molecular dynamics can also be used to find minimum energy structures (Fig. 4.6) and conformational analysis.

4.3.1 Conformational analysis

Each frame of the molecular dynamics 'movie' corresponds to a conformation of the molecule, which may be frozen in time and displayed on the monitor screen in any of the set formats. Since there are in theory an infinite number of conformers for a bond it is not possible for a molecular dynamics program to generate all the possible conformers for a molecular structure. However, it is possible to set the program to rotate each bond by a set number of degrees. The conformers corresponding to these set rotations may be displayed in a suitable format. The program is also able to compute the total energy of each of these conformations and plot a graph of energy against time or degree of rotation (Fig. 4.7). However, this can take some considerable time. For example, it can take several hours of computing time to find the conformations corresponding to a significant set of degree interval changes for a simple molecule containing six bonds if energy calculations are made at a rate of 10 determinations per second. This time may be considerably increased for molecules containing larger numbers of bonds that can rotate.

The ability of molecular dynamics programs to display different conformers of a molecule is an important aspect of the use of computers in drug discovery. It is believed that when a drug binds to its target it usually adopts a conformation that does not necessarily correspond to its global minimum. Similarly, it is also thought that the target usually adjusts its conformation to accept the drug (see section 8.5). Consequently, for a particular receptor system, it appears that a flexible drug molecule will often be more effective than a more rigid drug molecule since a flexible drug will more easily adjust its shape to fit the receptor and make the appropriate interactions. As a result, the ability of programs to display conformers other than global minima is important in many uses of molecular modelling (see sections 4.6 and 4.7).

4.4 Quantum mechanics

Unlike molecular mechanics, the quantum mechanical approach to molecular modelling does not require the use of parameters similar to those used in molecular mechanics. It is based on the realisation that electrons and all material particles exhibit wave-like properties. This allows the well defined, parameter free, mathematics of wave motions to be applied to electrons, atomic and molecular structure. The basis of these calculations is the Schrodinger wave equation, which in its simplest form may be stated as:

$$H\Psi = E\Psi \tag{4.8}$$

where ψ is a mathematical function known as the state function or time-dependent wave function that defines the state (nature and properties) of a system. In molecular modelling terms $E\psi$ represents the total potential and kinetic energy of all the particles (nuclei and electrons) in the structure and H is the Hamiltonium operator acting on the wave function ψ. Operators are mathematical methods of converting one function into another function in order to find a solution or solutions to the original function. For example, differentiation is an operator that transforms an equation representing a function into its first derivative.

Schrodinger equations for atoms and molecules use the sum of the potential and kinetic energies of the electrons and nuclei in a structure as the basis of a description of the three-dimensional arrangements of electrons about the nucleus. Equations are normally obtained using the Born–Oppenheimer approximation, which considers the nucleus to be stationary with respect to the electrons. This approximation means that one need not consider the kinetic energy of the nuclei in a molecule, which considerably simplifies the calculations. Furthermore, the form of the Schrodinger equation shown in equation (4.8) is deceptive in that it is not a single equation but represents a set of differential wave equations (Ψ_n) each corresponding to an allowed energy level (E_n) in the structure. The fact that a structure will only possess energy levels with certain specific values is supported by spectroscopic observations.

The precise mathematical form of $E\psi$ for the Schrodinger equation will depend on the complexity of the structure being modelled. Its operator H will contain individual terms for *all* the possible electron–electron, electron–nuclei and nuclei–nuclei interactions between the electrons and nuclei in the structure needed to determine the energies of the components of that structure. Consider, for example, the structure of the hydrogen molecule with its four particles, namely, two electrons at positions r_1 and r_2 and two nuclei at positions R_1 and R_2. The Schrodinger equation (4.8) may be rewritten for this molecule as:

$$H\Psi = (K + U)\Psi = E\Psi \tag{4.9}$$

where K is the kinetic and U is the potential energy of the two electrons and nuclei forming the structure of the hydrogen molecule. Consequently, the Hamiltonian operator for this molecule will contain operator terms for all the interactions between these particles and so may be written as:

$$H = -\frac{1}{2}\nabla_1^2 - \frac{1}{2}\nabla_2^2 + 1/R_1R_2 - 1/R_1r_1 - 1/R_1r_2 - 1/R_2r_1 - 1/R_2r_2 + 1/r_1r_2 \tag{4.10}$$

where $\frac{1}{2}\nabla_1^2$ and $\frac{1}{2}\nabla_2^2$ are terms representing the kinetic energies of the two electrons, and the remaining terms represent all the possible interactions between the relevant electrons and nuclei. The more electrons and nuclei in the structure, the more complex the H becomes and as a direct result the greater the computing time required to obtain solutions of the equation. Consequently, in practice it is not economic to obtain solutions for structures consisting of more than about 50 atoms.

It is not possible to obtain a direct solution of the Schrodinger equation for a structure containing more than two particles. Solutions are normally obtained by simplifying H using the Hartree–Fock approximation. This approximation uses the concept of an effective field V

to represent the interactions of an electron with all the other electrons in the structure. For example, the Hartree–Fock approximation converts the Hamiltonian operator (equation 4.10) for each electron in the hydrogen molecule to the simpler form:

$$H = -\frac{1}{2}\vec{V}^2 - 1/R_1 r - 1/R_2 r + V \qquad (4.11)$$

where r is the position of the electron. The use of the Hartree–Fock approximation reduces computer time and reduces the cost without losing too much in the way of accuracy. Computer time may be further reduced by the use of semi-empirical methods. These methods use experimentally determined data to simplify many of the atomic orbitals, which in turn simplifies the Schrodinger equation for the structure. Solving the Schrodinger equation uses a mathematical method that is initially based on proposing a solution for each electron's molecular orbital. The computer tests the accuracy of this trial solution and, based on its findings, modifies the trial solution to produce a new solution. The accuracy of this new solution is tested and a further solution is proposed by the computer. This process is repeated until the testing of the solution gives answers within acceptable limits. In molecular modelling the solutions obtained by the use of these methods describe the molecular orbitals of each electron in the molecule. The solutions are normally in the form of sets of equations that may be interpreted in terms of the probability of finding an electron at specific points in the structure. Graphics programs may be used to convert these probabilities into either representations like those shown in Figures 4.1 and 4.2 or electron distribution pictures (Fig. 4.8). However, because of the computer time involved, it is not feasible to deal with structures with more than several hundred atoms, which makes the quantum mechanical approach less suitable for large molecules such as the proteins that are of interest to medicinal chemists.

Quantum mechanics is useful for calculating the values of ionisation potentials, electron affinities, heats of formation, dipole moments and other physical properties of atoms and molecules. It can also be used to calculate the relative probabilities of finding electrons (the electron density) in a structure (Fig. 4.8). This makes it possible to determine the most likely points at which a structure will react with electrophiles and nucleophiles. A knowledge of the shape and electron density of a molecule may also be used to assess the nature of the binding of a possible drug to a target site (see next section).

Pyrrole   Pyrrole, orientation
                                               in the model

Figure 4.8 The stick picture of pyrrole on which is superimposed the probability of finding electrons at different points in the molecule obtained using quantum mechanics

4.5 Docking

The three-dimensional structures produced on a computer screen may be manipulated on the screen to show different views of the structures (4.1.3). It is also possible to superimpose one structure on top of another. In other words it is possible to superimpose the three-dimensional structure of a potential drug (*ligand*) on its possible target site. This process is known as *docking* (Fig.4.9). If the structure of a compound is complementary to that of its target site the compound is more likely to be biologically active. Furthermore, the use of a colour code to indicate the nature of the atoms and functional groups present in the three-dimensional structures also enables the medicinal chemist to investigate the binding of a drug to its target site.

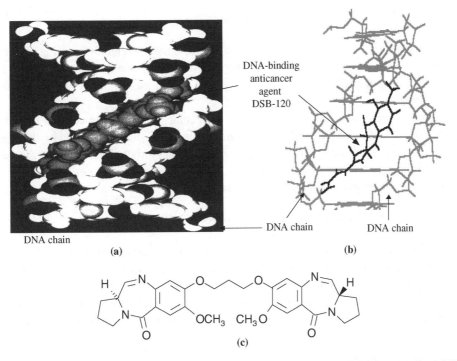

DNA-binding anticancer agent DSB-120

DNA chain

DNA chain

DNA chain

(a)

(b)

(c)

Figure 4.9 The docking of DBS-120 to a fragment of DNA. **(a)** CPK model and **(b)** Dreiding model (both courtsey of Professor D. Thurston, University of London). **(c)** DSB-120

Docking programs operate by placing the ligand in the target area and then attempting to orientate the ligand so that its binding groups line up with the complementary groups of the target with which they are likely to form bonds. For example, an electron donor group of the ligand lines up with an electron acceptor group of the target site. Not only does the program line up the appropriate complementary groups of the ligand and target site but it also attempts to separate them by a suitable bond distance. The simplest programs use specific conformers of both the ligand and target site, that is, so-called 'locked' conformations. They do not take into account the fact that the ligand and target site can exist in a number of conformers, in other words, these simple programs does not take into account the flexibility of the structures. This means that when using these molecular modelling programs in drug discovery studies the medicinal chemist should check the fit of a number of conformers of the potential drug molecule to different conformations of the target. The most used programs treat the ligand as being flexible but regard the target as being a rigid structure. More complex programs that treat both the ligand and target as being flexible are available. However, this type of program does require a considerable amount of computer time and so is not popular.

Molecular mechanics also enables the medicinal chemist to calculate the binding energy of a ligand. This is the energy *lost* when the ligand binds to its target site, that is:

$$E_{\text{binding}} = E_{\text{target}} + E_{\text{ligand}} - E_{\text{target plus bound ligand}} \qquad (4.12)$$

All the quantities on the right-hand side of the equation may be calculated using molecular mechanics force fields. However, it should be remembered that in the majority of cases the binding of a drug to its target should be weak because in most cases it has to be able to leave the target after it has activated that site.

A major problem with docking and other molecular modelling procedures is that the conformation adopted by a ligand when it binds to its target site will depend on the energy of the molecular environment at that site. This means that although a ligand may have the right pharmacophore, its global minimum energy conformer is not necessarily the conformation that binds to the target site, that is:

$$\text{Global minimum energy conformer} \rightleftharpoons \text{Bioactive conformer}$$

However, it is normally assumed that the conformers that bind to target sites will be those with a minimum potential energy. Since molecules may have large numbers of such metastable conformers a number of techniques, such as the Metropolis Monte Carlo method and Comparative Molecular Field Analysis (CoMFA), have been developed to determine the effect of conformational changes on the effectiveness of docking procedures.

4.5.1 *De novo* design

De novo design is the use of docking programs to design *new lead structures* that fit a particular target site. Two strategies are normally followed, the first is to use a template structure and the second is to construct a new molecule from component fragments of structure. These approaches both involve fitting fragments of structure into the target site. Consequently, both approaches can only be followed if the structure of the target site is reasonably well established.

The *de novo* design methods treat all the structures they use as being rigid, that is, bonds that can exhibit a degree of free rotation are locked into one conformation. Consequently, *de novo* programs do not usually take into account the flexibility of molecules and target sites. In other words, the software does not take into account that ligands and the target site can exist in more than one conformation so that those structures used in the program are not necessarily those that they assume in real life. In addition, in fully automated procedures the design is normally limited by the extent of the library of fragments, conformations and links held in the software's data base (Fig. 4.10) For these, and other reasons, *de novo* design is normally used to find new leads.

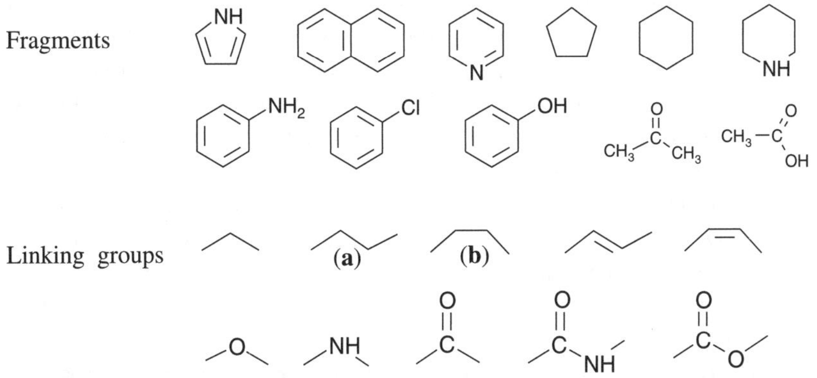

Figure 4.10 Examples of the types of fragment and linking structures found in *de novo* software programs such as the LUDI program. Since the structures are all regarded as being rigid, different conformations of the same fragment (structures a and b) are entered separately in the data base of the software

It should be appreciated that the nature of the activities and the ADME characteristics of the leads designed by *de novo* design methods may only be determined by synthesis and testing. Furthermore, the information obtained from the biological tests carried out on the lead will probably indicate that further modification to its structure is necessary. This is most likely to take the form of a SAR investigation (see Chapter 3).

The template method

The template method is based on finding a so-called template structure that approximately fits the target site and modifying that structure to improve its fit and binding characteristics. The method usually starts with a database search for suitable structures to act as a template. Software programs, such as CAVEAT, have been designed to act in conjunction with docking programs, such as DOCK, so that the structures are only listed if they fit the target site. Template structures found in this way are commonly referred to as *hits*.

The component fragment method

The groups and atoms of the binding site that can form bonds, such as hydrogen bonds and van der Waals' forces of attraction, with a suitable ligand are identified. An appropriate software program is used to fit suitable structural fragments (Fig.4.10) into the target area. These fragments are selected on the basis of both their shape and the type of interaction they could have with the target site. The program is used to place the fragments in the target site space at a suitable distance from a complementary group to which they could bind and is used to find the best fit for the fragments by trying different orientations of them in space. In this respect, *de novo* design normally treats all the structures it uses as being rigid, that is, bonds that can exhibit a degree of free rotation are locked into one conformation. The molecular modeller would need to try each of these conformations separately in order to find which one was the most suitable. In order to minimise the conformation problem, programs often contain a number of rigid conformations of the same structure in their data base. Once the fragments have been placed in position they are joined together by suitable linking structures (Fig. 4.10) to form a molecule that fits the docking site (Figs 4.11b and 4.11c). This molecule may be further modified by attaching groups that would allow the finished analogue to better fill the target area. Some software programs such as LUDI carry out the complete process automatically once they have had the required criteria loaded into the program. However, it should be appreciated that the activity, its nature and ADME characteristics of the designed molecule may only be determined by synthesis and testing.

4.6   Comparing three-dimensional structures by the use of overlays

Molecular modelling software programs are available that allow the investigator to compare the three-dimensional structures of two or more compounds. The structures are entered into the program, which automatically attempts to achieve the best fit between the three-dimensional structures according to criteria laid down by the operator. This normally consists of the operator specifying that certain *pairs of atoms*, one from each compound, should be placed as close to each other as possible when the structures are compared with each other. The program automatically carries out this instruction and the result is displayed using a suitable graphics package. It should be noted that the process treats both compounds as having rigid structures and so does not take into account any conformers that might give a better fit. Where a compound could exist in different conformational forms each individual conformer would have to be compared separately. Some programs are fully automatic. These programs define the reference pairs of atoms without any intervention by the operator, who only has to specify which structures are to be compared. Comparison of structures using three-dimensional overlays is usually more accurate than using two-dimensional comparisons.

Three-dimensional overlays are a useful tool in drug design, For example, once the three-dimensional structure of the active conformation and/or pharmacophore of an active compound is known it is possible to check whether proposed analogues are likely to have similar shaped conformations and/or pharmacophores. Analogues with conformations and/or pharmacophores that are similar to those of an active compound are more likely themselves to be active. This makes it easier for the medicinal chemist to decide which analogues to synthesise.

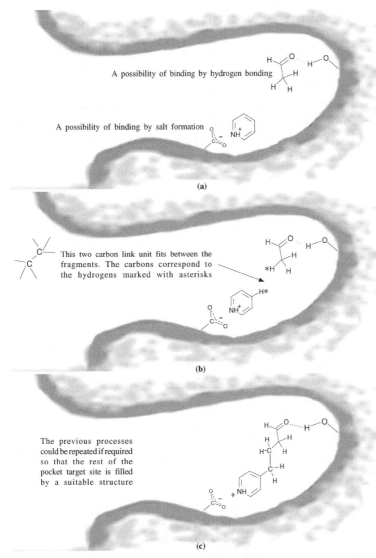

Figure 4.11 **(a)** A simulation of the binding of ethanal and pyridine fragments to complementary groups of a target site. The software program has decided which orientation of these fragments gives the best fit at the relevant points of the receptor site. It should be remembered that the software treats all structures as being rigid and does not allow for the fragments being able to exist in different conformations. **(b)** The selection of a suitable link group between ethanal and a benzene ring fragment to give an analogue that would probably bind to the target site. The link is selected according to the relative positions of the fragments it has to join. In this respect, it must be remembered that the link is treated as a rigid structure and that **(b)** is a two-dimensional arrangement of a three-dimensional situation. In the simulation a saturated two-carbon chain has been found to be suitable as the bridge between the fragments. **(c)** The completed structure. At this point the software may be used to incorporate appropriately shaped groups into the structure to fill empty spaces in the target site and, as a result, increase the possibility that the analogue binds in the designed manner. This gives the medicinal chemist a choice of analogues that might have similar binding characteristics to synthesise

4.6.1 An example of the use of overlays

DuPont carried out a SAR study in the early 1980s using *N*-benzylimidazole-5-ethanoic acids (Fig. 4.12a) that had been shown to be weak angiotensin II antagonists (see section 9.12.2). It resulted in the discovery of the highly active antihypertensive drug Losartan. This stimulated research based on a 2'-tetrazolyl-biphenyl template. At the same time Eli-Lilly carried out a SAR investigation to discover new angiotensin II antagonists based on *N*-imidazole-2-octanoic acid (Fig. 4.12b). It was originally thought that the analogues from this study had a different three-dimensional shape and size to Losartan and its analogues.

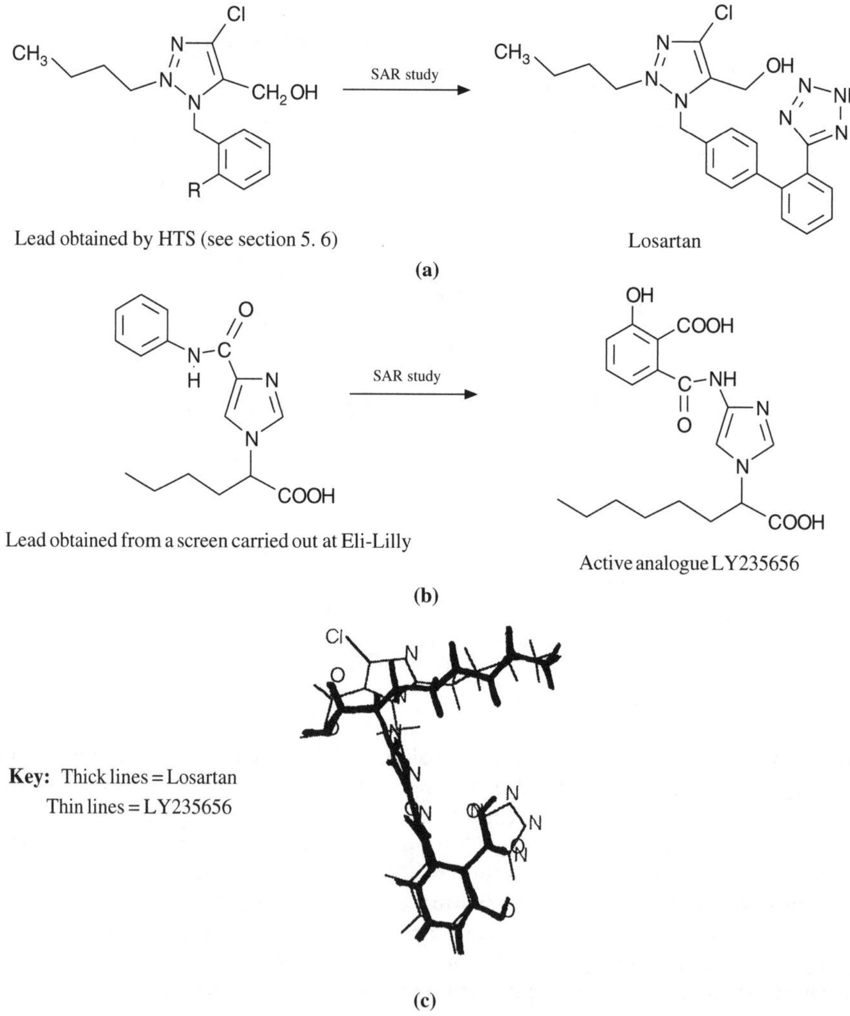

Figure 4.12 **(a)** The discovery of the antihypertensive Losartan by DuPont. **(b)** The discovery of the active analogue LY235656 by Eli-Lilly. **(c)** The fitting of the global minimum conformations of Losartan and LY235656. Reprinted with permission from *Computer-aided Molecular Design*, edited by C.H Reynolds, M. K. Hollaway and H. K. Cox, © 1995 The American Chemical Society.

However, a comparison of their global minimum energy conformations using molecular modelling techniques showed DuPont's Losartan to have a very similar shape and size to Eli-Lilly's LY235656 analogue (Fig. 4.12c).

4.7 Pharmacophores and some of their uses

The pharmacophore of a biologically active ligand is the three-dimensional geometric positions of the groups (pharmacophoric centres) of the ligand that form a unique pattern recognisable by the receptor, which is believed to be responsible for the ligand binding to and activating the receptor. These groups may be some distance apart in the structure of the ligand. Pharmacophores are frequently used as a tool for searching data bases for compounds with similar pharmacophores. The coordinates of the pharmacophore and other relevant details are fed into the software program, which searches the appropriate data bases for molecules with similar pharmacophores. Pharmacophores are also used to model binding sites in much the same way that a negative is used to produce a print of a photograph. However, in the case of pharmacophores the process uses a group of structurally different molecules with similar pharmacophores but different activities. These compounds are analysed using a software program to produces a 'three-dimensional map' of the regions, common to all of the molecules, that may be important in their binding to a receptor. This map is obtained using a similar method to that used to obtain three-dimensional QSAR maps (see section 4.9). It shows the relative positions of these regions and type of interaction (such as hydrogen bonding, hydrophobic interaction and ionic interaction) found in these areas. This is the so-called negative. It is converted into a model of the receptor site by comparing the types of interactions required for binding with the nature and position of the amino acid residues available in the target receptor. For example, phenylalanine is a candidate for hydrophobic interactions while aspartic acid would be suitable for ionic interactions. These amino acid residues are converted into a molecular model of the receptor site and the validity of the model is checked by examining the docking of molecular models of compounds known to bind to that receptor site. Once validated, the model may be used to discover new leads and drugs.

Pharmacophores of active compounds may be found using either high-resolution X-ray crystallography or NMR or obtained from data bases by analysis of a series of compounds with the same type of activity. However, the structure of the pharmacophore is usually a *perceived structure*, that is, a deduction based on what the researchers think is the most likely three-dimensional shape of the pharmacophore.

4.7.1 High-resolution X-ray crystallography or NMR

The advent of high-resolution X-ray and NMR techniques has allowed the three-dimensional structures of receptors and receptor–ligand complexes to be determined. These structures may be downloaded into molecular modelling programs from data bases such as the Brookhaven National Protein Data Bank in the USA. A study of these receptor and receptor–ligand structures using these molecular modelling programs enables the medicinal chemist to determine the possible nature of the binding of the ligand to its receptor. This allows the medicinal chemist to suggest with a reasonable degree of accuracy which groups of the ligand are involved in this binding and their relative positions, that is, the pharmacophore of the ligand. It also allows the investigator to determine the conformation adopted by the active form of the ligand at the binding site. In addition, a study of the X-ray structures of a receptor and a series of its receptor–ligand complexes involving different ligands will, if available, give a better picture of the shape and electrostatic fields of the target site. Ligand structures may be downloaded from suitable data bases such as the Cambridge Structural Database (CSD) in the UK.

4.7.2 Analysis of the structures of different ligands

Deduction from the analysis of the structures of different ligands is used when there is little or no information available about the target site. This is the most frequently encountered situation in medicinal chemistry. The approach consists of analysising the three-dimensional structures of a range of active compounds with the same type of activity but different potencies. It is assumed that because the compounds have the same activity they bind to the same receptor. The analysis consists of identifying the binding groups (their three-dimensional orientations in space) and conformations that the structures have in common. Additional information may be obtained from a comparison of active compounds with inactive compounds that are believed to bind to the same receptor. The three-dimensional structures of both the active and inactive compounds selected for the study are either individually modelled or obtained from a data base such as the CSD. The analysis of the structures allows the medicinal chemist to understand (perceive) the structural group and conformation requirements for that type of activity in a drug.

The analysis may be made manually using three-dimensional overlays (see section 4.6) or automatically by the use of specialised software, such as DISCO and HipHop. The software usually identifies potential binding groups such as benzene rings, hydrogen bond donor and acceptor groups, acidic and basic groups, etc. in the compound being studied. It records the relative three-dimensional positions of these groups in what is effectively a 'three-dimensional map'. Each compound of the series being studied is treated in the same way. Where the compound being analysed can exist in more than one conformer the software can usually be used to generate any significant conformers and separately record the new relative three-dimensional positions of the binding groups in each of these new structures. The software compares all the recorded three-dimensional maps and gives a combined visual three-dimensional display of all the groups it has identified. In essence, it is a three-dimensional map of all the groups identified from all the compounds analysed by the computer program to produce a three-dimensional map of the pharmacophore.

Pharmacophores determined by these methods are referred to as *perceived pharmaco-phores*. They are easier to determine for more rigid structures. Once they have been established, data bases are searched using a software program for suitable drug candidates with the same perceived pharmacophores. Some of these programs will also suggest possible alignments (*active conformers*) for the molecules found by the search with the receptor. These compounds are tested and if any are found to be active they are used as leads in further investigations.

4.8   Modelling protein structures

Many molecular modelling drug discovery procedures require a knowledge of the structure of the receptor, most of which are proteins. Unfortunately experimental determination of the three-dimensional structure determination of proteins by X-ray crystallography can only be carried out if it is possible to obtain crystalline samples of the protein. In cases where it is not possible to obtain a crystalline sample of a receptor protein it is sometimes possible to construct a model of that protein if the amino acid sequence of the protein is known. The method is based on the use of a suitable template (see section 7.5.1) that must meet certain criteria, namely:

- the template should have a similar structure to that of the receptor of interest;

- the template protein must have had its structure determined by either X-ray crystallography or NMR;

- sufficient sections of the template's primary structure should match a significant number of the sections of the primary structure of the receptor of interest.

Bacteriorhodopsin (Fig. 4.13) has been used as the template for models of the transmembrane proteins of *superfamily Type 2*, which are G-protein-coupled receptors (see section 8.4.2), since bacteriohodopsin is also a transmembrane protein and a significant number of the sections of its structure have similar amino acid sequences to those of G-protein-coupled receptor proteins. However, bacteriorhodopsin is not the best template for these proteins as it does not have a receptor site. Rodopsin, whose crystal structure was determined by X-ray crystallography in 2000, is a better template because it has a G-protein-coupled receptor.

Construction of the three-dimensional structure of a receptor protein is started by searching data bases of the structures of three-dimensional structures of proteins to find one that has sufficient sections of its primary structure in common with the primary structure of the protein whose structure is to be constructed. Once a template is found, the primary sequences of amino acids that the receptor protein has in common with the template are given the same three-dimensional structure as the template. For example, the fragment of a protein with a sequence that is the same as that found in an α-helix of the template would be modelled as an α-helix, the same as it is in the template. This procedure is carried out with all the sections that the protein and template have in common. These fragments are linked together by the appropriate sections of the amino acid sequence of the protein. The

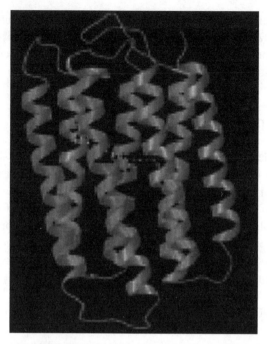

Figure 4.13 The structure of bacteriorhodopsin. Reproduced with permission from D.Voet, J. G. Voet and C Pratt, *Fundamentals of Biochemistry*, Fig. 10.4, (1999) © John Wiley and Sons Ltd.

link sections are obtained as three-dimensional structures, either by modelling the sequence using a separate molecular modelling program or using equivalent sections from other proteins with known three-dimensional structures. Any side chains are now added using low energy combinations sourced in the same way as the link sections. Once the model has been completed, its minimum energy conformation is determined using a suitable molecular dynamics program to give the final structure. The validity of the completed structure is tested experimentally. Once a protein structure has been validated it may be used to determine the active conformations and pharmacophores of ligands (see section 4.7) and also used in *de novo* design to design new leads (see section 4.5.1)

4.9   Three-dimensional QSAR

Traditional QSAR as discussed in sections 3.7–3.7.4 does not directly take into account the three-dimensional nature of molecules. However, it is now accepted that ligand–target interactions leading to biological activity depend on the three-dimensional structures of the ligand and target and the type of bonds responsible for binding the ligand to the target. These bonds are usually non-covalent, often electrostatic in nature. Three-dimensional QSAR uses steric and electrostatic parameters in an attempt to define the three-dimensional shape, and electrostatic fields of a molecule responsible for binding to produce QSAR type equations for predicting the activity of potential drug candidates. This approach attempts to give a target's eye view of the ligand.

A number of 3D-QSAR methodologies have been developed. They include the HASI (Dowyeko 1988), the CoMFA (Comparative Molecular Field Analysis, Cramer *et al.* 1988), the REMOTDISC (Ghose *et al.* 1989) and the GRID/GOLPE (Cruciani and Wilson 1994) methods. The most popular technique is the CoMFA method of Cramer, Patterson and Bunce. It has been developed and manufactured by Tripos and is based on the assumption that the bonding of the ligand to its target is electrostatic in nature. GRID is also popular and uses a similar approach as CoMFA, however this chapter will concentrate on the use of CoMFA.

3D-QSAR, like traditional QSAR, uses a group of compounds, the *training set*, with either *similar structures* or having a *common pharmacophore* and the *same type of activities* but ideally with *different potencies* in an investigation. A 3D-QSAR investigation is started by selecting one member of the training set as a reference compound and identifying its pharmacophore (see section 4.7). However, if the compound selected has a rigid structure it will not be necessary to identify the active conformer. The three-dimensional structure of the reference compound is locked into a regular three-dimensional lattice of so-called grid points (Fig. 4.14) that are usually set at a finite distance apart, typically 2 angstrom units (0.2 nm), in the x, y and z directions. Each grid point is located

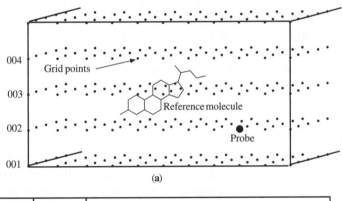

(a)

Compound	Activity	Calculated grid point energies							
		S001	S002	....	S999	E001	E002	....	E999
Compound 1	a								
Compound 2	b								
Compound 3	c								
etc.	etc.								

(b)

Figure 4.14 **(a)** The three-dimensional box of grid references containing the molecule under examination. The box is deeper than displayed. **(b)** A simulation of the data matrix for storing the results of the steric and electrostatic force field calculations at specific grid points. The prefixes S and E refer to steric and electrostatic field values, respectively, at the three-digit-numbered grid point at the top of the column

by a three- or four-digit number. A suitable molecular mechanical probe, such as an sp^3-hybridised carbon atom with a charge of $+1$, is placed in turn at each of these grid points. At each grid point the probe is used by the software to automatically calculate the energy of the steric and electrostatic forces of attraction and repulsion operating between the probe and the *entire* molecule. This is accomplished by using mathematical relationships similar to those used in molecular mechanics for bonded and non-bonded interactions. For example, equations *similar* to the Lennard–Jones equation for non-bonded interactions (see Table 4.1) may be used to calculate steric forces of attraction and repulsion, while electrostatic forces of attraction and repulsion are calculated using the equations *similar* to that given for coulombic forces (see Table 4.1). The results of these calculations are expressed as the potential energy of the probe at the relevant grid point, with forces of *attraction* having *negative* energy values and forces of *repulsion* having *positive* values. The nature of the equations used for the energy calculation means that steric field energies increase as a probe approaches the surface of the molecule, while electrostatic force fields increase as a positively charged probe nears electron deficient areas of the molecule but decrease as the same probe approaches electron-rich areas of the molecule. As the probe approaches very close to the surface or enters the volume of space of the reference molecule, the values of the potential energies of the field will rapidly increase. This would distort the statistical analysis used to obtain the 3D-QSAR equation and so, after reaching a certain magnitude chosen by the program operator, all grid points with this or a greater magnitude for their potential energies, irrespective of whether they are positive or negative, are either recorded as having that value or ignored as in the case of CoMFA in order to avoid this problem.

The grid may contain several thousand grid points and the results of the calculations for the reference molecule are stored in a data file in one line of a very large matrix under the relevant grid point reference number (Fig. 4.14). The complete set of energy values is referred to as a force field or simply a field. The next stage of the analysis is to align the other molecules of the training set in the lattice and measure their steric and electrostatic fields. A number of protocols exist for carrying out this alignment. For example, aligning the pharmacophore of the set molecule with that of the reference molecule usually gives good analysis results. Similarly, alignments based on matching the positions of common sections of the structure of the set molecule, such as a steroid ring system, have also been found to give good analysis results. Furthermore, computer programs, such as HipHop and DISCO, have been developed to automatically align training set structures with those of the reference structure. However, in all cases the method chosen will depend on the experience of the analyst. The success of this choice will also only become apparent if a viable 3D-QSAR model is obtained at the end of the analysis. The results of the calculations for the training set of compounds in the study are stored in the same data file matrix as the reference compound (Fig. 4.14).

The data from all the calculations are converted into a QSAR equation for *a molecule representative of the training set* using the statistical method of partial least-squares (PLS). QSAR equations obtained in this manner have the general form:

$$A = y + a(S001) + b(S002) + \ldots m(S999) + n(E001) + o(E002) \ldots m(E999) \quad (4.13)$$

where the coefficients *a*, *b*, etc. are numerical constants calculated by the PLS program for the field indicated by S (steric) or E (electrostatic) at the grid point specified by the three-digit number. The PLS method generates the traditional QSAR r^2 and s values (see section 3.7.1), which can be used to assess the validity of the equation. It also generates a more accurate test value called the *cross-validated correlation coefficient* q^2. Equations with a q^2 value greater than 0.3 are normally regarded as usable for activity prediction.

It is possible to predict the activity of a compound that is not a member of the training set by measuring the relevant fields for that compound and entering them into the 3D-QSAR equation, but it does require extensive use of a computer because of the large number of terms in the 3D-QSAR equation. However, it is possible to analyse a 3D-QSAR equation by considering the relative importance of the individual force fields at all the grid points Each grid point will have coefficients for the appropriate force field. For example, at grid point 001 the steric coefficient is *a* and the electrostatic coefficient is *n*. They are parameters representing a *consensus of the importance of the appropriate field of all the members of the training set* at each of the grid points. The analysis is carried out by the analyst selecting a so-called *cut-off* value. Values below the *cut-off* value are ignored. The *cut-off* value used is normally based on the experience of the analyst and several attempts may be made before a good result is obtained. It results in clusters of grid points that indicate the importance of the region as regards the type of interaction being considered, in other words, whether the molecule used as representative of the training set has an attractive or repulsive interaction with the probe in that region of space. The volumes of space in which these attractive and repulsive forces are found are either marked by colour-coded contour lines or a colour-coded solid filling (Fig. 4.15). For example, in many steric force field maps yellow is used to denote negative (attractive interaction) and green to denote positive (repulsion interaction) areas. Similarly, in many electrostatic force fields red is used to denote negative areas (attractive interaction) and blue to denote positive areas (repulsion interaction). These maps define areas where the steric and electrostatic fields

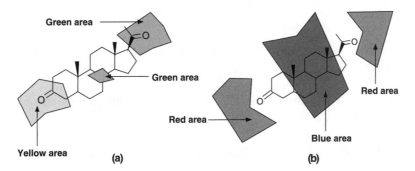

Figure 4.15 A two-dimensional simulation of the three-dimensional contour maps used to represent 3D-QSAR data. Normally both maps would be superimposed on one another but for clarity they have been separated in this black and white representation. **(a)** An imaginary simulation of a steric interaction contour map. The important steric areas are usually indicated by the green (positive) and yellow (negative) areas. **(b)** An imaginary simulation of an electronic interaction contour map. The important electrostatic areas are shown by red (negative) and blue (positive) areas.

play an important part in binding to the target site and hence determining the activity of the molecules in the training set.

3D-QSARs can be used to design new analogues with similar steric and electrostatic fields. However, it is important to realise that 3D-QSAR is normally used in conjunction with other approaches to drug discovery such as X-ray crystallography, molecular modelling, conformational analysis and traditional QSAR equations, amongst others The use of 3D-QSAR will depend on the nature of the investigation. For example, it may be used to suggest additions to the structure of an analogue that is already under investigation. A 3D-QSAR would be obtained using a training set of active compounds that bind to the target of interest. The structure of the analogue would be modelled in a similar alignment in the 3D-QSAR and its electrostatic and steric fields compared with those obtained for the training set for that grid. Examination of the results could enable the investigators to suggest structural modifications to the analogue in order to improve its binding to the target site.

In its original form CoMFA only handled steric and electrostatic probes but now programs such as GRID have been developed to use any type of probe to obtain a 3D-QSAR. Three probes in common use are a proton (H^+), a methyl carbonium ion ($^+CH_3$) and water (H_2O). H^+ is used for electrostatic, $^+CH_3$ for steric and H_2O for hydrophobic interactions. The last results in a series of columns in the data matrix for hydrophobic interactions, where the prefix H is usually used for the grid coordinates.

4.9.1 Advantages and disadvantages

The main advantages of the 3D-QSAR method are:

- The structure of the target site is not required.

- It does not require the input of either experimentally determined or calculated parameter values.

- It gives a visual picture that is easier to interpret than the mathematical formula of traditional QSAR.

- It is not restricted to a study of molecules with similar structures; it only requires similar pharmacophores.

- It allows the activities of new molecules to be predicted without synthesising them.

- The same general method of generating the 3D-QSAR model is followed for all studies.

The main disadvantages are:

- It is limited to predictions in the three-dimensional space covered by the training set.

- A model cannot give reliable predictions for substituent structures that do not lie in the structural confines of the original model. For example, if the model was obtained using a training set where only methyl substituents were present in a certain position it is not possible to get a reliable prediction for compounds in which that methyl is replaced by, say, a propyl group.

- It may be difficult to correctly align members of the training group, especially if it contains members with diverse structures.

- The members of the training set must interact with the target in similar ways.

- The accuracy of the analysis is dependent on the grid spacing employed.

4.10 Other uses of computers in drug discovery

Information concerning the structures and physical, chemical, biological and toxicological properties of compounds is mainly held in books and computer data bases. Computer data bases are a good source of this information as they are quick and readily easy to access. They are held and regularly updated by universities, government agencies and commercial firms. Government agencies will maintain some scientific data bases that are not generally available for security and economic reasons. Similarly, most major chemical and pharmaceutical companies will also maintain confidential scientific data bases of commercially sensitive information. The type of data stored will vary from data base to data base. Access to a data base may usually be made via the internet using the search engine Google, which has subdirectories with database listings at *directory. google.com/ Top/Science/Chemistry/Chemical_Databases/?tc = 1/*. These data bases may either be accessed free or require the payment of a fee. The latter is usually expensive and so these data bases are usually only available to industrial users. Searching a data base for the required information usually involves using parameters such as the name of a compound, its structure, its Chemical Abstracts Service (CAS) number or its molecular formula. The parameters used will depend to a certain extent on the nature of the search. For example, if the structures and properties of as many compounds as possible with a particular type of structural feature are required then a search based on the name of that feature may be used. This type of search is not specific and will probably turn up several thousand or more compounds meeting the structural requirement. For example, a search for compounds containing a quinoline fragment in the ChemIDplus database listed more than 4000 compounds. If information concerning the physical properties of one compound is required, the name of the compound might be used because it is more specific. However, variations in chemical nomenclature can lead to the researcher missing the compound altogether if the compound is listed in the data base under a different name to that used in the search. It follows that the parameters selected for a search need careful thought in order to prevent the researcher retrieving too much or too little data.

A wide range of computer programs have been developed to predict the numerical values of the physical properties of known and unknown compounds, the latter only existing on paper (Table 4.2). The accuracy of the values obtained from these computer programs is variable and each calculation has to be treated on its merits. Values that are inaccurate may be due to the structure of the compound concerned being outside the domain of the training set used to develop the program. However, many of these properties, such as partition coefficients, solubility and pK_a, are of considerable use to the medicinal chemist because they are directly involved in the discovery of new drugs but it should always be remembered that the values are calculated and so *may be* subject to considerable error.

Table 4.2 Examples of the programs used to predict the physical properties of known and unknown compounds. Programs are indicated by the initials by which they are normally known. Many more programs may be found on the International QSAR and Modelling Society's home page (www.qsar.org)

Property	Programs	Supplier	Notes
Melting point	ChemOffice	www.cambridgesoft.com	Calculations usually accurate
	MPBPWIN	EPA	to within $+$ 20 K
Boiling point	ADME Boxes	www.ap-algorthms	
	MPBPWIN	EPA	
Partition coefficients	ACD/log P	www.acdlabs.com	These are only a selection of
	Clog P	www.biobyte.com	over 30 programs available
	SCILOGP	www.scivision.com	for calculating P values
	VLOGP	www.accelrys.com	
Water solubility	ACD/Solubility	www.acdlabs.com	Little information available
	EPIWIN	EPA	concerning accuracy of water
	ToxAlert	Multicase Inc.	solubility programs
pK_a	ACD/pK$_a$	www.acdlabs.com	Values where molecules have
	Jaguar	Schrodinger Inc.	multiple ionisable groups are
			questionable

Computer programs such as GastroPlus and iDEA have been developed that predict aspects of the ADME of potential drug candidates in an attempt to help screen large numbers of compounds and reduce the use of live animals. They require the input of suitable physicochemical properties of the compound being investigated. Many of the predictions for first-pass metabolism (see section 11.5) are based on the metabolism of the compound involving enzyme systems such as cytochrome P-450 found in the liver (see section 12.4.1). They are based on the fit of the ligand to the active site of the enzyme and the associated energy changes incurred by this process. However, accurate prediction is complicated by the presence of inter-individual isoforms of enzyme systems. Currently the prediction of the metabolites of a potential drug candidate is of very limited value in drug discovery.

4.11   Questions

1 Describe, by means of notes and sketches, in outline only the following types of molecular model representations: (a) CPK model, (b) ball and stick model and (c) ribbon structure.

2 (a)   Describe the assumptions that form the basis of the molecular mechanics approach to molecular modelling.

 (b)   What is meant by the term force field?

 (c)   Construct an expression for $E_{Coulombic}$ for a molecule of ethane, ignoring all the possible long-range interactions.

 (d)   What are the advantages and disadvantages of using molecular mechanics to model molecular structures?

3 Outline the steps you would take in order to create a molecular mechanics model of aspirin.

4 Explain the significance of the terms global and local minimum energy conformations.

5 Describe, in outline only, what is meant by (a) molecular dynamics and (b) docking.

6 (a)   What is the principle that forms the basis of the quantum mechanical approach to molecular modelling?

 (b)   State and name the equation that is the starting point of the wave mechanical approach to molecular modelling.

 (c)   What are the major limitations of the quantum mechanical approach to molecular modelling?

7 Explain the meaning of the term pharmacophore.

8 Explain why a molecule with a flexible structure would be more likely to be more active than its analogue with a rigid structure. The compounds A and B both bind to the same target site, which of them would you expect to be the most active? State the basis of your decision.

(A)   (B)

9 Explain the meaning of the terms (a) training set, (b) probe and (c) grid coefficient in the context of 3D-QSAR.

5

Combinatorial chemistry

5.1 Introduction

The rapid increase in molecular biology technology has resulted in the development of rapid, efficient, drug testing systems. The techniques used by these systems are collectively known as *high-throughput screening* (HTS). High-throughput screening methods give accurate results even when extremely small amounts of the test substance are available (see section 5.6). However, if it is to be used in an economic fashion as well as efficiently, this technology requires the rapid production of a large number of substances for testing that cannot be met by the traditional approach to organic synthesis, which is usually geared to the production of one compound at a time (Fig. 5.1). Using this slow, labour-intensive traditional approach, a medicinal chemist is able to produce about 25 test compounds a year. Consequently, the production of the large numbers of compounds required by HTS would be expensive, both economically and in time, if this approach was used.

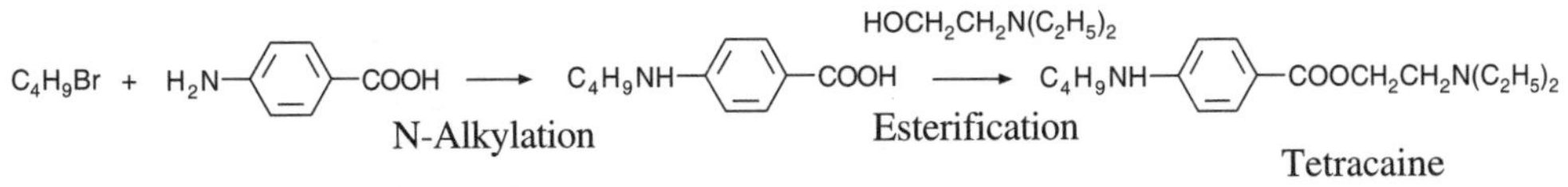

Figure 5.1 A traditional stepwise organic synthesis scheme illustrated by the synthesis of the local anaesthetic tetracaine

Combinatorial chemistry was developed to produce the large numbers of compounds required for high-throughput screening. It allows the simultaneous synthesis of a large number of the possible compounds that could be formed from a number of building blocks. The products of such a process are known as a *combinatorial library*. Libraries may be a collection of individual compounds or mixtures of compounds. Screening the components

Medicinal Chemistry, Second Edition   Gareth Thomas
© 2007 John Wiley & Sons, Ltd

of a library for activity using high-throughput screening techniques enables the development team to select suitable compounds for a more detailed investigation by either combinatorial chemistry or other methods.

The basic concept of combinatorial chemistry is best illustrated by an example. Consider the reaction of a set of three compounds (A_1–A_3) with a set of three building blocks (B_1–B_3). In combinatorial synthesis, A_1 would simultaneously undergo separate reactions with compounds B_1, B_2 and B_3, respectively (Fig. 5.2). At the same time compounds A_2

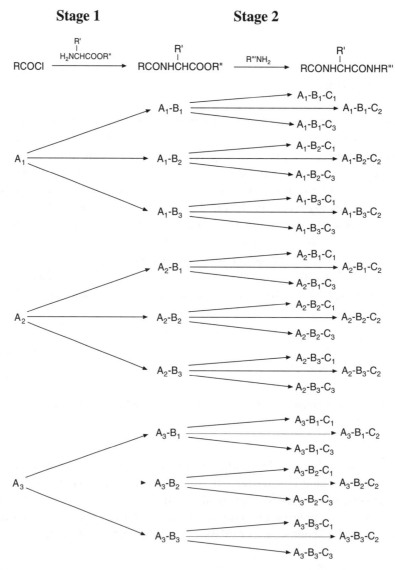

Figure 5.2 The principle of combinatorial chemistry illustrated by a scheme for synthesis of a hypothetical polyamide using three building blocks at each stage

and A_3 would also be undergoing reactions with compounds B_1, B_2 and B_3. These simultaneous reactions would produce a library of nine products. If this process is repeated by reacting these nine products with three new building blocks (C_1–C_3), a combinatorial library of 27 new products would be obtained.

The reactions used at each stage in such a synthesis normally involve the same functional groups, that is, the same type of reaction occurs in each case. Very few libraries have been constructed where different types of reaction are involved in the same stage. In theory this approach results in the formation of all the possible products that could be formed, but in practice some reactions may not occur. However, the combinatorial approach does mean that normally large libraries of many thousands of compounds can be formed rapidly in the same time that it takes to produce one product using the traditional approach to synthesis. A number of the techniques used in combinatorial chemistry to obtain this number of products are dealt with in sections 5.2 and 5.4.

5.1.1 The design of combinatorial syntheses

One of two general strategies may be followed when designing a combinatorial synthesis (Fig.5.3a). In the first case the building blocks are successively added to the preceding structure so that it grows in only one direction (*linear synthesis*, see section 15.2.3). It usually relies on the medicinal chemist finding suitable protecting groups so that the reactions are selective. This design approach is useful if the product is a polymer (*oligomer*) formed from a small number of monomeric units. Alternatively, the synthesis can proceed in different directions from an initial building block known as a *template* provided that the template has either the necessary functional groups or they can be generated during the course of the synthesis (Fig. 5.3b). Both routes may require the use of protecting groups (see section 15.2.4).

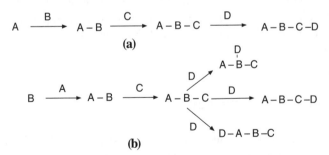

Figure 5.3 (a) Linear synthesis. The sequential attachment of building blocks. (b) Template Method. The non-sequential attachment of building blocks using B as a template

The reactions used when designing a combinatorial sequence should *ideally* satisfy the following criteria:

1. The reactions should be specific, relatively easy to carry out and give a high yield.

2. The reactions used in the sequence should allow for the formation of as wide a range of structures for the final products as possible, including all the possible stereoisomers.

3. The reactions should be suitable for use in automated equipment.

4. The building blocks should be readily available.

5. The building blocks should be as diverse as possible so that the range of final products includes structures that utilise all the types of bonding (see section 8.2) to bind to or react with the target.

6. It must be possible to accurately determine the structures of the final products.

In practice it is not always possible to select reactions that meet all these criteria. However, criterion 6 must be satisfied otherwise there is little point in carrying out the synthesis.

The degree of information available about the intended target will also influence the selection of the building blocks. If little is known a random selection of building blocks is used in order to identify a lead. However, if a there is a known lead, the building blocks are selected so that they produce analogues that are related to the structure of the lead. This allows the investigator to study the SAR/QSAR and/or determine the optimum structure for potency.

5.1.2   The general techniques used in combinatorial synthesis

Combinatorial synthesis may be carried out on a solid support (see section 5.2) or in solution (see section 5.4). In both cases, synthesis usually proceeds using one of the strategies outlined in Figure 5.3. Both solid support and solution synthetic methods may be used to produce libraries that consist of either individual compounds or mixtures of compounds. Each type of synthetic method has its own distinct advantages and disadvantages (Table 5.1).

5.2   The solid support method

The solid support method originated with the Merrifields (1963) solid support peptide synthesis. This method used polystyrene–divinylbenzene resin beads as a solid support for the product of each stage of the synthesis. Each bead had a large number of monochlorinated methyl side chains. The C-terminal of the first amino acid in the peptide chain was attached to the bead by an S_N2 displacement reaction of these chloro groups by a suitable amino acid (Fig. 5.4). The large number of chlorinated side chains on the bead meant that *one bead acts as the solid support for the formation of a large number of peptide*

Table 5.1 A comparison of the advantages and disadvantages of the solid support and in solution techniques of combinatorial chemistry

On a solid support	In solution
Reagents can be used in excess in order to drive the reaction to completion	Reagents cannot be used in excess, unless addition purification is carried out (see section 5.4.6)
Purification is easy, simply wash the support	Purification can be difficult
Automation is easy	Automation may be difficult
Fewer suitable reactions	In theory any organic reaction can be used
Scale up is relatively expensive	Scale up is relatively easy and inexpensive
Not well documented and time will be required to find a suitable support and linker for a specific synthesis	Only requires time for the development of the chemistry

molecules of the same type. Additional amino acids were added to the growing peptide chain using the reaction sequence shown in Figure 5.4. This sequence, in common with other peptide syntheses, uses protecting groups such as *t*-butyloxycarbonyl (Boc) to control the position of amino acid coupling. To form the amide peptide link the N-protected amino acids were converted to a more active acylated derivative of dicyclohexylcarbodiimide (DCC), which reacted with an unprotected amino group to link the new amino acid residue to the growing peptide (Fig. 5.4). At the end of the synthesis the peptide was detached from the bead using a mixture of hydrogen bromide and trifluroethanoic acid.

Merrifields' original resin bead has been largely superseded by other beads, such as the TentaGel resin bead. This bead is more versatile as it can be obtained with a variety of functional groups (X) at the end of the side chain (Fig. 5.5). These functional groups are separated from the resin by a polyethylene glycol (PEG) insert. As a result, the reacting groups of the side chains are further from the surface of the bead, which makes reactant access and subsequent reaction easier. Like the Merrifield bead, each TentaGel bead contains a large number of side chain functional groups. For example, the number of amine groups per bead is about 6×10^{13}. This means that in theory each bead could act as the support for the synthesis of up to 6×10^{13} molecules of the *same* compound. In peptide synthesis the amount of peptide found on one bead is usually sufficient for its structure to be determined using the Edman thiohydantoin microsequencing technique.

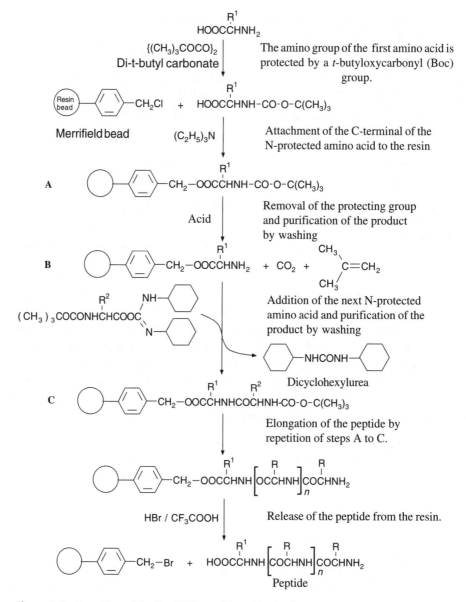

Figure 5.4 An outline of the Merrifield peptide synthesis where R is any amino acid side chain

5.2.1   General methods in solid support combinatorial chemistry

Solid support combinatorial chemistry has been carried out on a variety of supports that include polymer beads, arrays of wells, arrays of pins, glass plates, spatial arrays on microchips and cellulose sheets. However, most syntheses are performed using polymer beads. The group that anchors the compound being synthesised to the bead is known either

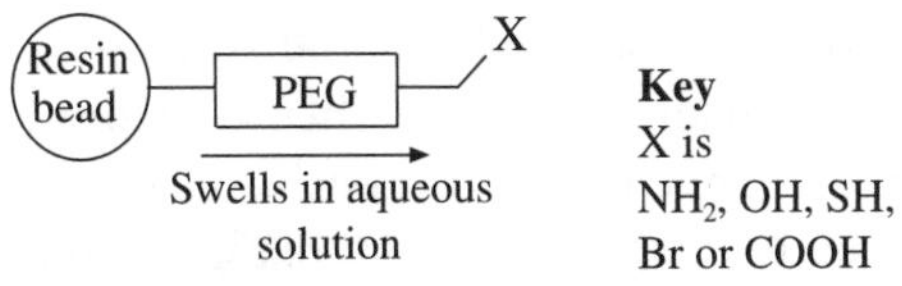

Figure 5.5 TentaGel resin beads

as a *handle* or a *linker* (Fig. 5.6). As well as modifying the properties of the bead they move the point of substrate attachment further from the bead, making reaction easier. The choice of linker will depend on the nature of the reactions used in the proposed synthetic pathway. For example, an acid-labile linker, such as HMP (hydroxymethylphenoxy), would not be suitable if the reaction pathway contained reactions that were conducted under strongly acidic conditions. Consideration must also be given to the ease of detaching the product from the linker at the end of the synthesis. The method employed must not damage the required product but must also lend itself to automation.

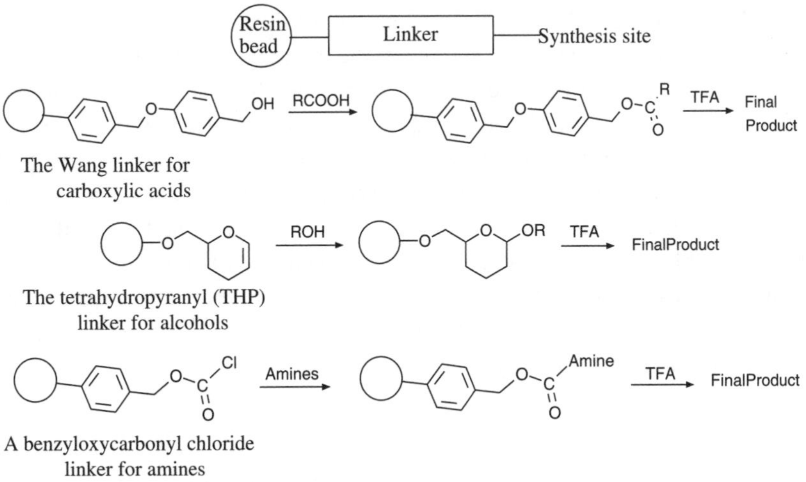

Figure 5.6 Examples of linkers and the reagents used to detach the final product. TFA is trifluoroacetic acid

Combinatorial synthesis on solid supports is usually carried out either by using parallel synthesis (see section 5.2.2) or Furka's mix and split procedure (see section 5.2.3). The precise method and approach adopted when using these methods will depend on the nature of the combinatorial library being produced and also the objectives of the investigating team. However, in all cases it is necessary to determine the structures of the components of the library by either keeping a detailed record of the steps involved in the synthesis or giving beads with a label that can be decoded to give the structure of the compound attached to that bead (see section 5.3). The method adopted to identify the components of the library will depend on the nature of the synthesis.

The reactions used in a solid support combinatorial syntheses are, of necessity, those that follow a relatively simple procedure. They are often developed from existing traditional organic chemistry reactions. This can take considerable time and effort to adapt a traditional organic reaction to use under solid phase conditions. Furthermore, many modified reactions cannot be used as they give too low a yield under solid phase conditions. This is a serious limitation for solid phase combinatorial chemistry. Reactions are normally carried out by mixing the reagents with the solid support to bring about reaction and, after reaction, washing the support with reagents and solvents to purify the product. Any heating required is usually carried out by placing the solid support in an oven. Recently, microwave heating has also been employed. However, multistep reactions and reactions that involve the use of high or low pressure, extreme temperature and inert atmospheres are avoided.

5.2.2   Parallel synthesis

This technique is normally used to prepare combinatorial libraries that consist of separate compounds. It is not suitable for the production of libraries containing thousands to millions of compounds. In parallel synthesis the compounds are prepared in separate reaction vessels but at the same time, that is, in parallel. The array of individual reaction vessels often takes the form of either a grid of wells in a plastic plate or a grid of plastic rods called pins attached to a plastic base plate (Fig. 5.7) that fits into a corresponding set of wells. In the former case the synthesis is carried out on beads placed in the wells whilst in the latter case it takes place on so-called plastic 'crowns' pushed on to the tops of the pins, the building blocks being attached to these crowns by linkers similar to those found on the resin beads. Both the well and pin arrays are used in the same general manner; the position of each synthetic pathway in the array and hence the structure of the product of that pathway is usually identified by a grid code.

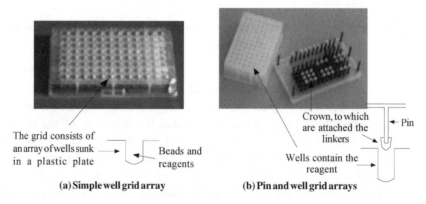

Figure 5.7 Examples of the arrays used in combinatorial chemical synthesis

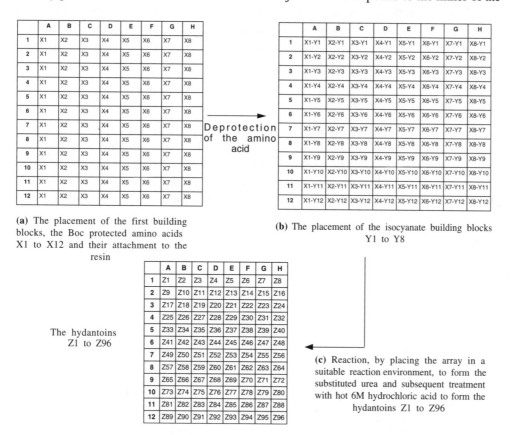

Figure 5.8 The reaction of amino acids with isocyanates to form hydantoins

The technique of parallel synthesis is best illustrated by means of an example. Consider the general theoretical steps that would be necessary for the preparation of a combinatorial library of hydantoins by the reaction of isocyanates with amino acids (Fig. 5.8) using a 96-well array. At each stage in this synthesis the product would be purified by washing with suitable reagents.

Eight N-protected amino acids ($X1, X2 \cdots X8$) are placed in the well array so that only one type of amino acid occupies a row, that is, row A will only contain amino acid $X1$, row B will only contain amino acid $X2$, and so on (Fig.5.9a). Beads are added to each well and the array placed in a reaction environment that will join the X compound to the linker of the

	A	B	C	D	E	F	G	H
1	X1	X2	X3	X4	X5	X6	X7	X8
2	X1	X2	X3	X4	X5	X6	X7	X8
3	X1	X2	X3	X4	X5	X6	X7	X8
4	X1	X2	X3	X4	X5	X6	X7	X8
5	X1	X2	X3	X4	X5	X6	X7	X8
6	X1	X2	X3	X4	X5	X6	X7	X8
7	X1	X2	X3	X4	X5	X6	X7	X8
8	X1	X2	X3	X4	X5	X6	X7	X8
9	X1	X2	X3	X4	X5	X6	X7	X8
10	X1	X2	X3	X4	X5	X6	X7	X8
11	X1	X2	X3	X4	X5	X6	X7	X8
12	X1	X2	X3	X4	X5	X6	X7	X8

(a) The placement of the first building blocks, the Boc protected amino acids X1 to X12 and their attachment to the resin

Deprotection of the amino acid →

	A	B	C	D	E	F	G	H
1	X1-Y1	X2-Y1	X3-Y1	X4-Y1	X5-Y1	X6-Y1	X7-Y1	X8-Y1
2	X1-Y2	X2-Y2	X3-Y2	X4-Y2	X5-Y2	X6-Y2	X7-Y2	X8-Y2
3	X1-Y3	X2-Y3	X3-Y3	X4-Y3	X5-Y3	X6-Y3	X7-Y3	X8-Y3
4	X1-Y4	X2-Y4	X3-Y4	X4-Y4	X5-Y4	X6-Y4	X7-Y4	X8-Y4
5	X1-Y5	X2-Y5	X3-Y5	X4-Y5	X5-Y5	X6-Y5	X7-Y5	X8-Y5
6	X1-Y6	X2-Y6	X3-Y6	X4-Y6	X5-Y6	X6-Y6	X7-Y6	X8-Y6
7	X1-Y7	X2-Y7	X3-Y7	X4-Y7	X5-Y7	X6-Y7	X7-Y7	X8-Y7
8	X1-Y8	X2-Y8	X3-Y8	X4-Y8	X5-Y8	X6-Y8	X7-Y8	X8-Y8
9	X1-Y9	X2-Y9	X3-Y9	X4-Y9	X5-Y9	X6-Y9	X7-Y9	X8-Y9
10	X1-Y10	X2-Y10	X3-Y10	X4-Y10	X5-Y10	X6-Y10	X7-Y10	X8-Y10
11	X1-Y11	X2-Y11	X3-Y11	X4-Y11	X5-Y11	X6-Y11	X7-Y11	X8-Y11
12	X1-Y12	X2-Y12	X3-Y12	X4-Y12	X5-Y12	X6-Y12	X7-Y12	X8-Y12

(b) The placement of the isocyanate building blocks Y1 to Y8

The hydantoins Z1 to Z96

	A	B	C	D	E	F	G	H
1	Z1	Z2	Z3	Z4	Z5	Z6	Z7	Z8
2	Z9	Z10	Z11	Z12	Z13	Z14	Z15	Z16
3	Z17	Z18	Z19	Z20	Z21	Z22	Z23	Z24
4	Z25	Z26	Z27	Z28	Z29	Z30	Z31	Z32
5	Z33	Z34	Z35	Z36	Z37	Z38	Z39	Z40
6	Z41	Z42	Z43	Z44	Z45	Z46	Z47	Z48
7	Z49	Z50	Z51	Z52	Z53	Z54	Z55	Z56
8	Z57	Z58	Z59	Z60	Z61	Z62	Z63	Z64
9	Z65	Z66	Z67	Z68	Z69	Z70	Z71	Z72
10	Z73	Z74	Z75	Z76	Z77	Z78	Z79	Z80
11	Z81	Z82	Z83	Z84	Z85	Z86	Z87	Z88
12	Z89	Z90	Z91	Z92	Z93	Z94	Z95	Z96

(c) Reaction, by placing the array in a suitable reaction environment, to form the substituted urea and subsequent treatment with hot 6M hydrochloric acid to form the hydantoins Z1 to Z96

Figure 5.9 The pattern of well loading for the formation of a combinatorial library of 96 hydantoins

bead. The amino acids are deprotected by hydrogenolysis and 12 isocyanates (Y1, Y2···Y8) added to the wells so that each numbered row at right angles to the lettered rows contains only one type of isocyanate. In other words, compound Y1 is only added to row one, compound Y2 is only added to row two, and so on (Fig. 5.9b). The isocyanates are allowed to react to form substituted ureas. Each well is treated with 6M hydrochloric acid and the whole array heated to simultaneously form the hydantoins and release them from the resin. Although it is possible to simultaneously synthesise a total of 96 different hydantoins (Z1–Z26, Fig.5.9c) by this technique, in practice it is likely that some of the reactions will be unsuccessful and a somewhat smaller library of compounds is normally obtained.

A well array combinatorial synthesis can consist of any number of stages. Each stage is carried out in the general manner described for the previous example. However, at each stage only the numbered or lettered rows are used, not both, unless a library of mixtures is required. Finally, the products are liberated from the resin by the appropriate linker cleavage reaction (see Fig. 5.6) and the products isolated. The structures of these products are usually determined by following the history of the synthesis using the grid references of the wells and confirmed by instrumental methods (mainly NMR, GC, HPLC and MS).

The pin array is used in a similar manner to the well array except that the array of crowns is inverted so that the crowns are suspended in the reagents placed in a corresponding array of wells (Fig. 5.7b). Reaction is brought about by placing the combined pin and well unit in a suitable reaction environment. The loading of the wells follows the pattern described in Figure 5.9.

Fodor's method for parallel synthesis

In theory almost any solid material can be used as the solid support for parallel combinatorial synthesis. Fodor *et al.* (1991) produced peptide libraries using a form of parallel synthesis that could be performed on a glass plate. The plate is treated so that its surface is coated with hydrocarbon chains containing a terminal amino group. These amino groups are protected by the UV-labile 6-nitroveratryloxycarbonyl (NVOC) group.

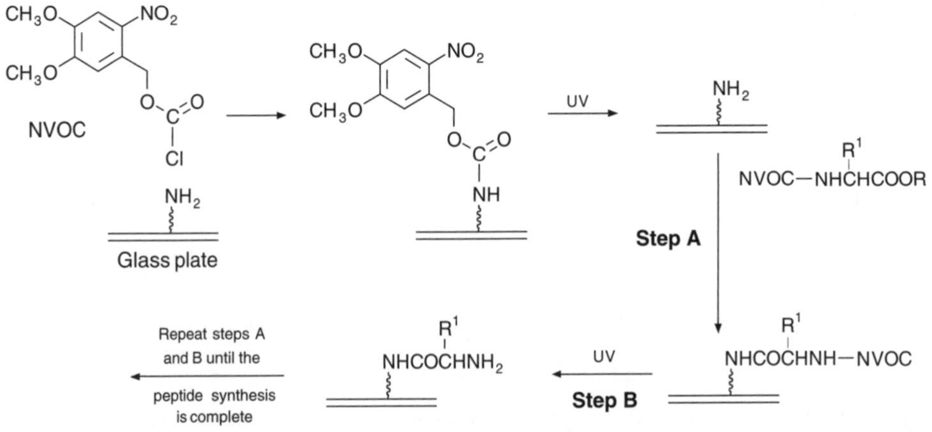

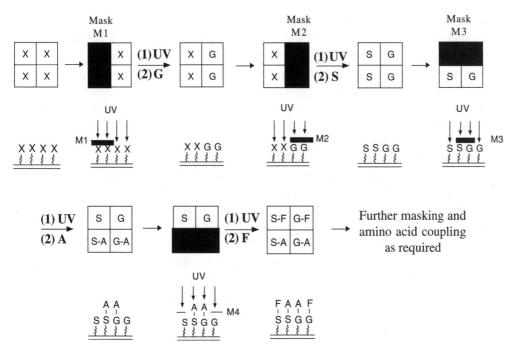

Figure 5.10 A schematic representation of the Fodor approach to parallel synthesis. X represents an NVOC-protected amino group attached to the glass plate. The other letters correspond to the normal code used for amino acids. Each of these amino acids is in its NVOC-protected form

A photolithography mask (M1) is placed over the plate so that only a specific area of the plate can be irradiated with UV light (Fig. 5.10). This results in removal of the NVOC protecting group from the amino groups in the irradiated area. The entire plate is exposed to the first activated NVOC-protected amino acid. However, it will only bond to the amino groups exposed in the irradiated area (Step A). The process is repeated using a new mask (M2) and a second activated NVOC-protected amino acid attached to the exposed amino groups (Step B). This process is repeated using different masks (M3, etc.) until the desired library is obtained, the structure of the peptide occupying a point on the plate depending on the masks used and the activated NVOC-protected amino acid used at each stage in the synthesis. The technique is so precise that it has been reported that each compound occupies an area of about 50 μm × 50 μm. A record of the way in which the masks are used will determine both the order in which the amino acids are added and, as a result, the structures of each of the peptides at specific coordinates on the plate.

5.2.3 Furka's mix and split technique

The Furka method was developed by Furka and co-workers from 1988 to 1991. It uses resin beads and may be used to make both large (thousands) and small (hundreds) combinatorial libraries. Large libraries are possible because the technique produces one type of compound on each bead, that is, all the molecules formed on one bead are the same but different from those formed on all the other beads. Each bead will yield up to 6×10^{13} product molecules, which is sufficient to carry out high-throughput screening procedures. The technique has the advantage that it reduces the number of reactions required to produce a large library. For example, if the synthetic pathway required three steps, it would require 30,000 separate reaction vessels to produce a library of 10,000 compounds if the reactions were carried out in separate reaction vessels using orthodox chemical methods. The Furka mix and split method reduces this to about 22 reactions.

The Furka method produces the library of compounds on resin beads. These beads are divided into a number of equally sized portions corresponding to the number of initial building blocks. Each of the starting compounds is attached to its own group of beads using the appropriate chemical reaction (Fig.5.11). All the portions of beads are now mixed and separated into the number of equal portions corresponding to the number of different starting compounds being used for the first stage of the synthesis. A different reactant building block is added to each portion and the reaction is carried out by putting the mixtures of resin beads and reactants in a suitable reaction vessel. After reaction, all the beads are mixed before separating

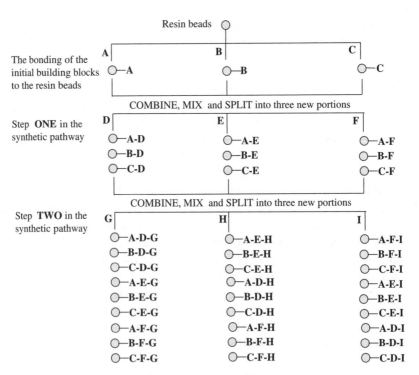

Figure 5.11 An example of the Furka approach to combinatorial libraries using a two-step synthesis involving three building blocks at each stage

them into the number of equal portions corresponding to the number of building blocks being used in the second stage of the synthesis. A different second stage building block is added to each of these new portions and the mixture is allowed to react to produce the products for this stage in the synthesis. This process of mix and split is continued until the required library is synthesised. In peptide and similar polymer library formation where the same building blocks are used at each step, the maximum possible number of compounds that can be synthesised for a given number of different building blocks (b) is given by:

$$\text{Number of compounds} = b^x \tag{5.1}$$

where x is the number of steps in the synthesis.

Unlike in parallel synthesis the history of the bead cannot be traced from a grid reference, it has to be traced using either a suitable encoding method (see section 5.3) or deconvolution (see section 5.5). Encoding methods use a code to indicate what has happened at each step in the synthesis. They range from putting an identifiable tag compound on to the bead at each step in the synthesis to using computer-readable silicon chips as the solid support. If sufficient compound is produced, its identity may also be confirmed using a combination of analytical methods such as NMR, MS, HPLC and GC.

5.3 Encoding methods

A wide variety of encoding methods have been developed to record the history of beads used in the Furka mix and split technique. This section outlines a selection of these methods.

5.3.1 Sequential chemical tagging

Sequential chemical tagging uses specific compounds (tags) as a code for the individual steps in the synthesis. These tag compounds are sequentially attached in the form of a polymer-like molecule to the same bead as the library compound at each step in the synthesis, usually by the use of a branched linker (Fig. 5.12). One branch is used for the library synthesis and the other for the encoding. At the end of the synthesis both the library compound and the tag compound are liberated from the bead. The tag compound must be

Figure 5.12 Chemical encoding of resin beads. Branched linkers, with one site for attaching the library compound and another for attaching the tag, are often used for encoding

produced in a sufficient amount to enable it to be decoded to give the history and hence the possible structure of the library compound.

Compounds used for tagging must satisfy a number of criteria:

(1) The concentration of the tag should be just sufficient for its analysis, that is, the majority of the linkers should be occupied by the combinatorial synthesis.

(2) The tagging reaction must take place under conditions that are compatible with those used for the synthesis of the library compound.

(3) It must be possible to separate the tag from the library compound.

(4) Analysis of the tag should be rapid and accurate using methods that could be automated.

Many peptide libraries have been encoded using single-stranded DNA oligonucleotides as tags. Each oligonucleotide acts as the code for one amino acid (Table 5.2). Furthermore, a polymerase chain reaction (PCR) primer is usually attached to the tag site so that at the end of the combinatorial synthesis the concentration of the completed DNA oligonucleo-tide tag may be increased using the *Taq* polymerase procedure. This amplification of the yield of the tag makes it easier to identify the sequence of bases, which leads to a more accurate decoding.

Table 5.2 The use of oligonucleotides to encode amino acids in peptide synthesis

Amino acid	Structure	Oligonucleotide code
Glycine (Gly)	$\overset{NH_2}{CH_2COOH}$	CACATG
Methionine (Met)	$CH_3SCH_2CH_2\overset{NH_2}{C}HCOOH$	ACGGTA

At each stage in the peptide synthesis a second parallel synthesis is carried out on the same bead to attach the oligonucleotide tag (Fig. 5.13). In other words, two alternating parallel syntheses are carried out at the same time. On completion of the peptide synthesis the oligonucleotide tag is isolated from the bead and its base sequence is determined and decoded to give the sequence of amino acid residues in the peptide.

Peptides and individual amino acids have also been used to code for the building blocks in a synthesis because they can be sequentially joined. Syntheses are usually carried out using a branched linker so that the synthesis of the encoding molecule can be carried out in parallel to that of the combinatorial library molecule it encodes. For example, the Zuckermann approach uses a diamine linker protected at one end by an Fmoc group and at

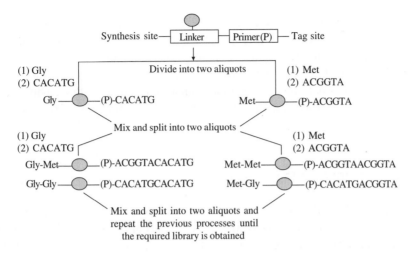

Figure 5.13 The use of oligonucleotides to encode a peptide combinatorial synthesis for a library based on two building blocks

the other end by a Moz group (Fig. 5.14). The Fmoc group was cleaved under basic conditions and the suitably protected building block was joined to the linker. The Moz group was removed under acid conditions and a suitably protected peptide was attached.

The process was repeated for the coupling of each building block to each portion of beads as the mix and split procedure progressed. At the end of the synthesis each bead is separated from its fellows and the product and its encoding peptide were liberated from that

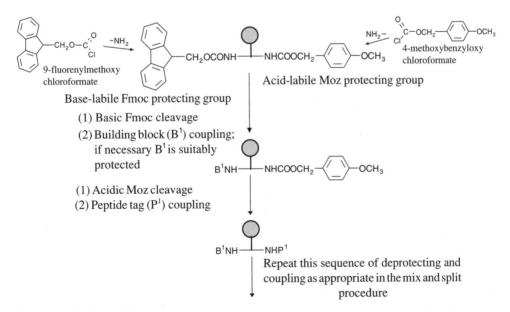

Figure 5.14 An outline of the Zuckermann approach using peptides for encoding

bead. It should be remembered that each bead will yield up to 6×10^{13} product molecules, which is suffcient to carry out high-throughput screening procedures. In addition, each bead will produce sufficient of the tagging compound to deduce the structure of the product molecule. The detached product and tag are separated and the sequence of amino acids in the encoding peptide is determined using the Edman sequencing method. This sequence is used to determine the history of the formation and hence the structure of the product found on that bead.

5.3.2   Still's binary code tag system

A unique approach by Still was to give each building block its own chemical equivalent of a binary code for each stage of the synthesis using inert aryl halides (Fig. 5.15a). One or more of these tags are directly attached to the resin using a photolabile linker at the appropriate points in the synthesis. They indicate the nature of the building block and the stage at which it was incorporated into the solid support (Table 5.3). Aryl halide tags are used because they can be detected in very small amounts by GC. They are selected on the basis that their retention times were roughly equally spaced (Fig. 5.15b). At the end of the synthesis all the tags are detached from the linker and are detected by GC. The gas chromatogram is read like a bar code to account for the history of the bead. Suppose, for example, that the formation of a tripeptide using six aryl halide tags allocated as shown in the tagging scheme outlined in Table 5.3 gave the tag chromatogram shown in Figure 5.15c. The presence of T1 shows that in the first stage of the synthesis the first amino acid residue is glycine. This residue will be attached via the C-terminal of the

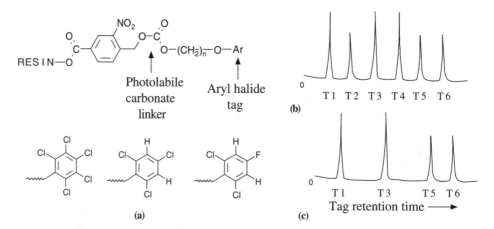

Figure 5.15   (a) Molecular tags used by Still; ⌇⌇⌇⌇ indicates the point at which the tag is attached to the linker. (b) A hypothetical representation of the GC plots obtained for some aryl halide tags. (c) The tag chromatogram for a hypothetical tagging scheme

Table 5.3 A hypothetical tagging scheme for the preparation of tripeptides using binary combinations of six tags

Stage	Tag		
	Glycine (Gly)	Alanine (Ala)	Serine (Ser)
1	T1	T2	T1 + T2
2	T3	T4	T3 + T4
3	T5	T6	T5 + T6

peptide if a linker with an amino group was used or via its N-terminal if a linker with an acid group was used. The presence of T3 shows that the second residue is also glycine, whilst the presence of T5 and T6 indicates that the third amino acid in the peptide is serine.

5.3.3 Computerised tagging

Nicolaou has devised a method of using silicon chips to record the history of a synthesis. Silicon chips can be coded to receive and store radio signals in the form of a binary code. This code can be used as a code for the building blocks of a synthesis. The silicon chip and beads are placed in a container known as a *can* that is porous to the reagents used in the synthesis. Each can is closed and treated as though it were one bead in a mix and split synthesis. The cans are divided into the required number of aliquots corresponding to the number of building blocks used in the initial step of the synthesis. Each batch of cans is reacted with its own building block and the chip is irradiated with the appropriate radio signal for that building block. The mix and split procedure is followed and at each step the chips in the batch are irradiated with the appropriate radio signal. At the end of the synthesis the prepared library compound is cleaved from the chip, which is interrogated to determine the history of the compound synthesised on the chip. The method has the advantage of producing larger amounts of the required compounds than the normal mix and split approach because the same compound is produced on all the beads in a can.

5.4 Combinatorial synthesis in solution

Solid phase combinatorial synthesis has a number of inbuilt disadvantages:

1. All the libraries have a common functional group at the position corresponding to the one used to link the initial building block to the linker or bead.

2. Syntheses are usually carried out using the linear approach.

3. Requires especially modified reactions with high yields (>98 per cent) if multistep syntheses are attempted.

4. Requires additional synthesis steps to attach the initial building block to and remove the product from the support.

5. The final product is contaminated with fragments (truncated intermediates) of the product formed by incomplete reaction at different stages and often needs additional purification steps.

Many of these disadvantages are eliminated or reduced when combinatorial syntheses are carried out in solution. For example, solution phase combinatorial chemistry does not have to have a common functional group at the position corresponding to the one used to link the synthesis substrate to the linker or bead. Both the linear, template and convergent synthesis routes (see sections 5.1.1 and 15.2.3) can be followed. Unmodified traditional organic reactions may be used but multistep syntheses will still require very efficient reactions. However, it does not require additional synthetic steps to attach the initial building block to and remove the product from a solid support. The product is not likely to be contaminated with truncated intermediates but unwanted impurities will still need to be removed at each stage in a synthesis. Disadvantages of solution phase combinatorial chemistry are given in Table 5.1.

Solution phase combinatorial chemistry can be used to produce libraries that may contain single compounds or mixtures (see section 5.4.2). Their production is usually by parallel synthetic methods (see sections 5.4.1 and 5.4.2). The main problem in their preparation is the difficulty of removing unwanted impurities at each step in the synthesis. Consequently, many of the strategies used for the preparation of libraries using solution chemistry are directed to purification of the products of each step of the synthesis (see sections 5.4.3–5.4.8). This and other practical problems have often limited the solution combinatorial syntheses to short synthetic routes.

5.4.1   Parallel synthesis in solution

Combinatorial synthesis in solution using the technique of parallel synthesis (see section 5.2.2) is used to prepare libraries of single compounds. Reactions are usually, carried out in microwave vials and 96-well plates. They are normally relatively simple, having one or two steps. For example, in 1996 Bailey et al. produced a library of 20 2-aminothioazoles by means of a Hantzsch synthesis. They used a 5 × 4 grid of glass vials (Fig 5.16). Five different substituted thioureas, one per row, were treated with four different α-bromoketones. Each of the α-bromoketones, only one per row, was added to a separate row. After reaction the products were isolated before being characterised by high-resolution MS and NMR. One of the compounds synthesised by this method was the anti-inflammatory fanetizole.

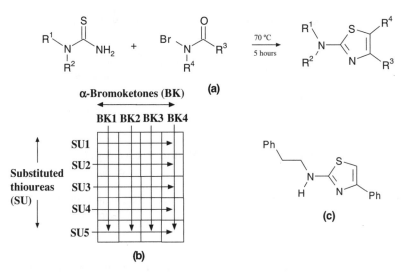

Figure 5.16 **(a)** The Hantzsch synthesis of 2-aminothioazoles. **(b)** The reaction grid. Each square corresponds to a glass vial. **(c)** Fanetizole

5.4.2   The formation of libraries of mixtures

Libraries of mixtures are formed by separately reacting each of the members of a set of similar compounds with the same mixture of all the members of the second set of compounds. Consider, for example, a combinatorial library of amides formed by reacting a set of five acid chlorides (A^1–A^5) with ten amines (B^1–B^{10}). Each of the five acid chlorides is reacted separately with an equimolar mixture of all ten amines and each of the amines is reacted with an equimolar mixture of all the acid chlorides (Fig. 5.17). This produces two sub-libraries, one consisting of a set of five mixtures based on individual acid halides and the other consisting of ten mixtures based on individual amines. This means that each compound in the main library is prepared twice, once from the acid chloride set in the acid chloride sub-library and once from the amine set in the amine sub-library. Consequently, determining the most biologically active of the mixtures from the acid halide set will define the acyl part of the most active amide and, similarly, identifying the most biologically active of the amine-based set of mixtures will identify the amine residue of that amide. Libraries prepared and used in this manner are often referred to as *indexed libraries*. They are restricted to two points of diversity. The approach is most successful when used to prepare small libraries.

This method of identifying the structure of the most active component of combinatorial libraries of mixtures assumes that inactive compounds in the mixture will not interfere with the active compounds in the bioassay used for determining activity. It depends on both of the mixtures containing the active compound giving a positive result for the assay procedure. However, it is not possible to identify the active structure if one of the sets of mixtures gives a negative result. Furthermore, complications arise if more than one mixture is found to be active. In this case all the active structures have to be synthesised and tested separately. However, it is generally found that the activity of the library mixture is usually

$$RCOCl \ + \ R'NH_2 \ \longrightarrow \ RCONHR' \ + \ Cl$$

The acid chloride-based set:

A^1 + $(B^1, B^2, B^3, B^4, B^5, B^6, B^7, B^8, B^9, B^{10})$ $\longrightarrow$ **Mixture 1** containing all the possible A^1—B compounds.

A^2 + $(B^1, B^2, B^3, B^4, B^5, B^6, B^7, B^8, B^9, B^{10})$ $\longrightarrow$ **Mixture 2** containing all the possible A^2—B compounds.

A^5 + $(B^1, B^2, B^3, B^4, B^5, B^6, B^7, B^8, B^9, B^{10})$ $\longrightarrow$ **Mixture 5** containing all the possible A^5—B compounds.

The amine-based set:

B^1 + $(A^1, A^2, A^3, A^4, A^5,)$ $\longrightarrow$ **Mixture 6** containing all the possible B^1—A compounds.

B^{10} + $(A^1, A^2, A^3, A^4, A^5,)$ $\longrightarrow$ **Mixture 15** containing all the possible B^1—A compounds.

Figure 5.17 A Schematic representation of the index approach to identifying active compounds in libraries formed in solution

higher than that exhibited by the individual compounds responsible for activity after they have been isolated from the mixture.

5.4.3 Libraries formed using monomethyl polyethylene glycol (OMe-PEG)

Polyethylene glycols are polymers with hydroxy groups at each end of the chain (Fig. 5.18a). These polymers are soluble in both water and organic solvents, the degree of solubility depending on the length of the polymer chain. Combinatorial syntheses in solution are carried out using monomethyl polyethylene glycol (OMe-PEG-OH), which tends to precipitate in diethyl ether. The synthesis is started by reacting the acid group of an acidic building block to the hydroxy group of OMe-PEG. The product is precipitated by adding diethyl ether and the excess reagent and other impurities are removed by washing (Fig. 5.18b). The solid product is redissolved in fresh solvent and the second stage of the synthesis is carried out using a similar reaction and washing procedure. At the end of the synthesis the product may be cleaved from the OMe-PEG, purified and assayed. In some cases the product is assayed when it is still attached to the OMe-PEG. This approach may be carried out using either the parallel synthesis or split and mix methods, the latter being carried out while the products of a stage are in solution.

5.4.4 Libraries produced using dendrimers as soluble supports

Dendrimers are branched oligomers (small polymers) with regular structures that have been used as soluble supports in a similar fashion to OMe-PEG. The synthesis initially involves

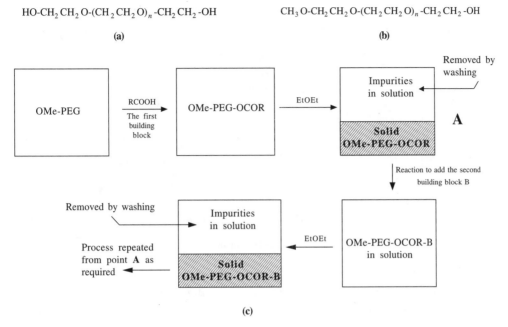

$HO\text{-}CH_2CH_2O\text{-}(CH_2CH_2O)_n\text{-}CH_2CH_2\text{-}OH$

(a)

$CH_3O\text{-}CH_2CH_2O\text{-}(CH_2CH_2O)_n\text{-}CH_2CH_2\text{-}OH$

(b)

Figure 5.18 (**a**) Polyethylene glycol (PEG). (**b**) Monomethyl polyethylene glycol (MeO-PEG-OH). (**c**) An outline of the use of OMe-PEG to produce combinatorial libraries

a functional group situated at the ends of each of the branches (X, Fig. 5.19). Standard chemical reactions and purification procedures based on size, such as gel filtration, are used in the synthesis. The final product is released from the dendrimer and purified, and its structure is determined using the history of the process and/or standard analytical techniques. An important advantage of this technique is that it usually gives high yields.

5.4.5 Libraries formed using fluorocarbon reagents

Polyfluorinated organic compounds are frequently insoluble in water and organic solvents but are soluble in liquid perfluoroalkanes. The polyfluoro compounds used in the production of combinatorial libraries usually have structures with long chains of CF_2 units attached via a short hydrocarbon chain to a silicon or tin atom (Fig. 5.20a). The library is produced by reacting the initial building block with the fluorous reagent. The product of this reaction may be separated from the reaction mixture by extraction into a perfluoroalkane solvent that is not miscible with water or the organic solvent used in the reaction. This process of purification by extraction is repeated at each stage of the synthesis. However, at the end of the synthesis the product is detached from the flurocarbon reagent prior to purification by extraction into a suitable organic solvent. In 1997 Studer *et al.* produced a small isoxazole library using this technique (Fig.5.20b).

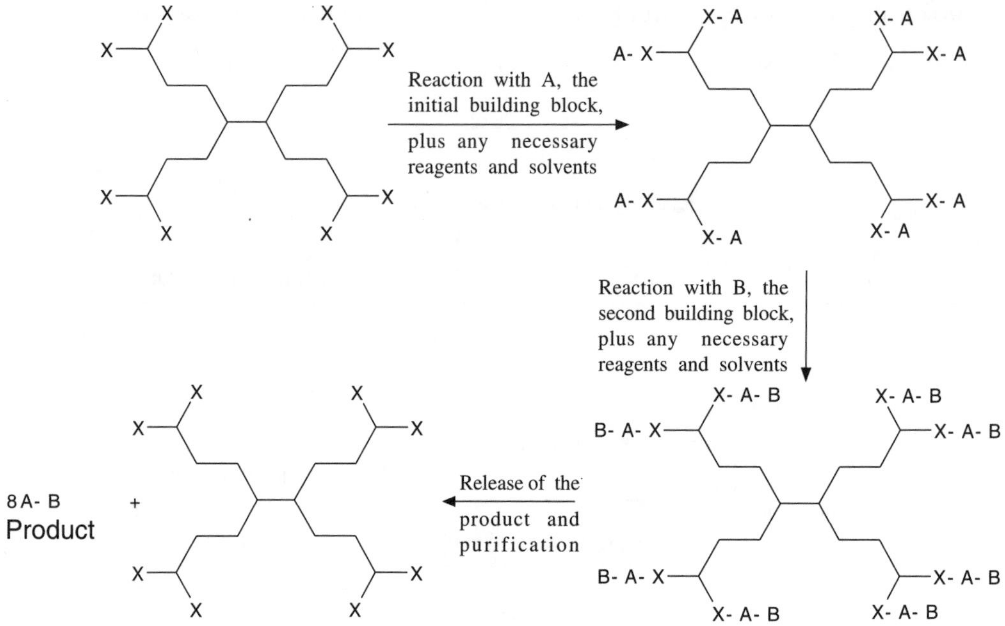

Figure 5.19 A schematic representation of the formation of a product using a dendrimer

5.4.6 Libraries produced using resin-bound scavenging agents

This approach depends on the removal of excess reagents and byproducts by the use of so-called solid phase *scavenging* or *sequestering agents*. These solid compounds are also referred to as complementary molecular reactivity sequestrants and polymer-supported quenching agents. They consist of resin beads to which is permanently attached a suitable residue with an organic functional group that will readily react with the relevant reactant (see Table 5.4) or byproduct. This reaction results in the formation of a solid complex

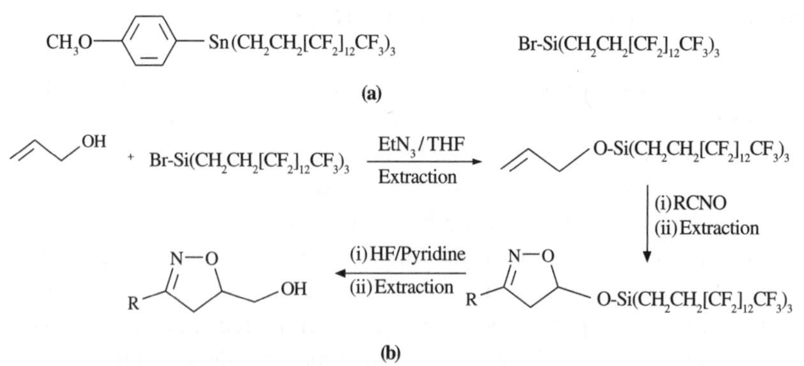

Figure 5.20 (a) Examples of fluorine compounds used in solution phase combinatorial chemistry. (b) The reactions used by Studer *et al.* to produce a small isoxazole library

Table 5.4 Examples of resin-bound reagent scavengers

Scavenger	Used to Remove	Example Reaction
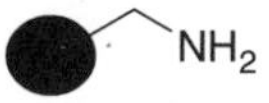 ●⁀NH$_2$	RCOCl, RSO$_2$Cl, RNCO, RNCS, RCHO and RCH$=$NR	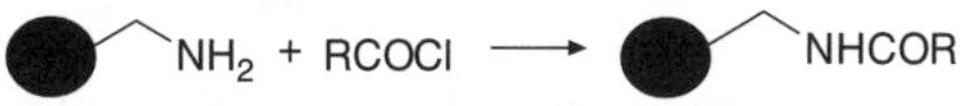 ●⁀NH$_2$ + RCOCl $\longrightarrow$ ●⁀NHCOR
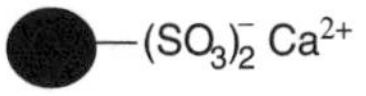 ●—(SO$_3$)$_2^-$ Ca^{2+}	Tetrabutyl ammonium fluoride (TBAF)	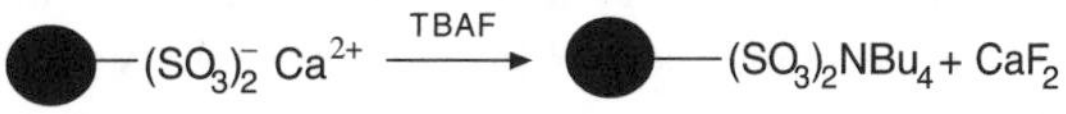 ●—(SO$_3$)$_2^-$ Ca^{2+} $\xrightarrow{\text{TBAF}}$ ●—(SO$_3$)$_2$NBu$_4$ + CaF$_2$
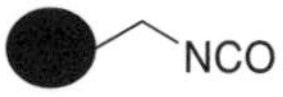 ●⁀NCO	Amines and hydrazines	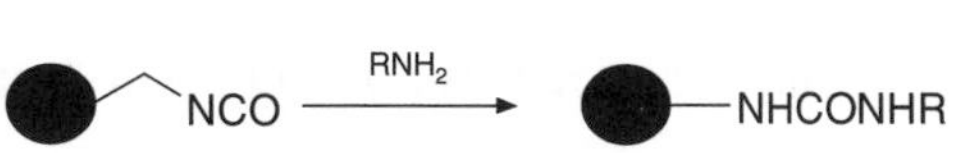 ●⁀NCO $\xrightarrow{\text{RNH}_2}$ ●—NHCONHR
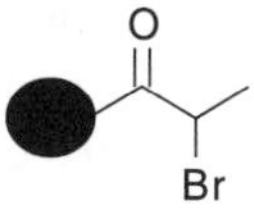 ●⁀(C=O)—CH(CH$_3$)—Br	Thioureas	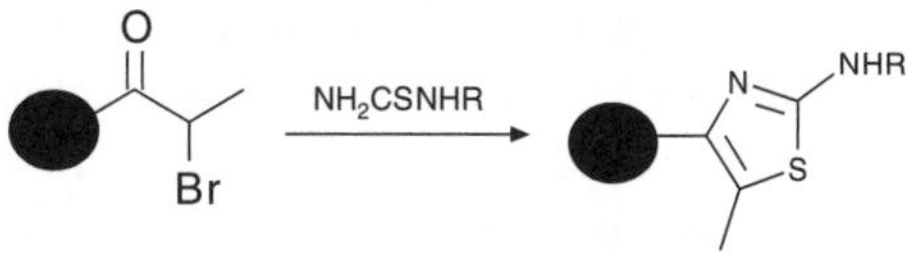 ●⁀(C=O)—CH(CH$_3$)—Br $\xrightarrow{\text{NH}_2\text{CSNHR}}$ ● thiazole NHR

containing the excess reagent/byproduct, etc., which can be removed from the reaction mixture by filtration through a cartridge containing an appropriate resin (Fig. 5.21).

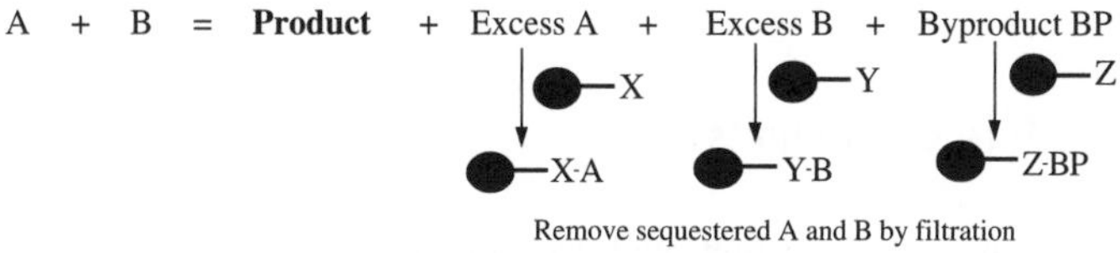

A + B = **Product** + Excess A + Excess B + Byproduct BP

●—X   ●—Y   ●—Z

●—X·A   ●—Y·B   ●—Z·BP

Remove sequestered A and B by filtration

Figure 5.21 A schematic representation of the use of resin-bound scavengers to remove excess reagents from a reaction mixture. *Note*: only relevant scavangers are used in a specific reaction

Reactant scavengers are used after a reaction to remove excess reagents. The byproducts of some of these scavenger reactions, such as the calcium sulphonate resin (Table 5.4) used to remove excess of tetrabutyl ammonium fluoride (TBAF) from a number of desilylation reactions, are also solid and may be removed by filtration. Resin-bound scavengers may also be used as mixtures where several different reactants need to be removed. This is because the functional groups of the scavengers are effectively isolated on their resins and so cross-resin interactions are unlikely to occur.

Resin-bound scavengers for the removal of byproducts of both the reaction and the reagents operate in the same way as those used for the removal of excess reagents, except they are normally used *during* the reaction. For example, carboxylic acids have been sequestered by the use of the anionic Amberlite-68 resin and 4-nitrophenol has been removed by ion exchange with a quaternary ammonium hydroxide resin.

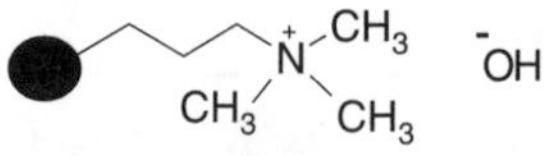

The use of these sequestering agents during the reaction usually helps to drive the reaction to completion.

A number of modifications to the scavanging agent concept are available where straightforward removal of excess reagents and byproducts is not possible. They are discussed in more detail in specialised texts such as: *Solid-Phase Organic synthesis*, edited by K. Burgess, published by Wiley-Interscience (2000).

5.4.7   Libraries produced using resin-bound reagents

Resin-bound reagents are used to remove reagent byproducts. They mediate the reaction sequestering the byproducts rather than acting as a source of part of the product. These sequestered byproducts and excess of the resin-bound reagent may be removed at the end of the synthesis by filtration. For example, a polymer-bound base has been used to produce aryl ethers.

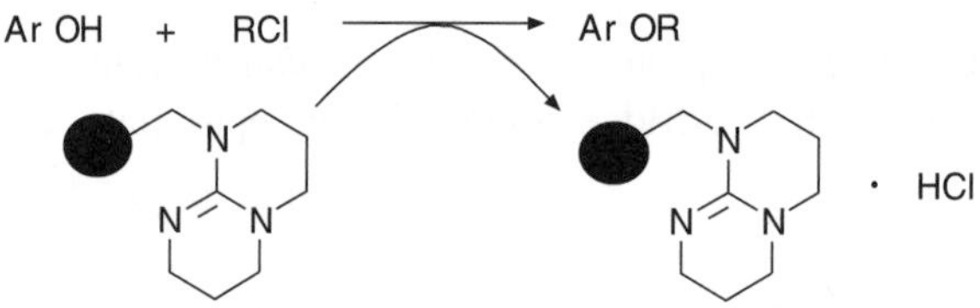

5.4.8   Resin capture of products

In this technique the resin has a functional group that can sequester the product. However, it must also be possible to break the bond linking the product to the resin to form the original functional group of the product. At the end of the reaction the product is captured on the resin and the excess reagents, and reagent byproducts are washed away with suitable solvents. The product is released from the resin, dissolved in a suitable solvent and the resin removed by filtration. For example, Blackburn *et al.* synthesised a library of 3-aminoimidazo[1,2-α]pyridines and pyrazines using parallel synthesis and this technique. They used a cation exchange resin to capture the products.

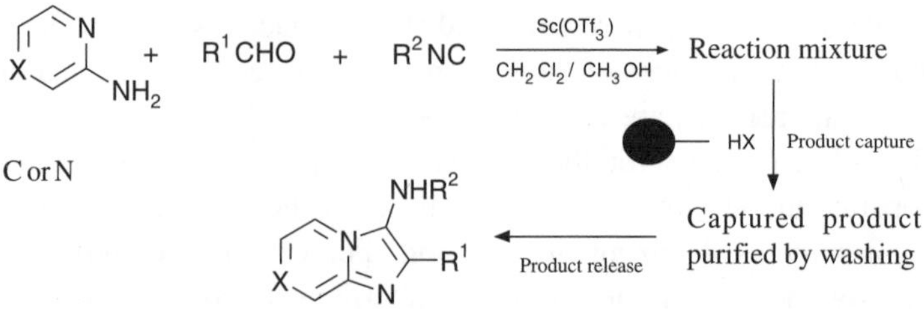

5.5   Deconvolution

The success of a library depends not only on it containing the right compounds but also the efficiency of the screening procedure for assessing the components of that library. A key problem with very large combinatorial *libraries of mixtures* is the large amount of work required to screen these libraries.

Deconvolution is a method, based on the process of elimination, of reducing the number of screening tests required to locate the most active member of a library consisting of a mixture of all the components. It is based on producing and biologically assaying similar secondary libraries that contain one less building block than the original library. It is emphasised that the biological assay is carried out on a mixture of all the members of the secondary library. If the secondary library is still as active as the original library, the missing building block is not part of the active structure. Repetition of this process will eventually result in a library that is inactive, which indicates that the missing building block in this library is part of the active structure. This procedure is carried out for each of the building blocks at each step in the synthesis. Suppose, for example, one has a tripeptide library consisting of a mixture of 1000 compounds. This library was produced from ten different amino acids (A^1–A^{10}) using two synthetic steps, each of which involved ten building blocks (Fig. 5.22). The formation of a secondary library by omitting amino acid A^1 from the initial set of amino acids but reacting the remaining nine amino acids with all ten amino acids in the first and second steps would produce 900 compounds. These compounds will not contain amino acid residue A^1 in the first position of the tripeptide. If the resulting library is biologically inactive the active compound must contain this residue at the first position in the tripeptide. However, if the mixture is active the process must be repeated using A^1 but omitting a different amino acid residue from the synthesis. In the worst scenario it would mean that the 900-member library would have to be prepared ten times in order to determine the first residue of the most active tripeptide. Repeating this process of omission, combinatorial synthesis and biological testing but using groups of

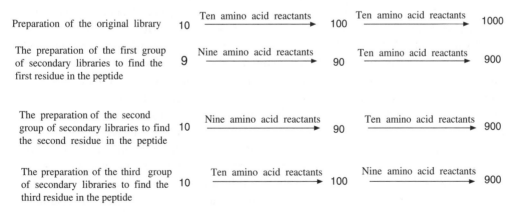

Figure 5.22   A schematic representation of deconvolution. The figures indicate the number of components in the mixture

nine reactants for the first step will give the amino acid that occupies the second place in the peptide chain. Further repetition but using groups of nine amino acid reactants in the second step will identify the third amino acid in the chain.

In order to be effective, deconvolution procedures require that both the synthesis and assay of the library be rapid. The procedure is complicated when there is more than one active component in the library. In this case it is necessary to prepare and test all the possible compounds indicated by deconvolution in order to identify the most active compound in the library.

5.6   High-throughput screening (HTS)

High-throughput screening (HTS) is the name given to rapid semi-automated simultaneous *primary screening* of large numbers of compounds, mixtures or extracts for active compounds. The process is based on the use of biomicroassays that are rapid to carry out and require very small quantities of the reagents and test compound. These assays are carried out on 96- and bigger-well plates using specialised handling equipment (see Fig. 5.24). They are based on the test compound interacting with a target, such as an enzyme, a cell membrane receptor, hormone, nuclear receptors and DNA, that is related to the disease state under investigation. Consequently, it may be necessary to identify, purify and isolate this target before a library can be screened. Furthermore, assays are often specially designed for an investigation and so have to be validated. These preliminary investigations can be expensive both in money and time.

The involvement of biological targets in assays means that assays are normally carried out in aqueous media. Consequently, an assay will only be effective if a significant amount of the compound under test dissolves in water. Consequently, as most assays are carried out in aqueous solution dimethylsulphoxide (DMSO) is often added to assay mixtures in order to improve the water solubility of the test compound.

The cost of the specialised reagents used for screening large libraries of single compounds can be very expensive. Consequently, many companies and researchers reduce costs by screening large libraries as mixtures of compounds. However, this can lead to misleading results. For example, *false positive* results may occur when the mixture under test contains a large number of individual compounds with a weak activity. As a result, the mixture gives a good overall response to the test. This result would be incorrectly interpreted by the analyst as the mixture having a strongly active compound. A *false negative* may occur if the inactive members of a mixture bind in preference to the assay target. This prevents active compounds binding to the target and giving a good assay response. The concentration of the test compound used in the assay may also give rise to false positives and negatives. Using too high a concentration of the test compound can also give a false positive because of concentration-driven non-selective binding to the target. Conversely, using too low a concentration can give a false negative when an insufficient amount of an active test compound is present to bind to the target. Other sources of inaccuracies also occur in specific types of microassay.

The microassays used in HTS may be classified for convenience as either biochemical or cell-based assays. Biochemical assays are those that are based on the interaction of the test compound with defined chemical entities isolated from cells such as an enzyme, hormone or receptor, whereas whole cell assays are based on the use of intact cells. However, it is emphasised that HTS is used as a primary screen for active compounds. Any active compounds (*hits*) that appear worthy of further investigation *must* be subjected to a wider range of activity tests before they could be considered for clinical development.

5.6.1 Biochemical assays

Biochemical assays are also referred to as mechanism-based assays. They are usually based on the binding of a ligand to a receptor or the inhibition of an enzyme-catalysed reaction using a target that has been identified as being relevant to a specified disease state. This target has usually been isolated from a cell and is no longer part of a cell. The binding of the test compound to the target is measured by the use of radioactive isotopes and/or traditional analytical methods such as spectroscopy (see section 6.2.1) using a variety of protocols, such as by measuring fluorescence in scintillation proximity assays (SPA).

Scintillation proximity assays use resin beads (SPA beads) whose surface has been engineered so it is capable of binding to a wide variety of substances. The bead also contains a scintillant that only fluoresces when a low energy radioactive source comes within about 20 μm of the surface of the beads. The radioactive isotopes used in SPA assays emit low energy emissions that have very short pathways in aqueous media (Table 5.5).

Table 5.5 Examples of the radioactive isotopes used in SPA assays

Element	Isotope	Half-life	Mode of Decay	Pathway in Aqueous Solution
Carbon	^{14}C	5730 years	β^-	
Hydrogen	^{3}H	12.26 years	β^-	< 1 μm
Iodine	^{125}I	60 days	Electron capture	~17 μm
Phosphorus	^{32}P	14.3 days	β^-	

Many SPA enzyme-based assays are carried out using radiolabelled substrates or ligands. Consider, for example, an enzyme inhibition assay, based on the use of a substrate A-B for the enzyme where B is the part of the substrate that contains the radioactive isotope and A contains a so-called *capture group*, which contains structures that bind to the SPA beads. B does not contain a capture group. Treatment of the substrate A-B with the enzyme and any essential co-enzymes in the absence of the inhibitor results in cleavage of all of the substrate A-B. When the SPA beads are added no fluorescence is observed as only the non-radioactive A binds to the beads. The radioactive B remains in solution too far from the beads to cause fluorescence (Fig. 5.23a). However, when a test compound is present the extent of the cleavage

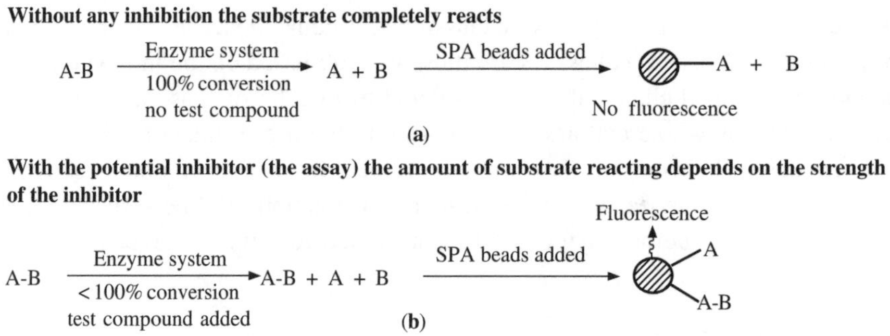

Without any inhibition the substrate completely reacts

A-B $\xrightarrow[\substack{100\% \text{ conversion} \\ \text{no test compound}}]{\text{Enzyme system}}$ A + B $\xrightarrow{\text{SPA beads added}}$ ⬤—A + B

No fluorescence

(a)

With the potential inhibitor (the assay) the amount of substrate reacting depends on the strength of the inhibitor

Fluorescence

A-B $\xrightarrow[\substack{< 100\% \text{ conversion} \\ \text{test compound added}}]{\text{Enzyme system}}$ A-B + A + B $\xrightarrow{\text{SPA beads added}}$ ⬤ ⟨A, A-B⟩

(b)

Figure 5.23 A schematic representation of an enzyme HTS microassay. Ideal situations are illustrated. (a) No fluorescence is observed as there are no A-B molecules to bind to the SPA beads. (b) The cleavage of only a fraction of the A-B substrate molecules results in the unreacted molecules binding to the SPA beads, resulting in fluorescence. The fragment A of the substrate will also bind to the beads

depends on the strength of the inhibition exhibited by the test compound: the greater the inhibition, the less the cleavage of the substrate A-B. Consequently, when the SPA beads are added, the unreacted radioactive substrate A-B and the non-radioactive A bind to the beads. The emissions from the unreacted A-B bound to the bead are close enough to cause the bead to fluoresce (Fig. 5.23b). However, B, the radioactive product of the enzyme-catalysed reaction, still remains in solution too far from the beads to cause fluorescence. Consequently, the intensity of the fluorescence is directly proportional to the bead-bound A-B. In other words, the greater the fluorescence, the greater the degree of enzyme inhibition of the substrate A-B by the test compound. This type of assay is ideal for HTS since it is easily automated and requires a small number of simple pipetting and shaking steps followed by the use of a suitable scintillation counter, such as the Microbeta produced by Wallac and the Topcount from Packard.

All the methods employed in bioassays depend for their success on producing a measurable effect. A wide range of techniques are used to measure those effects, including, for example fluorescence, absorption and luminescence spectroscopy. A discussion of these methods is beyond the scope of this text and readers are referred to more specialised texts, such as *High Throughput Screening, the Discovery of Bioactive Substances*, edited by J. P. Devlin, published by Marcel Decker Inc.,1997 and *High Throughput Screening in Drug Discovery* (2006), *Volume 35, Methods and Principles in Medicinal Chemistry*, published by John Wiley and Sons Ltd.

All HTS biochemical assays suffer from the disadvantage that they are carried out on material from incomplete cells. Consequently, compounds that exhibit a significant degree of activity under the conditions of a biochemical assay may not be active under normal physiological conditions. This loss of activity may be for a variety of reasons, such as they may be metabolised before reaching the target or they might not be able to cross the cell membrane. Furthermore, it is emphasised that a biochemical test should only be used if the target of the test can be related to the diseased state under investigation.

5.6.2 Whole cell assays

Whole cell assays are preferred when the nature of the steps in the mechanism of the disease state have not been well defined. They also offer a number of other advantages over biochemical tests. For example, whole cell tests may identify compounds that act at sites other than the target site. They are usually conducted under conditions that are more like those encountered if the test compound was used in a patient. Consequently, test compounds that are either too hydrophobic, and as a result bind too strongly to serum albumin (see section 1.7.1), or will not cross cell membranes will not usually be active. Therefore it is relatively easy to identify these compounds and eliminate them from the investigation. Furthermore, test compounds that are toxic are often readily identified because of their effect on the cells used in the test. These advantages mean that many medicinal chemists prefer whole cell assays to biochemical assays. A wide range of different types of whole cell assays are in use (Table 5.6).

Table 5.6 Examples of some of the types of whole cell assays used to determine activity

Type of assay	Notes
Reporter gene assays	A cell is transfected with a gene that produces a marker compound that is not normally present in the cell. This marker compound can be quantitatively measured. Its presence or absence gives a measure of the activity of a test compound
Cell proliferation assays	Changes in the rate of cell proliferation can be used as a measure of the activities of test compounds
Pigment-translocation assays	Used to identify new G-protein-coupled receptor ligands. Melanophores are cells that undergo a colour change when the concentration of cAMP in the intracellular fluid changes. At low cAMP levels the cells lighten but at high levels the cells-darken. Frog melanophores that have been transfected with the human receptor under test are exposed to the test compound for a limited time (ca. 30–60 min). The absorbance at 620 nm is a measure of the degree of activity of the test compound

5.6.3 Hits and hit rates

A *hit* occurs when the activity of a test compound has a value greater than an arbitrary minimum value set by the investigators using that assay. For example, in an enzyme inhibition assay a hit may be scored when the activity of the enzyme is inhibited by a preagreed value of 50 per cent. It is important to set the criteria for a hit before carrying out the assay since *hit rates* are often used as a measure of the validity of an assay procedure. Hit rates are defined as the number of active samples discovered by an assay expressed as a percentage of the total samples used in that screen. Assays with values of about 0.1–1 per cent hits are normally regarded as being valid. Higher values may occur for a number of reasons: for example, a series of compounds with similar structures and hence similar activities is being tested, the assay is not specific enough or the hit criteria are not rigorous enough. However, high hit rates may be of use when developing an assay.

Inaccurate hit values may be obtained because of false positives and negatives. In single-compound libraries false positives are due to inactive compounds giving a positive result because of an inappropriate interaction with the biological material and chemical reagents used in the assay Alternatively, they may be due to the compound passing a preliminary screen but not being active enough to pass more rigorous follow-up tests. Conversely, negative hits may also be due to compounds that are active but do not register as such in the assay. The screening of libraries of mixtures can also result in false positives and negatives (see section 5.6).

A hit is only converted into a lead when it exhibits a good activity profile in follow-up assays and its structure is confirmed.

5.7   Automatic methods of library generation and analysis

Traditional organic synthesis and assay are labour intensive. The advent of HTS and combinatorial chemistry has led to an increased use of automation in medicinal chemistry with a number of companies producing 'of the self' automatic compound synthesisers (Fig. 5.24) and HTS analysers, as well as custom-built machines. Synthesis is usually brought about in a series of computer-controlled steps using an appropriate array of reaction vessels at so-called stand-alone work stations. Manipulations in the synthesis such as loading the reaction vessels with reagents and solvents, pipetting, washing, heating, filtration, etc. are normally carried out using a robotic arm directed by the software controlling the synthesiser. Robotic arms and track systems are used to transfer microplates between work stations. Autoanalysis is carried out in the same general manner, that is, at work stations using robotic arms, etc. to transfer samples and reagents.

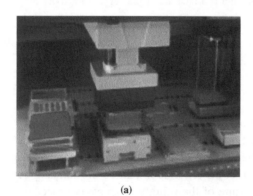
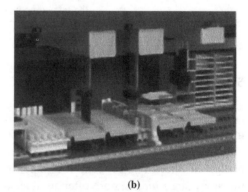

(a)   (b)

Figure 5.24   (a) Close up of an HTS workstation with two liquid handling arms and a robotic arm. (b) Close up of a 96 channel pipetting arm. Reproduced by permission of Tecan Deutschland GmbH.

5.8   Questions

1 List the general considerations that should be taken into account when designing a combinatorial synthesis.

2 Outline Merrifield's method of peptide synthesis.

3 A linker requires the use of oxidation to release its peptides from the solid support. Comment on the suitability of this linker when preparing peptides that contain (a) alanine, (b) methionine and (c) serine.

4 Describe Furka's mix and split technique for carrying out a combinatorial synthesis. How does this method differ from Fodor's parallel synthesis method?

5

$$\overset{R^1}{\underset{|}{}} \qquad \overset{R^2}{\underset{|}{}}$$
$$H_2NCHCONHCHCHCOOH$$

(B)

Design, in outline, a combinatorial synthesis for the formation of a combinatorial library of compounds with the general formula B using the Furka mix and split method. Outline any essential practical details. Details of the chemistry of peptide link formation are not required; it is sufficient to say that it is formed.

6 Outline the range of encoding methods used to deduce the structures of compounds produced in a Furka mix and split combinatorial synthesis.

7 Describe, in general terms, how the technique of deconvolution can be used to identify the most active component in a combinatorial library consisting of groups of mixtures of compounds.

8 Outline the technique of high-throughput screening.

9 (a) How do biochemical assays differ from whole cell assays. What are the advantages of whole cell assays over biochemical assays.

(b) What is the significance of hit rates when assaying the results of an assay?

(c) Suggest reasons for abnormally low and high hit rates in a high-throughput screen for a specific type of activity.

6

Drugs from natural sources

6.1 Introduction

The chemical diversity of the compounds found in nature makes plant, animal and marine materials important potential sources of new drugs, novel lead compounds and stereospecific structures for the synthesis of existing drugs. The most commonly used natural sources are plants and microorganisms, both land and marine. Selection of the plant or microorganism to be investigated may be based on *ethnopharmacology* or the current interest of the investigators. Ethnopharmacology is the investigation of the use of plants by an ethnic group. For example, the investigation of the coca leaves, which South American Indians chewed to alleviate fatigue, led to the discovery of cocaine. Development of this drug ultimately resulted in the discovery of the widely used local anaesthetics benzocaine and procaine (Fig.6.1).

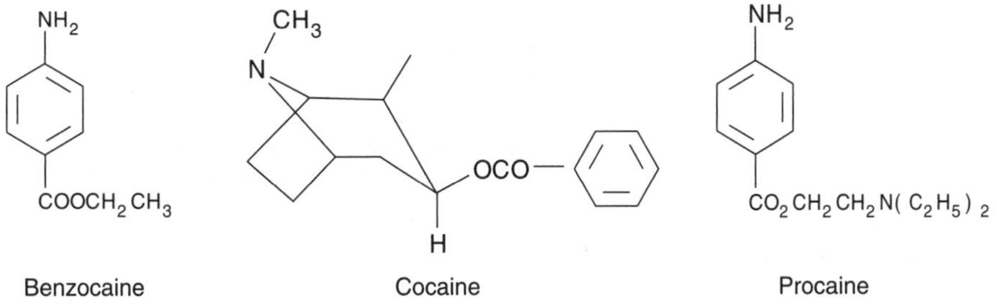

Figure 6.1 Local anaesthetics developed from cocaine

The first step in an investigation of a natural product is to decide its objective(s): for example, is it to be a general or specific search for active compounds in the natural material. In the latter case, is the investigation to concentrate on a search for compounds

Medicinal Chemistry, Second Edition   Gareth Thomas
© 2007 John Wiley & Sons, Ltd

active against a specific disease or is it to look for specific types of compounds that may be used as a lead? These decisions will dictate the scale of the investigation, the nature of the screening assay(s) required and the procedures followed. All these decisions have to take into account the budget allowed for the investigation. Once these issues have been decided the approach commonly used for the isolation consists of a series of steps (Fig. 6.2). These steps often use the same chemical methods but for different purposes. Active compound or compounds are followed through these steps by the use of appropriate bioassays (see section 6.2).

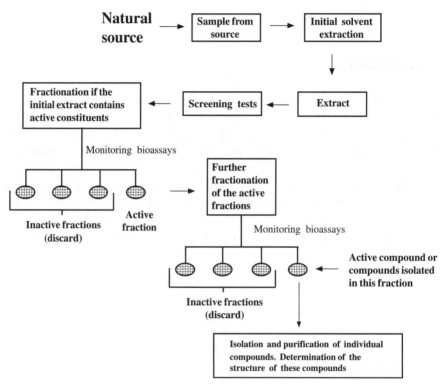

Figure 6.2   A schematic outline of the general approach to the isolation of active compounds from a natural source

The initial step in any investigation is to convert the chosen material into a form suitable for testing. This is normally done by extraction of the sample with a suitable solvent system. *Extraction* in this context is the separation of the active constituents from unwanted cellular materials (see section 6.6). Selection of the extraction solvent system will be influenced by the class of compounds that the investigators are aiming to extract and the screening test(s) that the investigators propose to use (see section 6.2.1). The extracted sample is tested for activity and if it shows the required level of activity it is fractionated. *Fractionation*, in the context of isolation of the compounds from natural sources, is the separation of an extract into groups of compounds (*fractions*) having the same physicochemical characteristics, such as solubility, size, electrical charge, acidic and basic nature. Separation methods normally used are solvent extraction, dialysis, distillation, precipitation, electrophoretic and chromatographic methods (see section 6.7). Each fraction is tested and the active samples fractionated normally undergo a second fractionation using different methods to those previously employed. The cycle of fractionation and testing for active samples is repeated until the active compound or compounds are isolated. This approach suffers from the drawback that a great deal of work can result in the isolation of a known compound. However, the chance of this happening can be reduced by *dereplication procedures* (see section 6.3). Once an active compound has been isolated, it is purified and its structure determined (see section 6.4). If the molecule's activity is of commercial interest it will be synthesised by medicinal chemists and evaluated by the development team (see Chapter 16). This will involve the synthesis of analogues and appropriate SAR and QSAR studies (see Chapter 3). A successful outcome of the development process will probably lead to either the large-scale synthesis of the original active compound or an analogue. However, in the former case it may be possible to obtain sufficient quantities of the source material to enable a viable commercial extraction of the original drug. This has the advantage of producing a drug with the correct stereochemistry, which is likely to be more cost effective than a difficult fully synthetic manufacturing process (see Chapter 15). For example, although the important anticancer drugs vincristine, etoposide and Taxol (see section 6.8) have been synthesised, they all have structures with chiral centres that make it more economic to extract the drug, or compounds from which the drug can be synthesised, from plant material that can be regularly harvested. It is important that the environmental impact of isolating a compound from a natural source be taken into account before commercial exploitation of that source is allowed. For example, the commercial isolation of the anticancer drug Taxol from the bark of the Pacific yew tree would have resulted in the tree's extinction. Taxol is now obtained using a less destructive semisynthetic route from the needles of the more common European and American yew trees.

6.2 Bioassays

Bioassays have two major roles in the isolation of active compounds from natural sources, namely screening and monitoring. *Screening tests* are used to detect the presence of active constituents in the initial extracts whilst *monitoring tests* are used to follow the path of the active constituents through the isolation process.

The amount and degree of activity of a constituent found in a naturally occurring sample will depend on the environment in which the organism grew, the time of collection and the way in which the sample was prepared and stored. It is imperative that detailed records of these events are accurately recorded at the time as the growing and collection conditions will vary. Consequently, it is useful to use screening and monitoring tests that can give an estimate of the concentration of the active constituent in the extract.

6.2.1   Screening tests

The tests used for screening must be rapid, accurate, reproducible, have a high capacity for samples, be cheap and easy to carry out. They should be usable for both tarry and poorly water-soluble extracts. Investigators either set up a limited range of tests to detect specific forms of activity, such as cytotoxic activity and antibiotic activity, or a wide range of screening tests to detect as many types of activity as possible. However, it is still possible to miss active extracts because of either the specific nature of the screening tests employed or their lack of sufficient sensitivity to respond to the amounts of active constituents in an extract. Consequently, in a general screening programme, it is essential to set up as wide a range of screening tests as possible. In both approaches high-throughput screening (HTS) is often the method of choice in the pharmaceutical industry as it is more cost effective than other forms of screening test.

Screening tests may be classified, for convenience, as broad and specific bioassays. *Broad screening tests* detect many different types of activity. They are used when the researchers are simply identifying active samples rather than samples that are active against specific diseases. *Specific bioassays* are used when the investigation is aimed at identifying compounds that are active against a specific disease or organism. Ideally, they should only give a positive response to that disease or organism. However, both general and specific screening tests are usually based on the response of whole organisms, cultured cells, isolated tissue samples and enzymes to test samples of an extract.

Whole organism screening tests

The organisms used in whole organism screening tests range from microorganisms through animals to human volunteers. Two of the most commonly used bioassays are the brine shrimp lethality test (BSLT) and the crown gall tumour inhibition test. The BSLT uses the brine shrimp (*Artemis salina*). Tests are normally carried out in triplicate by adding a range of concentrations of an extract to identical 5 ml brine solutions containing ten shrimps. The number of surviving and dead nauplii are counted after 24 hours and the results expressed as an LC_{50} or LD_{50} value, the dose required to kill 50 per cent of the nauplii. The LD_{50} values are calculated using set statistical methods. Active fractions have a significant LD_{50} value.

The crown gall tumour inhibition test is based on inhibition of the growth of crown gall tumours on 0.5 cm thick discs cut from potato tubers. These tumours are induced in the potato discs by the Gram-negative bacterium *Agrobacterium tumefaciens*. The discs are placed on 1.5 per cent agar contained in Petri dishes and a mixture of the extract and a broth of *Agrobacterium tumefaciens* is spread over the surface of the disc. The plates are incubated at room temperature and after 12 days the number of tumours are counted. Blanks are carried out and the results expressed as a plus or minus (inhibition) percentage of the tumours found on the blank discs. An inhibition of >20 per cent is regarded as a significant activity.

A number of simple screening tests utilise microorganisms such as bacteria, virus, fungi, yeasts and amoebae, amongst others. The tests are often based on growing the microorganism in agar and comparing the effect of the addition of various dilutions of an extract to the growing cultures with an untreated control. These tests may be carried out using a single microorganism in a Petri dish or a number of organisms in a 96-well microtitre plate. In the latter case, one or more extracts can be tested at different dilutions against a range of microorganisms responsible for human infections.

A disadvantage of whole organism tests is that members of the same species often have a different response to a drug. Consequently, it is often necessary to repeatedly test the extract using a large enough sample of the organisms to obtain statistically viable results. As a result, screening tests using larger mammals are very expensive. This, together with the question of the ethical use of animals, prohibits the use of higher mammals in initial (primary) screening tests for natural products.

Cultured cell tests

It is now possible to produce mammalian cell lines that can be sustained indefinitely in a suitable culture medium. These cells often contain receptors that are specific to a particular physiological process. This allows them to be used as the basis of *cultured cell tests*. These tests are used to study the binding of the constituents of an extract to these receptors as well as the inhibition of the action of that receptor. The cells can be produced in the wells of microtitre plates, which makes them suitable for HTS (see section 5.6). The small amounts of extract required mean that it is possible to check reproducibility by carrying out several identical tests at the same time. In addition, the effect of extract concentration can also be studied. Treatment of the cell culture with the extract is normally followed by reaction with a suitable reagent to produce a colour, fluorescence, luminescence or radioactivity (see section 5.6.1) related to the quantity of compound bound to the receptor or a metabolite produced by the cell.

Isolated enzyme tests

Isolated enzyme tests are used to determine whether an extract either activates or inhibits an enzyme system. This information can be related to the type of activity exhibited by the components of the extract. For example, activation of the enzyme trypsin could indicate debriding agents used to clean necrotic wounds, ulcers and abscesses while inhibition could show the presence of antithrombotic agents, anti-inflammatory agents and male antifertility agents. Tests are usually quantitative and are normally based on the conversion of a known amount of a substrate by a standard quantity of an enzyme to a product that is detectable by a suitable quantitative analytical method. Analysis is normally carried out after separation of the products of the enzyme-catalysed reaction by a suitable chromatographic technique. Product detection methods commonly used include radioactivity (see section 5.6.1), fluorescence, UV and visible light absorption. The results are compared with a control reaction containing no extract to determine whether the extract has caused activation or inhibition of the enzyme. For example, the action of trypsin on *N*-benzoylarginine-4-nitroanalide produces 4-nitroaniline, which can be assayed using visible spectroscopy by measuring its absorption at 400 nm (Fig. 6.3). Several control assays are carried out using *N*-benzoylarginine 4-nitroanalide and the absorption of the 4-nitroaniline produced by the enzyme-catalysed reaction in the absence of the test compound is measured. The average amount of 4-nitroaniline produced is calculated and used as a base line. The procedure is repeated using both *N*-benzoylarginine-4-nitroanalide and the test compound. A decrease in the amount of 4-nitroaniline produced indicates inhibition of trypsin by the test compound, while an increase indicates activation of the trypsin.

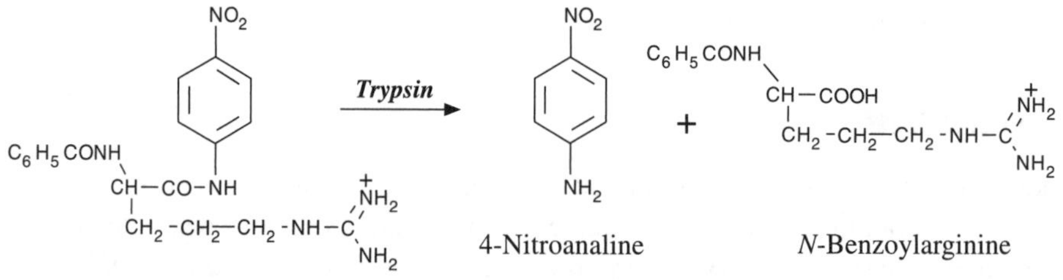

4-Nitroanaline

N-Benzoylarginine

N-Benzoylarginine-4-nitroanalide

Figure 6.3 Isolated enzyme test: action of trypsin on *N*-benzoylarginine-4-nitroanalide to produce 4-nitroanaline

Isolated enzyme tests can be carried out on 96-well microtitre plates using HTS (see section 5.6). However, the number of enzymes that have been used in screening tests is limited.

Isolated tissue tests

Classical *isolated tissue tests* are normally carried out by measuring either the contraction (Experiment 1, Fig 6.4) or inhibition of the contraction (Experiment 2, Fig 6.4) of a suitable tissue sample, such as guinea-pig ileum, suspended in a nutrient bath when the extract is added to the bath (see also section 8.5.1). This is a non-specific broad screening test. It is quick and easy to perform and economic on raw material.

Isolated tissue tests are often very sensitive and a wide range of different types of tissue sample have been used in this type of test. However, variations in the response of some similar tissue samples mean that some types of isolated tissue test can require large amounts of extract. Furthermore, it should be noted that the bioassays used in screening only provide preliminary information about the plant extract and must be followed up, if required, by more extensive and precise clinical tests on the isolated compounds.

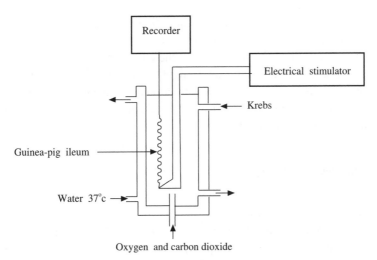

Figure 6.4 A diagrammatic representation of an isolated tissue test apparatus using guinea-pig ileum. Experiment 1: the contraction of the tissue is measured when the extract is added to the nutrient bath. Experiment 2: electrical stimulation is used to induce contraction of the ileum. The extract is added and any relaxing of the contraction is measured

6.2.2 Monitoring tests

The screening tests described in section 6.1.1 may also be used as monitoring tests to track the active extracts through the isolation process. For example, the guinea-pig ileum test is suitable as a monitoring test as it is quick and easy to perform and economic on raw material. However, as fractionation progresses the concentration of the active constituents will increase and this must be taken into account when preparing samples for testing.

The activity of fractions may fall or disappear during fractionation. This may be due to the decomposition of the active constituents during the fractionation process, removal of compounds that have a synergic effect on the activity of the bioactive constituents or the active constituent may simply be lost in the fractionation process.

Activity change due to decomposition

Decomposition of the active compounds may be due to either the nature of the fractionation process or because of the inherent instability of the compounds themselves. The most common routes for decomposition are oxidation, hydrolysis and polymerisation. One reason for decomposition during fractionation is that the conditions under which fractionation is carried out may be too vigorous. This cause of decomposition can be minimised by keeping the conditions as mild as possible. For example, temperatures should be kept low, solvent acidity and alkalinity should be kept as near neutral as possible and solvents removed by either freeze drying or distillation under a low vacuum. Other sources of decomposition during fractionation are the removal of inhibiting compounds such as naturally occurring antioxidants, an increase in the concentration of enzymes that catalyse the decomposition processes and exposure to light and air. The last two can be reduced by protecting from light and carrying out the fractionation in an atmosphere of nitrogen. Microorganisms can also cause decomposition, so all equipment should be sterilised if this type of decomposition is significant.

Activity change due to synergy

Synergy occurs when the activity of mixtures of two or more compounds is greater than the sum of the activities of each of the components of the mixture. Consequently, removal of one or more of the components of a mixture during fractionation can significantly reduce the activity of the remaining extract. Suppose, for example, that an extract with an activity of x contains components A, B, C, etc. with individual activities a, b, c, etc. If some or all of these components exert a synergic effect on each other then:

$$\text{Total activity of fraction } x > \Sigma(a + b + c + \ldots) \tag{6.1}$$

The mechanism by which synergy works is not understood. However, the phenomenon is used to explain why the activity of some drugs is significantly enhanced by the presence of a compound that is inactive or has a different activity from the drug. For example, the activity of thalidomide used in the treatment of multiple myeloma is considerably enhanced by the presence of dexamethasone, which on its own has no activity against myeloma.

Thalidomide   Dexamethasone

Activity change due to fractionation loss

The active compounds may remain 'locked up' in the procedure used to effect the fractionation. For example, if chromatography is used as part of the fractionation process the active compounds could be either very strongly or permanently absorbed on a solid stationary phase. In both cases, the active compounds will not show up in the fractions produced by the process. Furthermore, if distillation or sublimation is used as a technique to concentrate fractions in the fractionation process there is a loss of the active constituents by co-distillation or co-sublimation, leaving the inactive concentrate.

6.3   Dereplication

Dereplication is the use of chemical techniques to eliminate extracts that contain active constituents that have already been isolated and characterised. It is essentially chemical screening of the extracts using chemical procedures such as chromatography, NMR, UV and visible spectroscopy and mass spectrometry, and comparing the results to a data base to identify active compounds that have already been investigated. This analytical process does not always give a successful identification since it depends on the parameters on which the dereplication procedure is based. Consequently, it is not the complete answer to the problem of wasting time and resources isolating known compounds. Furthermore, it should be noted that some of the techniques used in dereplication are also used in the cleaning up procedures (see section 6.6.3). Consequently, it is not always possible to define the boundary between dereplication and cleaning up.

Each dereplication procedure is specifically designed for the extracts involved in an investigation. For example, from 1987 to 1992 the American National Cancer Institute (NCI) tested nearly 40 000 natural product extracts, both aqueous and organic, in a primary anti-HIV screen. Nearly 15 per cent of these extracts exhibited some degree of activity in this screen. This large number suggested that many of these active extracts contained similar classes of active compound. In view of the large number of extracts, Cardellina *et al.* devised a dereplication protocol for active *aqueous extracts* to eliminate those containing known classes of active compounds and to identify extracts that were likely to contain novel compounds that could be used either as drugs or more probably leads (Fig. 6.5). Their dereplication procedure was based on the molecular size, mass and polarity of the constituents of the extracts.

The first step in the procedure was to precipitate anti-HIV active sulphated polysaccharides, which are already known to be HIV active, using ethanol. The precipitate was separated from the aqueous supernatant liquid. This liquid was put through a second screen, the route through this screen depending on the source of the original extract. Plant extracts were passed through a chromatography column containing a polyamide resin. Polyamide resins have been shown to irreversibly retain tannins. Some of these polyhydroxyphenols have been shown to inhibit HIV. However, compounds with fewer phenolic groups can be eluted from these resins. This reduced the number of plant extracts by more than 90 per cent. Samples of each of the surviving plant extracts and the extracts from other sources were separately chromatographed through different types of chromatography columns, namely, Sephedex G-25 resin, Baker C_{18} and WPC_4 bonded-phase cartridges. These chromatography systems separate the constituents of the extracts according to molecular size, molecular mass and polarity. Fractions taken from these columns were tested for activity and a unique chromatographic profile of the active constituents of an extract was obtained, which was used to compare the nature of the constituents of different extracts and determine recurring elution patterns of extracts. This information, together with that concerning the molecular size, molecular mass and polarity of the constituents, which was also obtained from the dereplication screen, allowed Cardellina *et al.* to decide which were the most profitable extracts

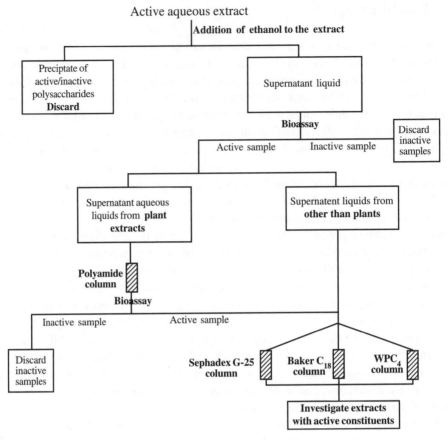

Figure 6.5   An outline of the dereplication protocol devised by Cardellina *et al.* for aqueous extracts that could possibly contain HIV-inhibiting constituents

to investigate. For example, the chromatographic profiles of six sponge extracts showed a recurring pattern of activity. Investigation of these extracts using NMR suggested the presence of sulphated sterols (Fig. 6.6). This led to the isolation of ibisterol from these extracts.

It is important to realise that dereplication procedures are tailored to suit the investigation in progress. Consequently, it is possible to miss compounds because of the design of the dereplication procedure. Furthermore, the cleaning up of extracts (see section 6.2) prior to dereplication may also result, in some cases, in the loss of novel active compounds.

6.4   Structural analysis of the isolated substance

The purity of isolated compounds must be assessed prior to their structure elucidation. As the quantities of isolated compounds are usually very small it is difficult to obtain pure

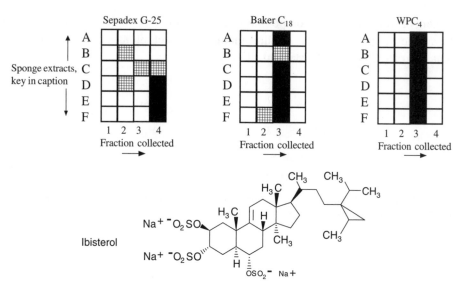

Figure 6.6 The chromatography profile of extracts from six sponges: **A**, *Topsentia* sp. 2; **B**, *Topsentia* sp. 4; **C**, *Pseudaxynissa* sp. A; **D**, *Aaptos* sp.; **E**, *Pseudaxynissa* sp. B, **F**, *Axinella* sp. Shaded boxes represent the elution of anti-HIV active constituents; the darker the shading, the higher the activity of the fraction

crystalline samples by repeated recrystallisation. Consequently, a compound is usually deemed to be pure if a single peak is obtained when the compound is subjected to a range of qualitative chromatographic techniques such as gas (GC), liquid (LC) and thin-layer chromatography (TLC).

The structure of a pure compound isolated by the fractionation process is determined by a combination of standard chemical procedures such as mass spectrometry (MS), nuclear magnetic resonance (NMR), infrared (IR), visible or ultraviolet spectroscopy (UV) used in conjunction with appropriate data bases. Unless there are some obvious clues, this could be a difficult process and may require considerable teamwork. However, the final isolated active fraction often consists of a mixture of compounds. The structures of the components of these mixtures are normally determined by the use of so-called *hyphenated techniques*. Typical combinations are GC-MS, LC-UV, LC-MS, LC-NMR and LC-DAD (photodiode array detection). In these procedures the mixture is initially separated into its components by either a gas or a liquid chromatography technique. In the latter case high-pressure liquid chromatography (HPLC) (see section 6.7.2) is one of the most popular techniques. The separated compounds are automatically passed, in succession, from the chromatography column into the specified analytical machine via a suitable interface. For example, three interfaces are commonly used with HPLC machines for separating and determining the structures of the components of plant extracts. They are known as thermospray (TS or TSP), continuous-flow fast atomic bomdardment (FAB) and electrospray (ES). Provided a well-defined LC or GC peak is obtained, combinations of these coupled techniques in conjunction with suitable data bases usually allow the assignment of structures to the compounds in an extract For example, Garo *et al.* identified the flavanone (Fig. 6.7) in the dichloromethane extract of *Monotes engleri* using a combination of LC-NMR, LC-MS and LC-UV.

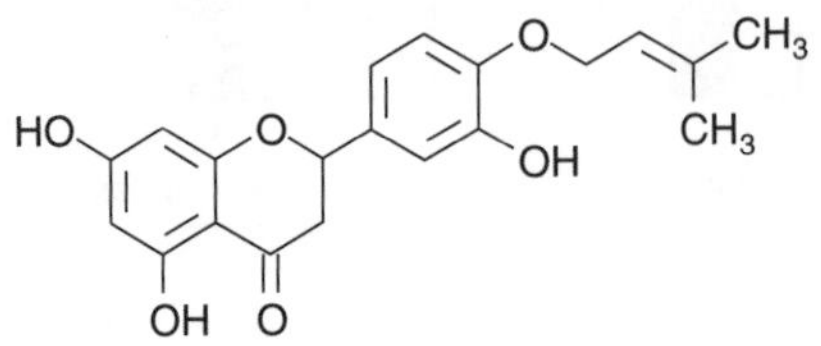

Figure 6.7   A flavone isolated by Garo *et al.* from the dichloromethane extract of *Monotes engleri* using a combination of LC-NMR, LC-MS and LC-UV

6.5   Active compound development

The development of naturally occurring active compounds into commercially viable drugs follows the lines outlined in Chapter 16. However, the stereochemistry of many naturally occurring drugs and the analogues derived from these compounds is so complicated that their large-scale commercial synthesis from synthetically manufactured starting materials is not economically viable. In these instances it is sometimes possible to grow sufficient quantities of a plant or microorganism that contains a compound whose structure contains the required stereochemistry. This naturally occurring compound is used as the starting point for the required drug. For example, in the mid-twentieth century it was discovered that cortisone and hydrocortisone (cortisol) (Fig. 6.8) were highly effective in cases of rheumatoid arthritis. Both of these drugs, like most steroids, have a large number of stereochemical centres that make their synthesis difficult (see section 15.3). Cortisone was originally extracted from the adrenal glands of cattle and eventually, using a laborious 30-step synthesis, from deoxycholic acid isolated from ox bile. These production methods could not meet the demand for these drugs and so alternative more efficient semisynthetic methods were developed. Originally hecogenin obtained from sisal leaves was used as the starting material but this has now been superseded by semisynthetic methods based on progesterone produced from diosgenin isolated from Mexican yams (Fig. 6.8). Progesterone has the same stereochemistry of rings A, B and C as cortisone and hydrocortisone. The synthesis starts with the conversion of diosgenin to progesterone. Progesterone is fermented with *Rhizopus arrhizus* or *R. nigricans* to form 11α-hydroxyprogesterone in greater than 85 per cent yield. A series of reactions convert this compound into compound A, the precursor of both cortisone ethanoate and hydrocortisol ethanoate. It is interesting to note that cortisone is metabolised in the liver to hydrocortisol, which is the active agent.

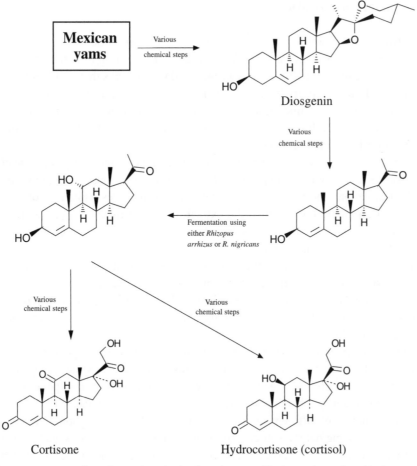

Figure 6.8 An outline of a semisynthesis of cortisone and hydrocortisone from Mexican yams

6.6 Extraction procedures

Extraction in this context is the initial step is the separation of the desired chemicals from the raw material. Most extracts are obtained in a liquid form. The resultant mixture is usually very complex and will normally require *cleaning up* (see section 6.6.3). This involves separating the desired chemicals from unwanted cellular debris and chemicals such as lipids, proteins and polysaccharides. The design of an extraction procedure will need to take into account the objective of the extraction, its scale, the nature of the screening assay(s) and the chemical nature of the compounds in the extract.

6.6.1   General considerations

The objective of an extraction will influence its general design. The first step is to select the type and number of bioassays needed for screening. For example, if the investigation is looking for compounds that will be active against HIV, bioassays that will identify possible compounds with this activity would be selected. However, if the extracts are being investigated in the hope of finding any active compounds, then broad spectrum bioassays would be selected. In both cases it is necessary to choose an extraction procedure that is compatible with the screening assay.

It is also necessary to consider the source of the extract. If the extracts are taken direct from the source it is only necessary at this stage to choose appropriate screening bioassays for the investigation. However, if the investigation involves isolation of the active constituents of an ethnic remedy then the extraction should follow the preparation of that remedy as closely as possible, to allow for any chemical and physical changes that occur during the preparation of that remedy.

It is also necessary to consider the scale of the extraction. Small-scale methods will usually only yield amounts sufficient for a preliminary screen. For more extensive work larger scale methods will need to be employed. These will require a different type of equipment, larger amounts of raw materials and it is possible that the proportions of compounds obtained may change.

The type of compound extracted from the natural material will depend on the polarity, acidity and alkalinity of the solvent. Polar solvents will tend to extract polar compounds while non-polar solvents will tend to favour the extraction of non-polar compounds (Table 6.1). Consequently, where there is no preconception of the nature of the compounds to be extracted, a number of different solvents with a range of polarities should be successively used to extract the same natural source. However, it may be possible to select a solvent that will preferentially extract the classes of compound that are the objective of an investigation.

Table 6.1 Examples of solvent polarity and the commonest classes of compound extracted by that solvent

Polarity	Solvent	Classes of compound extracted
Low	Hexane	Fats, waxes and essential and some volatile oils
	Chloroform	Alkaloids, aglycones and some volatile essential oils
	Diethyl ether	Alkaloids and aglycones
	Ethanol	Glycosides
High	Water	Amino acids, sugars and glycosides

The pH of the solvent will also affect an extraction. Acidic solvents will extract bases such as alkaloids, amines and amino acids, while basic solvents will tend to extract acidic compounds such as phenols, carboxylic acids, sulphonic acids and phosphorus oxyacids. The acids and bases in these solvents react with the acidic and basic groups of the constituents to form the corresponding salts. These salts are often sufficiently water soluble

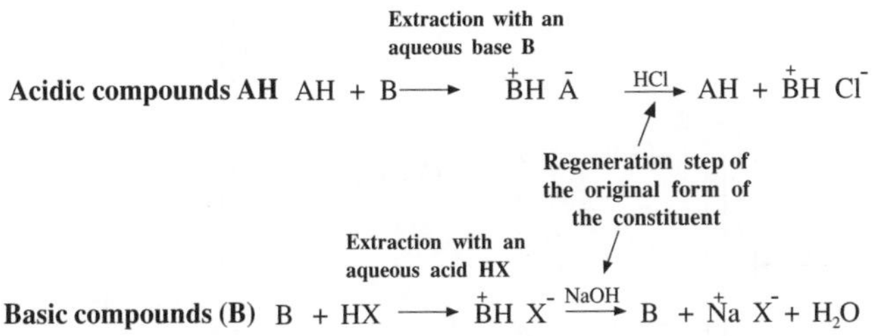

Figure 6.9 The extraction of acids and bases by acidic and basic aqueous solvents. Sodium hydroxide and hydrochloric acid may be used for B and HX, respectively. Other suitable acids and bases may be used in the regeneration step

to be extracted into an aqueous medium. Regeneration of the original acid or base is achieved by treating the salt with the appropriate acid or base (Fig. 6.9).

The use of acidic and basic solvents may affect the stability of the compounds being extracted. For example, acidic or basic conditions could cause hydrolysis of esters and amides, especially if the extraction is carried out above room temperature. Other solvents may also react with the constituents of an extract and so the temperature at which extraction is carried out should be kept as low as possible.

Bioassays are normally carried out in aqueous solution so after solvent removal the extracted constituents are dissolved in water. Constituents that are difficult to dissolve in water are sometimes dissolved in water miscible solvents, such as dimethylsulphoxide, acetone and ethanol, and the solution mixed with water. However, when water miscible solvents are used, a blank bioassay must be carried out to find out if the solvent is affecting the accuracy of the assay.

6.6.2 Commonly used methods of extraction

The most commonly used methods are those based on solvent extraction and supercritical fluid extraction.

Solvent-based methods

The natural material used in extraction processes is usually predried to remove water and, where relevant, the dried material broken up into small pieces for easy handling. In all methods the material is placed in direct contact with the solvent at a predetermined temperature for a set period of time. The extracts are often referred to as *infusions*. The equipment used to prepare an infusion depends mainly on the scale and temperature required. The simplest method is to stir the natural material with the solvent in a simple reflux distillation apparatus (Fig. 6.10a) at either room or an appropriate higher temperature. A disadvantage of using higher temperatures is that only pure solvents and azeotropic mixtures should be used. This is because the composition of a solvent mixture in

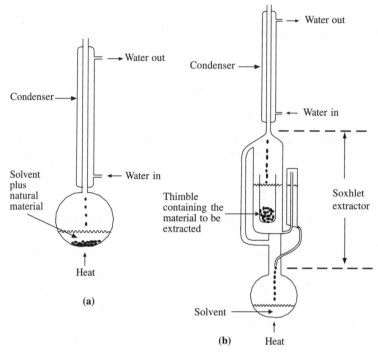

Figure 6.10   (**a**) A reflux distillation apparatus. (**b**) A Soxhlet extractor

the flask will change because the rates of distillation of the various components of the mixture will be different. Furthermore, the constituents of the extract will be more likely to undergo decomposition at higher temperatures. The advantage of this method are that it can be carried out on a large scale using industrial equipment (see section 16.2). This makes it suitable for use in the production of natural products from natural sources (see section 6.5). The main disadvantages of this method are that several portions of fresh solvent may be required if the solution becomes saturated with extract constituents and that the extract has to be separated from the remaining natural material. The latter problem can be alleviated by using a Soxhlet extractor (Fig. 6.10b).

The material to be extracted is placed in a cloth or cellulose thimble. Solvent is boiled in the round-bottomed flask and its vapour rises into the reflux condenser, where it condenses and the warm solvent runs back into the thimble and extracts the natural material. When the thimble and its container are full the solvent extract syphons back into the round-bottomed flask. This cycle can be repeated for as long as is required. Since each cycle effectively uses fresh solvent, the solvent in the flask becomes more concentrated as time goes by. This increase in concentration means that the solvent may become saturated with respect to some constituents, in which case these constituents could be precipitated in the round-bottomed flask and additional quantities of solvent will be required to remove these deposits at the end of the extraction. A further disadvantage is that the extract is in contact with hot solvent for long periods of time and so there is still a possibility of thermal decomposition of some of the constituents. In addition the extractor can only be used with pure solvents and azeotropic mixtures. However, the main advantage of Soxhlet extraction is that the debris from the natural material remains in the thimble and does not have to be removed by filtration or centrifugation.

Steam distillation　This is mainly used to extract essential oils from plants and can be on small and industrial scales. The extract is either boiled with water or has steam from an independent source passed through it. The total pressure of the immiscible liquid mixture of oil and water is the sum of the partial pressures of the components. Consequently, the mixture boils at a temperature lower than the boiling point temperature of each of the pure individual components. As a result, the oils co-distil with water. The oils are separated from the distillate by either allowing the oil and water layers to separate or by extraction into an organic solvent. However, steam distillation cannot be used to extract compounds such as esters, which are hydrolysed by water, or compounds that are thermally unstable.

Supercritical fluid extraction　This uses solvents that are kept in a physical state known as supercritical as the extraction solvents. They are formed when the solvent is kept above its critical point temperature (T_c) and pressure (P_c). A *supercritical fluid* normally has the density and appearance of a liquid but its viscosity is lower than would be expected for a liquid. Supercritical fluids are used for extraction because the diffusion coefficients of solutes are higher than would normally be expected for a liquid. Consequently, their penetration and mass transfer of constituents from natural materials is usually good. The extraction rate in the case of solids is governed by particle size: the smaller the particles, the better the rate of extraction. Liquid or wet natural material samples are often absorbed on inert drying agents such as magnesium sulphate prior to extraction.

Carbon dioxide is the most commonly used supercritical solvent (Table 6.2). This is because its low critical temperature reduces the risk of constituent decomposition, it provides an inert atmosphere for the extraction of thermally labile compounds, it can safely be allowed to evaporate into the atmosphere and it is relatively cheap for large-scale use. Furthermore, it has a good affinity for many non-polar compounds and, in the presence of modifiers such as ethanol, dichloromethane and water, for more polar compounds. Extracts may be analysed on-line by directly coupling the supercritical fluid extractor (SFE) to

Table 6.2 Examples of the supercritical temperatures and pressures of some common solvents

Solvent	T_c (K)	P_c (bar)	V_c (dm^3 mol^{-1})
Carbon dioxide	304	73.9	0.096
Dichlorodifluoromethane	385	41.1	0.217
Hexane	507	30.3	
Methanol	239	80.8	0.203
Water	374	217.8	0.056

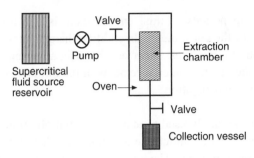

Figure 6.11   A schematic representation of an off-line SFE system

analytical instruments such as a gas chromatograph (SFE-GC), mass spectrometer (SFE-MS), high-pressure liquid chromatograph (SFE-HPLC) and a Fourier transform infrared spectrophotometer (SFE-FTIR). The equipment used in supercritical fluid extraction is more complex than that used in the simple forms of solvent extraction processes. A typical installation consists of a source of the fluid connected through a pump to an extraction chamber (Fig. 6.11). The pump delivers the fluid to the extraction chamber at the critical pressure or above. The extraction chamber is heated to either the critical temperature or above. This is carried out in an oven or by other means. On–off valves are use to control the flow of supercritical fluid into and out of the extraction chamber. The extract is either led to an on-line analytical instrument or transferred to a collection vessel where the solvent is removed and possibly recycled.

Supercritical carbon dioxide is used to extract volatile oils, lipids, triterpenes, steroids, alkaloids and carrotenoids from natural materials. It is used on an industrial scale to extract bitter acids from hops and to remove caffeine from coffee beans, avoiding the use of environmentally harmful solvents. Supercritical fluid extraction has also been used to extract the anticancer drug Taxol (see section 6.8), the alkaloid vindoline from *Catharanthus roseus* and parthenolide from feverfew (Fig. 6.12).

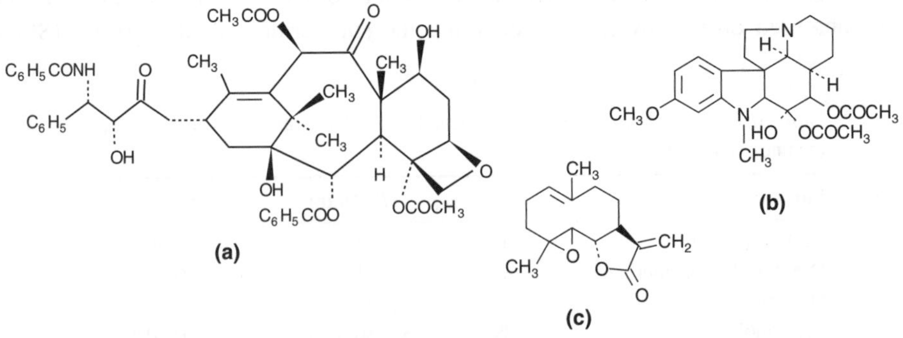

Figure 6.12   The structures of (**a**) Taxol, (**b**) vindoline and (**c**) partheolide

6.6.3 Cleaning up procedures

The extracts of natural products will contain a quantity of constituents that will make fractionation more difficult, either by interfering with the bioassay being employed or by making isolation procedures more difficult: for example, compounds such as tannins, carotenes and chlorophyll may cause problems with the bioassay(s) being used. Tannins are a particular problem in this respect as they cross-link with many proteins, thereby inhibiting bioassays involving proteins. This cross-linking also makes it more difficult to isolate proteins from an extract. Inorganic salts, such as sodium chloride, can be particularly troublesome in isolation procedures as they can be difficult to remove. These problem constituents are often removed by so-called *cleaning up procedures* (Table 6.3). The protocol used in an investigation will depend on the type of compounds that the investigators wish to remove. However, compounds removed in cleaning up procedures should not be rejected out of hand. Many are active and may warrant further investigation. For example, tannins are often biologically active.

Table 6.3 Examples of commonly used cleaning up procedures

Compound	Method	Notes
Chlorophyll	Precipitation with lead subacetate	Used with alcoholic extracts
	Partition	Dilute the extract with water and extract with diethyl ether
Polyphenols	Gel filtration using Sephadex LH-20®	Preferred method
(Tannins)	Partition	Hexane is used to remove the polyphenols
	Chromatography using a polyamide column	Polyphenols are retained on the column
Proteins	Precipitation with ammonium sulphate	
Lipids	Partition	Used for aqueous ethanolic extracts. Light petroleum is used to remove the non-polar constituents
Polysaccharides	Precipitation by the addition of ethanol or acetone	
Inorganic salts	Dialysis	Slow unless a dialyser is used. Can also remove small organic molecules

6.7 Fractionation methods

The techniques used in extraction can also be used in fractionation The most popular methods used are partition and chromatography. Other less popular methods include precipitation, fractional distillation, steam distillation and dialysis. The method selected depends on the nature of the constituents of the extract and the circumstances of the investigation. However, whatever the method, the conditions used should be kept as mild as possible to minimise decomposition. Selection of the solvents used in a fractionation must take into account their possible toxicity to any bioassay(s) being used to follow the fractionation, as well as the polarity of the constituents. In this respect it may be necessary to carry out a blank using the solvent in order to assess its effect on the bioassay being used.

6.7.1   Liquid–liquid partition

Liquid–liquid partition is the partition (distribution) of a solute between two immiscible liquids (see section 2.12). This distribution is governed by the partition law, which may be expressed mathematically as:

$$\text{Partition coefficient } (P) = \frac{[\text{Solute in solvent 1}]}{[\text{Solute in solvent 2}]} \tag{6.2}$$

where P is a constant at constant temperature for the specified system provided that ideal dilute solutions are formed. The mathematical nature of equation (6.2) means that in an extraction a 100 per cent transfer of solute from solvent 1 to an immiscible solvent 2 cannot occur. There will always be some solute left in solvent 1. Moreover, it can also be shown that the extraction of a solute dissolved in solvent 2 with a number of fresh samples of solvent 1 will result in the extraction of more of the solute into solvent 1 than a single extraction with solvent 1. Consequently, it is customary to carry out a series of extractions with an extract, each extraction being carried out with a fresh portion of extracting solvent. These portions of solvent are usually combined for further processing.

Small-scale extractions are usually carried out using a hand-operated glass separating funnel. Normally an aqueous solution of the extract is placed in the funnel and extracted by shaking with a suitable water-immisicible solvent. The mixture is allowed to equilibriate before the layers are separated and the solvent removed.

Extraction of an extract is not limited to one extracting solvent. An extract may be successively extracted by a series of solvents with either increasing or decreasing polarities. For example, a series with increasing polarity would be petroleum ether (low polarity), chloroform, ethyl ethanoate and ethanol (high polarity), whereas a series with decreasing polarity would be water (high polarity), water/ethanol (1:2), ethanol and diethyl ether (low polarity). The former series would give fractions containing constituents whose polarities range from low in petroleum ether to high in ethanol. Similarly the latter series would give fractions containing constituents whose polarities range from high in water to low in diethyl ether. When using a series of solvents, each layer is usually partitioned several times with fresh amounts of the appropriate solvents according to the type of scheme outlined in Figure 6.13. When all the partitions for a particular solvent in the scheme have been carried out, similar solvent layers are combined before the

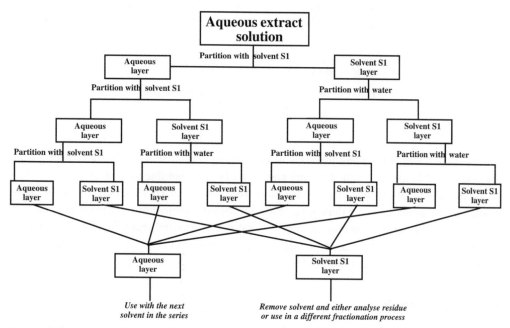

Figure 6.13 A three-step multiple liquid–liquid partition scheme for the partition of an extract by one of the solvents (S1) in a multiple solvent partition series. Each *T*-junction in the lines linking the boxes is a point where partition is carried out. The number of partition steps used in such a scheme depends on the researcher. At the end of the process similar layers are combined prior to partition of the combined extract layers by the next solvent in the series. Each solvent in the series uses the same partition scheme. If the extract was dissolved in a solvent other than water, this solvent would replace water in the appropriate partition steps

combined layer containing the residue of the original extract is partitioned with the next solvent in the series. This ultimately produces one fraction for each of the solvents used in the series. Finally all solvents are removed.

Acidic and basic compounds may be fractionated by adjusting the pH of the extract before separating by partition. For example an aqueous solution of a mixture of alkaloid salts may be fractionated in order of increasing basic strength by treating with increasing amounts of alkali. The careful initial addition of a small amount of alkali will liberate the weakest bases from their salts, allowing them to be separated from the other water-soluble components of the mixture by partition into a suitable organic solvent' leaving the more basic alkaloid salts in the aqueous solution (Fig. 6.14). Careful addition of more alkali will liberate the next most basic alkaloid, which can then be separated from the aqueous solution by partition into a suitable organic solvent. This process of neutralising the salts with alkali and separation of the liberated bases into an organic solvent is repeated until all the constituents of the extract have been fractionated. The same method can be used to fractionate aqueous solutions of mixtures of organic acid salts. However, in this case,

Mixtures of the salts $\qquad$ $\overset{+}{B}H$ $W^- \xrightarrow{\text{NaOH}} B + \overset{+}{Na} W^- + H_2O$
of organic bases ($\overset{+}{B}H$ W^-) $\qquad\qquad\qquad$ Base

Mixtures of the salts $\qquad$ $A^+Z^- \xrightarrow{\text{HCl}} AH + Z^+Cl^-$
of organic acids ($A^+ Z^-$) $\qquad\qquad\qquad$ Acid

Figure 6.14   The neutralisation of bases and acids

successive amounts of dilute aqueous acid are used to neutralise the salts. In both cases the method depends on the liberated acids and bases being insoluble in water.

Adjusting the pH of an aqueous solution may also be used to retain acidic or basic solutes in the aqueous layer during partition by forming their more water-soluble salts.

Liquid–liquid partition is a crude process that does not often result in the separation of the original extract into fractions containing pure compounds. However, it is frequently used as a preliminary fractionation process for other fractionation methods.

Mechanical methods

The liquid–liquid partition process is labour intensive but can be automated by the use of Craig counter-current distribution (CCCD) and steady-state distribution (SSD) apparatus. Both machines consist of a series of 'separating tubes' linked by tubes, the number depending only on the size of the machine. In CCCD machines the extract is dissolved in a suitable solvent or buffer solution and placed in the first tube with the immiscible partitioning solvent system The mixture is shaken mechanically for a set period of time before being allowed to stand for a set time and reach equilibrium. At the end of the standing time the machine automatically moves *all* the top layers from one separating tube to the next and also introduces a measured amount of fresh top-layer solvent from a reservoir into the first tube to replace the volume of solvent that has been moved into the next tube. This distributes the contents of the top layer from the first tube to the second tube. The whole process is repeated and the contents of both upper and lower layers are distributed over the number of tubes in the apparatus. Because the constituents of an extract have different partition coefficients' the counter-current technique will, if the machine has enough separating tubes, separate the extract into different fractions (see section 6.8). The SSD equipment operates in a similar manner, however the extract is fed or placed in the middle tube (A) and the machine programmed to move the layers in opposite directions. In this case a measured amount of the two solvent systems being used is automatically added to replace the layers moving in opposite directions along the bank of tubes. SSD will also separate an extract into fractions.

Both CCCD and SSD suffer from the disadvantages that they require large amounts of solvent and the machines can be rather bulky. They also require large amounts of extract. Consequently, health and safety concerns, cost and methods that require very much smaller quantities of extract have made their use less popular.

Salting out

The movement of many compounds from an aqueous layer to an immiscible non-aqueous layer such as diethyl ether or 2-butanone (ethyl methyl ketone) may be enhanced by salting out. Salting out is the addition of very soluble inorganic salts, such as sodium chloride and ammonium sulphate, to aqueous fractions. This treatment can considerably reduce the solubility of many non-ionic compounds in water and enhance their solubility in any less polar immiscible solvent that is present. If very large quantities (ca. 1 g cm^{-3}) of the inorganic salt are used and no other solvent is present, non-ionic compounds may precipitate from solution.

6.7.2 Chromatographic methods

Chromatographic separation techniques are the most widely used methods of fractionation. Both thin-layer chromatography (TLC) and all forms of column chromatography are used as fractionation techniques (Table 6.4). The selection of the method most appropriate for an investigation is usually based on the information available from the preliminary investigation of the extract and the experience of the investigators.

Table 6.4 Examples of the types of column chromatography commonly used for fractionation

Gas (GC)	Partition and adsorption (LC)
Gel filtration	High-pressure liquid chromatography (HPLC)
Ion exchange	Counter-current chromatography (CCC)
Reversed phase	

TLC is mainly used as an analytical tool but is useful for selecting solvent systems for use in preparative TLC and appropriate forms of column chromatography. It is only used to isolate small quantities (ca. 2–10 mg per plate) of extract components for structural determination. The positions of compounds on chromatograms are identified by the use of UV light, iodine adsorption or specific colour reactions (Table 6.5).

The most popular column chromatography methods are gas (GC) and high-pressure liquid chromatography (HPLC). Both of these methods are commonly used in tandem with other analytical techniques, the so-called hyphenated techniques (see section 6.4), to fractionate, identify and determine the structures of the components of an extract. For example, Wolfender *et al.* have used HPLC hyphenated techniques to isolate and determine the structures of xanthones in the dichloromethane extract of the Mongolian gentian *Halenia corniculata* (Fig. 6.15). Nine of these xanthones proved to be new compounds. Xanthones have been shown to inhibit monoamine oxidase (MAO), an enzyme that is known to play a role in the regulation of the neurotransmitters of the central nervous

Table 6.5 Examples of some of the reagents used to develop chromatograms. Note the conditions of use are not given and not all members of a class will give a positive result

Reagent	Common colours of spots	Class of compound detected
Anisaldehyde	Purple, blue and red	Some terpenoids, lignans and sugars
Antimony trichloride	Various colours and UV 365 nm	Some terpenoids
2,4-Dinitrophenylhydrazine	Yellow and orange	Aldehyde and ketone groups
Dragendorff's reagent	Orange on a yellow background	Alkaloids, heterocyclic amines and quaternary amines
Lieberman–Burchard reagent	Orange, red, blue and purple	Δ^5-3-Sterols and some steroids and triterpene glycosides
Ninhydrin	Purple (proline is yellow)	Amino acids, aminosugars, peptides and proteins
Van–Urk reagent	Blue and orange	Secale alkaloids

system. Consequently, it is believed that xanthones may have some potential as new antidepressant drugs.

6.7.3 Precipitation

It is sometimes possible to isolate compounds by selectively precipitating them from solution. Three general techniques are in common use, namely, salting out (see section 6.7.1), reducing solubility and insoluble salt formation. All of these techniques may be used at any appropriate stage of a fractionation. *Salting out* is widely used to isolate peptides and proteins from solution. *Solubility reduction* is achieved by the addition of a second miscible solvent in which the compound is less soluble. *Insoluble salt formation* relies on there being a way of regenerating and isolating the required compound from the salt after separation of the salt from the solution by either centrifuging or filtration. Some of the reagents used to form salts and the associated compound isolation methods are given in Table 6.6.

6.7.4 Distillation

Fractional distillation and steam distillation may be used to separate volatile compounds. These processes have a limited use and are normally only used to separate essential oils from plant material. Essential or volatile oils are used as pharmaceuticals, perfumery, flavourings and as the starting materials for the synthesis of other substances. These oils are commercially extracted on a large scale from many plants by steam distillation. For example, peppermint oil is obtained from two plant varieties, *Mentha piperita* var. *vulgaris* and *Mentha piperita* var. *officinalis*, by steam distillation.

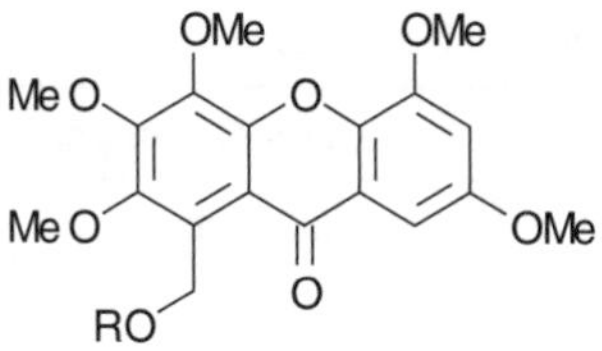

(1) R= H
(2) R = gentiobiosyl
(3) R = primeverosyl

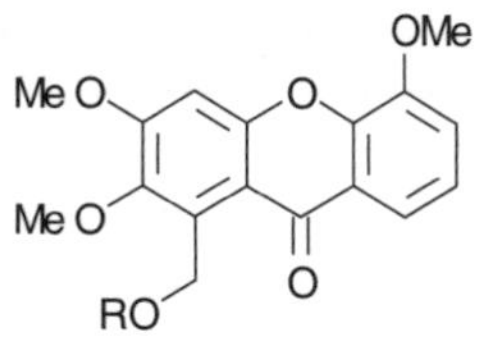

(4) R = H
(5) R = gentiobiosyl
(6) R = primeverosyl

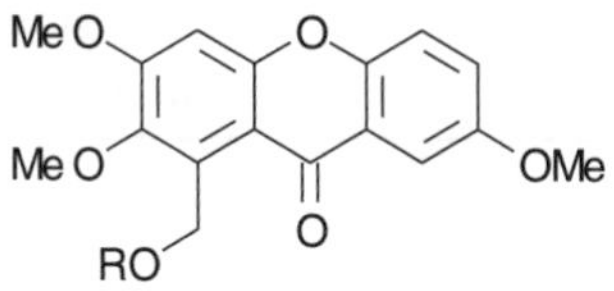

(7) R= H
(8) R = gentiobiosyl
(9) R = primeverosyl

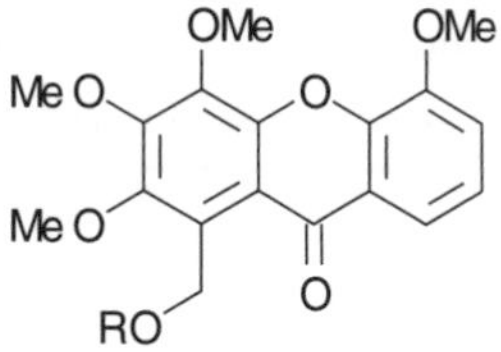

(10) R = H
(11) R = gentiobiosyl
(12) R = primeverosyl

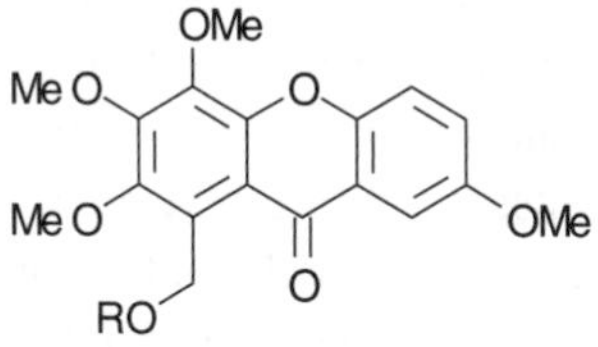

(13) R= H
(14) R = primeverosyl

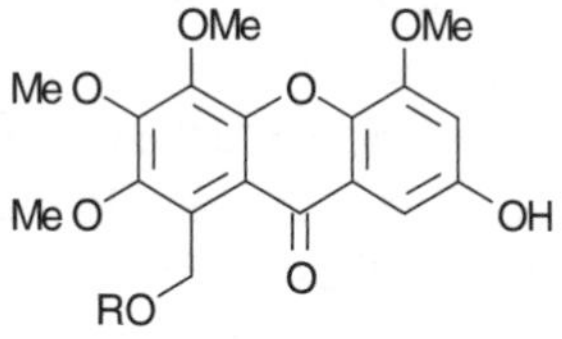

(15) R = H
(16) R = gentiobiosyl
(17) R = primeverosyl

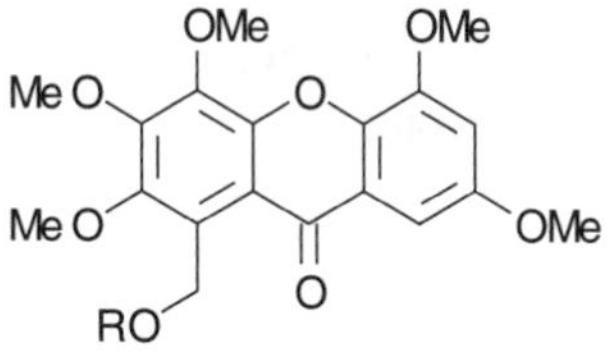

(18) R = H
(19) R = gentiobiosyl
(20) R = primeverosyl

Figure 6.15 The structures of the xanthones found by Wolfender *et al.* in the dichloromethane extract of *Halenia corniculata*

Table 6.6   Examples of reagents used to precipitate components of extracts

Precipitating agent	Regeneration	Class of compound isolated
Picric acid	Treat with aqueous alkali followed by extraction with a suitable solvent	Alkaloids
Dragondorff's reagent	Dissolve in a mixture of acetone, and water (6:2:1). Elute through an anion exchange column	Alkaloids
Ammonium reineckate	Dissolve in a mixture of acetone, methanol and water (6:2:1). Elute through an anion exchange column	Alkaloids
5% Gelatin solution	Treatment with aqueous alkali	Polyhydroxyphenols (tannins)
Lead subacetate solution	Treat with aqueous alkali	Flavonoids, tannins, chlorophyll

6.7.5   Dialysis

Dialysis is used to separate small (<1000 Da) from larger molecules in aqueous solution. Small molecules will pass through the pores in a semipermeable membrane when a concentration gradient exists across that membrane. In small-scale operations the sample is placed in a semipermeable membrane bag and suspended in pure water. Small molecules will diffuse through the membrane into the pure water until their concentration inside the bag equals their concentration outside the bag. Replacing the water outside the bag at regular intervals will eventually result in the transfer of nearly all the small molecules into the water from the extract.

Dialysis is used to separate small water-soluble compounds from large molecules. However, a range of semipermeable membranes with different pore sizes are available and so it is possible to fractionate a sample successively using a series of membranes with different pore sizes. The process is slow (days/weeks) unless dialysers are used. Dialysis is also used as a cleaning up procedure to remove inorganic salts from extracts (see section 6.6.3). It is often used for removing inorganic salts and other small molecules in the purification of polysaccharides and proteins.

6.7.6   Electrophoresis

Electrophoresis is used to separate components that carry an electrical charge. It is mainly used as an analytical tool. However, preparative electrophoresis can be carried out to fractionate extracts using a technique similar to that for preparative TLC.

6.8   Case history: the story of Taxol

The story of the discovery of Taxol (paclitaxel) is a good example of the time and persistence required to extract a new drug or lead from natural material. It also illustrates some of the difficulties encountered in this process and the multidisciplinary nature of the process. The story starts with a plant antitumour screening programme initiated by J. H. Hartwell of the American National Cancer Institute (NCI) in 1960. In 1962, 650 extracts were investigated by laboratories under contract to the NCI. Dr M. E. Wall of the Research Triangle Institute (RTI) requested that his laboratory investigate those extracts that exhibited activity in the 9KB cell culture assay as he had already noticed a good correlation between *in vivo* L-1210 mouse leukaemia activity and 9KB cytotoxicity in his studies on camptothecin. Amongst the samples sent to RTI in 1964 were those of the Pacific yew *Taxus brevifolia*. In the initial extraction, 12 kg of air-dried stem and bark were extracted with 95 per cent ethanol and the solution was concentrated. The isolation of active compounds was followed by biological testing by measuring the inhibition of the Walker-256 solid tumour (5WM) at each stage of the process. Activity was recorded as a T/C value, where T/C is defined as:

$$T/C = \frac{\text{Mean tumour mass of treated animals}}{\text{Mean tumour mass of control animals}} \times 100 \qquad (6.3)$$

and the smaller the T/C value relative to the dose used, the more active the sample. The concentrate was partitioned between water and a solvent consisting of a mixture of

(a)
Solids from the crude extract (166 g) dissolved in chloroform
↓
11 Tube CCCD. Solvent system hexane acetone *t*-butanol water (5:4:4:2)
↓
Tubes 6,7 and 8, 41 g solids, *T/C* 30% at 45 mg /kg^{-1}
↓
30 Tube CCCD. Solvent system methanol water tetrachloromethane/ chloroform (8:2:7:3)
↓
Tubes 8–18, 14 g solids, *T/C* 30% at 23 mg /kg^{-1}
↓
400 Tube CCCD. Solvent system methanol water tetrachloromethane/ chloroform (8:2:7:3)
↓
Tubes 170–189, 2.4 g solids, *T/C* 16% at 15 mg /kg^{-1}
↓
Titration with benzene, 0.5 g taxol, *T/C* 16% at 10 mg /kg^{-1}

(b)

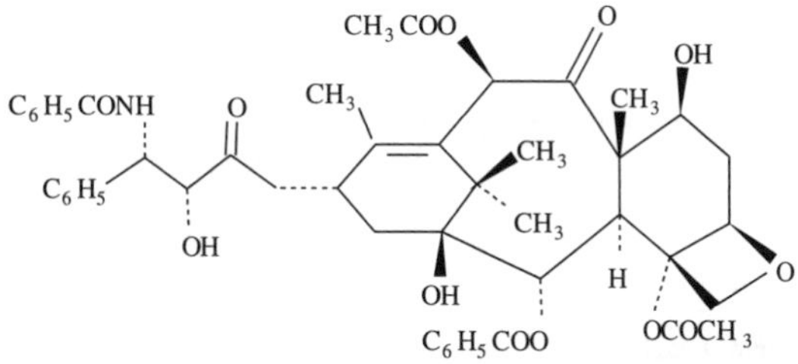

Figure 6.16 **(a)** The fractionation scheme to produce pure Taxol. CCCD is a Craig counter-current distribution procedure. Reprinted from M. E. Wall and M. C. Wani, Camptothecin and Taxol, from discovery to clinic, *J. Ethnopharmacol.*, **51**, 293–254, 1996, with permission from Elsevier. **(b)** The structure of Taxol

chloroform and methanol (4:1). The organic layer yielded 146 g of solids with a good activity against the solid tumour 5WM. These solids were fractionated using a series of three Craig counter-current distribution treatments (see section 6.7.1) using two different solvent systems (Fig. 6.16a). At each of these stages the most active tubes were passed on to the next stage. Finally titration with benzene gave 0.5 g of Taxol, an overall yield of 0.004 per cent. Isolation of the drug was followed by the determination of its structure in 1971 by Wani *et al.* (Fig. 6.16b).

Further development of the drug was slow because of the complex nature of the extraction, its low yield (0.004 per cent), the limited supply of bark, Taxol's molecular structure and its solubility. The trees had to be 100 years old and it required three mature trees to produce about 1g of Taxol. Harvesting also destroyed the trees with no prospect of reharvesting for a hundred or so years! The number of asymmetric centres in its structure meant that synthesis would be difficult. Taxol is almost insoluble in water, which made it difficult to obtain a formulation suitable for administration by intravenous infusion, the

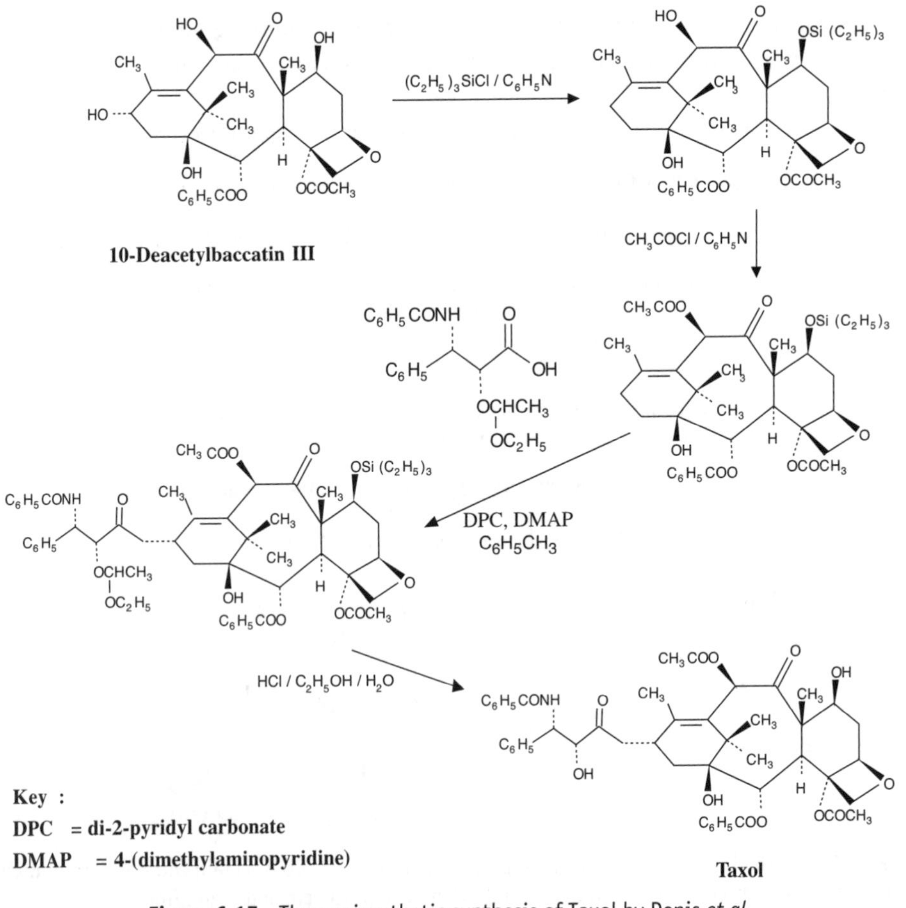

Figure 6.17 The semisynthetic synthesis of Taxol by Denis *et al.*

normal route for anticancer drugs. A solution to this problem was eventually found. A 50 per cent mixture of a polyethoxylated caster oil (Cremophor EL) and absolute alcohol was found to be suitable. This mixture is diluted before use with either 5 per cent dextrose solution or normal saline.

Preclinical development of Taxol started in 1977. It was given a boost when in 1979 Horwitz *et al.* showed that it had a unique mechanism of action. It prevents the dissembly of the microtubules formed during mitosis, the process of cell division. This uniqueness together with Taxol's activity against B16 melanoma and other cancers resulted in the drug going into clinical trials in 1982, Phase I trials in 1983 and Phase II trials in 1985. In 1988 Jean-Noel Denis *et al.* developed a synthesis of Taxol from 10-deacetylbaccatin III isolated from *Taxus baccata* (Fig. 6.17). In 1991 the NCI awarded Bristol-Myers Squibb a cooperative Research and Development Award (CRADA) to develop and market Taxol. This company went on to obtained FDA approval for the clinical use of Taxol in 1992, 32 years after the start of the story of discovery and isolation of Taxol. A full synthesis of Taxol was reported in 1994 by two groups of workers led by R.A. Holton and K. C. Nicolau, respectively. However, neither of these synthetic routes was a practical source of supply of Taxol.

Taxol is currently produced, using a semisynthetic pathway, from baccatin III and 10-deacetylbaccatin III. These compounds are isolated from the leaves of other *Taxus* species such as the common American and European yew, *Taxus baccata*, in yields up to 0.2 per cent. This source does not involve destroying the trees and the leaves are rapidly regenerated. Consequently, prudent regular harvesting gives a regular supply of baccatin III and 10-deacetylbaccatin III. Taxol has also been obtained from plant cell cultures: for example large-scale cultures based on cells from the Japanese yew *Taxus cuspidata* have produced about 3 mg of taxol in a litre of culture medium. Supercritical fluid extraction has also been used to extract Taxol from both *Taxus brevifolia* and *Taxus cuspidata*. This method is more selective than the ethanol used in the original extraction.

A wide range of analogues of Taxol have been prepared. SAR has shown that both the side chain and the four-membered oxetane ring are essential for Taxol's activity. The most promising analogue is Taxotere (docetaxel) (Fig. 6.18), which is also produced by a semisynthetic pathway from 10-deacetylbaccatin III. Taxotere is reported to be more active than Taxol. Its main advantages over Taxol are that it is more water soluble and a more potent antitumour agent. It is also active against a wider range of tumours than Taxol.

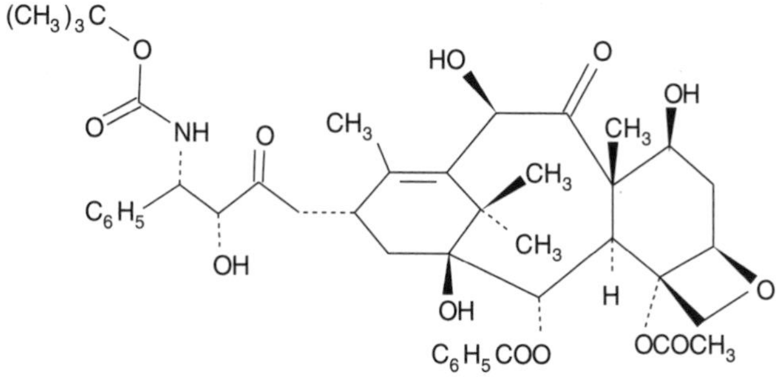

Figure 6.18 Taxotere (docetaxel)

6.9 Questions

1 Explain the meaning of the terms extraction and fractionation in the context of the isolation of active compounds from a plant extract.

2 Outline how the isolation of an active compound could be isolated from a naturally occurring source.

3 (a) Give one reason for not using aqueous acid for extracting peptides and proteins from natural sources.

   (b) Suggest a method that could be used to obtain a sample of water-insoluble organic acids from a mixture of chloroform and crushed plant leaves. Outline the chemistry behind the method using general equations and formulae.

   (c) Evidence suggests that the constituents of a marine sponge are thermally labile. What would be the best method to extract those constituents? Suggest a suitable solvent that could be used in the extraction.

   (d) An extract of the leaves of *Taraxacum officinale* (dandelion) is being prepared for fractionation. Outline a clean-up procedure for this extract.

4 Outline, with reasons, the considerations that need to be taken into account when deciding on an extraction procedure for an extract.

5 Outline a possible procedure for obtaining the proteins from an aqueous extract containing a mixture of proteins, chlorophyll and sodium chloride.

6 Describe the mode of action of a Soxhlet extractor. What are the advantages of supercritical fluid extraction over Soxhlet extraction?

7

Biological membranes

7.1 Introduction

All cells have a membrane, known as the *cytoplasmic* or *plasma membrane*, that separates the internal medium of a cell (*intracellular fluid*) from its surrounding medium (*extracellular fluid*). In the cells (*eukaryotes*) of higher organisms, membranes also form the boundaries of the internal regions that retain the intracellular fluids in separate compartments (Figs.7.1a and 7.1b). Those compartments that can be recognised as separate entities, such as the nucleus, mitochondria and lysosomes, are known collectively as *organelles*. Organelles carry out specialised tasks within the cell. However, in the *prokaryotic* cells (Fig. 7.1c) of simpler organisms where there are no organelles the plasma membrane is also involved in many of the functions of the organelles. The more fragile plasma membranes of plant cells and bacteria are also protected by a more rigid exterior covering known as the cell wall.

The primary function of the plasma membrane is to maintain the integrity of the cell in its environment. It is now also known that the membranes of all types of cell regulate the transfer of substances in and out of the cell and between its internal compartments. This movement controls the health, as well as the flow of information between and within cells. The plasma membrane of a cell is also involved in both the generation and receipt of chemical and electrical signals, cell adhesion, which is responsible for tissue formation, cell locomotion, biochemical reactions and cell reproduction. The internal cell membranes have similar functions and, in addition, are often actively involved in the function of organelles. Most drugs either interact with the receptors and enzymes attached to the membrane or have to pass through a membrane in order to reach their site of action.

The role of membranes and cell walls in maintaining cell integrity and their involvement in cellular function makes these areas of cells potential targets for drug action. However, in

Medicinal Chemistry, Second Edition   Gareth Thomas
© 2007 John Wiley & Sons, Ltd

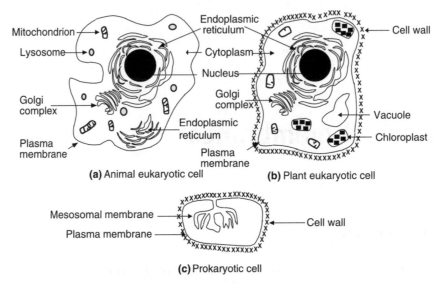

Figure 7.1 A diagrammatic representation of the structures of (**a**) animal and (**b**) plant eukaryote cells showing the principal cell organelles. These organelles can also have internal plasma membranes. (**c**) A diagrammatic representation of the structure of prokaryotic cells such as bacteria (see also Fig. 7.11)

order to design new and better drugs it is necessary to have a detailed picture of the structures of cell membranes and walls as well as a comprehensive knowledge of the chemistry of the biochemical processes that occur in these regions. This chapter attempts to give a broad picture of the current relevance of plasma membrane and cell wall structure to drug action and design.

7.2 The plasma membrane

The currently accepted structure of membranes (Fig. 7.2), based on that originally proposed by S. J. Singer and G. L. Nicolson in 1972, is a fluid-like bilayer arrangement of phospholipids with proteins and other substances such as steroids and glycolipids either associated with its surface or embedded in it to varying degrees. This structure is an intermediate state between the true liquid and solid states, with the lipid and protein molecules having a limited degree of rotational and lateral movement.

X-ray diffraction studies have shown that many naturally occurring membranes are about 5 nm thick. Experimental work has also shown that a potential difference exists across most membranes due to the movement of ions through ion channels (see section 7.2.2) and pumps in the membrane, the intracellular face of the membrane being the negative side of the membrane. For most membranes at rest, that is, not undergoing cellular stimulation, the potential difference (*resting potential*) between the two faces of the membrane varies from −20 to −200 mV.

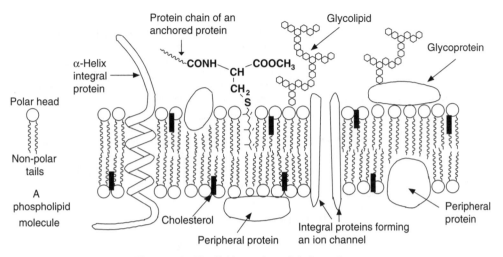

Figure 7.2 The fluid mosaic model of membranes

7.2.1 Lipid components

The lipid component of the plasma membrane of mammals is mainly composed of glyerophospholipids, sphingolipids and cholesterol. Each of the phospholipid molecules found in plasma membranes has a polar region (*hydrophilic head*) and a long non-polar hydrocarbon chain region (*hydrophobic tail*) (Fig. 7.3). The hydrocarbon chains usually contain between 14 and 24 carbon atoms, the most common being 16 and 18. They may be saturated or unsaturated and are usually unbranched. Living cells normally contain a higher proportion of unsaturated to saturated fatty acids. These lipid molecules are held together by weak hydrophobic bonding and van der Waals' forces, which give the structure liquid-like properties. The lipid molecules are aligned in the membrane so that their polar heads form the surfaces of the membrane that are in contact with either the extracellular or intracellular aqueous fluid. This means that the interior of a membrane is non-polar (hydrophobic) in nature. Consequently, non-polar compounds will diffuse into the membrane more readily than polar compounds. However, in order to be absorbed into the membrane, a compound must have some water solubility otherwise it will be repelled by the polar nature of the membrane's surface.

The glycosphingolipids are only present in small amounts in plasma membranes, but they are associated with a number of important cell functions. Specific glycosphingolipids appear to be involved in cell to cell recognition, tissue immunity blood grouping and also in the transmission of nerve impulses between neurons since large concentrations are found at nerve endings. The accumulation of certain gangliosides in tissue, due to a deficiency of the enzymes required for their degradation, is related to a number of genetically transmitted conditions, such as Tay-Sachs disease, Gaucher's disease, Krabbe's disease, Fabry's disease and Niemann-Pick disease.

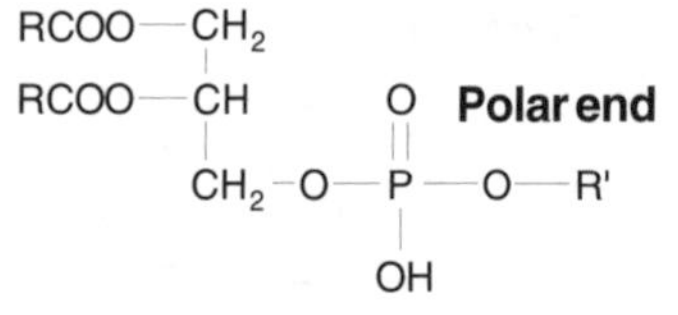

RCOO—CH$_2$
RCOO—CH   O   **Polar end**
       CH$_2$—O—P—O—R'
              OH

The phosphatidyl phospholipids

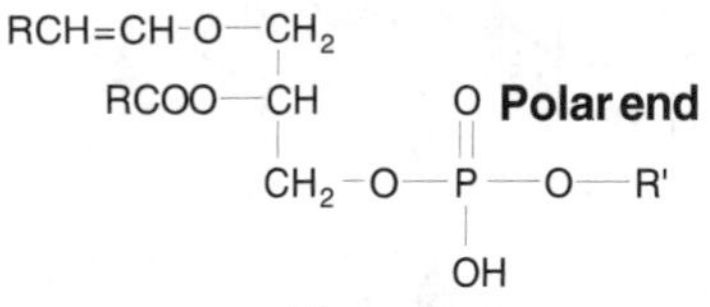

Key: R' =

a choline residue, α-Lecithins (phosphatidyl cholines [PC]), the main phospholipids found in mammals.

a glycerol residue, (phosphatidyl glycerol [PG]), the main plant phosphatidyl lipid. Phosphatidyl phosphate residues are attached to the 1,3 positions of the R' glycerol group (DPG).

an ethanolamine residue, α-Cephalins (phosphatidyl ethanolamine [PE]), a major mammal phospholipid.

a serine residue (phosphatidyl serine [PS]), small amounts in many tissues.

an inositol residue (phosphatidyl inositols [PI], small amounts present in all membranes.

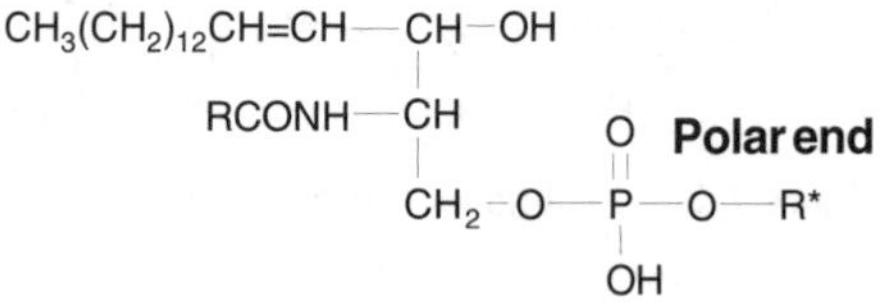

RCH=CH-O—CH$_2$
      RCOO—CH   O   **Polar end**
             CH$_2$—O—P—O—R'
                    OH

Plasmalogens

CH$_3$(CH$_2$)$_{12}$CH=CH—CH—OH
                  RCONH—CH   O   **Polar end**
                        CH$_2$—O—P—O—R*
                               OH

Sphingomyelins (nerve and muscle membranes of animals)

Figure 7.3 Some of the classes of phospholipids found in membranes. R groups are long-chain fatty acid residues, which may or may not be the same. R' can have one of the structures listed and R* is either a choline (SM) or an ethanolamine residue (SE). The C–C bonds of the hydrocarbon chains are mainly in the eclipsed conformation, giving these sections of the molecules a relatively straight structure

Cholesterol (Fig.7.4) molecules are found embedded in both the surfaces of animal plasma membranes and to a lesser extent in the membranes of their organelles. The molecule occurs in the membrane with its hydrocarbon side chain lined up alongside the hydrocarbon chains of the phospholipids. Unlike the phospholipids, the structure of cholesterol is relatively rigid and so it stiffens the membrane by hydrogen bonding to the adjacent oxygen atoms of the lipid esters. This helps to prevent the membrane from acting as a true liquid.

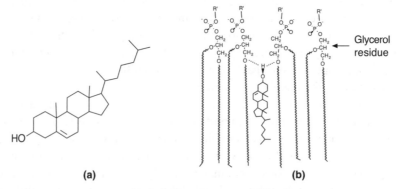

(a)   (b)

Figure 7.4 (a) Cholesterol and (b) a schematic representation of the hydrogen bonding of cholesterol in membranes. Note that the hydrocarbon chain attached to carbon-2 of the glycerol residue has an approximately 90° bend in its chain

7.2.2   Protein components

Cell membranes, with the exception of those of the Schwann cells of the myelin sheath of neurons (nerve cells), usually contain more protein than lipid in terms of the total dry mass of a membrane. These proteins are responsible for carrying out many of the active functions of membranes, such as acting as receptors and transportation routes for various substances across the membrane. They are normally classified as *integral* (*intrinsic*), *peripheral* (*extrinsic*) and *lipid-anchored* proteins. Those proteins whose chains either span the bilayer or are partly embedded in the bilayer usually have their N-terminal in the extracellular fluid and their C-terminal in the intracellular fluid.

Integral proteins are either deeply embedded in or pass right through the membrane. They can only be displaced from the membrane by disrupting its structure using solvents, disruptive enzymes and detergents. Integral proteins can be roughly divided into two types: those where most of the protein is embedded in the bilayer and those where part of the protein is embedded in the lipid layer but the greater part extends into either the extracellular or intracellular fluid or both. The former are often involved in the transport of species across the membrane (see sections 7.3.4 and 7.3.5). The latter usually has ogliosaccharides attached to the section protruding into the extracellular fluid. These oligosaccharides have a variety of functions. For example, the oligosaccharides of the protein glycophorin act as receptors for the influenza virus and also constitute the ABO and MN blood groups.

Both types of integral protein are *amphiphilic*, the surfaces of the protein segments in the extra- and intracellular fluids are hydrophilic in nature and the surfaces of the segment within the membrane are hydrophobic. The sections of proteins that cross a membrane are usually in the form of an α-helix with the polar groups orientated towards the centre of the helix and the non-polar groups on the outer surface of the helix.

Transmembrane integral proteins may cross a membrane several times. The sections of the protein that cross the membrane are referred to as *transmembrane domains* or *spans*. A single protein molecule may have several groups of transmembrane spans (Fig. 7.5a), which may be grouped together in a membrane in such a way that they form a water-filled pore through the membrane. However, it is more common for several transmembrane integral protein molecules to be grouped together to form a pore through the membrane. Each separate molecule in such a group is known as a *subunit*. For example, the protein molecules that form a pore large enough to allow the passage of K^+ ions have six transmembrane spans (Fig. 7.5b). Four of these subunits are grouped together to form the pore (Fig. 7.5c). Each subunit also has a smaller H5 domain that is embedded in the extracellular surface of the membrane so that it narrows the extracellular exit of the pore. The position of this domain explains why these pores are wider on the intracellular side of the membrane. This has a considerable bearing on the action of some drugs (see section 7.4.3). Pores that allow the passage of ions through a membrane are known as *ion channels*.

A wide variety of ion channels have been identified. Ion channels are usually fairly selective, allowing the passage of a specific ion but opposing the passage of other ions. Consequently, ion channels are referred to as Na^+, K^+ or Cl^-, etc. ion channels according to the nature of the ion allowed through the channel. Some channels are not permanently open

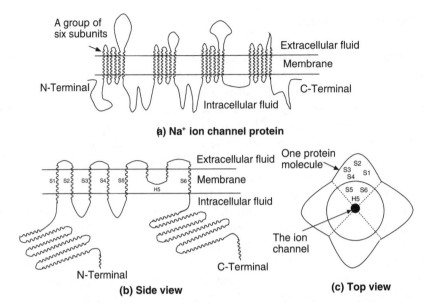

Figure 7.5 (**a**) A schematic representation of the transmembrane protein responsible for forming Na^+ ion channels in rat brain. The H5 domain in the membrane is not shown. The four groups of six transmembrane spans and one H5 unit form the Na^+ ion channel. (**b, c**) A schematic representation of the protein forming K^+ ion channels. Each subunit consists of six α-helical transmembrane spans plus a H5 domain. Four of these subunits are grouped together to form a K^+ ion channel.

but open briefly when changes occur in the conformations of the proteins forming the channel. These channels are referred to as *gated* channels. Opening of a channel may be initiated by either the binding of specific compounds to receptor (*ligand-gated channel*) or changes in the membrane's potential (*voltage-gated channels*). The movement of ions in and out of a cell through ion channels is an important process in cell function. The prevention of this movement is one of the ways in which drugs act on cell membranes (see section 7.4.3).

Peripheral proteins are probably attached to the surface of the membrane by electrostatic, hydrophobic and hydrogen bonding. These bonds are readily broken by metal chelating agents, changes in pH and ionic strength, which explains why many peripheral proteins are able to migrate over the surface of the membrane. Peripheral proteins have a variety of biological functions, including enzyme and antibody activity, whilst the intracellular peripheral proteins, actin and spectrin, form part of the cell's cytoskeleton. This is the network of protein filaments that is thought to determine the shape of the cell and controls its ability to move. It also controls the movement of organelles within the cell and is involved in cell division.

Lipid-anchored proteins are attached to the membrane by lipid molecules embedded in the lipid bilayer. These combined lipid–protein molecules, whose structures are made up of discrete protein and lipid regions, are often referred to and classified as *lipoproteins*. A wide variety of proteins are attached to membranes as lipid-anchored proteins. They include the *gag* proteins of certain retroviruses (see section 10.14.2), the transferrin receptor protein, yeast mating factors, the α-subunit of G proteins (see section 8.4.2), surface antigens (see section 10.15.2) and cell surface hydrolyases.

7.2.3 The carbohydrate component

Significant amounts of carbohydrate are associated with both the plasma and internal cell membranes. It takes the form of both small and large heterosaccharide chains consisting mainly of sialic acid, fucose, galactose, mannose, N-acetylglucosamine and N-acetylgalactosamine. The chains can also contain small amounts of glucose. Experimental evidence indicates that sialic acid is often the terminal sugar at the unattached end of the chain. These heterosaccharide chains, which are usually attached to lipids forming so-called *glycolipids* and surface proteins forming so-called *glycoproteins*, constitute an integral part of the structure of the membrane and are not easily removed from the cell surface. They have a variety of biological functions. For example, they are involved in cell–cell interaction and binding, tissue immunity receptor control and determine a person's blood group. The heterosaccharide chains also act as binding targets for infectious bacteria and as receptors for a variety of viruses. Consequently, carbohydrate chains are also potential targets for drug action.

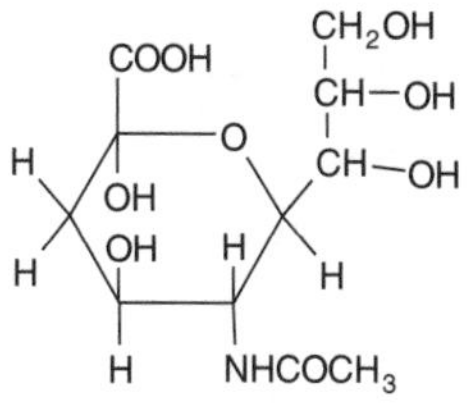

Sialic acid (N-acetylneuraminic acid) 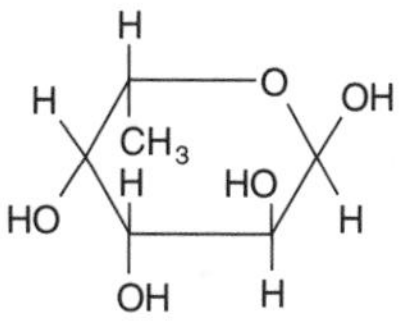   α-L-Fucose 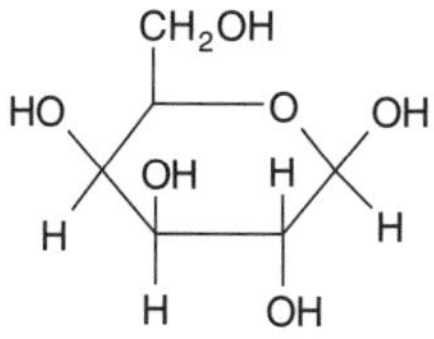   β-D-Galactose 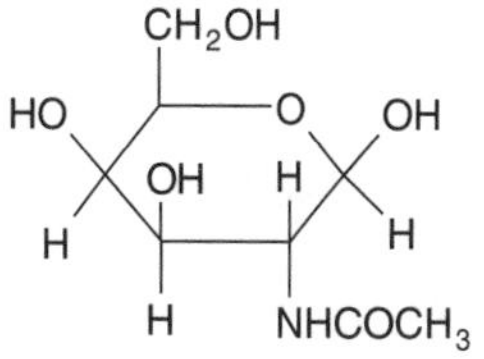   β-D-N-Acetylgalactose

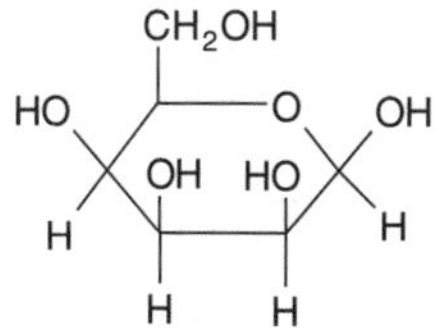

β-D-Mannose 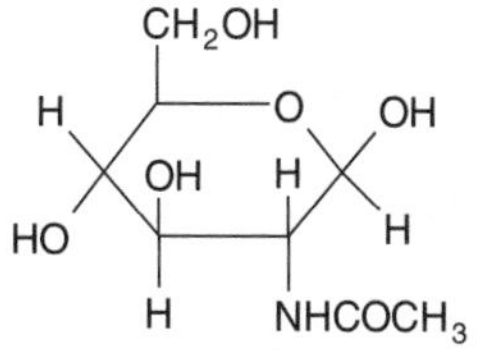   β-D-N-Acetylglucosamine 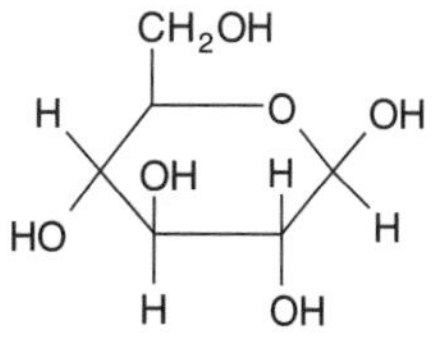   β-D-Glucose

7.2.4 Similarities and differences between plasma membranes in different cells

The chemical nature of the lipid components of the plasma membranes of different cells varies both in composition and concentration according to the cell function and the nature of the organism (Table 7.1). Differences in composition and concentration may also occur between plasma membranes and the organelle membranes in the same cell. For example,

Table 7.1 The most abundant components of membranes. The organism or organelle membrane is given in parentheses. (For abbreviations, see Fig. 7.3)

Compound	Animals	Bacteria	Fungi	Plants
Steroids	Cholesterol	Rare	Ergosterol (yeast and other fungi)	Stigmasterol Sitosterol Cytosterol
Phospholipids	PC, PE, DPG, SM	PC, PE, PI (cyanobacteria)		PC, PE, PI, DPG (chloroplasts)

some organelle membranes have a higher percentage of unsaturated fatty acyl and ether residues. However, the membranes of the organelles of plants and animals that have the same function have similar lipid compositions except that plant membranes contain stigmasterol or cytosterol whilst animal membranes contain cholesterol (Fig. 7.6). The various differences enable medicinal chemists to selectively target some types of cell.

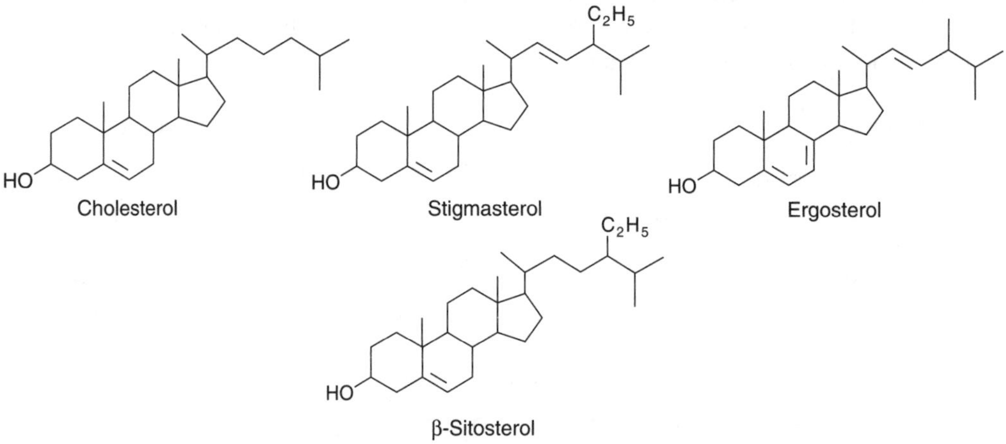

Figure 7.6 The principal steroids found in membranes

Membranes normally contain the same proteins. However, the amount of each protein shows considerable variation. For example, the concentrations of actin and myosin are much higher in muscle cells than in other types of eukaryotic cell.

7.2.5   Cell walls

Plants, fungi and most bacteria have a well-defined cell wall that covers the outer surface of the plasma membrane of the cell. This is a rigid structure that protects the fragile interior of the cell from damage by the surrounding environment as well as cementing cells together to form larger organisms such as plants. The cell wall consists mainly of a complex polypeptide–polysaccharide matrix generally referred to as a *peptidoglycan*. Its components depend on the nature of the organism: for example, in plants its polysaccharide components are mainly cellulose whilst in bacteria a wider variety of sugar residues are found. Cell walls are continually being renewed and so this process offers a potential target for drugs. For example, a number of antibiotics act by preventing cell wall synthesis (see section 7.4).

Bacterial cell walls

Bacteria have a high internal osmotic pressure. Their strong rigid cell walls maintain their shape and integrity by preventing either the swelling and bursting (lysis) or the shrinking of the bacteria when the osmotic pressure of the surrounding solution changes. It enables the bacteria to survive in hypotonic and hypertonic environments. The cell walls of bacteria are broken down by enzymes in their surrounding medium and so they are continuously being rebuilt. A number of antibiotics, such as penicillin, act by preventing this renewal of the cell wall, which results in the lysis of the bacteria because of its high internal osmotic pressure (see section 7.4.2).

Bacteria are commonly classified as being either Gram-positive or Gram-negative depending on their response to the Gram stain test. The cell walls of Gram-positive bacteria are about 25 nm thick and consist of up to 20 layers of the peptidoglycan (Fig.7.7). In contrast, the cell walls of Gram-negative bacteria are only 2–3 nm thick and consist of an outer lipid bilayer attached through hydrophobic proteins and amide links to the peptidoglycan. This lipid–peptidoglycan structure is separated from the plasma membrane by an aqueous compartment known as the *periplasmic space*. This space contains transport sugars, enzymes and other substances. The complete structure separating the cytoplasm from its surroundings is known as the *cell envelope*.

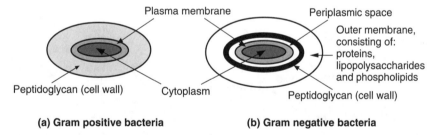

Figure 7.7 Schematic cross-sections of the cell envelopes of (**a**) Gram-positive and (**b**) Gram-negative bacteria

The peptidoglycans found in Gram-positive and Gram-negative bacteria are commonly known as mureins. They are polymers composed of polysaccharide and peptide chains that form a single, net-like molecule that completely surrounds the cell. The polysaccharide chains consist of alternating 1–4-linked β-*N*-acetylmuramic acid (NAM) and β-*N*-acetylglucosamine (NAG) units (Fig. 7.8). Tetrapeptide chains are attached through the lactic acid residues of the NAM units of these polysaccharide chains.

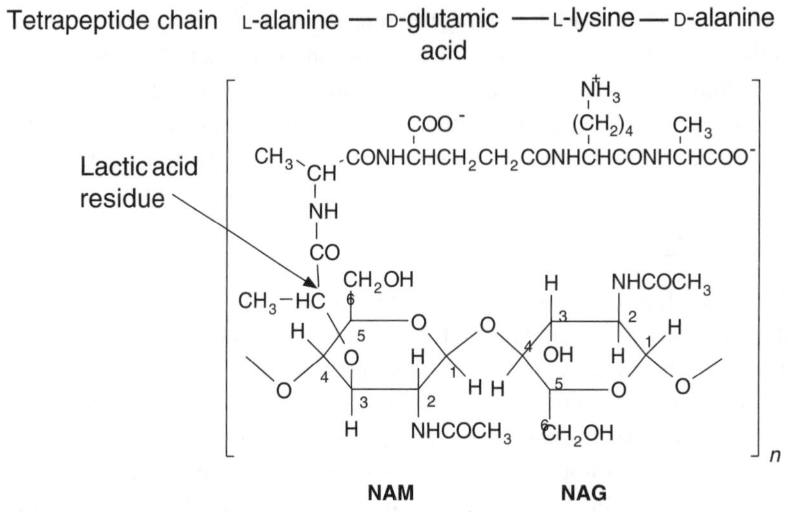

Tetrapeptide chain   L-alanine — D-glutamic ——L-lysine — D-alanine
                                    acid

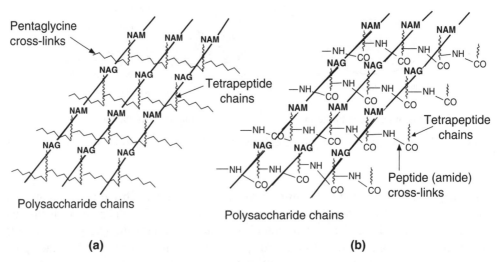

NAM   **NAG**

Figure 7.8 The structure of the monomer unit of the polyglycan of *Staphylococcus aureus*. The cross-links between chains are from the γ-amino group of the lysine to the carboxylate of the C-terminal D-alanine. The tetrapeptide can contain other amino acid residues

In Gram-positive bacteria, the tetrapeptide chains of one polysaccharide chain are cross-linked from the γ-amino group of lysine by pentaglycine peptide bridges to the terminal alanine of the tetrapeptide chains of a second polysaccharide chain to form a net-like polymer (Fig. 7.9a). The pentapeptide occasionally contains other residues. In Gram-negative bacteria the tetrapeptide chains are directly linked by peptide bond (amide group) bridges (Fig. 7.9b). The structure is denser than that found in the Gram-positive cell wall because the peptidoglycan chains are closer together.

Figure 7.9 A schematic representation of the structure of the peptidoglycan of (**a**) Gram-positive and (**b**) Gram-negative bacteria

7.2.6 Bacterial cell exterior surfaces

The exterior surface of Gram-positive bacteria is covered by *teichoic acids*. These are ribitol–phosphate or glycerol–phosphate polymer chains that are frequently substituted by alanine and glycosidically linked monosaccharides (Fig. 7.10). They are attached to the peptidoglycan by a phosphate diester link. Teichoic acids can act as receptors to bacteriophages and some appear to have antigenic properties.

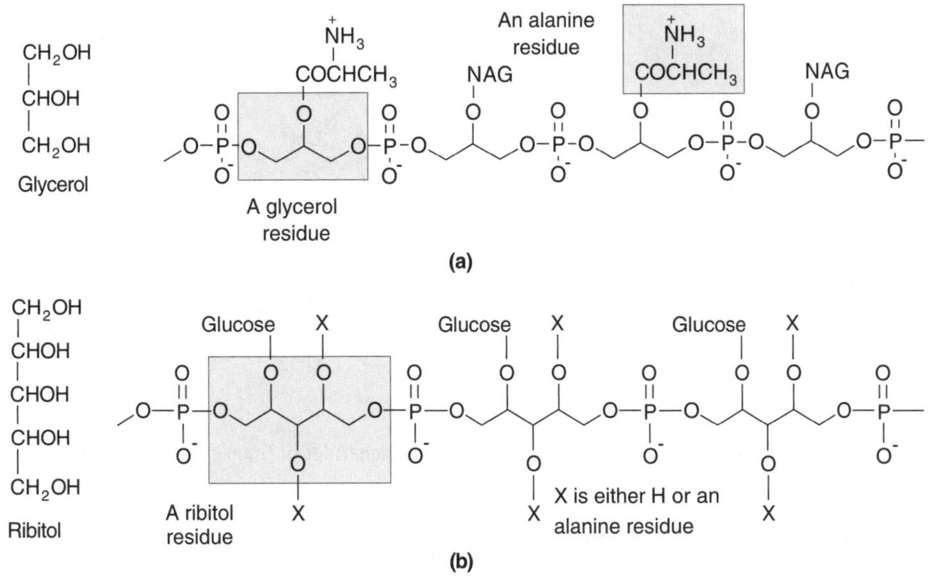

Figure 7.10 (a) A glycerol-based teichoic acid. (b) A ribitol-based teichoic acid. In many teichoic acids the monosaccharide residues are glucose and *N*-acetylglucosamine

The exterior surface of Gram-negative bacteria are more complex than those of the Gram positive bacteria. It is coated with *lipopolysaccharides*, which largely consist of long chains of repeating oligosaccharide units that are attached to the outer membrane by a core oligosaccharide (Fig. 7.11). These lipopolysaccharides often contain monosaccharide units such as abequose (Abe) and 2-keto-3-deoxyoctanoate (KDO) that are rarely found in other organisms. The repeating units, which are known at *O-antigens*, are unique to a particular type of bacteria. Experimental evidence suggests that they play a part in the bacteria's recognition of host cells. It is also believed that they enable the host's immunological system to identify the invading bacteria and produce antibodies that destroy the bacteria. However, a particular bacterial cell can have a number of different O-antigens and it is this diversity that allows some bacteria to evade a host's immune system.

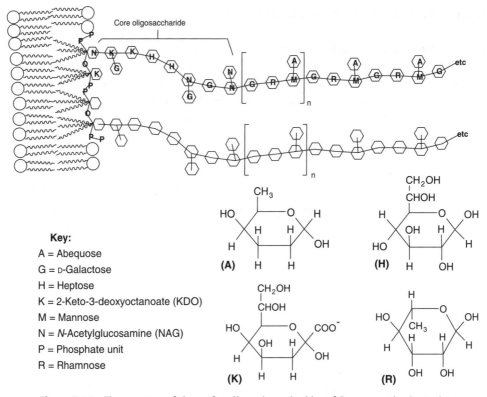

Figure 7.11   The structure of the surface lipopolysaccharides of Gram-negative bacteria

7.2.7   Animal cell exterior surfaces

The surfaces of animal cells play an essential part in the function of the cells. Numerous experiments have shown that cells are able to communicate with each other. For example, normal healthy cells stop growing when their surfaces come into contact. This is known as *contact inhibition*. One of the characteristics of cancerous cells is that they lose this contact inhibition and do not stop growing when they make contact with other cells. It is now thought that information is also passed by the interaction of glycoproteins on the cell surface with the proteins, especially the protoglycans, found in the extracellular fluid between the cells. Protoglycans are a family of glycoproteins in which the main carbohydrates are glycoaminoglycans. Surface carbohydrate moieties have also been shown to be involved in many cell processes ranging from the aggregation of cells to form organs to the infection of organisms by bacteria.

7.2.8   Virus

A discussion of the structure of virus is given in section 10.14.1.

7.2.9 Tissue

Cells are the basic building blocks for all known life forms. They occur in a huge variety of sizes and shapes and have a tremendously varied range of functions. Tissue is the biological structure formed by groups of cells adhering together. Its physical and biochemical properties will depend on the types of cell forming the tissue. However, all tissues have certain features in common, such as a supporting framework, blood vessels to supply nutrients and remove waste products and a nerve system to transmit relevant information to the cells forming the tissue. In addition, ancillary cells, such as macrophages, melanocytes and lymphocytes, enter the tissue from other sources, either during its formation or continuously during its life time. Furthermore, all the cells in a tissue have a limited life span and are continually dying and being replaced by new cells.

The space between cells varies depending on the type of tissue. For example, the gaps between the endothelial cells forming the stomach lining are very small in order to prevent the leakage of hydrochloric acid into the underlying tissues. These gaps are known as *tight junctions*. Tight junctions are also found between the endothelial cells lining the interior surfaces of many of the blood vessels of the circulatory system. This means that the gaps between these endothelial cells are too small to allow the passage of xenobiotics to their target sites in the underlying tissue. This forces drug molecules to pass through a large number of cell membranes in order to reach their sites of action. Consequently, it is important to ensure that potential drugs are able to cross plasma membranes and also, where necessary, penetrate cell walls. However, in some conditions, such as inflammation, the cells lining the blood vessels move apart and allow the leakage of unwanted fluid into the underlying tissues, causing them to swell (*oedema*). These conditions could also allow xenobiotics to penetrate to the underlying tissues.

The gaps between the endothelial cells that line the capillaries of the brain are extremely small. They form a structure that is known as the *blood–brain barrier* (*BBB*). The extremely small gaps in the BBB mean that almost all the substances entering the CNS and the brain from the blood have to pass through an endothelial cell. In other words, the substance must pass through an endothelial cell membrane, cross the cell and exit by passing across a second membrane. This makes it more difficult for polar substances to enter the brain unless they are actively transported (see section 7.3.5). In addition, the BBB also contains enzymes that protect the brain. Both of these factors must be taken into account when designing drugs to target the brain.

The gaps between endothelial cells can also be quite large. For example, the gaps (up to 1 µm) between the endothelial cells lining the capillaries of the liver and spleen are large enough to allow the passage of protein molecules.

7.2.10 Human skin

Human skin consists of three distinct regions: the epidermis, the dermis and a fatty subcutaneous layer (Fig. 7.12). The epidermis consists of dead cells that have migrated to

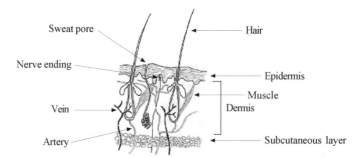

Figure 7.12 A representation of the basic of structure of skin

the surface, undergoing *apoptosis* en route. Apoptosis is the process by which cells undergo programmed cell death in order to maintain a healthy body. The outermost layers of the epidermis, the *stratum corneum*, consist of these dead cells laid down in a keratin matrix. This results in a strong structure with no gaps between the dead cells. It protects the body by preventing excess water loss and the entry of infectious organisms. However, both sweat glands and hair follicles penetrate the stratum corneum. The dermis underlying the epidermis consists of connective tissue in which are found various glands, hair roots and blood vessels. Beneath this lies a subcutaneous fatty layer.

The structure of skin affects the design of topically applied drugs in that the more lipid soluble a drug is, the more likely it is to penetrate the stratum corneum and be absorbed by the underlying blood vessels. Conversely, the more polar the drug, the less likely it is to penetrate the stratum corneum. Water-soluble drugs are only able to penetrate an intact stratum corneum through the hair follicles and sweat glands and so only very small amounts of these drugs can reach the underlying tissues. However, large quantities of water-soluble drugs will be rapidly absorbed through lesions of the skin because these drugs will now be in direct contact with the underlying tissue.

7.3 The transfer of species through cell membranes

7.3.1 Osmosis

Water is able to diffuse through membranes when there is a solute particle concentration gradient across the membrane. All the fluid-containing compartments of the body are either apparently or almost iso-osmotic. This iso-osmotic equilibrium is time dependent and so if there is a sudden change in composition of a fluid in a compartment a concentration gradient may be formed across any relevant membranes. This will result in a net movement of water from the area of lower solute particle concentration to the area of higher solute particle concentration, which can lead to cells either contracting (crenation) or swelling with perhaps subsequent lysis. Consequently, in sensitive areas of the body, such as the eye, the production of a solute particle concentration gradient by the introduction of a xenobiotic can result in unwanted tissue damage.

Changes in solute particle concentration may be brought about by the introduction of substances into a body compartment. However, metabolism of the substance may reduce the effect. For example, a 5 per cent solution of glucose is initially isotonic with plasma. Therefore, initially, introduction of this solution into a body compartment will not change the osmotic pressure of the system. However, as the glucose is metabolised the solution becomes hypotonic, water will flow into the compartment and the subsequent change in osmotic pressure could cause cell damage. Consequently, to reduce cell damage due to unwanted osmosis, it is important that liquid formulations of drugs are designed to be isotonic with the relevant body fluids.

7.3.2   Filtration

The channels formed by some integral proteins (see section 7.2.2) act like the pores in a filter paper and allow the passage of small molecules in and out of the cell. Filtration occurs when solute particles are forced through these channels in the membrane by the external pressure and no other agencies. Most pores have a diameter of 0.6–0.7 nm, which allows the passage of species with relative molecular masses up to about 100. Since drugs usually have relative molecular masses above this value, it means that most drugs cannot filter through a membrane. However, they may be carried through the gaps between cells by the pressure gradients produced by the heart. Even so, these gaps are narrow and so will not usually allow the passage of large protein molecules and so the passage of drugs bound to these proteins will also be restricted.

7.3.3   Passive diffusion

Passive diffusion is the process whereby a solute diffuses through a membrane from a *high to a low concentration* of solute without the membrane actively participating in that process. It is a major route for the transfer of *uncharged* and *non-polar* solutes, which readily dissolve in lipids, through membranes. The process initially requires the partition of the solute between the donating aqueous fluid and the lipids of the membrane, diffusion of the solute through the membrane from one side to the other and, finally, partition of the solute between the lipid membrane and the receiving aqueous fluid (Fig. 7.13). The

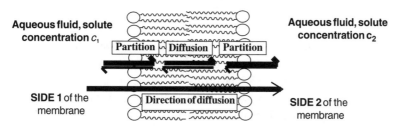

Figure 7.13   A diagrammatic representation of the mechanism of passive diffusion through a membrane from side 1 to side 2 where $c_1 c_2$

diffusion of a solute through a membrane is driven solely by that solute's concentration gradient between the aqueous fluids on either side of the membrane. It only ceases when the concentrations of the solute are the same on both sides of the membrane. However, in many cases, diffusion does not stop since the solute is carried away, metabolised or both when it has passed into the aqueous fluid on the receiving side of the membrane. This ensures that there is always a concentration gradient across the membrane.

The passage of more polar uncharged compounds by passive diffusion is restricted or prevented by most membranes since charged species are too polar to dissolve in the lipid membrane. However, small ions and small polar molecules can diffuse through the water-filled protein channels. Furthermore, the outer membranes of Gram-negative bacteria and mitochondria show a high non-discriminatory permeability to uncharged polar molecules because of the presence of large pores formed by the transmembrane protein porin. In addition, small anions will diffuse to positively and small cations to negatively charged areas, that is, down their respective electrical gradients. In this case, diffusion ceases when the electrical gradient is neutralised.

The ease of diffusion of polar compounds whose structures contain ionisable acidic and basic groups is dependent on both their degree of ionisation at physiological pH and the lipid solubility of the unionised form. The degree of ionisation may be calculated using the Henderson–Hasselbalch equation (see section 2.11). Since a membrane will not usually allow the passage of charged organic compounds, the lower the degree of ionisation, the more likely the compound will be transferred through a membrane. In addition, the compound must have an optimum solubility in the lipid membrane because if the solute is too soluble it will readily enter the membrane but will be reluctant to leave. An estimate of the lipid solubility of a compound may be obtained from its partition coefficient in a model system such as the octanol/water system (see section 2.12). In general, the lipid solubility of a solute will increase with an increase in the value of its partition coefficient.

The rate of diffusion, at constant temperature, of an uncharged species through a membrane is dependent on: the partition coefficients for the entry (K_1) and exit of the drug (K_2), the thickness of the membrane (x), the concentration gradient across the membrane ($c_1 - c_2$) and the surface area (S) of the membrane involved in the diffusion. The relationship between the rate of diffusion and these parameters is summarised in Fick's first law of diffusion:

$$J = \frac{DS(K_1c_1 - K_2c_2)}{x} \tag{7.1}$$

where J is the rate of appearance of the drug in the fluid on side 2 (Fig. 7.13) of the membrane, D is a constant known as the diffusion coefficient and c_1 and c_2 are the solute concentrations defined in Figure 7.13. The diffusion coefficient is a characteristic property of the diffusing substance/membrane system. It makes allowance for the chemical and physical states of the diffusing species and the membrane. The term D/x is known as the permeability coefficient (P) for the membrane and the diffusing substance. It is a measure of the solute's ability to move from the aqueous medium into the membrane. This deduction is supported by the fact that experimental evidence shows that for a number of compounds the value of P increases with an increase in the value of the non-polar solvent–water partition coefficient, that is, increases with increasing lipid solubility of the compound.

7.3.4 Facilitated diffusion

Facilitated diffusion involves specific carrier proteins known as *permeases*, which facilitate the transport of a chemical species through the membrane (Fig. 7.14). *It only occurs in the direction of the concentration gradient*, that is, from high to low concentration, and so does not require energy supplied by the cell. Consequently, it is a form of passive diffusion. The rate of transport will be rapid for low concentrations of the species but will slow down when its concentration reaches the point where it saturates all the available carriers. Facilitated diffusion appears to play a minor role in the transport of xenobiotics through a membrane but is the mechanism by which glucose is transported through most cell membranes.

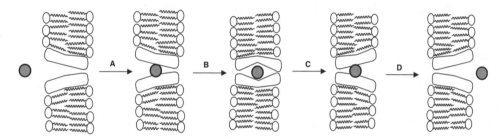

Figure 7.14 A schematic representation of carrier protein transport. A, the substrate binds to the carrier protein. B and C, the conformation of the carrier protein changes. D, the substrate is released on the other side of the membrane

7.3.5 Active transport

Active transport is the transport of a solute through a membrane by a so-called *carrier protein*. The solute combines with a specific protein, causing this protein to change its conformation. This change results in the transport of the solute from one side of the membrane to the other, where it is released into the aqueous medium. Active transport normally operates against the concentration gradient, the solute travelling from a low concentration to a high concentration. This process requires the cell to expend considerable quantities of energy.

The carrier proteins are highly selective, transporting solutes with specific chemical structures. They are involved in the transport of many naturally occurring compounds and so will also transport drugs with structures related to these natural products. For example, the drug levodopa, which is used to treat Parkinson's syndrome, is an amino acid. It is transported by the same active transport system as the naturally occurring amino acids tyrosine and phenylalanine. This type of structural relationship can be used in the approach to the design of new drugs.

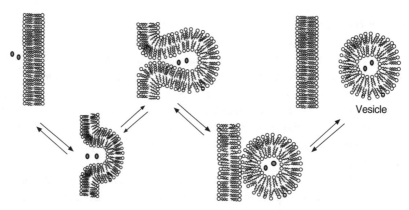

Phenylalanine   Levodopa   Tyrosine

The rate of active transport is dependent on the concentration of the solute at the absorption sites. Transport follows first-order kinetics at concentrations less than those required to saturate the available carriers but changes to zero order at concentrations above the carrier saturation point. Consequently, increasing a drug's concentration in the extracellular fluid above this saturation concentration does not increase the rate at which the drug is delivered to its site of action in the cell if the drug is transferred through the cell membrane by an active transport mechanism. One of the most important active transport systems is the sodium pump. This is a protein that moves Na^+ into a cell when the cell is deficient in those ions. A number of drugs, such as digitoxin (see Figure 1.5), act by interfering with active transport.

7.3.6 Endocytosis

Large molecules, fragments of dead tissue, whole bacteria and other particles that are visible under a microscope can pass through a membrane into a cell by a process known as *endocytosis* (Fig. 7.15). In endocytosis, contact of the substance being transported with the membrane causes the membrane to form a pocket (invaginate). The substance enters the pocket, which closes around the substance forming a vesicle (see section 2.13.3). The vesicle separates from the membrane and passes into the intracellular fluid. Once in the intracellular fluid the membrane of the vesicle has to disperse in order to release its

Figure 7.15 Schematic representations of endocytosis (left to right) and exocytosis (right to left). The extracellular fluid is on the left of each membrane whilst intracellular fluid is on the right of each membrane

contents into the intracellular fluid. Endocytosis has the potential to decrease the surface area of the membrane but this is balanced by coupling with exocytosis (see section 7.3.7).

Endocytosis can be classified into two general types: *phagocytosis* and *pinocytosis*. Phagocytosis (cell eating) is concerned with the transport of particles that are visible under a microscope through the membrane. Pinocytosis (cell drinking) is the same process, except the substances being transported are in solution. In both cases two different mechanisms have been discovered: *constitutive endocytosis* and *receptor-mediated endocytosis*. The former is a non-induced continuous process whilst the latter occurs mainly through receptors located in the membrane or in either clathrin (a polypeptide) or caveolin (a protein) coated indentations in the membrane. In this instance the cytoplasmic surface of the vesicle is coated with the appropriate protein. Receptor-mediated endocytosis is more specific and occurs more rapidly than constitutive endocytosis. It is responsible for the transport of low density lipoproteins, vitamins, insulin, iron, toxins, viruses and growth factors into a cell.

7.3.7   Exocytosis

Exocytosis is essentially the reverse of endocytosis. The substance that is to be transferred out of the cell is packaged in the form of a vesicle that fuses with the membrane. This fusion results in the formation of a pocket open to the extracellular fluid. The substance diffuses out of the pocket but on the extracellular side of the membrane (Fig. 7.15). The process requires the expenditure of energy by the cell and the presence of Ca^{2+} ions but details of the mechanism are not fully known. However, it has been shown to occur by either a *constitutive* or a *non-constitutive* pathway. The former pathway uses vesicles that are lined with the protein clathrin by the appropriate organelles, whilst the latter utilises unlined vesicles. Exocytosis has the potential to increase the total quantity of the membrane of a cell. This potential increase is balanced in healthy cells by the coupling of exocytosis with the removal of excess membrane by endocytosis.

Exocytosis is used to secrete digestive enzymes into the gastrointestinal tract and release many hormones and neurotransmitters into the extracellular fluid.

7.4   Drug action that affects the structure of cell membranes and walls

Most drugs act on the enzymes and receptors found in cell membranes and walls. These aspects of cell membranes are discussed in succeeding chapters. However, a number of drugs act by blocking ion channels, disrupting the structure of the cell membranes and walls or inhibiting the formation of cell membranes and walls. For example, the venoms of both the eastern diamondback rattlesnake and the Indian cobra contain the enzyme phospholipase A2. This enzyme catalyses the hydrolysis of the C2 fatty acid residue from phosphatidyl lipids. The phospholipid product of this hydrolysis acts as a detergent, breaking down the membranes of red blood cells and causing them to burst, usually with fatal results to the infected mammal.

$$
\underset{\substack{\text{RCO-O-CH} \\ |\\ \text{R'CO-O-CH}_2}}{\overset{\substack{\text{O} \\ || \\ \text{CH}_2\text{-O-P-O-X} \\ | \\ \text{O}^-}}{}} \xrightarrow[\text{Phospholipase A}_2]{\text{H}_2\text{O}} \underset{\substack{\text{HO-CH} \\ |\\ \text{R'CO-O-CH}_2}}{\overset{\substack{\text{O} \\ || \\ \text{CH}_2\text{-O-P-O-X} \\ | \\ \text{O}^-}}{}} + \text{RCOO}
$$

Acts as a detergent

In general, drugs that act by disrupting the structure of membranes and walls, or their synthesis, appear to act by:

• inhibiting the action of enzymes and other substances in the cell membrane involved in the production of compounds necessary for maintaining the integrity of the cell;

• inhibiting processes involved in the formation of the cell wall, resulting in an incomplete cell wall and leading to loss of vital cellular material and subsequent death of the cell;

• forming channels through the cell wall or membrane, making it more porous and resulting in the loss of vital cellular material and the death of the cell;

• making the cell more porous by breaking down sections of the membrane.

All microorganisms have plasma membranes that have characteristics in common. Consequently, drugs can act by the same mechanism on quite different classes of microorganism. For example, griseofulvin is both an antifungal and an antibacterial agent (see section 7.4.1). However, the membranes of prokaryotic cells exhibit a number of significantly different characteristics to those of eukaryotic cells (see section 7.1). It is these differences that must be exploited by medicinal chemists if they are to find new drugs to treat microbiological infestations. They also account for the selectivity of current drug substances when used on humans, animals and plants.

7.4.1 Antifungal agents

Fungal infections or *mycoses* may generally be divided into either superficial or systemic mycoses. Superficial mycoses affect the skin, nails, scalp and mucous membranes, while systemic mycoses affect internal tissues and organs. Since the middle of the twentieth century there has been an increase in both superficial and systemic mycoses. A significant degree of this increase is believed to be due to medical treatments, such as the use of antibiotics, radiotherapy, immunosuppressant drugs and steroids, which suppress a patient's immune system. It is believed that this supression allows the fungal microorganisms to flourish. Fungal infections caused in this manner are referred to as *opportunistic fungal infections*. Opportunistic infections also occur in conditions such as AIDS where the immune system is suppressed by the disease.

Fungal microorganisms are believed to damage the cell membrane, leading to a loss of essential cellular components. This may result in inflammation of the infected tissue, which in some cases may be severe. Antifungal agents counter myoses by both *fungistatic* and *fungicidal* action. Fungistatic action occurs when a drug prevents the fungi reproducing, with the result that it dies out naturally, whilst fungicidal action kills the fungi. The suffixes *-static* and *-cide* are widely used to indicate these general types of action.

Fungal microorganisms differ from other microorganisms in that they consist of eukaryotic cells with a chitin cell wall. This means that their chemical structures and biochemistry are similar to those of humans. Consequently, it is more difficult to design drugs that would selectively target these fungi. However, there are some differences that can be utilised. For example, human cell membranes contain cholesterol but those of fungi contain ergosterol. A number of antifungal drugs are believed to act by blocking the biosynthesis of ergosterol in fungi (see below).

Azoles

The azoles are a group of substituted imidazoles that exhibit fungistatic activity at nanomolar concentration and fungicidal activity at higher micromolar concentrations (Fig. 7.16). They are active against most fungi that infect the skin and mucous membrane. Azoles are also active against some systemic fungal infections, bacteria, protozoa and helminthic species.

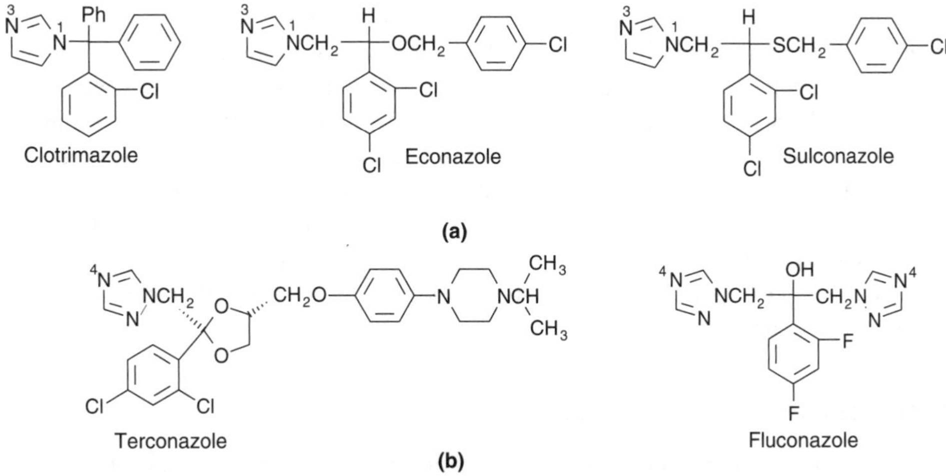

Figure 7.16 (a) Examples of the structures of some active 1,3-diazoles. Note the common structural features. (b) Examples of azoles based on 1,2,4-triazole ring systems

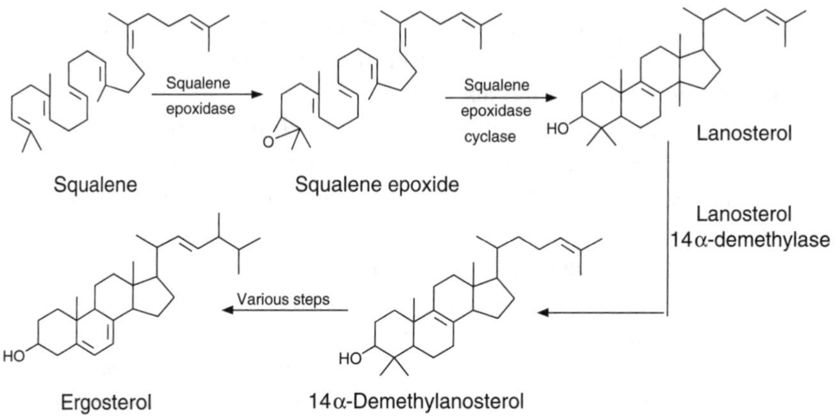

Figure 7.17 An outline of the biosynthesis of ergosterol

In common with most drugs, the azoles are believed to act at a number of different sites, all of which contribute to their fungicidal action. However, their main point of action is believed to be the inhibition of some of the cytochrome P-450 oxidases found in the membranes of the microorganisms. In particular, azoles have been linked to inhibition of the enzyme 14α-sterol demethylase (P-450$_{DM}$), which is essential for the biosynthesis of ergosterol, the main sterol found in the fungal cell membranes (Fig. 7.17). It is believed that nitrogen at position 3 of the imidazole rings (Fig. 7.16a) and nitrogen at position 4 of the triazole rings (Fig.7.16b) bind to the iron of the haem units found in the enzyme, thereby blocking the action of the enzyme. This appears to lead to an accumulation of 14α-methylated sterols such as lanosterol in the membrane, which is thought to increase the membrane's permeability, allowing essential cellular contents to leak causing irreversible cell damage and death. However, the precise details of the mode of action of azoles have yet to be fully elucidated. Azoles also inhibit the P-450 oxidases found in mammalian steroid biosynthesis, but in mammals much higher concentrations than those necessary for inhibition of the fungal sterol 14α-demethylases are usually required.

Structure–action studies have shown that a weakly basic imidazole or 1,2,4-triazole rings substituted only at the N-1 position are essential for activity. The substituent must be lipophilic in character and usually contains one or more five- or six-membered ring systems, some of which may be attached by an ether, secondary amine or thioether group to the carbon chain. The more potent compounds have two or three aromatic substituents, which in the more potent compounds are singly or multi-chlorinated or -fluorinated at positions 2, 4 and 6. These non-polar structures give the compounds a high degree of lipophilicity, and hence membrane solubility.

Allylamines and related compounds

Allylamines are synthetic derivatives of 3-aminopropene (Fig. 7.18) developed from naftifine. They are weak bases, their hydrochlorides being only slightly soluble in water. The allylamine group appears to be essential for activity.

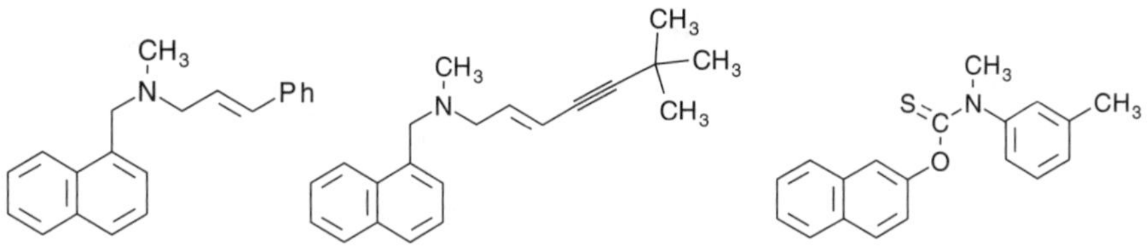

Figure 7.18 Examples of the structures of allylamines. Naftidine and terbinafine are used as fungicides to treat deratophytes and filamentous fungi but only have a fungistatic action against pathogenic yeasts

Allylamines are believed to act by inhibiting squalene epoxidase, the enzyme for the squalene epoxidation stage in the biosynthesis of ergosterol in the fungal membrane (Fig. 7.17). This leads to an increase in squalene concentration in the membrane with subsequent loss of membrane integrity, which allows loss of cell contents to occur. Tolnaftate, although it is not an allyl amine, appears to act in a similar fashion. However, allylamines do not appear to significantly inhibit the mammalian cholesterol biosynthesis.

Phenols

There are numerous phenolic antifungal agents (Fig. 7.19). They are believed to destroy sections of the cell membrane, which results in the loss of the cellular components and the death of the cell. The mechanism by which this destruction occurs is not known. Ciclopirox is not a phenol but appears to have a similar action. However, at low concentrations it has been shown to block the movement of amino acids into susceptible fungal cells.

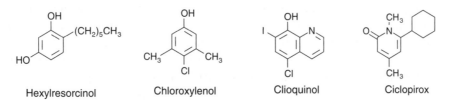

Figure 7.19 Examples of phenolic compounds used as antifungal agents

Antibacterial antifungal agents

A number of antibiotics are important antifungal agents. They are mainly polyenes such as amphopterin B, nystatin and natamycin (Fig. 7.20). However, the antibiotic griseofulvin (Fig. 7.20d) is also active. The smaller polyenes (26-membered ring) exhibit both fungistatic and fungicidal action at the same concentration. In contrast, the larger ring polyenes (38-membered ring) show fungistatic action at lower concentrations and fungicidal action at higher concentrations. This indicates some differences in their mode of action. All the polyenes are believed to act by binding to the cell membrane, causing

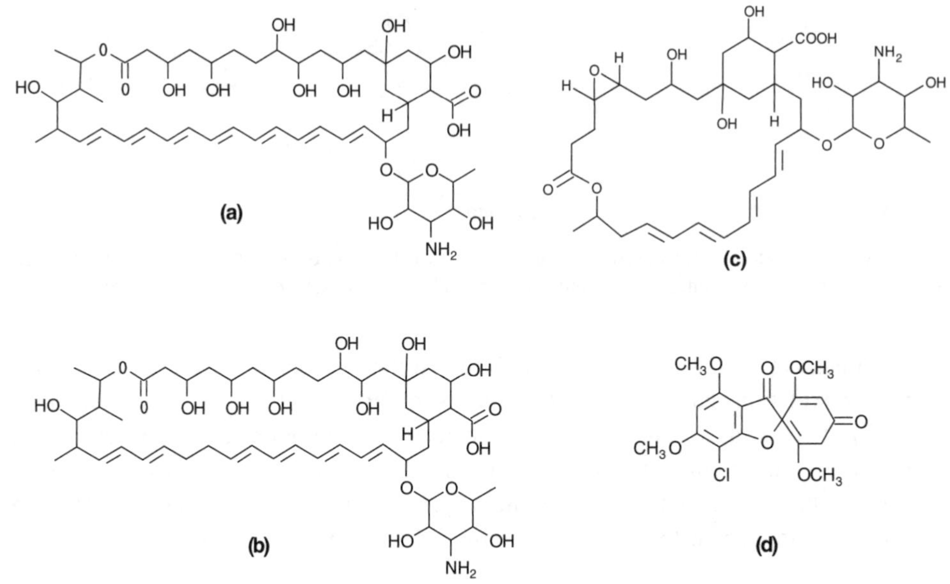

Figure 7.20 (**a**) Amphopterin B, first isolated from Streptomyces nodosus by Gold *et al.* (**b**) Nystatin, first isolated from Streptomyces noursei by Hazen and Brown. (**c**) Natamycin, first isolated from Streptomyces natalensis by Struyk *et al.* (**d**) Griseofulvin,first isolated from Penicillium griseofulvium by Oxford *et al.*

leakage of the cytoplasmic contents and cell lysis. It is thought that amphopterin B binds to the ergosterol found in the cell membranes of microfungi to form a transmembrane channel that allows the leakage of the cell contents. Unfortunately, it also appears to act in the same way with the cholesterol found in human cell membranes, which accounts for its toxic side effects in humans when administered by parenteral routes. The poor water solubility of polyenes means that they are difficult to administer by parenteral routes. However, amphopterin B may be administered parentally using micelle formulations (see section 2.13.1). Consequently, their difficulty of administration and often unpleasant side effects results in polyenes being used mainly in topical preparations.

Griseofulvin is a fungistatic agent. It acts by preventing the infestation of new tissue as that tissue is formed. This is a slow process and so its use is only successful if the patient sticks rigidly to the prescribed drug regimen. Griseofulvin is used to treat systemic infections and has few side effects. It has a poor water solubility and so its oral adsorption and hence its effectiveness will depend on how the dose is formulated.

7.4.2 Antibacterial agents (antibiotics)

Antibacterial antibiotics act at a variety of sites. However, in many cases, they act by either making the plasma membrane of bacteria more permeable to essential ions and other small molecules by ionophoric action or by inhibiting cell wall synthesis (see below). Those compounds that act on the plasma membrane also have the ability to penetrate the cell wall structure. In both cases, the net result is a loss in the integrity of the fungal cell envelope, which leads to irreversible cell damage and death.

Ionophoric antibiotic action

Ionophores are substances that can penetrate a cell membrane and increase its permeability to ions. They may be naturally occurring compounds such as the antibiotic gramicidin A produced by *Bacillus brevis* and valinomycin obtained from *Streptomyces fulvissimus*, or synthetic compounds like the crown and cryptate compounds (Fig. 7.21). However, ionophores transport ions in both directions across a membrane. Consequently, they will only reduce the concentration of a specific ion until its concentration is the same on both sides of a membrane. However, a number of drugs are believed to owe their action to the ionophoric transfer of essential substances out of the cell.

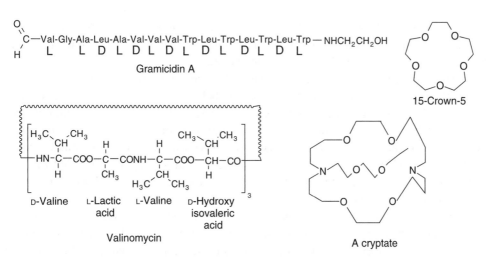

Figure 7.21 Examples of naturally occurring and synthetic ionophores

Ionophores operate in two ways:

1. They form channels across the membrane through which ions can diffuse down a concentration gradient (Fig. 7.22a).

2. They act as carriers that pick up the ion on one side of the membrane, transport it across, and release it into the fluid on the other side of the membrane (Fig. 7.22b).

The structure of each channel in a *Channel ionophore* is characteristic of the channel-former. For example, gramicidin forms a channel (tube) composed of two molecules whose N-terminals meet in the middle of the membrane. Each gramicidine molecule is in the form

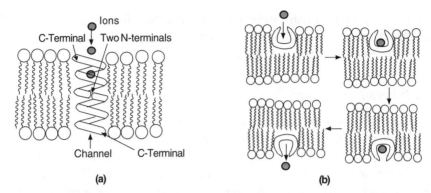

Figure 7.22 The general mode of action of ionophores in ion transport. (**a**) A channel formed by two gramicidin A molecules, N-terminal to N-terminal. (**b**) The sequence of events in the operation of a carrier ionophore such as valinomycin

of a left-handed helix, which results in the polar groups lining the interior of the channel. This facilitates the transfer of polar ions through the channel. A single gramicidin channel can allow the transport of up to 10^7 K$^+$ ions per second.

Carrier ionophores are specific for particular ions. For example, valinomycin will transport K$^+$ ions but not Na$^+$ or Li$^+$ ions. It is believed to form an octahedral complex with six carbonyl group oxygen atoms acting as ligands. The resulting chelation complex has a hydrophobic exterior that allows the complex to diffuse through the membrane. However, the rigid nature of the molecule coupled with its size makes the binding site of valinomycin too large to form octahedral complexes with the smaller Na$^+$ and Li$^+$ ions. Consequently, it is more energetically favourable for Na$^+$ and Li$^+$ ions to remain in solution as their hydrated ions.

Ionophores are mainly active against Gram-positive bacteria. However, until now, most of the compounds examined do not significantly differentiate between bacterial and mammalian membranes and so are of little clinical use. However, they are of considerable use as research tools in the investigation of drug action.

Cell wall synthesis inhibition

The cell walls of all bacteria are being continuously replaced because they are continuously being broken down by enzymes in the extracellular fluid. Antibacterial agents can inhibit this replacement biosynthesis of the cell wall at any stage in its formation. Investigations using *Staphylococcus aureus* have yielded a great deal of detail about the biosynthesis of its cell wall but there are still areas of the biosynthesis that have not yet been fully elucidated. Experimental investigations of the cell wall synthesis of other bacterial species suggest that similar routes are followed. A detailed knowledge of the route followed and the enzymes involved is an extremely useful prerequisite in the quest for new drug substances.

It is convenient to introduce the subject of antibacterial action due to inhibition of cell wall synthesis by dividing the synthesis into three stages:

1. The formation of precursor starting materials.

2. The formation of the peptidoglycan chains.

3. The cross-linking of the peptidoglycan chains.

However, it should be realised that not only can an antibiotic inhibit cell wall formation but it may also have other areas of action such as the plasma membrane of a bacterium. Furthermore, it is emphasised that the biochemical pathways discussed are a simplification based on experimental evidence. However, it is likely that the drugs act in the same manner on other susceptible bacteria.

Drugs that inhibit the formation of the starting compounds   A convenient starting point for cell wall synthesis is *N*-acetylglucosamine-1-phosphate (NAG-1-P), which is found in all life forms. This compound is believed to react with uridine triphosphate (UTP) to form uridine diphospho-*N*-acetylglucosamine (UDPNAG), one of the precursors of the peptido-glycan chain (Fig. 7.23). Some of the UDPNAG is further converted by a series of steps into the uridine diphospho-*N* acetylmuramic acid pentapeptide derivative (UDPNAM-pentapep-tide), the second precursor of the peptidoglycan polymer chain. Drug action can inhibit any of the steps in the formation of both UDPNAG and UDPNAM-pentapeptide. However, inhibition of the synthesis of the latter is likely to be potentially more rewarding since its formation requires a larger number of steps, which gives a wider scope for intervention. Drugs in clinical use that act by inhibiting different processes in this stage of cell wall synthesis are cycloserine and fosfomycin (Fig. 7.23).

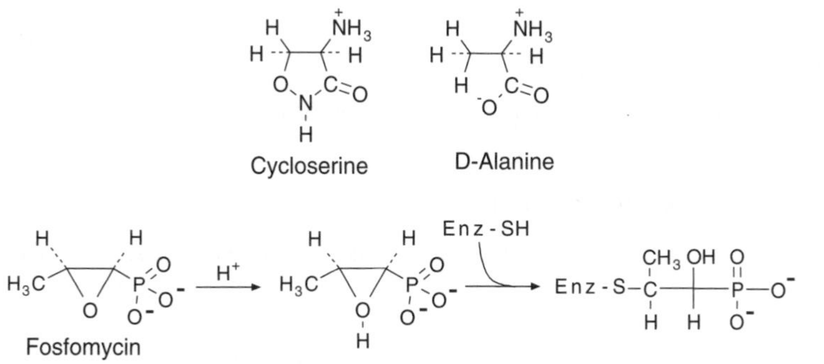

Figure 7.23   Cycloserine and fosfomucin

Cycloserine is a broad spectrum antibiotic produced by *Streptomyces orchidaceus*. The drug is used mainly as a second-line antitubercular agent. It enters the bacteria by active trans-port systems, which results in a high concentration in the bacterial cell, a primary require-ment for activity. D-Cycloserine inhibits both alanine racemase and D-alanyl-D-alanine

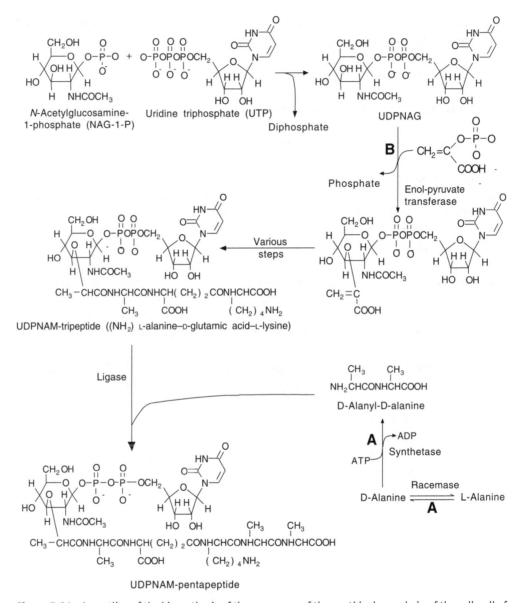

Figure 7.24 An outline of the biosynthesis of the precursors of the peptidoglycan chain of the cell wall of *Staphylococcus aureus*

synthetase, which blocks the conversion of the tripeptide to the pentapeptide at two places (A in Fig. 7.24). The affinity of the enzymes in *Staphylococcus aureus* for the drug has been found to be 100 times higher than its affinity for its natural substrate D-alanine. This affinity is believed to depend on the isoxazole ring, whose shape corresponds to one of the conformations of D-alanine. It is believed that the rigid structure of the isoxazole ring gives the drug a better chance of binding to the active sites of the enzymes than the more flexible structure of D-alanine.

Fosfomycin, produced by a number of Streptomyces species, is active against both Gram-positive and Gram-negative bacteria. However, it is used mainly to treat Gram-positive infections. The drug acts by inhibiting the enol-pyruvate transferase (B in Fig. 7.23) that catalyses the incorporation of phosphoenolpyruvic acid (PEP) into the UDPNAG molecule. However, the drug does not inhibit other enol-pyruvate transferases used to incorporate PEP in a number of other biosynthetic reactions. Consequently, it appears that the activity of the drug is due to it forming an inactive product with the enzyme. It has been suggested that this product is formed by the acid-catalysed nucleophilic substitution of the oxiran ring by the sulphydryl groups of the cysteine residues in the active site of the enzyme.

Drugs that inhibit the synthesis of the peptidoglycan chain   The sequence of reactions starting from UDPNAM-pentapeptide and UDPNAG to form the peptidoglycan chain (Fig. 7.25) is not completely known although the main stages have been identified. However, it is known that the reactions are catalysed by membrane-bound enzymes. A number of

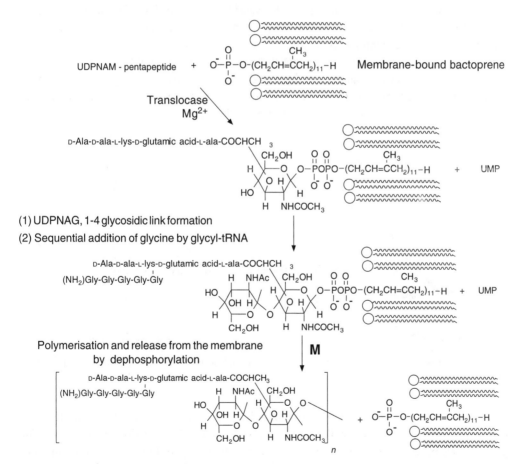

Figure 7.25 An outline of the formation of the peptidoglycan chains of *Staphylococcus aureus* from UDPNAM-pentapeptide and UDPNAG. UMP is uridine monophosphate

antibiotics, such as bacitracin, are believed to inhibit some of the stages of the biosynthesis of the peptidoglycan chains.

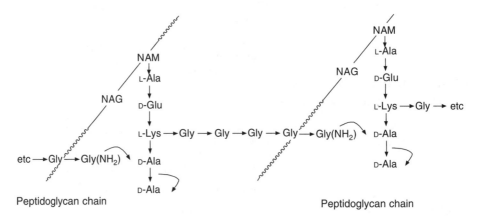

Bacitracin is a mixture of similar peptides produced from *Bacillus subtilis*. The main component of this mixture is bacitracin A, which is active against Gram-positive bacteria. However, its high degree of neuro- and nephrotoxicity means that the drug is seldom used, and then somewhat cautiously. Its main site of action appears to be inhibition of the dephosphorylation of membrane-bound phospholipid carrier bactoprene (step M in Fig. 7.25). Its action is enhanced by the presence of zinc ions

Drugs that inhibit the cross-linking of the peptidoglycan chains   The final step in the formation of the cell wall is the completion of the cross-links. This converts the water-soluble and therefore mobile peptidoglycans into the insoluble stationary cell wall. Investigations using *Staphylcoccus aureus* indicated that the cross-linking is brought about by a multistep displacement of the terminal alanine of the peptide attached to one peptidoglycan chain and its replacement by the terminal glycine of the peptide attached to a second peptidoglycan chain (Fig. 7.26). This reaction is catalysed by transpeptidases.

Figure 7.26   An schematic outline of the formation of the peptide cross-links in the formation of the cell wall of *Staphylcoccus aureus*

The β-lactam group of antibiotics inhibit cell wall synthesis by inhibiting the transpeptidases responsible for the cross-linking between the peptidoglycan chains. This group of antibiotics, named after the β-lactam ring that they all have in common, includes the widely used penicillins and the cephalosporins (Fig. 7.27). Both of these groups of

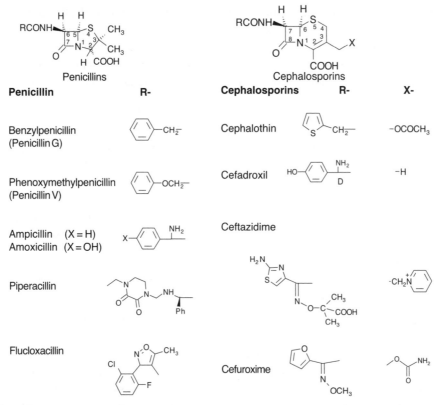

Figure 7.27 Examples of the range of penicillins and cephalosporins. The R residues of ampicillin, amoxicillin and ceftazidime have D configurations

β-lactam antibiotics are more effective against Gram-positive bacteria than Gram-negative bacteria. However, some cephalosporins, such as ceftazidime, which is administered intravenously, are very effective against Gram-negative bacteria.

The β-lactam antibiotics have to reach the plasma membrane of the bacteria before they can act. As the outer surfaces of Gram-positive bacteria are covered by a thin layer of teichoic acids (see section 7.2.6) they offer less resistance to drug penetration than Gram-negative bacteria, where the drug has to penetrate both the outer membrane and the periplasmic space (Fig. 7.7) before it can interfere with cell wall synthesis. In Gram-negative bacteria the drug diffuses through the outer membrane via *porin channels* formed by integral trimeric proteins. A large number of these channels, with diameters of about 1.2 nm, are found in each bacterial cell wall. However, not all porin channels are able to transport β-lactam drugs. Some bacterial genera such as *Pseudomonas* are resistant to β-lactam antibiotics because their porin channels will not allow the transport of these drugs. Once through the outer membrane the drug diffuses across the periplasmic space, which contains β-lactamases that can inactivate the drug. Gram-positive bacteria also produce these enzymes, which they release into the extracellular fluid. Finally, the drug penetrates to the outer surface of the plasma membrane where it binds to, and blocks the action of, the transpeptidases and other proteins involved in cell wall synthesis. The precise nature of the blocking mechanism has not yet been fully elucidated but appears to involve the β-lactam ring system. This ring system is very reactive and is easily decomposed by acid and base catalysed hydrolysis (Fig. 7.28), the rate of which depends on the structure of the penicillin.

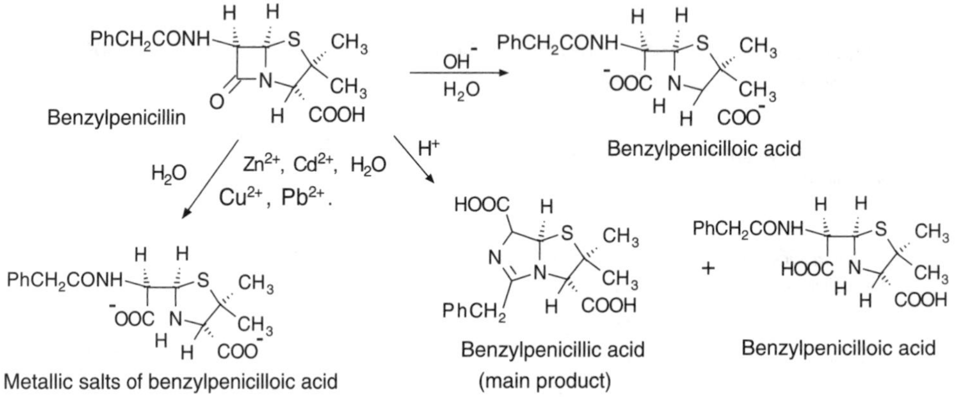

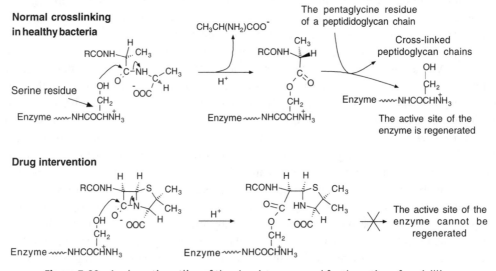

Figure 7.28   Some of the decomposition routes of the β-lactam ring of benzylpenicillin

The final stage in the formation of the cross-links between the peptidoglycan chains in bacteria is catalysed by a glycopeptide transpeptidase. It is believed that the hydroxyl group of a serine residue in this enzyme displaces the last alanine residue from the tetrapeptide chain. The displaced alanine diffuses away from the reaction site, allowing attack by the amino group of the terminal glycine of the pentaglycine chain on the alanine bound to the enzyme to complete the peptide linkage and regenerate the enzyme (Fig. 7.29). However, it is thought that since the geometry of the penicillins resembles that of the alanyl-alanyl unit

Figure 7.29   A schematic outline of the chemistry proposed for the action of penicillins

the bacteria mistakes the drug for its normal substrate. The β-lactam ring reacts with the enzyme to form a covalently bound acyl derivative. The 1,3-thiazolidine ring of this derivative is believed to prevent a pentaglycine unit from attacking the enzyme–acyl linkage and regenerating the active site of the enzyme.

Penicillins are unstable under acid conditions, the rate of decomposition depending on which penicillin is being considered. For example, piperacillin (Fig. 7.27) is so acid labile that it has to be administered by intravenous infusion. This means that the acid conditions of the stomach could reduce the amount of the drug reaching the general circulatory system and hence its effectiveness. The reactivity of penicillins under acid conditions can be attributed to the presence of a reactive four-membered lactam ring, the carbonyl group of which is readily attacked by nucleophiles under acid conditions.

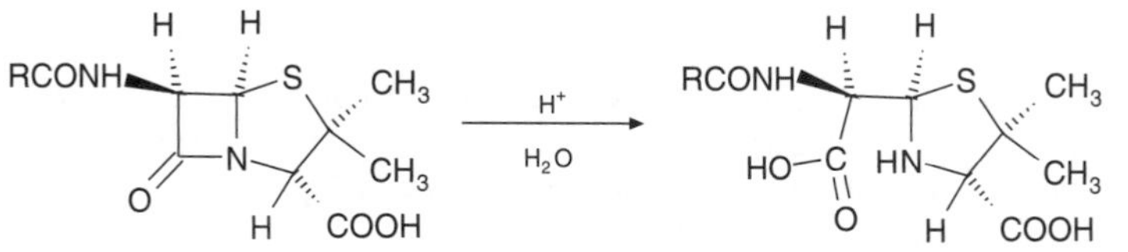

This reactivity is believed to be enhanced by the presence of the acyl side chain in a so-called *neighbouring group effect*. This group is believed to enhance the electronegative nature of the oxygen of the carbonyl group of the lactam, which makes the carbon of the lactam group more susceptible to nucleophilic attack (Fig. 7.30).

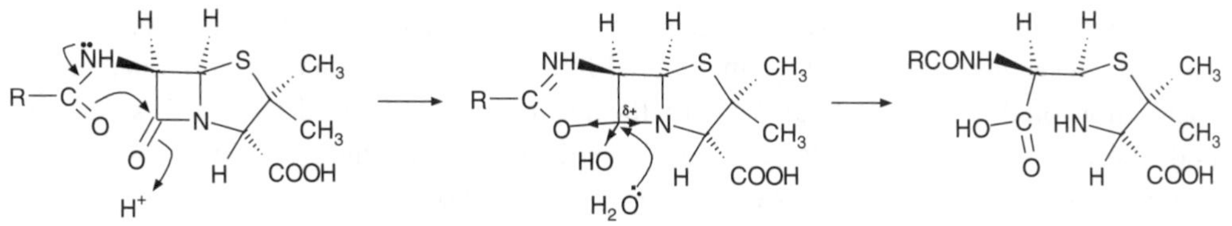

Figure 7.30 A possible mechanism for the enhancement of the reactivity of the carbonyl group of the lactam by a neighbouring group

Consequently, some acid-resistant penecillins have been produced by introducing an electron withdrawing group on the alpha carbon of the side chain. This reduces neighbouring group participation and, as a result, the reactivity on the lactam's carbonyl group (Fig. 7.31). This approach led to the development of penicillin V (phenoxymethyl-penicillin), amoxacillin and ampicillin.

Cephalosporins usually exhibit a greater resistance to acid hydrolysis than penicillins. However, the first generation of cephalosporins were not as potent as the penicillins but were active against a wider range of bacteria. However, their absorption from the GI tract is often poor and so they have to be given by injection. Consequently, cephalosporin C, first isolated by workers in Oxford University in the late 1940s, was used as the lead to develop more active analogues (Fig. 7.27). A large number of different cephalosporins are now in clinical use.

The relationship between the structures of β-lactams and their activity has been the subject of much discussion. Originally it was believed that the amide-linked side chain, the

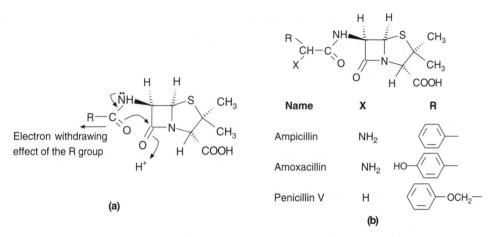

Figure 7.31 (a) It is believed that the electron withdrawing effect of the R group reduces the ability of the electrons of the carbonyl group of the amide link to influence those of the carbonyl group of the lactam. (b) Examples of penicillins in clinical use with an electron withdrawing R group

carboxylic acid at position 2 and the fused thio-ring systems were all essential for the pharmacological activity of β-lactam antibiotics. However, the discovery of β-lactams such as thienamycin, aztreonam (for synthesis see Fig. 15.13) and nocardin A (Fig. 7.32), whose structures do not contain all these functional groups, suggests that the β-lactam ring is the only essential requirement for activity.

The increasing number of bacteria resistant to β-lactam antibiotics is becoming a major problem. Resistance to penicillins and cephalosporins by some bacteria is mainly due to inactivation of the drug by hydrolysis of the lactam ring catalysed by the β-lactamases produced by that bacterium. However, in general, penicillins tend to be more susceptible than cephalosporins to hydrolysis catalysed by β-lactamases.

Both Gram-positive and Gram-negative bacteria produce β-lactamases. In the former case the enzyme is liberated into the medium surrounding the bacteria. This results in

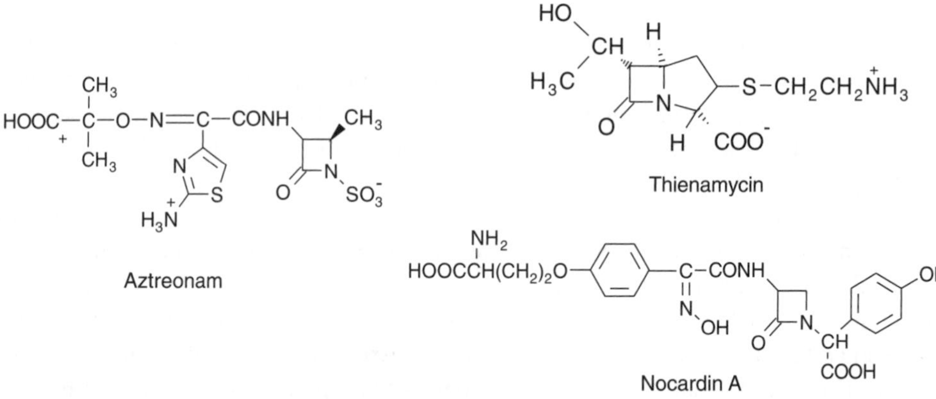

Figure 7.32 Examples of β-lactam antibiotics that do not contain a thio-ring system

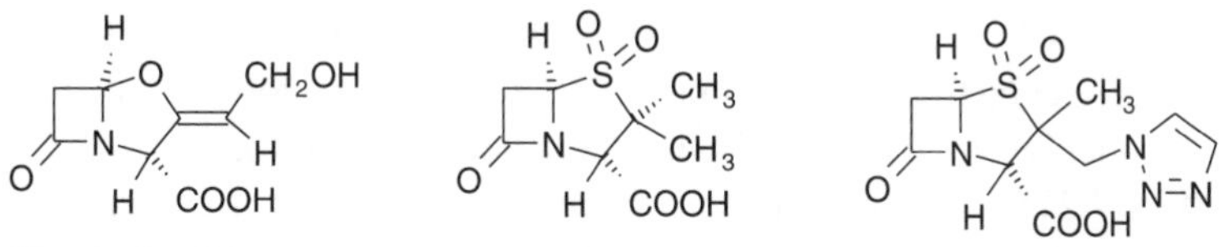

Figure 7.33 Examples of β-lactamase inhibitors used in penicillin dosage forms

inactivation of the penicillin, cephalosporin and other β-lactam drugs before the drug reaches the bacteria. However, with Gram-negative bacteria, the hydrolysis takes place within the periplasmic space. In addition some Gram-negative bacteria produce *Acylases* which can cleave the side chains of penicillins. Bacteria that have developed a resistance to β-lactam antibiotics are often treated using a dosage form incorporating a β-lactamase inhibitor such as clavulanic acid, sulbactam or tazobactam (Fig. 7.33) and Table 7.2).

Table 7.2 Examples of dosage forms containing β-lactamase inhibitors

Preparation	Penicillin	β-Lactamase Inhibitor
Co-amoxiclav	Amoxicillin	Clavulanic acid
Tazocin	Piperacillin	Tazubactum
Unasyn	Ampicillin	Sulbactam
Timentin	Ticarcillin	Clavulinic acid
Augmentin	Amoxicillin	Clavulanic acid

An alternative approach to the problem of bacterial resistance was the development of β-lactamase-resistant drugs. The strategy adopted by Beecham Research Laboratories for penicillins in 1961 was to use bulky substituents close to the labile lactam ring in order to use steric hindrance to prevent the drug binding to the enzyme's active site and undergoing lactam hydrolysis. This approach resulted in the discovery of methicillin. However, methicillin had a reduced potency compared with other penicillins and was acid labile to the extent that it could only be administered by injection. Further work by Beecham using an isoxazole ring in the side chain resulted in the discovery of flucloxacillin (Fig 7.27), oxacillin, cloxacillin and dicloxacillin. These drugs are used against Gram positive bacteria. They are inactive against Gram negative bacteria. However, oxacillin is also acid resistant.

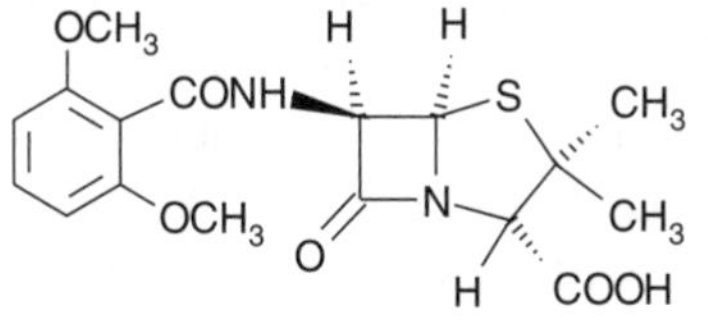

Methicillin

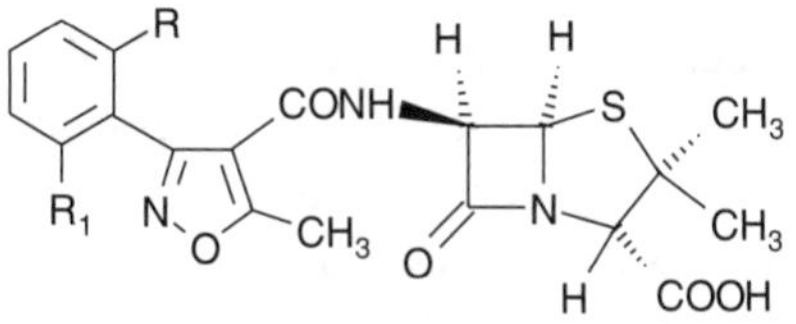

Cloxacillin   R = H, R₁=Cl
Dicloxacillin R = Cl, R₁=Cl
Oxacillin   R = H, R₁= H

The introduction of a bulky *syn* α-oximino group (-C=N-O-) side chain in so-called 'second-generation cephalosporins' improved their stability towards β-lactamases and esterases by probably sterically hindering the hydrolysis of the lactam. For example, cefuroxime (Fig. 7.27) is active against a wide range of Gram-positive and Gram-negative bacteria. Further development has resulted in the discovery of so-called 'third- and fourth-generation' cephalosporins that also exhibit enhanced resistance to β-lactamases. This increased resistance is believed to be partly due to the presence of a *syn* α-oximino side chain. In the third-generation cephalosporins, such as ceftazidine (Fig.7.27), the presence of an aminothiazole group is thought to increase the ease of transfer of the drug through the outer membrane of Gram-negative bacteria, which would account for their good Gram-negative activity and variable Gram-positive activity. The fourth-generation cephalosporins, such as cefepime and cefpirome, are zwitterions. They have a high potency against Gram-positive, Gram-negative and *Pseudomonas aeruginosa*.

Cefepime   Cefpirome

Polypeptide antibiotics A large number of polypeptide antibiotics have been discovered. They are active against a wide variety of microorganisms and operate by a range of mechanisms. Vancomycin and teicoplanin (teichomycin A_2) act by inhibiting the cross-linking of the peptidoglycan chains in bacterial cell walls. Other polypeptide antibiotics such as gramicidin (see section 7.4.2) and viomycin, which is used to treat TB, have different mechanisms of action.

Vancomycin (Fig. 7.34a) is a glycopeptide antibiotic that was isolated from *Streptomyces orientalis* in 1955 and inhibits the formation of the peptide links between the peptidoglycan chains. In spite of the extensive use of the drug, very little bacterial resistance to vancomycin has developed. It is mainly used for Gram-positive infections but is irritating on intravenous injection. Oral administration does not give useful blood levels. However, the drug is used orally to treat pseudomembranous enterocolitis caused by high concentrations of *Clostridium difficile* in the intestine.

The structure of vancomycin (Fig.7.34a) is based on a tricyclic ring system of aliphatic and aromatic amino acids with a disaccharide side chain. This structure is rigid with a peptide-lined pocket that has a strong affinity for D-ala-D-ala residues. Vancomycin inhibits cell wall synthesis by binding to the D-ala-D-ala end group of the pentapeptide chain of the peptidoglycan cell wall precursor. NMR spectroscopy and molecular modelling suggest that the D-ala-D-ala residue is multiple hydrogen-bonded to the vancomycin in this pocket. This inhibits the formation of the peptide cross-links between the polyglycan chains (Fig. 7.26), which results in a loss of bacterial cell wall integrity.

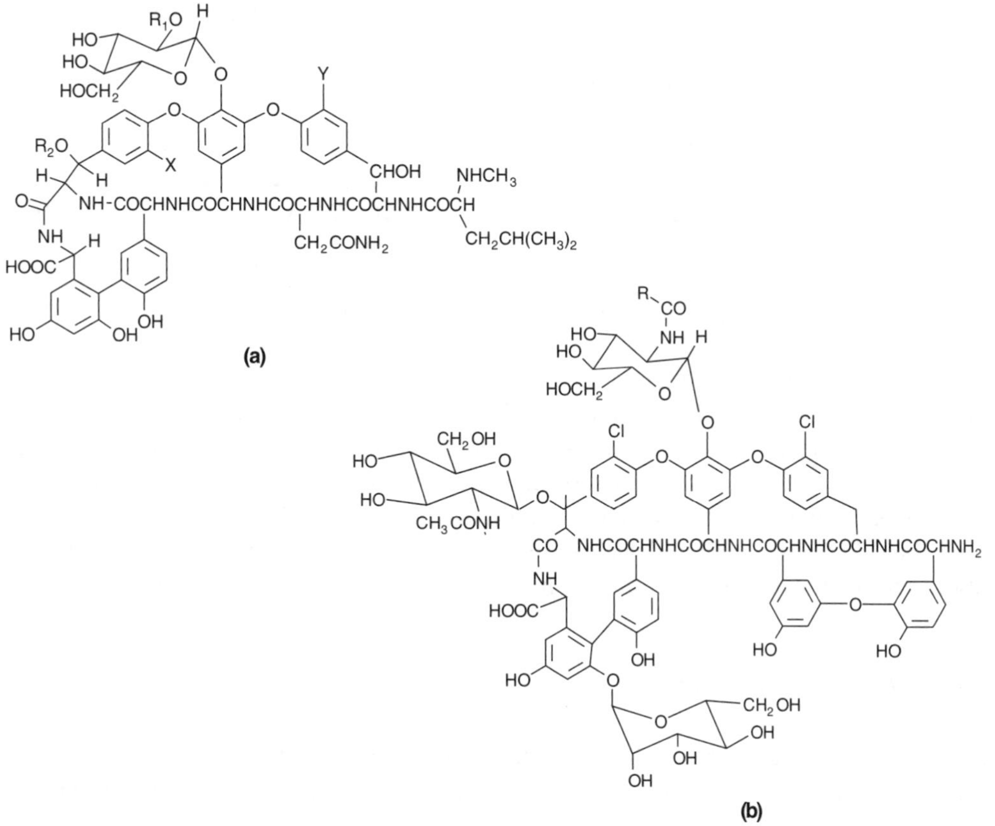

Figure 7.34   (**a**) Vancomycin and some related compounds. Vancomycin, X = Cl, Y = Cl, R_1 = α-vanco-saminyl, R_2 = H; Decaplanin, X = H, Y = Cl, R_1 = α-rhamnosyl, R_2 = α-epivancosaminyl and Orienticin A, X = Cl, Y = H, R_1 and R_2 = α-epivancosaminyl. (**b**) Teicoplanin: A_{2-1}, R = $CH_3(CH_2)_4CH=CHCH_2CH_2$-; A_{2-2}, R = $(CH_3)_2CH(CH_2)_5CH_2$-; A_{2-3}, R = $CH_3(CH_2)_7CH_2$-; A_{2-4}, R = $CH_3CH_2CH(CH_3)(CH_2)_5CH_2$-; A_{2-5}, R = $(CH_3)_2CH(CH_2)_6CH_2$-

Those bacteria that exhibit resistance to the drug appear to have replaced the D-ala-D-ala unit of the pentapeptide residue by a D-ala-D-lactate unit. This structural change is believed to reduce the number of hydrogen bonds between the drug and the D-ala-D-lactate of the peptidoglycan precursor by one. However, this small change is sufficient to prevent the drug operating efficiently.

Teicoplanin is a complex of five compounds (A_{2-1}, A_{2-2} to A_{2-5}) with similar structures (Fig.7.34b) that is active against a range of Gram-positive and Gram-negative bacteria. It was isolated in 1976 from *Actinoplanes teichomycetius*. The components of the mixture were separated in 1983 and their structures elucidated in 1984. Teicoplanin has good water solubility but significantly better lipid solubility than vancomycin. It is generally less toxic than vancomycin and its long half-life means that it need only be administered once a day. Its mechanism of action is believed to be identical to that of vancomycin.

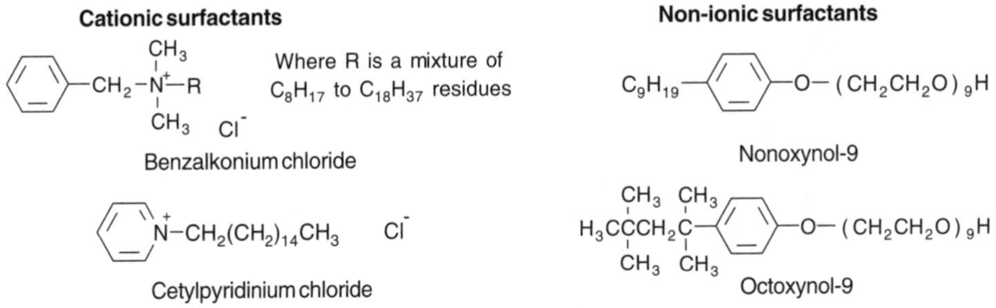

Cationic surfactants   **Non-ionic surfactants**

Where R is a mixture of
C_8H_{17} to $C_{18}H_{37}$ residues

Benzalkonium chloride   Nonoxynol-9

Cetylpyridinium chloride   Octoxynol-9

Figure 7.35   Surfactants used as antibacterial agents

Surface active agents (see section 2.13) disrupt cell membranes because they dissolve in both the aqueous extracellular fluid and the lipid membrane. This lowers the surface tension of the membrane, which allows water to flow into the cell and ultimately results in lysis and bactericidal action. In all cases a balance between the hydrophilic and lipophilic sections of the molecule is essential for action. Both cationic and non-ionic surfactants are used (Fig. 7.35). In addition, detergent surfactants, such as sodium dodecyl sulphate, are also used to remove proteins from cell membranes.

7.4.3   Local anaesthetics

Local anaesthetic agents block the nerves that transmit the sensation of pain. They mainly act on the cell membrane of the nerve cells (*neurons*, Fig. 7.36). This anaesthetic action is

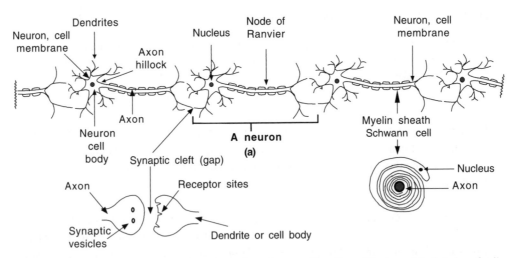

Figure 7.36   A schematic representation of a section of a nerve fibre. Nerve fibres consist of chains of cells known as neurons **(a)** where the axon of one neuron is separated from the end of either a dendrite or the cell body of a second neuron by the synaptic cleft or gap. The synaptic vesicles contain the chemical messengers (neurotransmitters) that transmit the nerve impulse by diffusing across the synaptic cleft to the receptors. Nerves consist of bundles of nerve fibres enclosed in various protective tissues

reversible. However, although local anaesthetic agents are administered locally they can also exert systemic effects since they can be transported from the point of application to other organs of the body by the blood stream.

Neurons are activated by chemical, electrical and mechanical stimuli. They transmit the information provided by these sources as an electrical signal known as the *action potential* or *nerve impulse* that which travels along the length of the cell at speeds of up to $120 \, ms^{-1}$ in myelinated axons but only $10 \, m \, s^{-1}$ in unmyelinated axons. At the synaptic knob the electrical signal releases chemical messengers, which cross the synaptic cleft and trigger the appropriate electrical signal in the next neuron.

The transmission of the signal along the neuron is due to successive changes in the potential difference across the membrane (*membrane potential*). For cells at rest, that is, cells that are undergoing no stimulation, the interior surface of the plasmic membrane is the negative side of the potential difference. The so-called *resting potential* of such cells can vary from -20 to about $-90 \, mV$. This potential is mainly due to the passive transport of small inorganic ions such as Na^+, K^+, Ca^{2+} and Cl^- ions through the membrane via ion channels (see section 7.2.2). The number per unit area of these ion channels varies: the highest density is at the nodes of Ranvier but there are considerably fewer in the myelinated sections of the neuron. Stimulation of a nerve is believed to result in the opening of the Na^+ gated channels in the plasma membrane. This allows a large number of Na^+ ions to flood into the neuron before the gated channels automatically close. These Na^+ ions neutralise the negative potential of the inner surface of the membrane, which allows K^+ ions to flow out of the neuron through ungated K^+ leak channels. This results in the membrane being depolarised to the extent that its potential reaches values of about $+40 \, mV$ relative to the exterior of the neuron. However, the automatic closure of the Na^+ channels after a brief time interval and the action of the sodium pump in removing the excess Na^+ ions from the neuron repolarise the membrane by deneutralising the negative charges on the inner surface of the plasmic membrane. These negative charges attract K^+ ions back into the axon so that the membrane returns to its resting state, the complete process of depolarisation and repolarisation usually being accomplished in about 4 milliseconds.

The initial action potential of a neuron either originates in the cell body or is generated by messages from the dendrites. It stimulates the sequence of changes described in the preceding paragraph. The action potential generated by these changes will trigger a similar series of changes in an adjacent section of the membrane, and so on. However, the automatic closing of the gated Na^+ channels means that the action potential usually moves along the axon away from the cell body. This also means that the nerve is ready to transmit a new action potential from the cell body within milliseconds of a previous transmission.

The main local anaesthetic agents in clinical use are basic compounds that can be broadly classified as either aromatic amide or ester derivatives (Fig. 7.37). Cocaine and other alkaloid local anaesthetics drugs are only used under strictly controlled conditions because of their addictive properties. However, both the alkaloid and basic aromatic amide and ester local anaesthetics agents are believed to act by blocking the Na^+ ion channels,

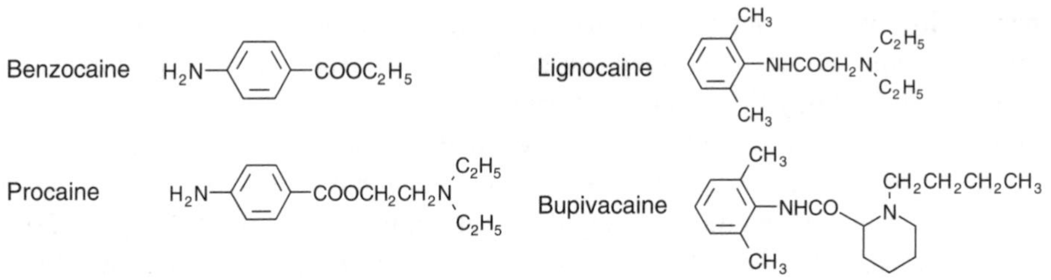

Benzocaine $H_2N-\langle\rangle-COOC_2H_5$

Lignocaine

Procaine $H_2N-\langle\rangle-COOCH_2CH_2N\begin{smallmatrix}C_2H_5\\C_2H_5\end{smallmatrix}$

Bupivacaine

Figure 7.37 Examples of ester- and amide-based local anaesthetic agents in clinical use

which prevents the initial influx of Na^+ ions into the neuron. This blocking action is thought to occur in three ways:

1. The agent blocks the external entry to the Na^+ ion channel.

2. The drug enters the channel and acts as a stopper, closing the channel about mid way along its length.

3. The drug binds to the proteins forming the channel and, as a result, distorts their structures to the extent that the channel will not allow the passage of Na^+ ions.

The structures of most of the local anaesthetic agents in clinical use contain basic groups such as tertiary amine groups, which are ionised at physiological pH:

$$RN(R')_2 + H^+ \rightleftharpoons RNH(R')_2$$

Since neutral molecules diffuse into the non-polar lipid membrane more easily than charged species, the local anaesthetics must enter the lipid membrane in its uncharged form. However, once inside the neuron, experimental evidence suggests that it is the charged form of the drug that binds to the receptor site(s). This is believed to occur by van der Waals' forces, dipole–dipole attractions and electrostatic forces (Fig. 7.38).

Alkaloid local anaesthetic agents are believed to act by blocking the external entries of the Na^+ ion channels, whilst it is thought that the main action of basic amide- and ester-based

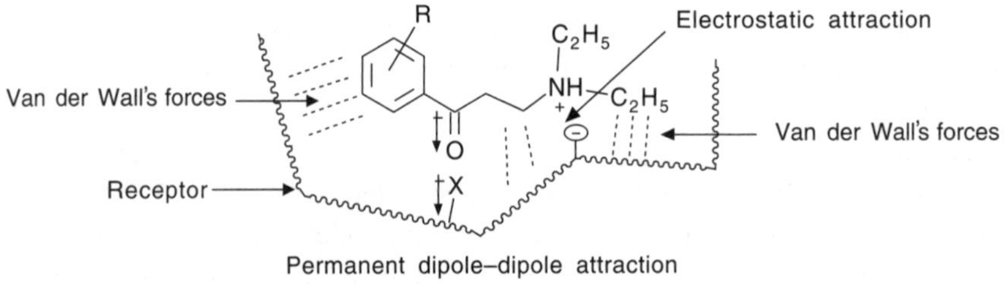

Figure 7.38 A schematic representation of the binding of ester-type local anaesthetic agents to a receptor site

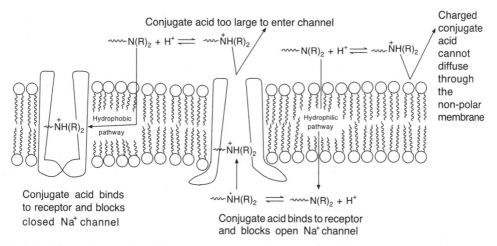

Conjugate acid too large to enter channel

$\text{www-N(R)}_2 + \text{H}^+ \rightleftharpoons \text{www-}\overset{+}{\text{N}}\text{H(R)}_2$

$\text{www-N(R)}_2 + \text{H}^+ \rightleftharpoons \text{www-}\overset{+}{\text{N}}\text{H(R)}_2$

Charged conjugate acid cannot diffuse through the non-polar membrane

$\text{-N}\overset{+}{\text{H}}\text{(R)}_2$ Hydrophobic pathway

Hydrophilic pathway

$\text{-}\overset{+}{\text{N}}\text{H(R)}_2$

Conjugate acid binds to receptor and blocks closed Na⁺ channel

$\text{www-}\overset{+}{\text{N}}\text{H(R)}_2 \rightleftharpoons \text{www-N(R)}_2 + \text{H}^+$

Conjugate acid binds to receptor and blocks open Na⁺ channel

Figure 7.39 A representation of the hydrophobic and hydrophilic routes for blocking both open and closed ion Na^+ channels

agents is to internally block these channels. In the latter case, experimental evidence shows that the drug follows either a hydrophobic or a hydrophilic route to its site of action. In the hydrophobic route, the drug passes into the membrane where it diffuses within the membrane to its site of action (Fig. 7.39). This route is particularly important since it offers a route for the blocking of closed Na^+ ion channels by local anaesthetic agents. In the hydrophilic route, the drug passes through the membrane into the intracellular fluid where it forms its conjugate acid, which is the active form of the drug. This species diffuses into the Na^+ ion channel to its site of action. Experimental evidence has shown that drugs following the hydrophilic route cannot enter Na^+ ion channels from the extracellular fluid since the channel entrances are too narrow. They have to enter via the wider interior entrances. Furthermore, experimental studies of the action of benzocaine on the giant squid axon have indicated that there are at least two sites of action of this drug, which implies that other local anaesthetic agents have more than one site of action inside the Na^+ channels.

The structure–activity relationships of local anaesthetic agents are well documented. Active amide- and ester-based drugs have hydrophilic and lipophilic centres separated by a structure containing an ester or amide group (Fig. 7.40). The hydrophilic centre normally contains a basic group, usually a secondary or tertiary amine, whilst the lipophilic centre usually consists of a substituted carbocyclic or heterocyclic ring system. As well as an ester or amide group, the structure linking the hydrophilic and lipophilic centres can contain a short hydrocarbon chain and nitrogen, oxygen and sulphur atoms. The most common links are based on hydrocarbon chains.

| Hydrophilic centre |—| Link group |—| Amide or ester group |—| Lipophilic centre |

Figure 7.40 The common features of the structures of most local anaesthetics in clinical use

The hydrophilic centre gives the local anaesthetic agent its water solubility whilst the lipophilic centre confers lipid solubility on the molecule. Water solubility is essential for transporting the drug to the membrane and, once in the neuron, to its site of action. Lipid solubility is also essential for the local anaesthetic agent to penetrate the non-polar lipid membrane and reach its site of action. It follows that the best local anaesthetic action will occur with compounds that have the correct balance between their hydrophilic and lipophilic centres. A measure of this balance is the degree of ionisation of the compound in water. Since local anaesthetic agents are bases, it is possible to relate the degree of ionisation to their pK_a values: the higher the pK_a value, the higher the compound's degree of ionisation. Compounds with a high degree of ionisation will be more soluble in water but less able to reach their site of action, because transport through the cell membrane will be more difficult due to a reduced lipid solubility. Most of the clinically useful local anaesthetic agents have pK_a values in the range 7.5–9.5.

The pH of the extracellular fluid can significantly affect the action of local anaesthetics whose mode of action depends on the molecule penetrating the membrane. Conditions, such as inflammation, that give the extracellular fluid an acidic pH will increase the ionisation of the basic local anaesthetic molecules, forming an equilibrium mixture with a higher proportion of charged molecules. These charged molecules do not penetrate the neuron, which reduces the proportion of neutral local anaesthetic molecules that can penetrate the membrane and act on the ion channels and reduces the effectiveness of the local anaesthetic action of the drug

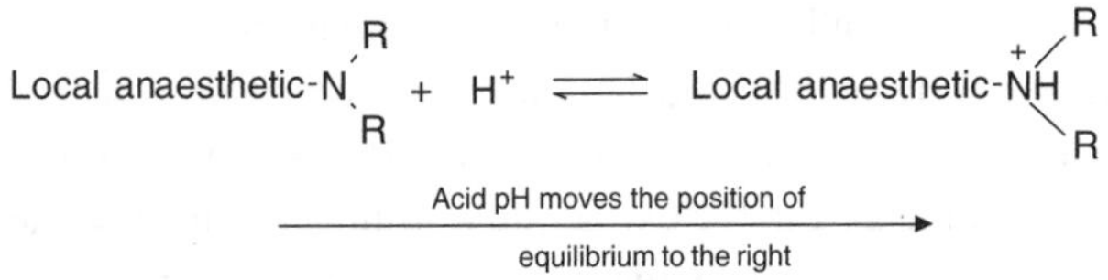

$$\text{Local anaesthetic-N} \overset{R}{\underset{R}{}} + H^+ \rightleftharpoons \text{Local anaesthetic-}\overset{+}{N}H \overset{R}{\underset{R}{}}$$

Acid pH moves the position of
_____→
equilibrium to the right

Substitution of the aromatic ring of the lipophilic centres of some ester-based local anaesthetics by electron donor groups resulted in increased local anaesthetic activity, whereas substitution by electron acceptor groups gave reduced activity. It has been suggested that electron donor groups increase the strength of the dipole of the carbonyl group by increasing the polarisation of the carbonyl group (Fig. 7.41). This subsequently increases the strength of the carbonyl–receptor binding. In contrast, electron acceptor substituents lower the strength of the carbonyl's dipole by decreasing the polarisation of the group. This reduces the strength of the carbonyl group's dipole, which in turn reduces the strength of the dipole–dipole binding to the receptor.

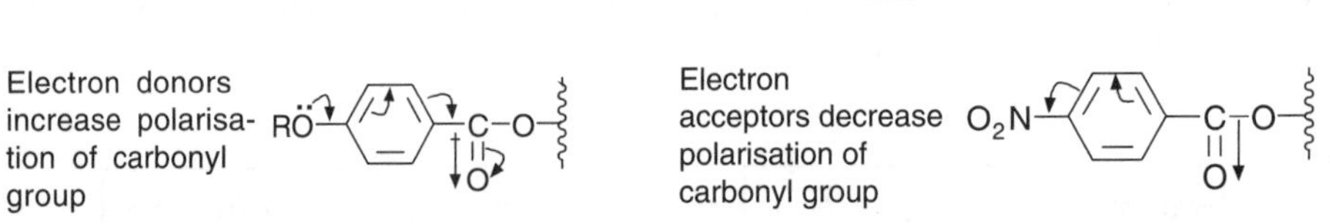

Electron donors increase polarisation of carbonyl group

Electron acceptors decrease polarisation of carbonyl group

Figure 7.41 The effect of electron donor and acceptor substituents on the binding of some local anaesthetics to their receptor sites

7.5 Questions

1 Briefly define the meaning of each of the following terms: (a) extracellular fluid, (b) cytoplasmic membrane, (c) organelle and (d) prokaryotic cell.

2 Outline the fluid mosaic model of the plasma membrane structure. Explain why the interior surfaces of most membranes are negatively charged with respect to the exterior surface.

3 Draw the general structural formulae of each of the following compounds: (a) SM; (b) PE; (c) PI (d) PC and (e) PS.

4 Distinguish carefully between the terms cell wall, plasma membrane and outer membrane in the context of bacteria. Ilustrate the answer by a description of the general chemical structures of each feature.

5 Describe the structure of the cell envelope of (a) Gram-positive bacteria and (b) Gram-negative bacteria.

6 Describe the main differences between each of the following:

(a) Fungicidal and fungistatic drugs.

(b) Exocytosis and endocytosis.

(c) Eukaryotic and prokaryotic cells.

(d) Active transport and facilitated diffusion.

7 Calculate the degree of ionisation of each of the following local anaesthetics at pH 7.4: benzocaine pK_a 2.5; cocaine pK_a 5.59 and procaine pK_a 8.15. Use the data to suggest a potential order of activity of these compounds. On what principle is this order of activity based?

8 Briefly explain how active transport could influence the design of a drug. Illustrate the answer by reference to one suitable example.

9 What are ionophores? How do they act and when does their action cease?

10 Suggest two reasons for Gram-negative bacteria being more resistant to treatment with penicillin than Gram-positive bacteria.

11 Describe the essential structural features that should be incorporated into the design of a potential local anaesthetic agent.

12 (a) Suggest a *feasible* chemical explanation of how a cephalosporin might act on bacteria cell walls.

(b) To what is the stability of third- and fourth-generation cephalosporins attributed?

13   A drug B is readily absorbed from the GI tract by passive diffusion. Predict the most likely effect of each of the structural changes on the absorption of the resulting analogues of B, assuming that each of the analogues is still absorbed by passive diffusion.

(a) Incorporating a carboxylic acid group into the structure of B.

(b) Incorporating an amino into the structure of B.

(c) Esterifying an existing carboxylic acid group in the structure of B.

8

Receptors and messengers

8.1 Introduction

Receptors are specific areas of certain proteins and glycoproteins that are found either embedded in cellular membranes or in the nuclei of living cells. Any endogenous or exogenous chemical agent that binds to a receptor is known as a *ligand*. The general region on a receptor where a ligand binds is known as the *binding domain*.

The binding of a ligand to a receptor depends on the ligand having a significant part of its stereoelectronic structure complementary to that of the receptor's stereoelectronic structure (see section 1.3). Ligands that fulfil this requirement will bind to the receptor and may cause either a positive or negative biological response. For example, positive responses can result in an immediate physiological response, such as the opening of an ion channel, or lead to a series of biochemical events that may result in the release of so-called *secondary messengers* (see section 8.4.2), such as cyclic adenosine monophosphate (cAMP). These secondary messengers promote a sequence of biochemical events that result in an appropriate physiological response (Fig.8.1). Alternatively, the binding of the ligand to the receptor may prevent a physiological response by either initiating the inhibition of an associated series of biological events (see section 8.4.2) or simply preventing the normal endogenous ligand from binding to a receptor. For example, β-blockers such as propranolol act by blocking the β-receptors for adrenaline. In all relevant cases, the mechanism by which any message carried by the ligand is translated through the receptor system into a tissue response is known as *signal transduction*.

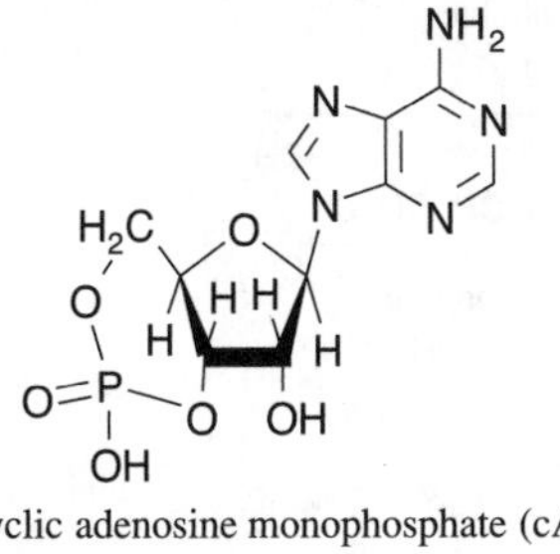

Cyclic adenosine monophosphate (cAMP)

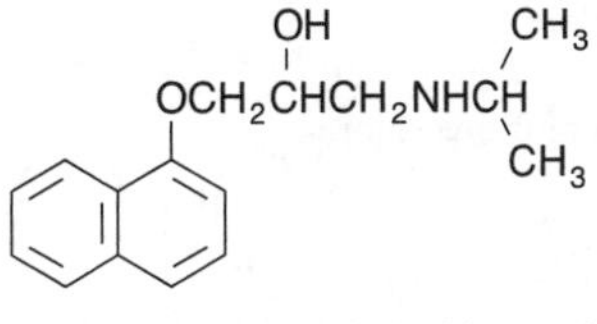

Propranolol (the *S* isomer is 100 times more active than the *R* isomer)

Medicinal Chemistry, Second Edition   Gareth Thomas
© 2007 John Wiley & Sons, Ltd

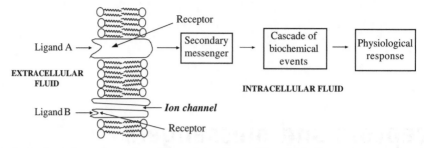

Figure 8.1 A schematic representation of some of the effects of ligands binding to receptors. The binding of ligand A to the extracellular surface of the receptor results in the activation of a secondary messenger, which is followed by a cascade of biochemical events and the appropriate physiological response. The binding of ligand B causes the ion channel to open

The binding of a drug to a receptor either inhibits the action of the receptor or stimulates the receptor to give the physiological responses that are characteristic of the action of the drug. Drugs that bind to a receptor and give a similar response to that of the endogenous ligand are known as *agonists*, whilst drugs that bind to a receptor but do not cause a response are termed *antagonists*. However, drugs are not the only xenobiotic that can bind to a receptor: viruses, bacteria and toxins can also bind to the receptor sites of specific tissues.

8.2   The chemical nature of the binding of ligands to receptors

The forces binding ligands to receptors cover the full spectrum of chemical bonding, namely: covalent bonding, ionic bonding and dipole–dipole interactions of all types, including hydrogen bonding, charge-transfer bonding, hydrophobic bonding and van der Waals' forces (Fig. 8.2). The bonds between the ligand and the receptor are assumed to be formed spontaneously when the ligand reaches the appropriate distance from its receptor

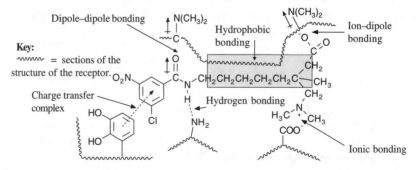

Figure 8.2   Theoretical examples of some of the common forms of bonding (excluding van der Waals' forces) found in drug–receptor interactions

for bond formation. To achieve this situation the ligand is transported by either diffusion or a transport protein to the receptor.

Covalent bonding is by far the strongest form of bond between a ligand and a receptor. It usually forms an irreversible link between the drug and the receptor. Consequently, except in cancer therapy and the inhibition of certain enzymes, covalent bond formation is seldom found in drug action. Ionic or electrostatic bonding is an important form of bonding between ligands and receptors since many of the functional groups on the receptor will be ionised at physiological pH. Ionic interactions are usually reversible. They are effective at distances that are considerably greater than those required by other types of bonding but their strength is inversely proportional to the square of the distance between the charges (the inverse square law). Electrostatic attractions in the form of ion–dipole attractive forces, dipole–dipole interactions and hydrogen bonding usually form weaker bonds than ionic bonds. However, they are important because of the large numbers of these bonds that are usually formed between the drug and the receptor.

Charge-transfer complexes may also be formed when an electron donor group is adjacent to an electron acceptor group. In this situation experimental evidence suggests that the donor may transfer a portion of its charge to the acceptor. As a result, one compound becomes partially positively charged with respect to the other and a weak electrostatic bond is formed (Fig. 8.3). The precise nature of this type of bonding has not been fully elucidated but donors are usually π-electron-rich species and chemical moieties with lone pairs of electrons.

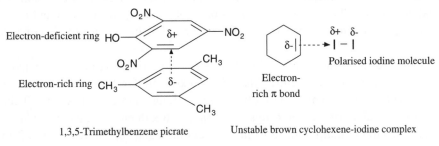

Figure 8.3 Examples of charge-transfer compounds

Hydrophobic bonding is a very weak form of bonding that occurs when non-polar sections of molecules are forced together by a lack of water solubility (Fig. 8.4a). The precise nature of hydrophobic bonding is not known but the formation of hydrophobic bonds leads to a fall in the energy of the system and a more stable structure. Some workers do not believe that hydrophobic effects exist.

London dispersive forces may also contribute to the bonding between a drug and a receptor. These forces are very weak dipole–dipole interactions due to the formation of transient dipoles within a structure. Transient dipoles are time dependent. They arise because the electron distribution in a molecule varies with time, giving an uneven distribution of electrons that results in a temporary charge distribution and a transient dipole within the molecule (Fig. 8.4b).

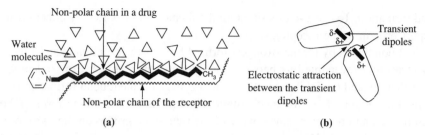

Figure 8.4   Hypothetical examples of **(a)** hydrophobic bonding and **(b)** London dispersive forces

The binding of many drugs to their receptors is by weak reversible interactions.

$$\text{Drug} + \text{Receptor} \rightleftharpoons \text{Drug--Receptor} \tag{8.1}$$

This means that the binding of a drug to its receptor is concentration dependent. As the concentration of the ligand in the extracellular fluid increases, the equilibrium (8.1) will move to the right and the drug will bind to the receptor. However, when the concentration of the drug in the extracellular fluid falls, the equilibrium (8.1) will move to the left and the drug–receptor complex will dissociate. Consequently, drugs and endogenous ligands become ineffective as soon as their concentrations fall below a certain limit because an insufficient number of receptors are being activated by these ligands. Reduction of both drug and endogenous ligand concentration is brought about by metabolism and excretion. Consequently, both of these processes will have a direct bearing on the duration of action of a drug.

Drugs that form strong bonds with their receptors do not readily dissociate from the receptor when their concentrations in the extracellular fluid fall. Consequently, drugs that act in this manner will often have a long duration of action. This is a particularly useful attribute for drugs used in the treatment of cancers, where it is particularly desirable that the drug forms irreversible bonds to the receptors of tumour cells.

8.3   Structure and classification of receptors

Receptors are classified according to function into four so-called *superfamilies* of receptors (Table 8.1). The members of a superfamily will all have the same general structure and general mechanism of action. However, individual members of a superfamily tend to exhibit variations in the amino acid residue sequence in certain regions and also the sizes of their extracellular and intracellular domains.

Each of the superfamilies is subdivided into a number of types of receptor whose members are usually defined by their endogenous ligand. For example, all receptors that bind acetylcholine (ACh) are of the cholinergic type and those that bind adrenaline and noradrenaline are of the adrenergic type. These sub-types are further classified either

Table 8.1 The four receptor superfamilies. The rectangles represent α-helices and the single lines represent polypeptide chains

Family	Endogenous ligands	Examples	General structure
1.	Fast neurotransmitters	nAChR, GABA$_A$ receptor, glutamate receptor	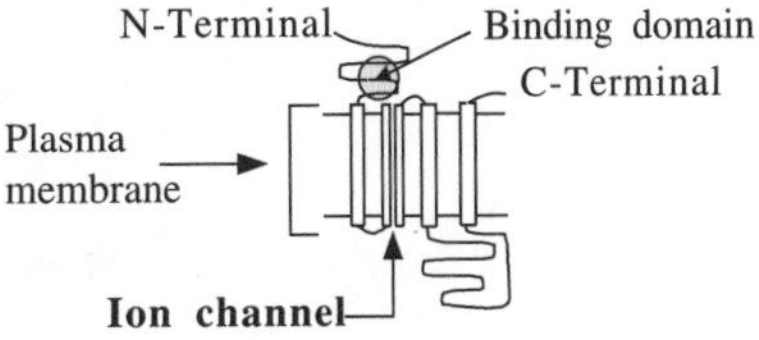 Receptors consist of four or five subunits of this type with a total of 16–20 membrane-spanning domains
2.	Hormones and slow transmitters. The receptor is coupled to the effector system by G-protein	mAChR and noradrenergic is receptors	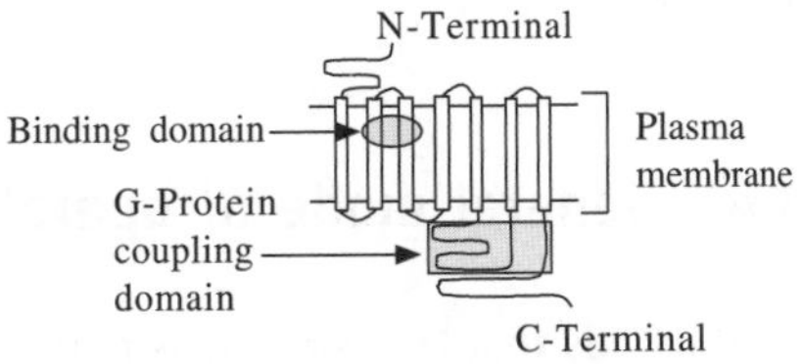
3.	Insulin and growth factors. The receptor is linked to tyrosine kinase	Insulin receptors	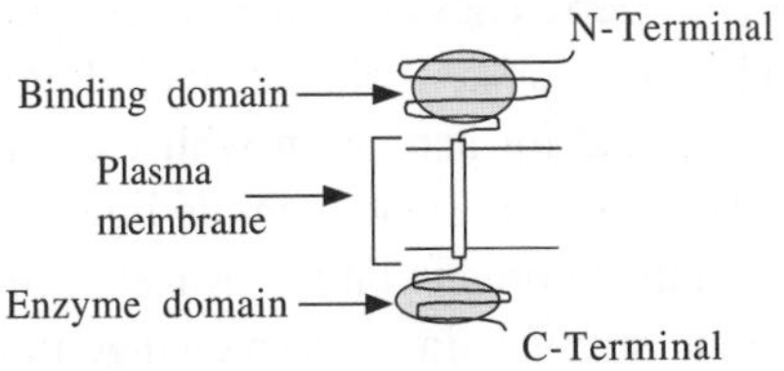
4.	Steroid hormones, thyroid hormones, vitamins such as vitamin D and retinoic acid	Antidiuretic hormone (ADH) or vasopressin receptors	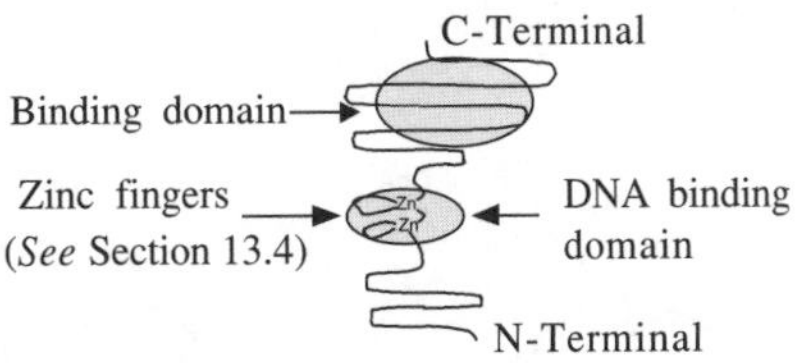

according to the type of genetic code responsible for their structure or after the exogenous ligands that selectively bind to the receptor. For example, the endogenous ligand acetylcholine will bind to all cholinergic receptors (AChR) but the exogenic ligand nicotine will only bind to nicotinic cholinergic receptors (nAChR). Similarly muscarine will only bind to muscarinic cholinergic receptors (mAChR). However, it is possible to differentiate between different types of receptor within a sub-type. For example, three different types of muscarinic cholinergic receptors have been detected and a further two predicted from a study of the genes that code for this type of receptor. These five mAChR receptors are classified using numerical subscripts as m_1AChR, m_2AChR, m_3AChR, m_4AChR and m_5AChR, respectively.

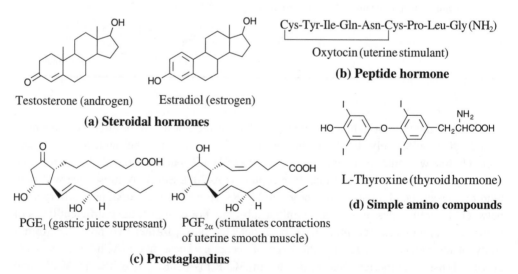

Adrenaline (epinephrine)

Acetylcholine (ACh)

Noradrenaline (norepinephrine)

Nicotine

Muscarine

Similar classifications are used for other receptors. For example, adrenoceptors are classified as α and β sub-types. These sub-types are further classified according to the nature of their exogenous ligands. These are further classified by the use of subscript numbers.

8.4 General mode of operation

The ligands that *activate* or *deactivate* (inhibit) a receptor are known as *primary* or *first* messengers. These first messengers may be hormones, neurotransmitters, other endogenous substances or xenobiotics, including drugs, bacteria and viruses. Some receptors require one ligand for activation whilst other receptors require two ligands for activation.

Hormones are endogenous ligands (Fig. 8.5) that are produced, stored and released by specialised organs either in response to a stimulus of the physical senses (such as light, odour, stress, pain) or to a change in the concentration of a compound in a body fluid. For

Testosterone (androgen)   Estradiol (estrogen)

(a) Steroidal hormones

Cys-Tyr-Ile-Gln-Asn-Cys-Pro-Leu-Gly (NH$_2$)

Oxytocin (uterine stimulant)

(b) Peptide hormone

L-Thyroxine (thyroid hormone)

(d) Simple amino compounds

PGE$_1$ (gastric juice supressant)   PGF$_{2\alpha}$ (stimulates contractions of uterine smooth muscle)

(c) Prostaglandins

Figure 8.5 Examples of four major classes of ligands that act as hormones in humans

example, an increase in glucose concentration in blood stimulates the release of insulin. Many hormones are released into the blood stream where they may be transported distances of up to about 20 cm in order to reach their target receptor. Others, such as the prostaglandins, operate where they are formed. These hormones are referred to as local hormones or *autocoids*. Autocoids differ from ordinary hormones in that they are continuously synthesised and released into the circulation.

Neurotransmitters are the endogenous ligands (Fig.8.6) released by a neuron when it communicates with other neurons across a synaptic cleft (gap) (see section 7.4.3). The neurotransmitter is released from the presynaptic site and diffuses to its target on the postsynaptic neuron or an effector cell where it binds to the appropriate receptor. This activates the receptor, which in turn transmits the chemical message carried by the ligand to the cell (signal transduction). The mechanisms by which both hormones and neurotransmitters activate a receptor have been found to be remarkably similar.

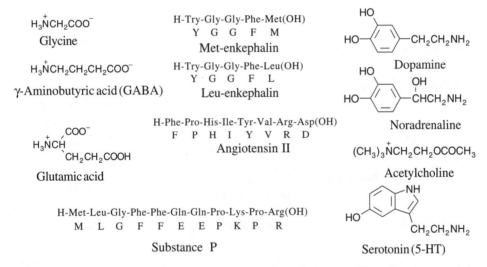

Figure 8.6 Examples of ligands that act as neurotransmitters in humans. These ligands are varied and include some amino acids, small peptides, acetylcholine, β-phenylethanolamines, catecholamines and their derivatives. The letters under each structure are those currently used for the corresponding amino acid

The mechanism by which chemical messages are received and delivered by a receptor is not fully understood. It is believed that some sections of the structure of the ligand are complementary in shape to sections of the receptor and so are able to bind to the receptor by appropriate bonds (Fig. 8.7). However, the complete structure of the ligand does not exactly match that of the receptor site. This means that other sections of the ligand can only form very weak bonds with the receptor because they are not quite correctly aligned to maximise their interaction with the receptor. However, it is thought that the receptor then changes its conformation to maximise its interaction with these sections of the ligand. It is as though these sections of the ligand were acting as a magnet for the receptor.

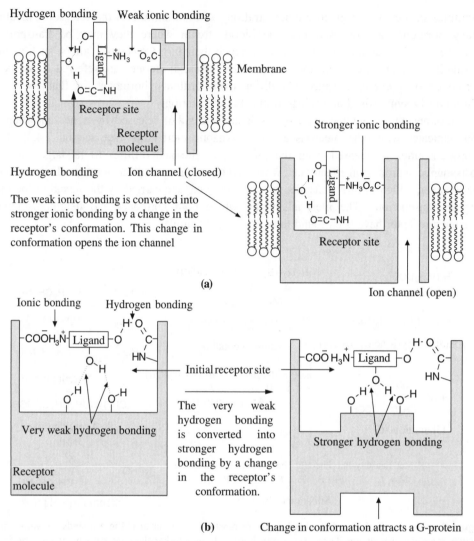

Figure 8.7 Diagrammatic representations of the general methods by which it is believed that a receptor transmits a chemical message when a ligand binds to the receptor site. The bonds shown are not the only types that can bind a ligand to a receptor. (a) The binding of the ligand causes a change in the conformation of the receptor molecule, which results in the opening of an ion channel (Superfamily Type 1). The change in conformation that opens the channel may be some distance from the receptor site for the ligand and not as close as shown in the diagram. (b) The binding of a ligand results in a change of conformation, which results in the receptor protein developing a strong affinity for a G-protein (Superfamily Type 2)

The change in conformation caused by this maximising of the binding between receptor and ligand is thought to affect the whole of the receptor molecule and result in conformational changes in another domain of the receptor molecule. These subsequent changes may, for example, result in the opening or closing of an ion channel (Fig. 8.7a), the activation or deactivation of an enzyme, the activation of a G-protein (Fig. 8.7b) or another biochemical event. The physiological significance of these events will depend on the nature of the process controlled by the receptor (see sections 8.4.1–8.4.4).

It is emphasised that it is not yet known whether the binding of the ligand occurs in a series of steps as described in the foregoing discussion or simultaneously in one step. However, although the binding must be strong enough to alter the conformation of the receptor, it must not be too strong to prevent the removal of the ligand from the receptor after it has delivered its message. This is particularly important for neurotransmitters. If they are not rapidly removed from their receptors the nerve receiving the message would be switched on or off for long periods of time, which could result in injury and death.

8.4.1 Superfamily Type 1

These receptors control ion channels. They are membrane-embedded proteins that have both their C-terminal and N-terminals in the extracellular fluid. It usually has four to five membrane spanning subunits surrounding a central pore. Each subunit consists of a sequence of 20–25 amino acid residues in the form of an α-helix. The different sequences are identified by a Greek letter and in some cases a suitable subscript. Sugar residues are attached to the extracellular N-terminal chain but these do not appear to take part in the receptor's biological function.

The ion channels of most ion channel receptors normally require two ligand molecules to activate the receptor and open the ion channel. For example, the pore of the nicotinic acetylcholine (nACh) receptor, which consists of five subunits (two α and one each of β, γ and δ subunits), is only opened when an acetylcholine molecule binds to each of the α subunits (Fig. 8.8). In other words, the ion channel only opens when two acetylcholine molecules have bound to the receptor. The way in which the channel opens is not fully understood. However, Unwin showed in 1993 that about halfway through the membrane the five subunits protrude into the channel,

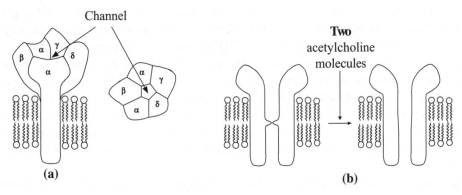

Figure 8.8 **(a)** The proposed general structure of the nACh receptor and **(b)** its proposed mode of action. Based on electron microscopy and other experimental data

effectively blocking it. He suggested that the binding of the acetylcholine residues caused a conformational change that resulted in the removal of this protuberance (see also Fig. 8.7b).

The physiological response to the binding of the appropriate ligand to an ion channel receptor occurs in microseconds. The opening of the channel allows the passage of ions into or out of a cell, which leads to a variety of cellular effects. For example, the opening of ion channels in a nerve cell increases the flow of Na^+ ions into the neuron. This depolarises the membrane, which results in the rapid transmission of a nerve impulse along the nerve (see section 7.4.3). The opening of the Ca^{2+} ion channels of endothelial cells results in a flow of Ca^{2+} ions into the cell, which, with the calcium binding protein calmodulin (CaM), activates the eNOS and nNOs enzymes to catalyse the production of nitric oxide (see section 14.4).

8.4.2 Superfamily Type 2

Most of the receptors in this family have been found to consist of a single polypeptide chain containing 400–500 amino acid residues. The N-terminal of this chain lies in the extracellular fluid whilst the C-terminal is found in the intracellular fluid. The lengths and sequences of the terminal polypeptide chains vary considerably between members of this family. All the receptors have a group of seven transmembrane subunits, which consist of 20–25 amino acid residues arranged in an α-helix. These transmembrane subunits are grouped around a central pocket that is believed to contain the receptor site.

The binding of a ligand to the receptor site results in a conformational change in the large intracellular polypeptide loop and C-terminal chain. These changes attract a protein known as *G-protein* because of its close association with guanine nucleotides. G-Proteins are a family of unattached proteins that are able to diffuse through the cytoplasm. They consist of three polypeptide subunits known as α, β and γ subunits. In its resting state the G-protein has guanosine diphosphate (GDP) bound to its α subunit. Experimental evidence suggests that when the G-protein reaches the receptor the GDP is exchanged for guanosine triphosphate (GTP) and in the course of this exchange the α-GTP subunit becomes detached and migrates to either the receptor of an ion channel (an effector protein) or the active site of an enzyme (an effector protein). The coupling of the α-GTP subunit to the receptor of the ion channel opens or closes the channel, whilst the binding of the α-GTP subunit to the enzyme either inhibits or activates the enzyme (Fig. 8.9). The action of the α-GTP subunit is terminated when the GTP is hydrolysed to GDP by the catalytic action of the α subunit. When the GTP is converted to GDP the α unit loses its affinity for the effector protein and migrates back to its β and γ subunits to complete the cycle. It should be noted that some researchers have suggested that ion channels are controlled by the β–γ subunit and not the α-GTP subunit.

The type of activity initiated by a G-protein is thought to depend on the nature of the G-protein involved. It is now believed that there are three major groups of G-proteins, namely: the G_s that stimulate adenylate cyclases, the G_i that inhibit adenylate cyclases and the G_o

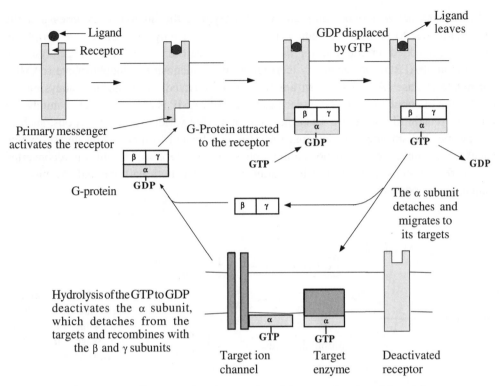

Figure 8.9 A diagrammatic representation of the general action of G-proteins

that mediate neurotransmission in the brain by a pathway that is not yet fully understood. It is believed that these G-proteins act through specific receptors activated by the appropriate agonist. For example, an enzyme would be stimulated by one receptor but inhibited by a different receptor (Fig. 8.10). It is interesting to note that cholera toxin bonds (conjugates) to a G_s-coupled receptor, resulting in the permanent activation of adenylyl cyclase. This induces excessive secretion of fluid from the GI epithelium, which leads to excessive diarrhoea and dehydration, the classic symptoms of cholera.

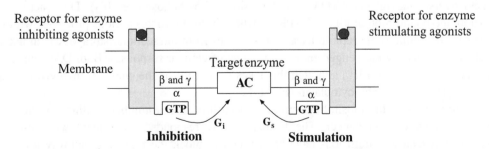

Figure 8.10 A diagrammatic illustration of the specific nature of receptors and G-proteins

G_s-Proteins activate two major classes of enzymes, the adenylate cyclases and the phospholipases, to produce secondary messengers. The adenylate cyclases are peripheral proteins embedded in the interior surface of the cell membrane. They catalyse the conversion of ATP to the secondary messenger cyclic adenosine monophosphate (cAMP). Adenylate cyclase (AC) acts as an amplifier, one activated AC molecule catalysing the conversion of about 10 000 ATP molecules to cAMP. cAMP is able to activate a number of different protein kinases (Fig. 8.11) that control a wide variety of cell functions, such as energy metabolism and cell division. cANP is deactivated by phosphodiesterase, which catalyses its conversion to adenosine monophosphate (AMP), which inturn is reconverted by a series of steps to ATP. It is interesting to note that both caffeine and theophylline inhibit phosphodiesterase.

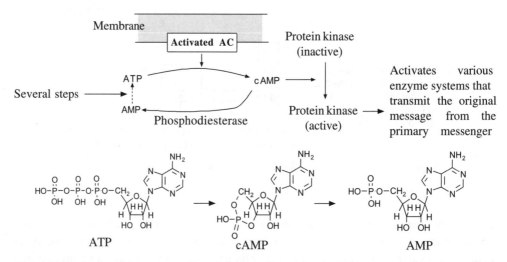

Figure 8.11   A schematic representation of the formation and action of the secondary messenger cAMP

Other members of the G-protein family activate phospholipase C. These enzymes catalyse the hydrolysis of phosphatidylinositol bisphosphate (PIP_2) to two secondary messengers, diacylglycerol (DAG) and inositol-1,4,5-triphosphate (IP_3). DAG activates membrane-bound protein kinase C (PKC), which in turn initiates various processes within the cell. IP_3 initiates the rapid release of Ca^{2+} ions from intracellular storage, which itself acts as a secondary messenger initiating a range of cellular responses. Both DAG and IP_3 are converted back to PIP_2, although this takes some time as the enzymes involved in this process are found in the cytoplasm (Fig. 8.12).

Malfunctions of the inositol system have been linked to a number of illnesses such as manic depression and cancer. The former usually responds to treatment with lithium carbonate because lithium ions have been found to block the recycling pathway for IP_3. Cancer formation is believed to occur because protein kinase is involved in cell division.

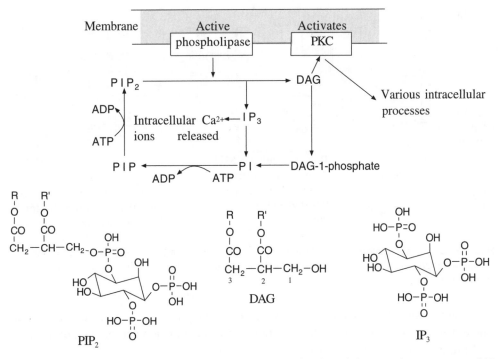

Figure 8.12 A schematic representation of the formation and action of the secondary messengers DAG and IP_3

8.4.3 Superfamily Type 3

Tyrosine–kinase linked receptors have a single helical transmembrane subunit. This subunit is attached to large extracellular and intracellular domains. The extracellular domains are varied and their nature depends on that of their endogenous ligand. However, the intracellular domains of all these receptors contain a tyrosine–kinase residue as an integral part of their structures. In addition, it is believed that all these intracellular domains have an ATP binding site near the surface of the membrane, whereas the substrate site is nearer to the end of the intracellular domain.

It is thought that the binding of the ligand to Type 3 receptors normally causes the receptors to dimerise (Fig. 8.13). This is believed to result in conformational changes that trigger the autophosphorylation of various tyrosine residues in the intracellular domain. These tyrosine-phosphorylated residues appear to act as high affinity binding sites for specific intracellular proteins. The structures of these proteins appear to have in common a conserved sequence of about 100 amino acids referred to as the SH2 domain, which binds to specific regions of the phosphorylated tyrosine–kinase domain. This binding is thought to trigger either enzyme activity or the binding of further functional proteins to the SH2 containing protein, which leads to a wide variety of biological activities.

Many cytokine receptors appear to act in a similar way to tyrosine–kinase receptors even though they do not have a tyrosine–kinase domain. It is believed that the dimerisation of

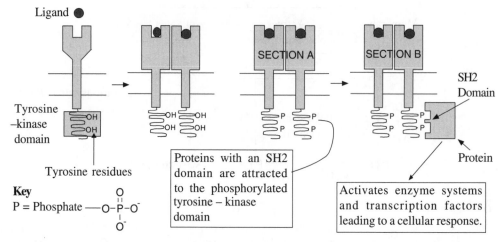

Figure 8.13 A diagrammatic representation of the mode of action of tyrosine–kinase linked receptors. A number of different proteins with SH2 domains exist in the intracellular fluid. These proteins control a wide variety of cell functions

the receptors attracts a tyrosine–kinase unit in the cytoplasm to the intracellular domain, where it is autophosphorylated This is followed by the attraction and binding of a specific SH2 protein, which ultimately leads to a cellular response.

The discovery of SH2 domains opens the way for the design of new drugs that will have structures that are either complementary to or mimic the action of the SH2 domain.

8.4.4   Superfamily Type 4

The members of this family of intracellular proteins are located within the nucleus of living cells, although they were first isolated from the cytoplasm of homogenised cells. This led to the erroneous conclusion that these receptors existed in the cytoplasm. They are activated by steroid and thyroid hormones, retinoic acid and vitamin D (Fig. 8.14).

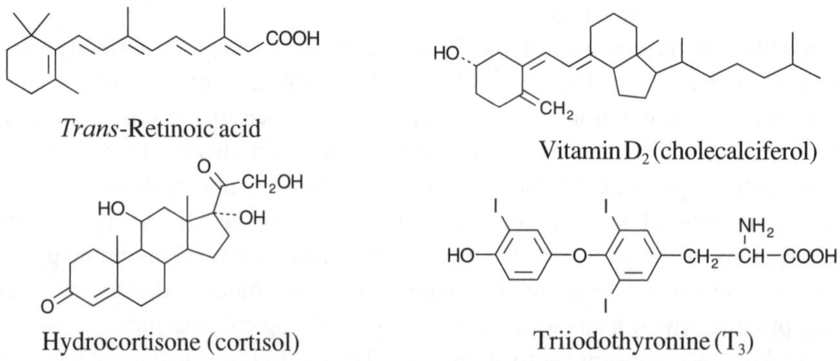

Trans-Retinoic acid

Vitamin D_2 (cholecalciferol)

Hydrocortisone (cortisol)

Triiodothyronine (T_3)

Figure 8.14 Examples of the agonists of the members of Superfamily Type 4

The members of the family are large proteins containing 400–1000 amino acid residues. Their structures have similar central sections of about 60 residue sequences that contain two loops of about 15 residues. These loops are known as zinc fingers, since each loop originates from a group of four cysteine residues coordinated to a zinc atom (see section 13.4). The hormone receptor lies on the C-terminal side of this central region, whilst located on the N-terminal side is a highly variable region that controls gene transcription. The conformational changes caused by the binding of the hormone to the receptor are believed to expose the DNA binding domain, which is normally hidden within the structure of the protein. This allows the DNA to bind to the protein. This is accompanied by an increase in RNA polymerase activity and within minutes the production of a specific mRNA. This mRNA controls the synthesis of a specific protein that produces the cellular response. However, this response can take from hours to days to develop. For example, the binding of glucocorticoids to the DNA binding domain is believed to increase the concentration of lipocortin, which acts as an anti-inflammatory by inhibiting the activity of phospholipase A_2 (PLA$_2$).

8.5   Ligand–response relationships

The binding of a ligand (L) to a receptor (R) results in a loss of energy. This loss is broadly described as the *affinity* of the ligand for the receptor. The greater the loss of energy, the greater the affinity of the ligand for the receptor, and the more easily it binds to that receptor. Clark (see section 8.6.1) in the 1920s envisaged this binding process as a dynamic equilibrium in which the binding of a ligand to a number of identical independent receptor sites activated those sites. He postulated that the magnitude of the corresponding response was proportional to the number of receptors occupied at equilibrium. Clark assumed that a maximum response would occur when all the receptors were occupied and that the response to the ligand would cease when the ligand dissociated from the receptor. As a result, the affinity for a receptor may be defined in terms of the equilibrium constant for the dissociation of the ligand–receptor complex, that is:

$$\underset{\text{(Ligand–receptor complex)}}{L-R} \quad \rightleftharpoons \quad \underset{\text{(Ligand)}}{L} \; + \; \underset{\text{(Receptor)}}{R} \tag{8.2}$$

and:

$$K_D = \frac{[L]\,[R]}{[L{-}R]} \tag{8.3}$$

where $K_D (= K_A$, the affinity constant) is the dissociation constant for the process. It follows that the smaller the value of K_D, the greater the affinity of the ligand for the receptor. Most agonists have K_D values in the order of 10^{-10}–10^{-6} mol dm^{-3}. These values are usually recorded as pD_2 values where:

$$pD_2 = -\log K_D = -\log EC_{50} \tag{8.4}$$

because from equation (8.20). (see section 8.6) $K_D = EC_{50}$, where EC_{50} is the molar concentration of the drug that produces half the maximum biological response observed when a ligand binds to a receptor. According to Clark's theory the EC_{50} corresponds to half the receptors being occupied by the ligand. The EC_{50} of a drug is normally determined *in vitro* and may be used to estimate the value of K_D but the correlation is not always reliable. However, a low EC_{50} value will indicate a high affinity for the receptor. Both EC_{50} and pD_2 are used to compare the relative effects of the members of a series of analogues in the preclinical development of new drugs.

The affinity of a ligand may also be measured in terms of the tendency of a ligand to associate with the receptor, that is:

$$L + R \rightleftharpoons L-R \tag{8.5}$$

and:

$$K_a = \frac{[L-R]}{[L]\,[R]} \tag{8.6}$$

where K_a is the association constant and is the reciprocal of K_D. High affinity ligands have large K_a values.

A high proportion of the amount of a ligand in general circulation will bind to receptor sites for which it has a strong affinity. However, not all the ligand binds to these sites; some will bind to secondary sites for which the ligand has a lower affinity. The binding of the ligand to these secondary sites could give rise to unwanted side effects, although this is not the only source of unwanted side effects.

8.5.1   Experimental determination of ligand concentration–response curves

Quantitative information concerning the dose–response relationship of a drug is usually obtained from a study of the effect of the drug on an isolated tissue, such as guinea-pig ileum and rabbit jejunum. These studies are normally carried out using an organ bath (Fig. 8.15). The lower end of the tissue is attached to the air line whilst the top is connected

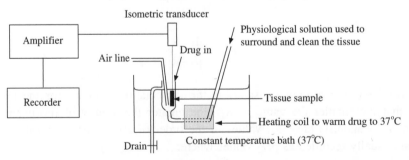

Figure 8.15   A schematic representation of a typical organ bath experiment. The drug is injected directly onto the tissue

to an isometric transducer that detects changes in the tension in the tissue. A dose of the drug is administered and the response of the tissue is detected, amplified and recorded as a trace on a chart.

The appearance of the trace will depend on the nature of the study (Fig. 8.16). It will depend on both the drug and the tissue. For example, prenalterol acts as a full agonist (see section 8.5.2) on thyroxine-treated guinea-pig atria, a partial agonist on rat atria and an agonist in canine coronary artery, whereas isoprenaline acts as a full agonist in all three tissues.

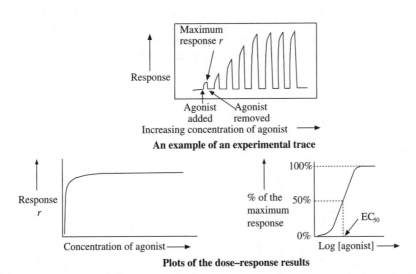

An example of an experimental trace

Plots of the dose–response results

Figure 8.16 A simulation of the results of a typical molar concentration–response investigation for a full agonist. If the amount of the drug is measured in terms of the dose administered, the plots will have the same general appearance but the EC_{50} will now be recorded as the ED_{50}. The appearances of the trace will vary: the example shown is for an ideal set of experimental results

8.5.2 Agonist concentration–response relationships

The response due to an agonist increases with increasing concentration of the agonist until it reaches a maximum (Fig. 8.16). At this point, further increases in dose have no further effect on the response. It should be noted that the scale on the x-axis of response curves may be either in terms of the molar concentration or the dose (total amount used) of the agonist. The latter is referred to as a *dose–response* curve. Agonists with similar structures acting on the same receptor will usually exhibit similar log molar concentration–response plots (Fig. 8.17). However, the slopes of these plots are not always similar (Fig. 8.18). Furthermore, some agonists in the same structural series may show a lower maximum response value than the other members of the series. These compounds are referred to as *partial agonists* (see section 8.5.4) whilst those that show the maximum response are known as *full agonists*. Most drugs act as partial agonists.

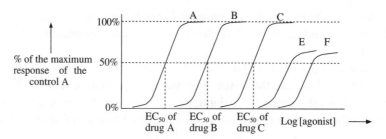

Figure 8.17   Ideal representations of the percentage response against log molar concentration plots for several agonists that cause the same response. Plot A is used as the control to calculate the percentage response. Plots B and C are for drugs acting on the same receptor site but with different EC_{50} values. The dose-response curves of these drugs are similar and their ED_{50} values can be read off the log dose axis of these plots. Plots E and F are for drugs in the same series that are partial agonists

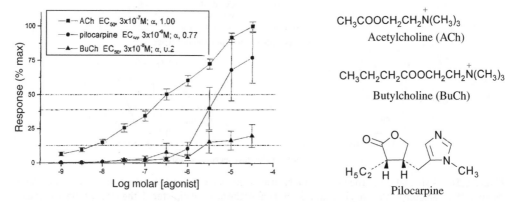

Figure 8.18   Experimental results showing the effect of different agonists on guinea-pig ilium. (Courtsey of Dr S Arkle and the students of the 1998 cohort of the ABMS, BMS and MPharm degree courses)

Agonist response plots may be used during preclinical drug development to compare the potency of the different analogues of a lead compound. These comparisons may be made in terms of the EC_{50} and ED_{50} values (Fig. 8.16) of the potential drugs. However, the values of both the EC_{50} and ED_{50} of a drug will depend on the biological effect being studied. For example, the EC_{50} value for aspirin used as an analgesic is significantly different from the EC_{50} value of aspirin when it is used as an anticoagulant.

8.5.3   Antagonist concentration–receptor relationships

Antagonists inhibit the action of an agonist (Fig. 8.19). They may be classified as either *competitive* or *non-competitive*. Competitive antagonists are more common than non-competitive antagonists.

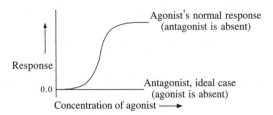

Figure 8.19 The effect of an ideal antagonist on the response of a receptor

A competitive antagonist binds to the same receptor molecule as an agonist but does not cause a response (Fig. 8.19). As the concentration of the antagonist increases, the response due to the agonist decreases. However, for competitive antagonists this decrease can be reversed by increasing the concentration of the agonist (Fig. 8.20).

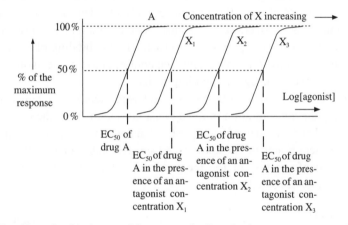

Figure 8.20 The effect of an ideal competitive antagonist X on the dose–response curves for an agonist A. Plot A is the dose–response curve for the agonist A in the absence of the antagonist X. Plots X_1 to X_3 are the dose–response curves for the agonist A in the presence of three different constant concentrations (X_1, X_2 and X_3) of the antagonist X. The value of the EC_{50} of A will depend on the concentration of the antagonist

This behaviour can be explained if the binding of both the agonist and competitive antagonist to the receptor are in dynamic equilibria (Fig. 8.21). When the concentration of the competitive antagonist is increased, the position of the equilibrium K_1 moves to the left. However, as the concentration of agonist increases, the antagonist is displaced and the equilibrium moves to the right. This behaviour means that in the presence of a competitive antagonist a higher concentration of the agonist is required in order to obtain the same degree of response as that observed in the absence of the antagonist (Fig. 8.21). In other words, the effect of a competitive antagonist will depend on the relative concentrations of the agonist and antagonist.

Originally, *competitive antagonists* were thought to compete for the same receptor site as the agonist. However, in most cases the structures of an agonist and its antagonists are not

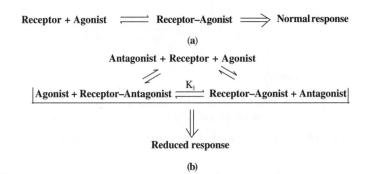

Figure 8.21 An outline of the mode of action of a competitive antagonist. **(a)** With no antagonist and **(b)** with an antagonist

terribly similar. Consequently, it is now thought that the antagonist can also bind to a receptor at a different site close to the agonist receptor site. This site may be close enough for the bound antagonist to overlap the agonist receptor site and so sterically, hinder the binding of the agonist to its receptor site. Alternatively, the binding of the antagonist at a site near the agonist receptor site may change the conformation of the agonist receptor site, which would also prevent the agonist binding to its site.

The action of *non-competitive antagonists* is not dependent on the concentration of the agonist. Increasing the concentration of the agonist does not restore the degree of response (Fig. 8.22a). It is believed that noncompetitive antagonists bind irreversibly by strong bonds such as covalent bonds to allosteric sites (see section 9.3 page 318) on the receptor. This changes the conformation of the receptor site, which prevents the binding of the agonist to the receptor (Fig. 8.22b).

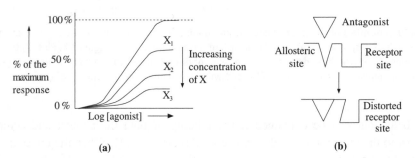

Figure 8.22 **(a)** The effect of a non-competitive antagonist on the dose–response curves for a drug A. Plot A is the dose–response curve for the agonist A in the absence of the antagonist X. Plot X_1 is the dose–response curve for the agonist A in the presence of a constant concentration (X_1) of the antagonist X. The effect of increasing the constant concentration of X is shown in the succeeding plots. **(b)** A representation of the general mode of action of a non-competitive antagonist

The magnitude of the response to an agonist in the presence of an antagonist will depend on the relative affinities of the receptor for the antagonist and the agonist. The concentration at which an antagonist exerts half its maximum effect is known as

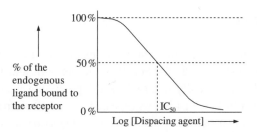

Figure 8.23 A displacement curve for the displacement of a ligand antagonist or agonist from a receptor. The IC_{50} is for the displacing agent

its IC_{50} (Fig. 8.23). This measurement may either refer to an antagonist displacing an agonist or an agonist displacing an antagonist. In both cases, the IC_{50} is a measure of the affinity of the drug for the receptor under the appropriate conditions. The smaller the value of the IC_{50}, the stronger the affinity of the displacing drug for the receptor.

IC_{50} values may be used in drug development to compare the potencies/affinities of a series of drugs that bind to the same receptor molecule and inhibit the same biological response. The values may be determined *in vitro* or *in vivo* and so can also be of use in estimating the dose of a drug required in preclinical trials.

8.5.4 Partial agonists

Partial agonists are compounds that act as both agonists and antagonists. They are believed to act as antagonists by preventing the endogenous ligand binding to the receptor but at the same time weakly activating the receptor. This is thought to cause a weak signal to be sent to the appropriate domain of the receptor. The net result of these opposing effects is that a much higher dose of the agonist is required to obtain the maximum response. Furthermore, this response is less than that of pure agonists with similar structures. Most drugs are partial agonists.

Partial agonism has been explained in two ways:

1. The structure of the partial agonist is such that it can bind in two different ways to the receptor. In one orientation it acts as an agonist whilst in the alternative orientation it acts as an antagonist. As a result, the degree of activity of the receptor would depend on the relative numbers of molecules binding in each orientation.

2. An alternative explanation is that the partial agonist is a reasonable but not perfect fit to the receptor site and so its binding does not cause a great enough change in the receptor molecule's conformation to allow a full transmission of the signal. However, the conformational change is large enough to allow a weak transmission of the signal.

8.5.5   Desensitisation

The repeated exposure of a receptor to identical doses of a drug can in some cases result in a reduction of the response (Fig. 8.24). For example, smooth and voluntary muscle will become insensitive when repeatedly exposed to a depolarising agent. It appears that the drug starts by acting as a full agonist but its repeated use results in partial agonistic action. This phenomenon is known as *desensitisation* or *tachyphylaxis*. At present there is no universally accepted explanation for desensitisation, however rate theory does account for the phenomenon (see section 8.6.2).

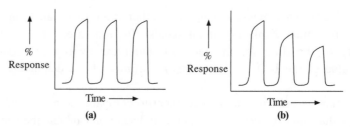

Figure 8.24   **(a)** No desensitisation. **(b)** Desensitisation. The start of each peak corresponds to the administration of an identical dose of the agonist

8.6   Ligand–receptor theories

A number of theories have been advanced to explain the action of ligands on receptors. They attempt to explain desensitisation: the relative potencies of different drugs acting on the same receptor and why a drug may act as an agonist in one tissue, a partial agonist in another and an antagonist in a third tissue even though it is acting on the same receptor.

8.6.1   Clark's occupancy theory

Clark in the 1920s visualised the drug–receptor interaction as being a bimolecular dynamic equilibrium with the drug molecules continuously binding to and leaving the receptor (see also section 8.5), that is:

$$\textbf{Drug (D)} + \textbf{Receptor (R)} \rightleftharpoons \textbf{Drug–Receptor Complex (DR)} \Rightarrow \textbf{Response} \quad (8.7)$$

Clark stated that the intensity of the response at any time was proportional to the number of receptors occupied by the drug: the greater the number occupied, the greater the pharmacological effect, that is:

$$\text{Response effect } E \propto [\text{DR}] \quad (8.8)$$

According to Clark a maximum response would be obtained when all the receptors were occupied, that is:

$$\text{Maximum response effect } E_{\max} \propto [R_T] \tag{8.9}$$

where R_T is the total number of receptors. It follows from equations (8.8) and (8.9) that for a given dose of a drug the fraction of the maximum response is given by:

$$\text{Fraction of the maximum response } \frac{E}{E_{\max}} = \frac{[DR]}{[R_T]} \tag{8.10}$$

The dissociation of the drug–receptor complex may be represented as:

$$D-R \rightleftharpoons D + R \tag{8.11}$$

and applying the law of mass action:

$$K_D = \frac{[D]\,[R]}{[DR]} \tag{8.12}$$

where K_D is the dissociation constant for the drug–receptor complex.
But the total receptor concentration is:

$$[R_T] = [R] + [DR] \tag{8.13}$$

Substituting equation (8.13) in equation (8.12):

$$K_D = \frac{[D]([R_T] - [DR])}{[DR]} \tag{8.14}$$

Rearranging equation (8.14) gives:

$$K_D = \frac{[D][R_T]}{[DR]} - \frac{[D]\,[DR]}{[DR]} \tag{8.15}$$

therefore:

$$K_D = \frac{[D][R_T]}{[DR]} - [D] \tag{8.16}$$

and:

$$K_D + [D] = \frac{[D][R_T]}{[DR]} \tag{8.17}$$

$$\frac{K_D + [D]}{[D]} = \frac{[R_T]}{[DR]} \tag{8.18}$$

Substituting equation (8.18) in equation (8.10):

$$\frac{E}{E_{max}} = \frac{[DR]}{[R_T]} = \frac{[D]}{K_D + [D]} \tag{8.19}$$

Equation (8.19) shows that the relationship between E and molar drug concentration [D] is in the form of a rectangular hyperbola, whilst that between E and log [D] is sigmoidal. These theoretical relationships derived using Clarks' theory are often in good agreement with the experimental results (Fig. 8.25) obtained in a number of investigations. Furthermore, substituting the value of $E/E_{max} = 1/2$ in equation (8.19) gives the relationship:

$$K_D = EC_{50} \tag{8.20}$$

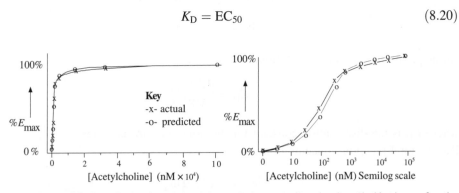

Figure 8.25 The correlation of experimental results and those predicted using Clark's theory for the stimulated contraction of guinea-pig ileum by acetylcholine

where EC_{50} is the molar concentration of the drug that produces half the maximum biological response observed when a ligand binds to a receptor. However, in practice this theoretical relationship does appear to be the exception rather than the rule.

The value of the dissociation constant K_D is a measure of the affinity of the drug for the receptor. Drugs with small K_D values have a large affinity for the receptor whilst those with high values have a low affinity. As a result, K_D values are used to compare the activities of a series of analogues during drug development. The value of K_D may be determined experimentally from tissue binding experiments using a radioactive form of the drug. The data obtained from this type of experimental work may be analysed using a Scatchard plot of the ratio of bound to free ligand against bound drug (Fig. 8.26). This gives a straight line with a slope of $-1/K_D$ provided that the drug binds to only one type of receptor. In industry the data are now analysed by the use of a computerised method of least squares.

Although Clark's occupancy theory is still a cornerstone of pharmacodynamics, a number of its assumptions have now been shown to be incorrect. It is now known that:

- the formation of many drug–receptor complexes is not reversible;

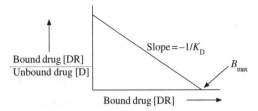

Figure 8.26 Scatchard plots for an ideal ligand binding to only one type of receptor. The steeper the slope of the line, the higher the affinity of the drug for the receptor. B_{max}, the intercept on the x axis, gives an estimate of the maximum number of receptors to which the drug can bind. Any units that measure quantity may be used, but the units for the x axis are usually femto- or pico-moles per mg of protein.

- the receptor sites are not always independent;

- the formation of the complex may not be bimolecular: for example, two acetylcholine molecules bind to nACh receptors of ion channels (see section 8.4.1);

- a maximum response may be obtained before all the receptors are occupied;

- the response is not linearly related to the proportion of receptors occupied, especially in the case of partial agonists.

In the 1950s. Ariens and Stephenson separately modified Clark's theory to account for the existence of agonists, partial agonists and antagonists. They based their modifications on a proposal by Langley in 1905, which visualised the action of a receptor as taking place in two stages. The first stage was the binding of the ligand to the receptor, which was controlled by the ligand's affinity for the receptor. The second stage was the initiation of the biological response. Ariens said that this second step was governed by the ability of the ligand–receptor complex to initiate a response. Ariens called this ability the *intrinsic activity* (α), whilst Stephenson referred to it as the *efficacy* (e) of the ligand–receptor complex.

Intrinsic activity may be defined as:

$$\alpha = \frac{E_{max} \text{ of a drug}}{E_{max} \text{ of the most active agonist in the same structural series}} \qquad (8.21)$$

Using the concept of intrinsic activity Clark's equation (equation 8.19) becomes:

$$\frac{E}{E_{max}} = \frac{\alpha[D]}{K_D + [D]} \qquad (8.22)$$

When $\alpha = 1$ for ligands with identical affinities for a receptor, equation (8.22) reverts to the original form of Clark's equation (equation 8.19). This means that a normal response curve is obtained and the drug acts as a full agonist. However, when $\alpha = 0$ the percentage

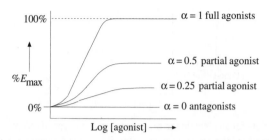

Figure 8.27   A pictorial representation of the variation of dose–response curves with the value of α. Values for α between 1 and 0 correspond to drugs that act as partial agonists, the degree of partial agonism depending on the value of α

response is zero and the drug is a full antagonist. Intermediate values between 1 and 0 for α indicate a partial agonist (Fig. 8.27).

In 1950s, Stephenson discovered that a maximum response was obtained when only a proportion of the available receptors were occupied. This discovery was in direct conflict with Clark's occupancy theory and led Stephenson to independently propose a two-stage route for receptor action. Independently of Ariens he proposed that the binding of a ligand to a receptor produced a stimulus (S) that was related to tissue response. The magnitude of the stimulus depends on both the affinity of the ligand for the receptor and its efficacy (e).

$$D + R \xrightarrow{\text{Affinity}} D\text{–}R \xRightarrow{\text{Efficacy}} \text{Tissue response}$$

As a result, Clark's equation (8.19) was modified to:

$$S = \frac{E}{E_{\text{max}}} = \frac{e[D]}{K_D + [D]} \tag{8.23}$$

Equation (8.23) shows that ligands with an e value of zero will have no biological response. Consequently, full antagonists will have an e value of zero. To obtain a positive response e must have a positive value and so agonists and partial agonists will have positive e values. Moreover, the higher the positive value, the greater the response (Fig. 8.28) and the lower the dose of agonist [D] required to achieve the maximum response. This means that agonists with a high efficacy will produce a maximum response even though they do not

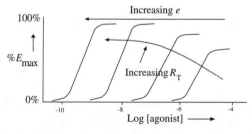

Figure 8.28   The effects on the dose–response curves of increasing e and R_T

occupy all of the available receptor sites. Unoccupied receptors are known as *spare receptors*. Their presence increases the sensitivity of a receptor to other ligands.

It is now known that cells can contain several thousand receptors of a particular type. This number can increase (*upregulation*) or decrease (*downregulation*). These changes may be brought about by both pathological and physiological cell stimuli. They can affect drug response. For example, an increase (upregulation) in the number of receptors (R_T) moves the drug response curve to a lower concentration, whilst a decrease will move it to a higher concentration (Fig. 8.28).

Stephenson observed that the magnitude of the response was not linearly related to the stimulus. This lead to a further modification of Clark's equation to:

$$S = \frac{E}{E_{max}} = f\left(\frac{e[D]}{K_D + [D]}\right) \tag{8.24}$$

where f is a function known as the *transducer function*. The transducer function represents the properties of the signal transducer mechanism that links the signal from the ligand to the tissue response and is a characteristic of the responding tissue. As a result, the same ligand could have different transducer functions when it is bound to different tissues. This difference would explain why a ligand may act as an agonist in one tissue but as a partial agonist in a different tissue even though it is acting on the same receptor. Furthermore, differences in the transducer functions of different ligands acting on the same receptor in the same tissue would also explain why their relative potencies may be different.

Potency depends on both the ligand–receptor complex and its efficacy. It is used to compare the relative effectiveness of different drugs and is defined as:

$$\text{Potency} = \varepsilon/K_D \tag{8.25}$$

where ε is the *intrinsic efficacy*, which is the efficacy per receptor, that is:

$$\varepsilon = e/R_T \tag{8.26}$$

Since intrinsic efficacy is independent of the total number of available receptors, a drug with the same value for ε in different tissues is likely to be acting on the same receptor in those tissues. Conversely, if the values are different, the drug is likely to be acting on different receptors in the different tissues.

8.6.2 The rate theory

This theory was proposed by Paton in 1961 as an alternative to the occupancy theory. Paton proposed that the stimulus was produced only when the ligand first occupied the receptor site. Stimulation does not continue even though the site was still occupied. This is because the receptor undergoes a second conformational change, which results in its inactivation. It is also likely that this second conformational change enhances the binding of the agonist to that site, resulting in a more stable drug–receptor complex. Consequently, as long as the ligand is bound to the receptor, the receptor is unable to produce a further stimulus. As soon as the ligand disengages from the receptor it returns to its original conformation. As a result, further stimulation of the receptor can now occur. Stimulation, according to the rate theory, is analogous to playing a piano: the note (stimulation) only occurs when the key is struck. Holding the key down does not produce a continuous note. In contrast the occupancy theory is like playing an organ: the note (stimulation) is maintained when the key is held down.

The rate theory suggests that the general type of activity exhibited by a drug is independent of the number of receptors, which explains the existence of spare receptors. Instead it is a function of the rates at which the drug binds to and is released from the receptor. Full agonists rapidly bind and even more rapidly dissociate from the receptor. Partial agonists dissociate more slowly, whilst antagonists will dissociate very slowly from the receptor. This theory, like Clark's theory, results in the establishment of a dynamic equilibrium between the ligand and the ligand–receptor complex. Consequently, the mathematics of rate theory is similar to that of the occupancy theory at equilibrium. However, according to rate theory, efficacy is related to the association and dissociation rate constants. Unfortunately, correlation between practical observations and this theory tends to be poor.

The rate theory also offers an explanation of desensitisation (see section 8.5.5). Ligands with a strong affinity for a receptor will still be bound when the second conformational change occurs. This second inactive conformation will be maintained for as long as the agonist occupies the receptor site. Consequently, as these receptors cannot respond to any further dose of the drug the degree of response is smaller because the overall number of potentially active receptors is reduced. However, once the agonist leaves the receptor its structure reverts to its original conformation and the receptor is able to function normally. Obviously the rate of dissociation of the agonist will influence the degree of desensitisation. Furthermore, intrinsic efficacy becomes a measure of a ligand's ability to stabilise the ligand–receptor complex after binding.

8.6.3   The two-state model

In its simplest form, this model is based on the concept that receptors can exist in either an active or an inactive state. The active state is known as the *relaxed* or *R state* whilst the inactive state is referred to as the *tensed* or *T state*. Receptors in the R state can provide a stimulus but those in the T state are unable to produce a stimulus.

The two-state model postulates that in the absence of any ligands a population of receptors of the same type will consist of an equilibrium mixture of receptors in the R and T states.

$$\text{T} \underset{k_{-1}}{\overset{k_1}{\rightleftharpoons}} \text{R} \qquad\qquad (8.27)$$

where k_1, and k_1 are the rate constants for the forward and reverse processes, respectively. In the absence of ligands there will be no receptor stimulation even though some of the receptors are in the R state. This situation may be regarded as being the 'at rest' form of the receptor system. However, when the equilibrium of equation (8.27) is to the right, by the interaction of a suitable ligand with the receptor the number of receptors in the R state will increase. This increase results in stimulation, which is followed by a tissue response. Conversely, ligands that cause equilibrium (8.27) to move to the left will inhibit stimulation and tissue response. This model offers a simple explanation of the general mechanism of the action of agonists, antagonists and partial agonists.

Full agonists cause maximum receptor stimulation and so will have a strong affinity for the R state. This affinity will move the position of equilibrium to the right, with a subsequent stimulation. For a drug to act as an agonist $k_1 > k_{-1}$. The larger the ratio k_1/k_{-1}, the greater the efficacy of the drug. Antagonists do not cause any stimulation and so have a strong affinity for the T state. This affinity shifts the position of equilibrium to the left, resulting in no stimulation and no subsequent tissue response. For a drug to act as an antagonist $k_1 < k_{-1}$; the smaller the ratio k_1/k_{-1}, the greater the efficacy of that drug. Partial agonists have an affinity for both states of the receptor, the degree of partial agonism exhibited depending on the relative values of k_1 and k_{-1}. This picture is a simplification and more detailed models, which are beyond the scope of this text, are discussed in a review by Kenakin (1987).

8.7 Drug action and design

Drugs can act at any stage of the signal transduction process. However, it is convenient to consider the drugs acting on receptors separately to those acting at other stages in the transduction process.

8.7.1 Agonists

Agonists often have structures that are similar to that of the endogenous ligand (Fig. 8.29). Consequently, the normal starting point for the design of new agonists is usually the structure of the endogenous ligand or the structure of those parts of the ligand (its *pharmacophore*) that interact with the receptor. This information is normally obtained from a study of the binding of the endogenous ligand to the receptor using X-ray crystallography, nuclear magnetic resonance (NMR) and computerised molecular modelling techniques. However, it is also emphasised that many agonists have structures that are not directly similar to those of their endogenous ligands.

The usual approach to designing a new drug is to carry out a SAR study using a series of compounds with similar structures to that of either the endogenous ligand or its pharmacophore (see section 3.6). This approach is based on the assumption that a new agonist is more likely to be effective if its structure contains the same binding groups and bears some resemblance to the endogenous ligand. It should be remembered that the

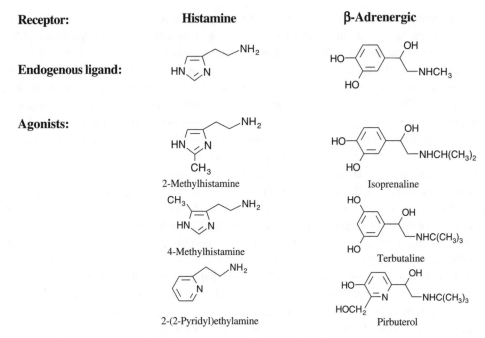

Figure 8.29 Examples of the structures of the agonists of some common receptors

groups involved in binding the endogenous ligand to the receptor are responsible for controlling both the binding of the endogenous ligand to the receptor and the ease with which it disengages from the receptor after its message has been received by the receptor. In other words, the nature of the binding groups controls the length of time a ligand remains bound to a receptor. Consequently, using mainly the same binding groups in the drug ensures that it has a good chance of fulfilling this requirement.

The nature of the binding groups is not the only structural factor that influences the activity of an agonist. The drug must also be of the correct size and shape to bind to and activate the receptor. Once again the initial approach is to use the structure of the endogenous ligand as a model. In the first instance the size of the pharmacophore of the potential drug should approximate to that of the pharmacophore of the endogenous ligand. However, it may be difficult to determine what section of the structure of the drug acts as its pharmacophore.

Information concerning the best shape for a new agonist may be obtained from a study of the conformations and configurations of a number of active analogues of the endogenous ligand. For example the torsion angle τ_2 of the endogenous mACh receptor agonist acetylcholine chloride (Fig. 8.30) is $+85°$. Experimental work has shown that many agonist drugs acting on mACh receptors have τ_2 angles between $+68°$ and $+89°$. Consequently, a proposed new drug will be more likely to exhibit agonistic cholinergic activity if its pharmacophore has an τ_2 angle within this range.

The configuration of a potential new agonist will also affect its shape and so should also be taken into account. For example, the effect of the cholinergic agonist $(-)$ acetyl-2S-

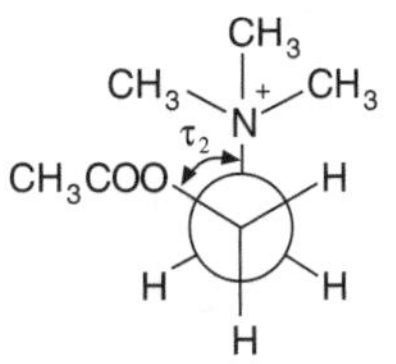

Structure	Torsion	angle	τ_2
$CH_3COOCH_2CH_2\overset{+}{N}(CH_3)_3$ Cl⁻ Acetylcholine chloride		+ 85	
$CH_3COOCH(CH_3)CH_2\overset{+}{N}(CH_3)_3$ I⁻ (+)Acetyl-2S-methylcholine iodide		+ 85	
$CH_3COOCH_2CH(CH_3)\overset{+}{N}(CH_3)_3$ I⁻ (−)Acetyl-2R-methylcholine iodide		+ 89	
$CH_3COOCH(CH_3)CH(CH_3)\overset{+}{N}(CH_3)_3$ I⁻ Acetyl-1R,2S-dimethylcholine iodide		+ 76	

Figure 8.30 The conformers of some acetylcholine receptor agonists. A torsion angle is positive when the bond of the specified front group of the drawn structure is rotated to the right to eclipse the bond of the specified rear group. The angle is negative when the front group is rotated to the left

methylcholine chloride on guinea-pig ileum is about 24 times greater than its $R(+)$ analogue, whilst $(-)2S, 3R, 5S$-muscarine iodide has about a 400 times greater effect than $(+)2R, 3S, 5R$-muscarine iodide on guinea-pig ileum.

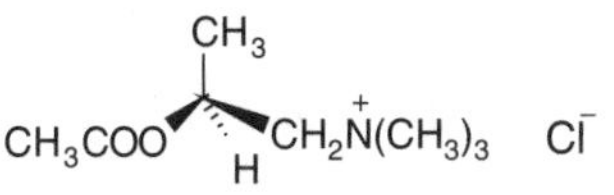

(−) Acetyl-2S-methylcholine chloride

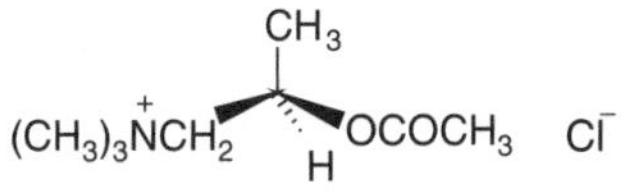

(+) Acetyl-2R-methylcholine chloride

2S,3R,5S-muscarine iodide

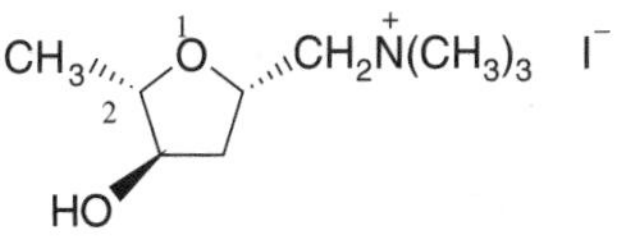

2R,3S,5R-muscarine iodide

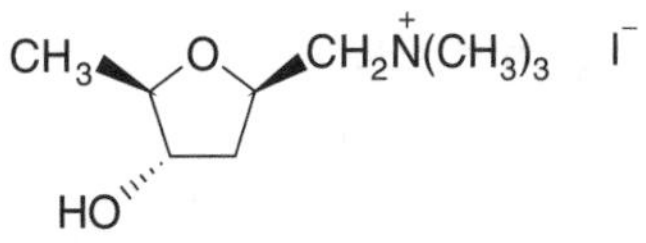

8.7.2 Antagonists

Antagonists act by inhibiting a receptor and so are used to reduce the effect of an endogenous ligand. Their action is classified as being either competitive or non-competitive (see section 8.5.3), depending on the nature of their binding to the receptor.

The structures of the antagonists acting on a receptor usually have little similarity to those of the endogenous ligands for the receptor (Fig. 8.31). Consequently, the structure of the endogenous ligand cannot be taken as a good starting point for the design of a new antagonist. The ideal starting point for the design of a new antagonist would be the structure of the receptor. However, it is often difficult to identify the receptor and also obtain the required structural and stereochemical information. Consequently, although it is not the ideal starting point, many developments start with the structure and stereochemistry of either the endogenous ligand or any other known agonists and antagonists for the

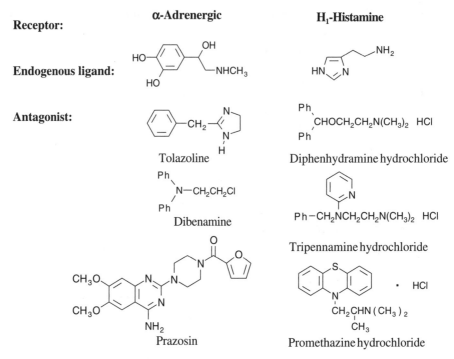

Figure 8.31 Examples of some of the antagonists of some common receptors

receptor. Since antagonists exert a stronger affinity for the receptor than its natural agonist, the binding groups selected for the new drug are often groups that could form stronger bonds with the receptor. As in the case of agonist drug design (see section 8.7.1) the relative sizes and shapes, that is, the conformations and configuration of the new antagonist and its binding site, should also be taken into account as this can have a significant effect on activity. For example, the antihistamine $S(+)$-dexchlorpheniramine is 200 times more potent than its $R(-)$-isomer. If the structure of the receptor site is known in sufficient detail, molecular modelling (see section 4.1.2) can be of some use in the design of new antagonists.

8.7.3 Citalopram, an antagonist antidepressant discovered by a rational approach

Depression is a state of mind characterised by lethargy, loss of appetite, melancholia and suicidal tendencies, amongst other similar symptoms. It is believed by many researchers to arise from a depletion of the neurotransmitters serotonin and noradrenaline (Fig. 8.6). These neurotransmitters are stored in vesicles in the presynaptic knob of a neuron. The arrival of an action potential at the presynaptic knob stimulates the release of these neurotransmitters. They diffuse across the synaptic cleft and bind to a postsynaptic receptor. This binding triggers an appropriate biological response (Fig. 8. 32). Once the response has occurred the

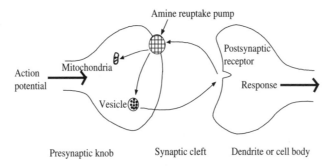

Figure 8.32 The arrows show the route of the neurotransmitters when they are released from the presynaptic knob

serotonin and noradrenaline are released from the receptor and diffuse back across the synaptic cleft to be reabsorbed into the presynaptic knob by the action of the amine reuptake pump. Once they have been reabsorbed they are stored in vesicles (see section 2.13.3) in the neuron. When all the available vesicles are full, any remaining neurotransmitters in the neuron are thought to be metabolised by the mitochondrial enzyme monoamine oxidase (MAO). Any noradrenaline that is not reabsorbed into the neuron is believed to be metabolised by catechol-O-methyltransferase(COMT) in the neighbouring tissue.

In the 1970s a number of pharmaceutical companies adopted similar rational approaches to discover antidepressant drugs that were more effective, less toxic and had an increased therapeutic index (see section 1.3) when compared with existing drugs. These approaches were based on an increased knowledge of the biochemistry involved in depression. This knowledge allowed the various teams to decide to develop drugs, now referred to as selective serotonin reuptake inhibitors (SSRIs), that selectively inhibited the amine pump that was responsible for the reuptake of serotonin in the presynaptic sites in the brain. This would increase the concentration of serotonin in contact with the postsynaptic receptor and, as a result, should reduce the patient's depression. In 1971 researchers at H. Lundbeck A/S selected talopram (Fig.8.33a) as the lead structure for a project to develop selective serotonin reuptake pump inhibitors. Talopram had been investigated by them as a potential antidepressant, however it had been found to be clinically unsatisfactory. A SAR investigation eventually revealed citalopram as the most clinically acceptable analogue. This was after less than 50 analogues of talopram had been synthesised. It was marketed by Lundbeck in 1989. Other companies soon followed suite and a number of SSRIs are now regularly used to treat mild and medium depression (Fig. 8.33b).

Citalopram was first marketed as its racemate but it was found that the S-isomer was very much more active than the R-isomer (Fig. 8.34). The former is now marketed as escitalopram. SSRIs appear to exhibit fewer detrimental side effects than the earlier tricyclic antidepressants (TCAs), such as amitryptiline (Fig. 12.2), imipramine (Fig. 2.17) and doxepin. However, TCAs are more effective in treating severe depression.

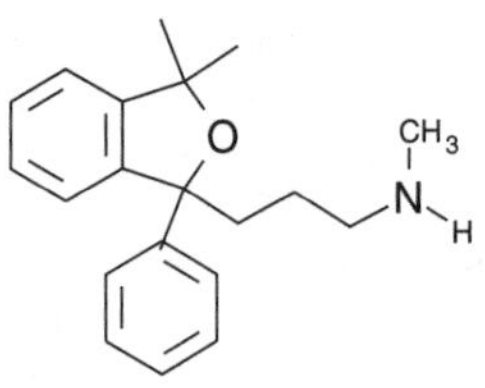

Talopram

(a)

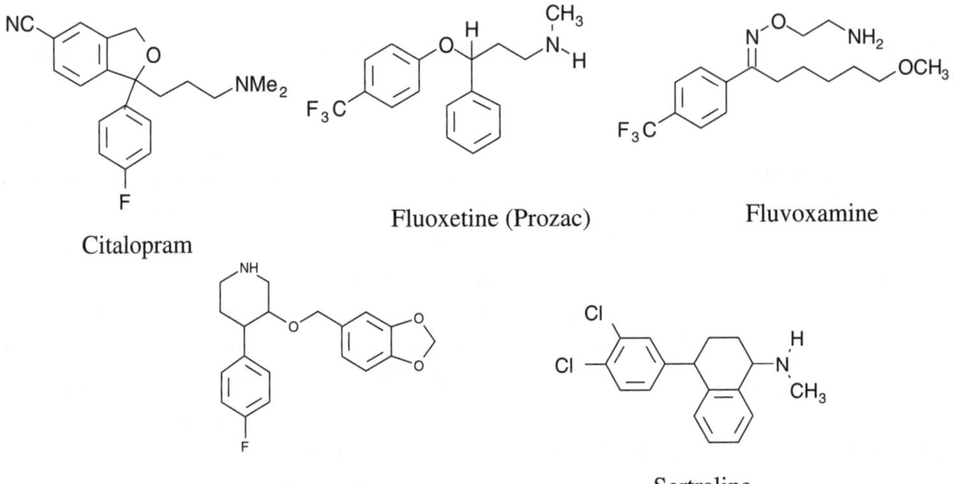

Citalopram

Fluoxetine (Prozac)

Fluvoxamine

Paroxetine

Sertraline

(b)

Figure 8.33 **(a)** Talopram. **(b)** Citalopram and other related drugs developed using a similar strategy to that for selective serotonin reuptake inhibitors (SSRIs)

A number of syntheses of citalopram were published by Klause *et al.* in 1977 (Fig. 8.35). All of the routes were very versatile, allowing a large number of similar analogues to be synthesised using variations of the starting materials and some of the intermediate reagents.

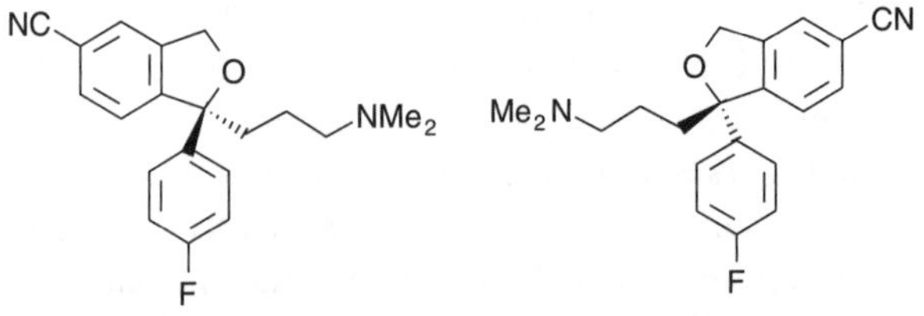

S-Citalopram (Escitalopram)

R-Citalopram

Figure 8.34 *R* and *S*-citalopram

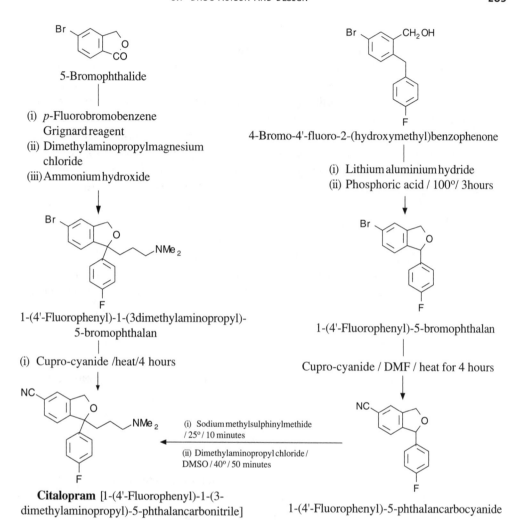

Figure 8.35 Outline examples of the synthetic pathways used to prepare citalopram. The isobenzofruan ring system is also referred to as the phthalan ring system. The formulae represent the points in the syntheses where an intermediate was isolated, purified and characterised before proceeding to the next step

8.7.4 β-Blockers

Heart disease is one of the Western world's biggest killers. A class of drug that is particularly useful in the treatment of heart disease are the so-called *β-blockers*, which act as antagonists for β-adrenoceptors in the heart. The first of these drugs, like many others, was discovered by luck. In the mid-1950s, Irwin Slater of Eli Lilly in Indianapolis was testing analogues of isoprenaline in an attempt to discover long acting bronchodilators. The bioasssay used was based on measuring the relaxation of tracheal strips that had been contracted to simulate the effect of bronchoconstriction caused by asthma (see section

6.2.1.4). These strips were used for a number of tests and so were treated with adrenaline between tests to check whether they were still responsive. Slater noticed that the strips that had been treated with dichloroisoprenaline did not relax when they were treated with adrenaline, that is, the dichloroisoprenaline was acting as an antagonist to adrenaline but was also found to some have partial agonist activity, mimicking the action of adrenaline. He reported this work in 1957 and it was followed up by Neil Moran at the Emory University in Atlanta who showed that dichloroisoprenaline antagonised changes in heart rate produced by adrenaline. The story was taken up by James Black at the ICI Pharmaceuticals Division, Cheshire, who was looking for a way to treat patients with angina by controlling heart rate. He realised that it should be possible to develop an analogue of dichloroisoprenaline that would protect patients against fluctuations in heart rate caused by changes in their adrenaline and noradrenaline levels. The first effective β-blocker, pronethalol, was synthesised by John Stephenson for James Black in 1960. It was used successfully to treat angina. It was also found to act as an antiarrhythmic and to reduce blood pressure in hypertensive patients. However, it was found that a long-term toxicity test could cause cancer of the thymus in mice As a result, its use was restricted to seriously ill patients.

Key: * = chiral centre

Isoprenaline   Dichloroisoprenaline   Pronethalol

   The similarity of the structures of isoprenaline, dichloroisoprenaline and pronethalol resulted in pharmaceutical company drug discovery teams retaining an aromatic ring system and the main structure of the side chain in the race to discover new, safer β-blockers. The first approaches were to increase the length of the side chain between the secondary alcohol group and the aromatic ring system and to vary the nature of that system. This resulted in the discovery and marketing of the aryloxypropanolamines (Fig. 8.36a) propranolol (ICI), oxprenolol (Ciba), alprenolol (AB Hassle), timolol and nadolol (Squibb). These compounds are relatively easy to synthesise (Fig. 8.36b), the reactions involved in their preparation being adaptable to large-scale use. They have all been used to treat angina, hypertension and cardiac dysrhythmias and normally have mild adverse side effects, such as headaches, diarrhoea, nausea, vivid dreams and insomnia.
   Most of the first generation of aryloxypropanolamines are not specific for the heart. They act on the α_1-, β_1- and β_2-adrenoceptors (Table 8.2) in a variety of tissues. This lack of specificity does not usually cause any serious problems except in the case of patients with asthma and restricted air way diseases. The bronchoconstriction caused by the binding of a drug to β_2-adrenoceptors in the lung could precipitate bronchospasm in some patients. Propranolol is a pure antagonist. It was the first aryloxpropanolamine to be used clinically

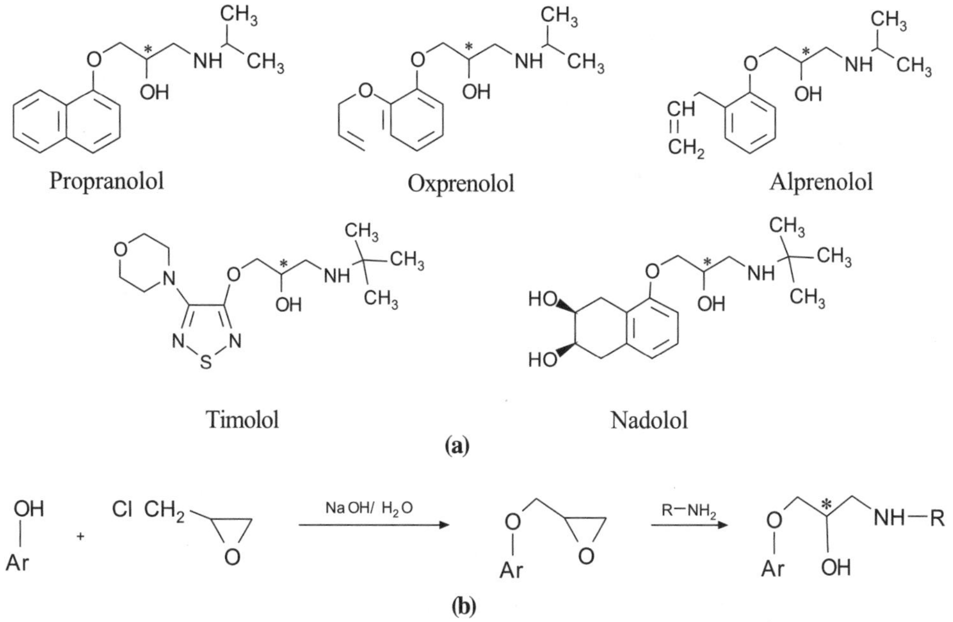

Figure 8.36 **(a)** Aryloxypropanolamines used as β-blockers. Chiral centres are marked with an asterisk. **(b)** An outline of the synthetic pathway to produce aryloxypropanolamines. The final product is a racemate because of the chiral centre (*)

(1964) and was found to be non-carcinogenic and very much more potent than pronethalol. Oxprenolol and alprenolol have considerable partial agonist activity. This may be of some use in patients with heart disease where it is necessary to maintain some degree of β_1-adrenoceptor activity in order to prevent partial cardiac failure. Timolol is mainly used, in the form of eye drops, to treat glaucoma. Nadolol does not cause the nightmares and sleep disturbances that occur with other β-blockers. This is probably because it is more polar than other β-blockers and so does not enter the CNS.

A further extension of the side chains of aryloxypropanolamines beyond the amino group using alkylaryl groups resulted in the discovery of a number of compounds that

Table 8.2 Antagonist activity related to the binding of β-blockers to α_1-, β_1- and β_2-adrenoceptors

α_1	β_1	β_2
Vasodilation	Decreases heart rate	Bronchoconstriction
Decreases blood glucose concentration	Decreases blood pressure	Decreases blood glucose
Pupil constriction	Increase in gut secretions and motility	
Causes impotence		

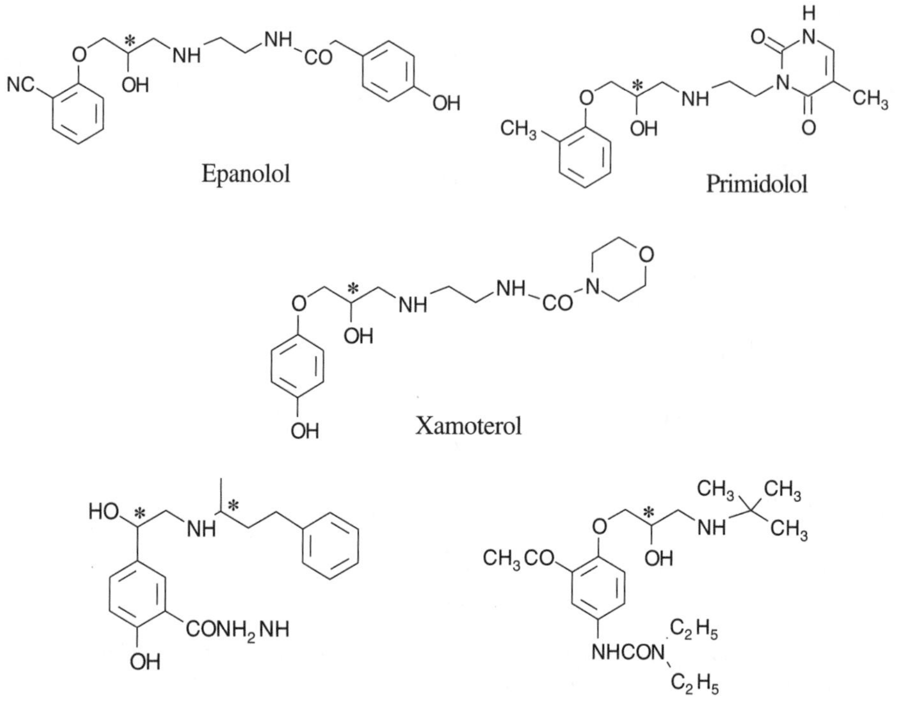

Figure 8.37   Examples of β-blockers with side chains extended beyond the amino group. *Note*: Chiral centres are indicated by an asterisk

acted as adrenoceptor antagonists (Fig. 8.37). Epanolol and primidolol are very selective β_1-adrenoceptor antagonists. Epanolol has been used to treat angina pectoris while primidolol has been used to treat hypertension. Xamoterol is a selective β_1 partial agonist that acts as a β-blocker when large amounts of adrenaline and noradrenaline are being produced during strenuous exercise. The related compounds, labetalol and celiprolol, also act as antagonists to both α- and β-adrenoceptors. These compounds are used to treat all forms of hypertension but they can exhibit a number of adverse side effects and are not well tolerated by patients.

β-*Blockers that selectively act on the heart*

In 1964 A. A. Larsen and P. M. Lish reported the synthesis of sotalol, a potential β-blocker in which the two phenolic hydroxy groups of isoprenaline were replaced by one sulphonamide group (Fig. 8.38). It was found to block tachycardia caused by the action of adrenaline on the heart. Furthermore, its polar nature prevented it entering the CNS and causing vivid dreams. ICI chemists used sotalol as a lead and in a SAR study synthesised a series of analogues. This resulted in the discovery of practolol, a β-blocker that acted mainly on β_1-adrenoceptors of the heart. Unfortunately, its long-term oral use in some

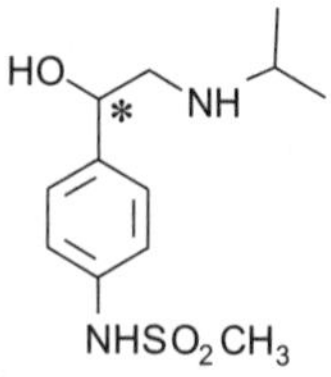

Sotalol

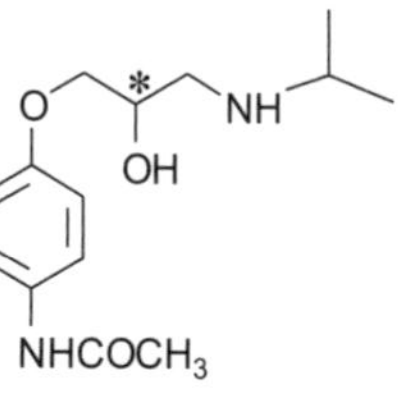

Practolol

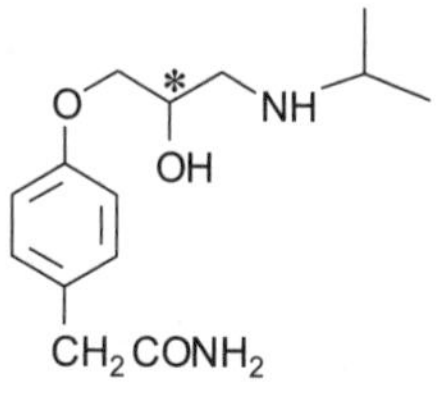

Atenolol

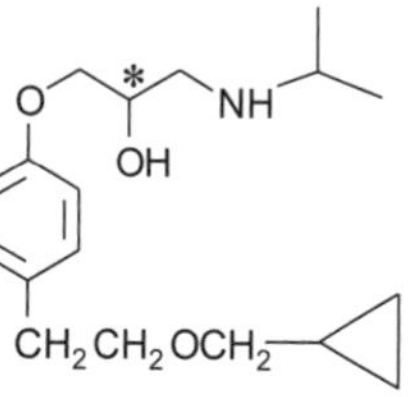

Betaxolol

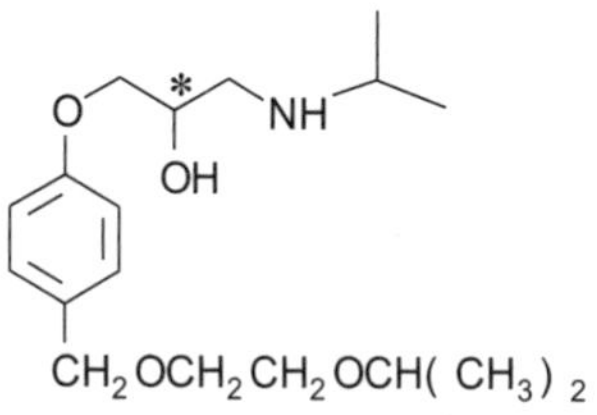

Bisoprolol

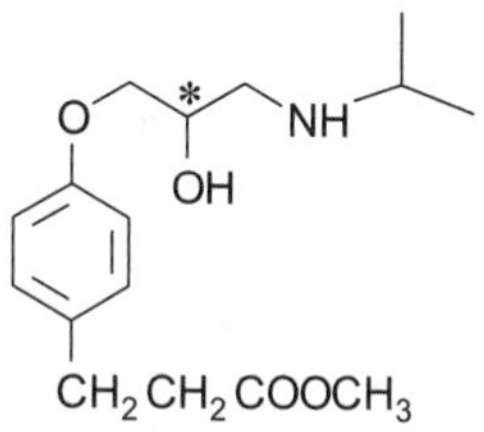

Esmolol

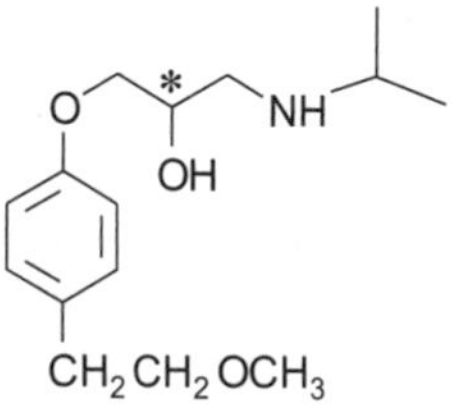

Metoprolol

Figure 8.38 Examples of cardioselective β-blockers. Chiral centres are marked by an asterisk

patients caused a serious oculomucocutaneous reaction, which in some cases resulted in blindness. Consequently, it has now been withdrawn from clinical use. However, SAR studies showed that the selectivity of the β_1-adrenoceptor activity of practolol depended on the amido group being in the *para* position. This suggests that hydrogen bonding between the amido group and the receptor plays a part in practolol's β_1 selectivity. Consequently, replacement of the amido group by other groups that could bind to the receptor and further SAR studies resulted in the discovery of a number of other drugs (cardioselective drugs) that had a relatively selective action on the β_1-adrenoceptors of the heart (Fig. 8.38). Since these drugs mainly block β_1-adrenoceptors of the heart, their use reduces the possibility of potentially life threatening situations, such as bronchospasm, caused by the blocking of β_2-adrenoceptors.

8.8 Questions

1 Explain the meaning of the terms (a) ligand, (b) endogenous molecule, (c) agonist, (d) antagonist, (e) first messenger and (f) secondary messengers in the context of medicinal chemistry.

2 Explain the meaning of the abbreviation m_2AChR in the context of the classification of receptors.

3 Suggest where the compound A would form (a) strong bonds and (b) weak bonds with a receptor. The receptor and compound A are drawn to the same scale. Compound A approaches the receptor site from in front of the paper

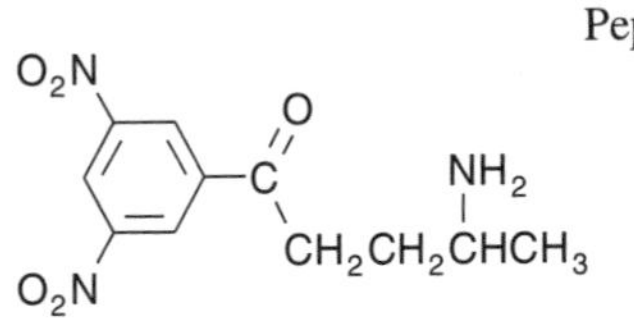

(A) 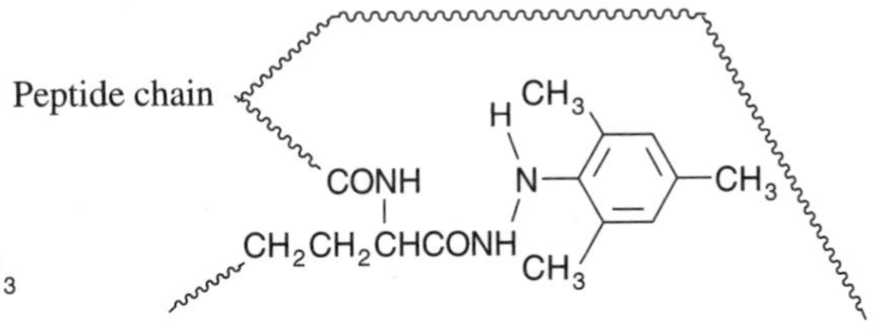   The structure of the receptor site

4 Distinguish between the terms hormone, neurotransmitter and autocoid.

5 Explain the meaning of the term *affinity* in the context of ligand–receptor relationships. What is the relationship between K_A, K_a and K_D.

6 Outline how cAMP is generated within a cell by the action of a G-protein.

7 Draw structural formulae for (a) DAG and (b) IP_3. What is the cellular function of these compounds?

8 Distinguish between competitive and non-competitive antagonists.

9 Outline the main features of the rate theory for explaining the relationships between ligands and receptors. How does this theory explain desensitisation?

10 Outline a strategy for designing a new agonist for histamine receptors. The answer should include both the structures of possible analogues and an outline of the experimental work that would be needed to support the drug development.

9

Enzymes

9.1 Introduction

Enzymes act as catalysts for almost all of the chemical reactions that occur in all living organisms. Their important general characteristics are:

- mild conditions are required for the enzyme action;

- they have a big capacity, a minute amount of enzyme rapidly producing a large amount of a product;

- they usually exhibit a high degree of specificity;

- their activities can be controlled by substances other than their substrates.

Enzymes are usually large protein molecules that are sometimes referred to as *apoenzymes*. However, some RNA molecules, known as *ribozymes* (see section 9.14), can also act as enzymes. The structures of a number of enzymes contain groups of metal ions, known as *clusters* (see section 13.2.4), coordinated to the peptide chain. These enzymes are often referred to as *metalloenzymes*.

Some enzymes require the presence of organic compounds known as *coenzymes* (Fig. 9.1) and/or metal ions and inorganic compounds referred to as *cofactors* before the enzyme will function. Coenzymes and cofactors are separate chemical species that are bound to the apoenzyme by electrostatic bonds, hydrogen bonds and van der Waals' forces. These composite active enzyme systems are known as *holoenzymes*:

$$\text{Apoenzyme} + \text{Coenzyme and/or Cofactor} = \text{Holoenzyme}$$
$$(\textit{inactive}) \qquad\qquad (\textit{inactive}) \qquad\qquad\qquad (\textit{active})$$

Medicinal Chemistry, Second Edition   Gareth Thomas
© 2007 John Wiley & Sons, Ltd

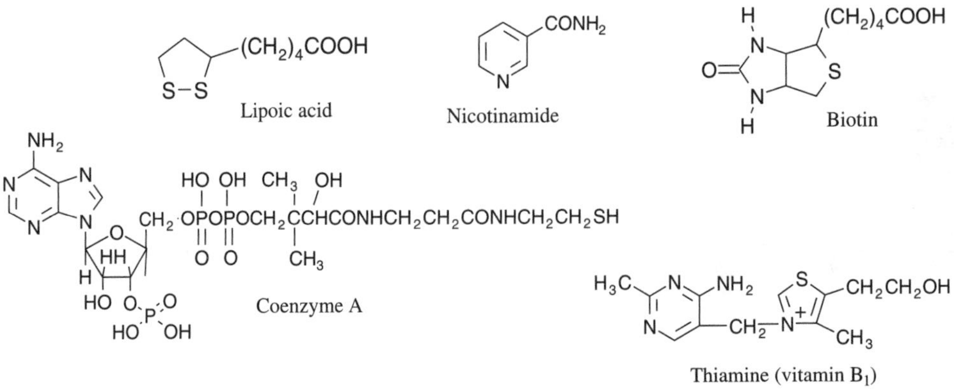

Figure 9.1 Examples of the varied nature of coenzymes

However, the term enzyme is commonly used to refer to both holoenzyme systems and those enzymes that do not require a coenzyme and/or cofactor.

Enzymes are widely distributed in the human body. It has been reported that there are over 3000 in a single cell. They are found embedded in cell walls and membranes as well as in the various biological fluids in living organisms. All enzymes are produced by cells and mainly function within that cell. It is often difficult to isolate and purify enzymes, however several thousand enzymes have been purified and characterised to some extent.

A number of enzymes are produced by the body from inactive protein precursors. These precursors are known as *proenzymes* or *zymogens*. The formation of the enzyme often requires the removal of part of the peptide chain of the precursor protein. For example, the active form of enzyme trypsin is produced from the proenzyme trypsinogen by the loss of a six-amino-acid residue chain from the N-terminal of trypsinogen. This loss is accompanied by a change in the conformation of the resultant protein to form the active form of the enzyme. Proenzymes allow the body to produce and control enzymes that could be harmful in the wrong place. For example, trypsin catalyses the breakdown of proteins. Consequently, the formation of active trypsin in the body could pose a problem and so the body produces the inactive form that is only converted to the active form when it reaches the intestine, where it catalyses the breakdown of proteins in our food.

Enzymes with different structural forms can catalyse the same reactions. These enzymes are usually isolated from different tissues but can be found in the same tissue. Where enzymes have the same catalytic activities they are referred to as *isoenzymes* or *isozymes*. Isoenzyme structures often contain the same protein subunits but either in a different order or ratio. For example, the structure of lactate dehydrogenase (LDH), which catalyses the conversion of lactate to pyruvate and vice versa, is a tetramer consisting of four so-called H and M subunits. The ratio of H to M subunits in the enzyme will depend on its source. Lactate dehydrogenase isolated from the heart contains mainly H subunits whilst that from the liver and skeletal muscle contains mainly M subunits. Although isoenzymes catalyse the same reaction they can exhibit significantly different properties, such as thermal stability, the values of pharmacokinetic parameters (see Chapter 11), electrophoretic mobilities and the effects of inhibitors.

Isoenzymes may also be used as a diagnostic aid. For example, the presence of H-type LDH in the blood indicates a heart attack, since heart attacks cause the death of heart muscle with a subsequent release of H-type LDH into the circulatory system.

Enzymes that catalyse the same reactions but have significantly different structures are known as *isofunctional* enzymes. Isofunctional enzymes are produced by different tissues and different organisms. Their different origins and structures mean that they can act as selective targets for drugs. For example, the enzyme dihydrofolate reductase catalyses the interconversion of dihydrofolate to tetrahydrofolate in both humans and bacteria. The antibiotic trimethoprim is more effective in inhibiting this conversion in bacteria than in humans, which makes it useful for the treatment of bacterial infections in humans. Nevertheless, resistance to this drug is found in some species of bacteria (see section 9.13).

Variations in the structures of enzymes can also occur between and within ethnic groups of the same species. For example, a number of different dehydrogenase isoenzymes have been observed in some groups of Asians.

9.2 Classification and nomenclature

Enzymes are broadly classified into six major types (Table 9.1).

The International Union of Biochemistry has recommended that enzymes have three names, namely: a systematic name that shows the reaction being catalysed and the type of reaction based on the classification in Table 9.1; a recommended trivial name; and a four-figure Enzyme Commission code (EC code) also based on the classification in Table 9.1. Nearly all systematic and trivial enzyme names have the suffix *-ase*.

Table 9.1 The Enzyme Commission classification of enzymes

Code	Classification	Type of reaction catalysed
1	Oxidoreductases	Oxidations and reductions
2	Transferases	The transfer of a group from one molecule to another
3	Hydrolases	Hydrolysis of various functional groups
4	Lyases	Cleavage of a bond by non-oxidative and non-hydrolytic mechanisms
5	Isomerases	The interconversion of all types of isomer
6	Ligases (synthases)	The formation of a bond between molecules

Systematic names show, often in semi-chemical equation form, the conversion the enzyme promotes and the class of the enzyme. For example, L-aspartate: 2-oxoglutarate aminotransferase catalyses the transfer of an amino group from L-aspartic acid to 2-oxoglutaric acid, whilst D-glucose ketoisomerase catalyses the conversion of D-glucose to D-fructose.

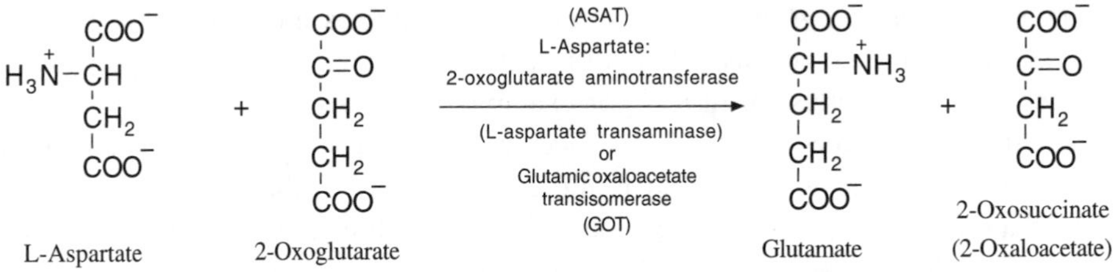

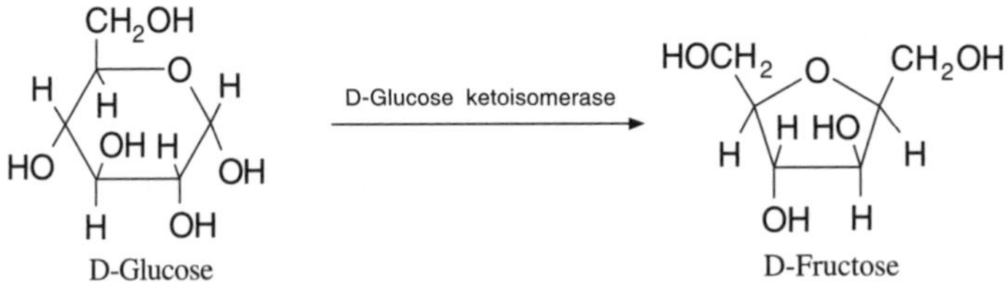

Trivial names are usually based on the function of the enzyme but may also include or be based on the name of the substrate. For example, alcohol dehydrogenase is the trivial name for alcohol NAD$^+$ oxidoreductase, the enzyme that catalyses the oxidation of ethanol to ethanal, whilst glucose-6-phosphatase catalyses the hydrolysis of glucose-6-phosphate to glucose and phosphate ions.

$$NAD^+ + \underset{\text{Ethanol}}{CH_3CH_2OH} \xrightarrow{\text{Alcohol dehydrogenase}} \underset{\text{Ethanal}}{CH_3CHO} + NADH + H^+$$

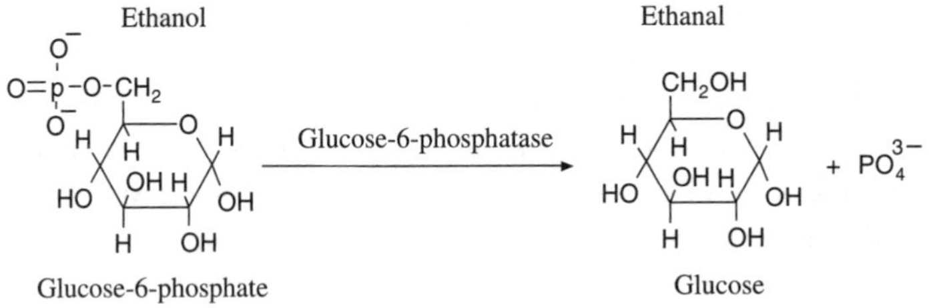

However, some trivial names in current use are historical and bear no relationship to the action of the enzyme or its substrate, for example pepsin and trypsin are the names commonly used for two enzymes that catalyse the breakdown of proteins during digestion.

The Enzyme Commission's code is unique for each enzyme. It is based on the classification in Table 9.1 but further subdivides each class of enzyme according to how it functions. For example, the Enzyme Commission code for the lactate dehydrogenase that catalyses the oxidation of L-lactate to pyruvate is EC 1.1.1.27. The letters EC show that the numbers are an Enzyme Commission code for an enzyme. The first number indicates that the enzyme is an oxidoreductase, the second that it is acting on the CH–OH bond of electron donors and the third that NAD^+ is the electron acceptor. The fourth number identifies the enzyme as a specific oxidoreductase, namely lactate dehydrogenase. The full code is given in the International Union of Biochemistry (IUB) publication *Enzyme Nomenclature*.

$$NAD^+ + HO\!-\!\overset{\displaystyle COO^-}{\underset{\displaystyle CH_3}{C}}\!-\!H \xrightarrow{\text{Lactate dehydrogenase}} \overset{\displaystyle COO^-}{\underset{\displaystyle CH_3}{C}}\!=\!O + NADH + H^+$$

L-Lactate   Pyruvate

This text uses trivial names as they are usually easier to pronounce. In addition, some letter abbreviations will also be used.

9.3 Active sites and catalytic action

The reactant(s) whose reaction is catalysed by an enzyme is known as the *substrate(s)*. All naturally occurring enzyme-controlled processes can be regarded as equilibrium processes. However, under body conditions many processes appear only to proceed in one direction because as soon as the products are formed they are either immediately used as the substrates of a subsequent enzymecatalysed reaction or physically removed from the vicinity of the enzyme.

In its simplest form, the catalytic action of enzymes is believed to depend on the substrate or substrates binding to the surface of the enzyme. This binding usually occurs on a specific part of the enzyme known as its *active site*. Once the substrate (S) or substrates are bound to the surface of the enzyme (E) a reaction takes place and the products (P) are formed and released whilst the enzyme is recycled into the system.

$$E + S \rightleftharpoons E\text{–}S \text{ complex} \rightleftharpoons E\text{–}P \text{ complex} \rightleftharpoons P + E$$

Recycled

The mechanism by which enzymes act is probably more sophisticated than that suggested by this simple active site model.

Active sites are usually visualised as pockets, clefts or indentations in the surface of the enzyme. These physical features are the result of the conformations of the protein's peptide chain. Consequently, the amino acid residues forming the site can be located some distance apart in the peptide chain but are brought together by the folding of the peptide chain. For example, the amino acid residues that form the active site of lactate dehydrogenase occupy positions 101, 171 and 195 in the peptide chain. The amino acid residues most often involved in forming active sites are serine, histidine, arginine, cysteine, lysine, aspartic acid and glutamic acid.

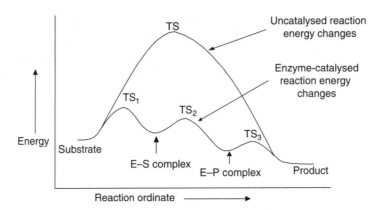

Figure 9.2 The effect of an enzyme on the minimum energy pathway of a simple enzyme-catalysed reaction involving a single substrate (TS = transition state). The heights of the energy barriers TS_1, TS_2 and TS_3 will vary depending on which step in the enzyme controlled route is the rate controlling step

It is believed that the enzyme reduces the activation energy required for each of the stages in an enzyme-controlled reaction (Fig. 9.2).

The specific nature of most enzyme action has been explained by a number of models, the earliest being the lock and key. In this model the enzyme acts as the lock and the reactants as the key. Just like the key, which has to be of the correct shape to turn in the lock, the reactant molecules must have a shape that fits the configuration of the active site in order to react (Fig. 9.3). This model assumes that the enzyme structure is rigid and that the active site will always have the correct shape for that reaction. However, it is now known that the structures of enzymes are not rigid and can change to accommodate the substrate. In spite of this the lock and key model is still a useful concept, even though it is an oversimplification that does not satisfactorily explain all enzyme behaviour.

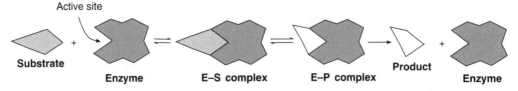

Figure 9.3 The lock and key model of enzyme action

The discovery that the binding of a substrate to an active site changes the conformation of the site led Daniel Koshland to propose the *induced fit hypothesis* of enzyme action. He proposed that the enzyme and substrate interact with each other, mutually adjusting their conformations to allow the substrate to fit the active site (Fig.9.4) in the exact position necessary for reaction This concept explains why enzymes are more effective than chemical catalysts in increasing the rates of reactions.

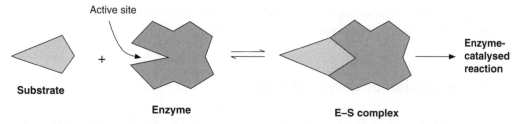

Figure 9.4 A schematic representation of the induced fit hypothesis of enzyme action

Coenzymes are essential components of the active sites of many enzymes. They appear to act by binding to the apoenzyme to form the active site of the enzyme (Fig.9.5). This binding is often very strong and can involve covalent bonds as well as the full range of electrostatic forces.

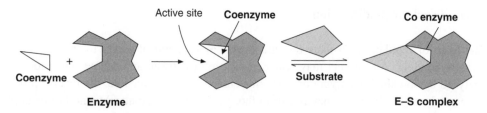

Figure 9.5 A schematic representation of the action of the coenzymes using the induced fit model

The active site is not the only place where compounds may bind to an enzyme. These alternative sites are known as *allosteric sites*. The binding of a compound to the allosteric site of an enzyme may cause a change in its conformation (Figure 9.6). These changes may result in the activation of inhibition of an enzyme or have no effect (see section 9.4.2).

9.3.1   Allosteric activation

The change in conformation caused by a substrate binding to the active site of an enzyme can result in the formation of another active site on the enzyme (Fig. 9.6). This behaviour is

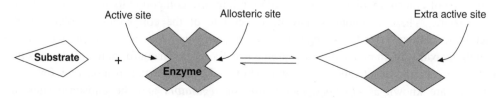

Figure 9.6 A schematic representation of allosteric activation

known as allosteric activation. It occurs with increasing concentration of the substrate and increases the capacity of the enzyme to process substrate.

9.4 Regulation of enzyme activity

The chemical activity of a cell must be controlled in order for the cell to function correctly. Enzymes offer a means of controlling that activity in that they can be switched from active to inactive states by changes induced in their conformations. Compounds that modulate the action of enzymes and other molecules by these conformational changes are known as *regulators* or *effectors*. These compounds may switch an enzyme on (activators) or off (inhibitors). They may be generally classified for convenience as compounds that covalently modify the enzyme and as allosteric regulators.

9.4.1 Covalent modification

This form of regulation involves the attachment of a chemical group to the enzyme by covalent bonding. Attachment may either inactivate or activate an enzyme. The process is catalysed by so-called *modifying* or *converter* enzymes. It is reversible, the reverse process being catalysed by different enzymes from those required for the attachment. For example, glycogen phosphorylase, the enzyme that catalyses the formation of glucose from glycogen, is inactive until the hydroxy group of a serine residue is phosphorylated. This conversion is itself catalysed by kinases that are themselves usually activated by Ca^{2+} ions.

$$\text{Glycogen} \xrightarrow{\text{Glycogen phosphorylase}} \text{Glucose-1-phosphate}$$

$$\text{Active enzyme} -CH_2-O-\overset{\overset{\displaystyle O}{\|}}{\underset{\underset{\displaystyle O^-}{|}}{P}}-O^-$$

Phosphorylation by protein kinase ↑ | ↓ Dephosphorylation by glycogen phosphoryl phosphatase

$$\text{Inactive enzyme} -CH_2OH$$

9.4.2 Allosteric control

This type of control involves the reversible binding of a compound to an allosteric site on the enzyme. These compounds may be either one of the compounds involved in the metabolic pathway (feedback regulators) or a compound that is not a product of the metabolic pathway. In both cases the binding usually results in conformational changes that either activate or deactivate the enzyme (Fig. 9.7). Compounds responsible for these changes are known as *effectors, modulators* or *regulators* and the allosteric sites are referred to as *regulatory sites*.

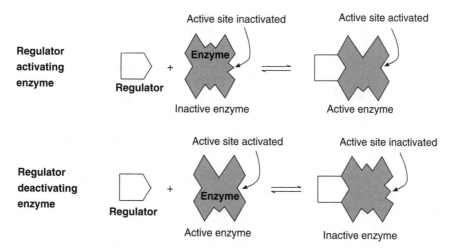

Figure 9.7 A schematic representation of allosteric control

Feedback control is an automatic form of allosteric regulation. It often, but not always, utilises the end product of the metabolic pathway to regulate its production. Consider, for example, the metabolic pathway:

where E, E_1, to E_n are the enzymes used in the metabolic pathway. When the concentration of D is low the enzymes operate to increase the production of D. As the concentration of D increases it increasingly inhibits the action of E_1. Eventually the enzyme stops acting and no more B and therefore no more D is produced. As D is utilised by the body its concentration decreases and, as a consequence, its inhibiting effect on E_1 also decreases.

Some regulators that activate an enzyme need to be activated themselves before they can act. For example, calmodulin (CaM) is a small dumb-bell-shaped protein with four binding sites that has a high affinity for Ca^{2+} ions. It activates a number of enzymes such as ATPase and nitric oxide synthase (see section 14.4). CaM is inactive until the concentration of Ca^{2+} ions in the surrounding medium increases. As the concentration of Ca^{2+} ions in the medium increases, the number of sites occupied by ions increases until all four are occupied. Binding of Ca^{2+} ions to all four sites changes CaM's conformation, which allows it to bind to and activate the enzyme. Conversely, a decrease in the calcium ion concentration of the surrounding medium results in the release of Ca^{2+} ions from CaM, which results in it becoming inactive with a subsequent deactivation of the enzyme system.

$$CaM + Ca^{2+} \rightleftharpoons CaM{-}Ca^{2+} \rightleftharpoons Enzyme{-}CaM{-}Ca^{2+}$$

Calmodulin   Active enzyme

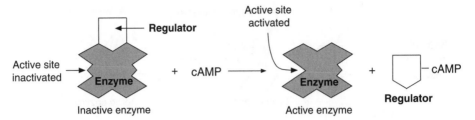

Figure 9.8 A schematic representation of the action of cAMP as a positive modulator

Where a regulator inhibits an enzyme by binding to a regulatory site it may be removed by a second modulator. For example, the regulator of protein kinase, an enzyme that catalyses the transfer of phosphate groups, is a protein molecule that binds to the regulatory site of the kinase to form an inactive complex. This complex is activated by cyclic adenosine monophosphate (cAMP), which removes the regulator (Fig. 9.8). cAMP is known as a *positive modulator* since it activates the enzyme.

9.4.3   Proenzyme control

Proenzymes (zymogens) are precursors of enzymes. They are also a form of enzyme control since the active form of the enzyme is produced from the proenzyme where and when it is required. This often requires the removal of a section of the protein chain by a reaction controlled by another enzyme. This second enzyme is known as an *activator* because, unlike a modulator, it does not bind to the enzyme and cannot switch off an enzyme. For example, the active form of the digestive enzyme chymotrypsin is produced from its inactive proenzyme chymotrypsinogen by cleavage of an arginine–isoleucine peptide bond by trypsin to form the active π-chymotrypsin (Fig.9.9). π-Chymotrypsin autocatalytically removes two dipeptides from itself to form α-chymotrypsin, which is the active form normally referred to as chymotrypsin. The bond fission and the removal of the two dipeptides allow the protein molecule to assume its active conformation.

9.5   The specific nature of enzyme action

The specific action of enzymes can be broadly described in terms of the nature and functional groups of compounds and their stereochemistry. Most enzymes will only act on a specific functional group in groups of structurally similar compounds of roughly similar size. For example, alcohol dehydrogenase will catalyse the oxidation of a number of small primary and secondary alcohols (such as methanol, ethanol and propan-2-ol) to the corresponding aldehydes and ketones. The rates and yields of these reactions vary considerably. However, some enzymes are selective in that they mainly catalyse reactions involving one particular compound, whilst others will catalyse the reactions of large numbers of different substrates with the same functional group. For example, the digestive

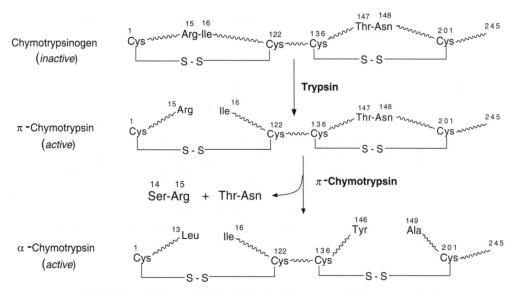

Figure 9.9   The production of active chymotrypsin from its proenzyme chymotrypsinogen. The enzymes for each stage of the process are in bold type

enzyme carboxypeptidase will remove all types of C-terminal amino acid residues from proteins except arginine, lysine and proline. Enzymes are also often stereospecific because they can form asymmetric active sites. This means that an active site will have a strong affinity for the complementary-shaped compound (Fig. 9.10) but not its stereoisomers. For example, trypsin will catalyse the hydrolysis of L-Arg–L-Ile peptide links in proteins but

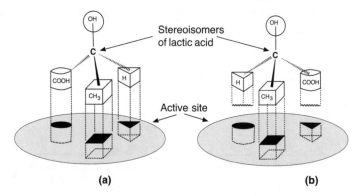

Figure 9.10   A schematic illustration of the stereospecific nature of enzymes. The groups of the two stereoisomers of lactic acid are represented by the three-dimensional shapes shown and the sections of the active site that bind to these groups are represented as the correspondingly shaped sockets. In **(a)** the relevant groups fit into these sockets, that is, the structures are complementary and the groups of L-lactic acid are correctly aligned for binding. However, in **(b)** two of the groups will be misaligned. The wedge and cylinder are unable to fit their sockets and so the groups of D-lactic acid will not be correctly aligned for binding to the active site

will not catalyse the hydrolysis of D-Arg–D-Ile peptide links. Most enzymes exhibit this type of stereospecific behaviour.

9.6 The mechanisms of enzyme action

The mechanisms of the reactions at the active sites of enzymes have been widely investigated. There is now no doubt that the increased speed of enzyme-catalysed reactions is due in part to the close proximity and the correct alignment of the reactants at the active site. A detailed knowledge, obtained from X-ray crystallography, of the three-dimensional structures of the enzyme and bound substrate plus other experimental evidence means that in many cases the details of enzyme action have been established. This has resulted in the proposal of reaction mechanisms for a number of processes. However, it should be realised that each mechanism is specific for the enzyme–substrate system under consideration.

9.7 The general physical factors affecting enzyme action

The main physical factors affecting enzyme action are the relative concentrations of the enzyme and substrate, the pH and the temperature. If the concentration of the substrate is kept constant, the rate of the reaction increases with an increase in enzyme concentration. However, if the concentration of the enzyme is kept constant, as is likely in biological processes, and the concentration of the substrate is increased, the rate of reaction increases to a maximum value known as the *saturation value* (Fig. 9.11a). This maximum rate corresponds to the saturation of all the available active sites of the enzyme. In the case of enzymes that are under feedback control (see section 9.4.2) the curve is sigmoidal (Fig. 9.11b), as against hyperbolic for enzymes not under feedback control but still regulated by a modulator, that is, under allosteric control.

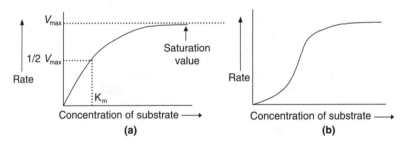

Figure 9.11 The effect of concentration on the rate of allosteric (**a**) modulated and (**b**) feedback-regulated enzyme-catalysed reactions

Enzymes are only usually effective over specific ranges of pH characteristic of the enzyme (Fig. 9.12). These ranges usually correspond to the pH of the environment in which the enzyme occurs. Outside this range the enzyme may undergo irreversible denaturation

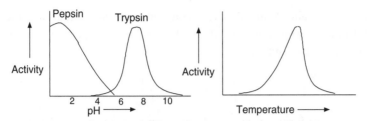

Figure 9.12   The effect of pH and temperature on the rates of enzyme-catalysed reactions

with subsequent loss of activity. Initially an increase in temperature will usually result in an increase in the rate of an enzyme-controlled reaction. However, enzymes are sensitive to temperature and once it increases beyond a certain point the protein irreversibly denatures and the enzyme becomes inactive. An enzyme's operating temperature range will normally correspond to about that of its normal environment.

9.8   Enzyme kinetics

9.8.1   Single substrate reactions

In these reactions a single substrate (S) is converted by the enzyme (E) to the products (P):

$$E + S \rightleftharpoons E\text{--}S \rightleftharpoons E\text{--}P \longrightarrow E + P$$

The mathematical relationship between the rate of the enzyme-catalysed reaction and the concentration of the single substrate depends on the nature of the enzyme process. Many processes exhibit first-order kinetics at low concentrations of the substrate, which changes to zero order as the concentration increases, that is, a hyperbolic relationship between rate and substrate concentration (Fig. 9.11a). Processes under allosteric control have a sigmoid relationship between reaction rate and the substrate concentration (Fig. 9.11b). This suggests second and higher orders of reaction, that is, the rate is proportional to $[S]^n$ where $n > 1$. Only the kinetics of enzyme processes that exhibit hyperbolic rate/substrate concentration relationships will be discussed in this text.

The kinetics of enzyme processes with a single substrate that exhibit a hyperbolic rate/ substrate concentration relationship may often be described by the Michaelis–Menten equation:

$$V = \frac{V_{max}[S]}{K_m + [S]} \tag{9.1}$$

where K_m is a constant known as the *Michaelis constant*, $[S]$ is the concentration of the substrate, V is the rate of the reaction and V_{max} is the rate of the reaction when the concentration of the substrate approaches infinity. When $V = V_{max}/2$, it follows that

$K_m = [S]$, that is, the value of the Michaelis constant is the concentration of the substrate at $V_{max}/2$ (Fig.9.11a). Rearrangement of the Michaelis–Menten equation (9.1) gives the more useful Lineweaver–Burk equation:

$$\frac{1}{V} = \frac{K_m}{V_{max}[S]} + \frac{1}{V_{max}} \tag{9.2}$$

This equation is of the form $y = mx + c$. Consequently, a graph of $1/V$ against $1/[S]$ will be a straight line with a slope of K_m/V_{max} and an intercept on the y axis of $1/V_{max}$ (Fig. 9.13). Furthermore, when $V = 0$ the intercept on the x axis will be $-1/K_m$.

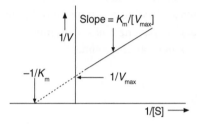

Figure 9.13   The Lineweaver–Burk double reciprocal plot of an enzyme process with a hyperbolic rate curve

The values of K_m and V_{max} are characteristic properties of the enzyme and substrate system under the specified conditions (Table 9.2). The value of K is a measure of the affinity of the substrate for the enzyme.

Table 9.2   Examples of K_m values for some enzymes and their substrates

Enzyme	Substrate	K_m (mM)
Chymotrypsin	Acetyl-L-tryptophanamide	5
	N-Acetyltyrosinamide	32
	Glycyltyrosinamide	122
Glutamate dehydrogenase	Glutamate	0.12
	NADH	0.018
Penicillinase	Benzylpenicillin	0.05

The use of reciprocal values leads to an increased error for low values when plotting any data. Consequently, several other rearrangements of the Michaelis–Menten equation have been developed to convert this equation into a straight line format. Two popular rearrangements are the Eadie–Hofstee and Hanes–Wolf plots (Fig. 9.14) that are used to obtain accurate values of K_m and V_{max}. The Eadie–Hofstee plot is particularly useful for detecting deviations from a straight line, that is, deviations from Michaelis–Menten kinetics. These deviations may indicate that the enzyme is under feedback control. However, the Lineweaver–Burk plot will be used in this text.

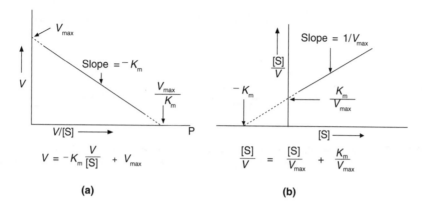

Figure 9.14 The **(a)** Eadie–Hofstee and **(b)** Hanes–Wolf plots for an enzyme process with a hyperbolic rate curve

9.8.2 Multiple substrate reactions

Enzymes often catalyse reactions that involve more than one substrate. The kinetics of these reactions will depend on the number of substrates and the general mechanism of enzyme action. However, the kinetics of many of these reactions will still follow the mathematics discussed in section 9.8.1 provided that the concentrations of all the substrates except the one being studied are kept constant. For example, reactions involving two substrates A and B (bisubstrate reactions) often proceed by one of two general routes: the sequential and the double displacement routes.

The sequential or single-displacement reactions

These reactions can follow two general routes. In the first case the order in which the substrates A and B bind to the enzyme has no effect (random order) but in the second case one substrate, known as the leading substrate, must bind to the enzyme before the other substrate can bind to the enzyme (ordered process). Once both substrates are bound to the enzyme, reaction to the products occurs.

$$E + A + B \rightleftharpoons A\text{--}E\text{--}B \rightleftharpoons P\text{--}E\text{--}N \longrightarrow E + P + N$$
$$E + A \rightleftharpoons A\text{--}E + B \rightleftharpoons A\text{--}E\text{--}B \rightleftharpoons P\text{--}E\text{--}N \longrightarrow E + P + N$$

Both of these types of sequential reaction routes exhibit the characteristic type of Lineweaver–Burk plot shown in Figure 9.15a when the concentrations of one of the substrates and the enzyme are kept constant and the concentration of the other substrate is varied.

Double-displacement or ping-pong reactions

In reactions following this route a substrate binds to the enzyme and forms a complex, which forms a product P and a modified form of the enzyme. At this point the second

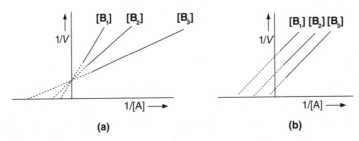

Figure 9.15 The Lineweaver–Burk plots of reactions for **(a)** the sequential and **(b)** double-displacement routes for enzyme-catalysed reactions. The lines correspond to different constant concentrations of substrate B where $[B_3] > [B_2] > [B_1]$

substrate binds to the modified enzyme and reacts to form a second product N and regenerate the original form of the enzyme.

$$E + A \rightleftharpoons E\text{-}A \rightleftharpoons E'\text{-}P \xrightarrow[\text{B}]{\text{P}} E' \longrightarrow E'\text{-}B \rightleftharpoons E'\text{-}N \longrightarrow E + N$$

Reactions proceeding by this type of route exhibit characteristic Lineweaver–Burk plots of the type shown in Figure 9.15b when the concentration of one substrate is varied and the concentrations of the enzyme and other substrate are kept constant.

9.9 Enzyme inhibitors

The use of compounds to inhibit enzyme action is an important possibility for therapeutic intervention. This approach has been used to disrupt essential steps in a key metabolic pathway responsible for a pathological condition (see section 10.12) as well as the cell wall and plasma membrane synthesis of microorganisms (see section 7.4). Compounds used in this way have a wide range of structures. Those compounds whose structures closely resemble that of the normal substrate for an enzyme are called *antimetabolites* (see section 10.13).

Inhibitors (I) may have either a reversible or irreversible action. Reversible inhibitors tend to bind to an enzyme (E) by electrostatic bonds, hydrogen bonds and van der Waals, forces and so tend to form an equilibrium system with the enzyme. A few reversible inhibitors bind by weak covalent bonds but this is the exception rather than the rule. Irreversible inhibitors usually bind to an enzyme by strong covalent bonds. However, it should be realised that in both reversible and irreversible inhibition the inhibitor does not need to bind to all of the active site in order to prevent enzyme action. It is only necessary for the inhibitor to partially block the active site in order for it to inhibit the reaction.

$$\textbf{Reversible:}\quad E + I \rightleftharpoons E\text{-}I \text{ complex}$$
$$\textbf{Irreversible:}\quad E + I \longrightarrow E\text{-}I \text{ complex}$$

9.9.1 Reversible inhibitors

These are inhibitors that form a dynamic equilibrium system with the enzyme. Reversible inhibitors are normally time dependent because the removal of unbound inhibitor from the vicinity of its site of action will disturb this equilibrium. This allows more enzyme to become available, which decreases the inhibition of the process. Consequently, the inhibitory effect of reversible inhibitors is time dependent. Furthermore, drugs acting as reversible enzyme inhibitors will only be effective for a specific period of time.

Most reversible inhibitors may be further classified as being competitive, non-competitive or uncompetitive.

Competitive inhibition

In competitive inhibition the inhibitor *usually* binds by a reversible process to the same active site of the enzyme as the substrate. This competition between the inhibitor and substrate for the same site results in some enzyme molecules being inhibited whilst others function as normal. The net result is an overall reduction in the rate of conversion of the substrate to the products (Fig. 9.16). Since the binding of the inhibitor is reversible, its effect will be reduced as the concentration of the substrate increases until, at high substrate concentrations, the inhibitory effect is completely prevented. This allows the V_{max} for the process to be reached. However, it also follows that as the concentration of the inhibitor increases its inhibiting effect increases. Consequently, for an inhibitor to be effective a relatively high concentration has to be maintained in the region of the enzyme.

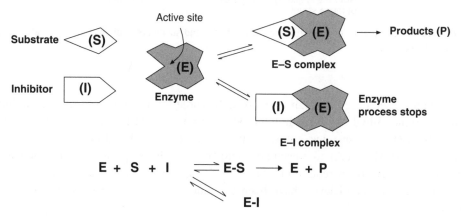

Figure 9.16 A schematic representation of competitive inhibition

For single substrate enzyme processes exhibiting simple Michaelis–Menten kinetics the Lineweaver–Burk plots for competitive inhibition show characteristic changes when different concentrations of the inhibitor (I) are used. These changes do not affect the value of V_{max} but can be used to diagnose the possible presence of a competitive inhibitor (Fig. 9.17).

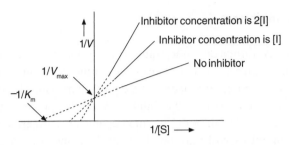

Figure 9.17   The Lineweaver–Burk plot for competitive inhibition

Since the substrate and inhibitor compete for the same active site it follows that they will probably be structurally similar. For example, succinate dehydrogenase, which catalyses the conversion of succinate to fumarate, is inhibited by malonate, which has a similar structure to succinate (Fig. 9.18). The structural relationship between substrates and competitive inhibitors offers a rational approach to drug design in this area (see section 9.11).

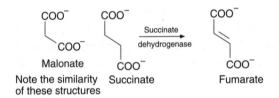

Figure 9.18   The similarity of the structures of malonate and succinate explains why malonate inhibits succinate dehydrogenase

The statins, used to reduce high levels of LDL-cholesterol, act as competitive inhibitors of the enzyme 3-hydroxy-3-methylglutaryl coenzyme A (HMG-CoA) reductase (see section 9.12.3). This enzyme catalyses the conversion of 3-hydroxy-3-methylglutaryl coenzyme A (HMG-CoA) to mevalonic acid, one of the steps in the biosynthesis of cholesterol. Cholesterol is required by the body for membrane construction and hormones, however high cholesterol levels have been associated with an increase in cardiovascular disease.

Both methotrexate and trimethoprim are competitive inhibitors of dihydrofolate reductase (DHFR). This enzyme, which occurs in organisms ranging from bacteria, to plants and mammals, catalyses the conversion of dihydrofolic acid to tetrahydrofolic acid (see section 10.13.1). Methotrexate is used to treat a variety of cancers while trimethoprim is used in combination with sulphamethoxazole to treat bacterial infections (see section 10.13.2). Trimethoprim has a high affinity for bacterial DHFR, which makes it particularly useful as an antibacterial.

Non-competitive inhibition

Non-competitive inhibitors bind reversibly to an allosteric site on the enzyme (Fig. 9.19). In *pure* non-competitive inhibition, the binding of the inhibitor to the enzyme does not influence the binding of the substrate to the enzyme. However, this situation is uncommon and the binding of the inhibitor usually causes conformational changes in the structure of the enzyme, which in turn affects the binding of the substrate to the enzyme. This is known as *mixed* non-competitive inhibition.

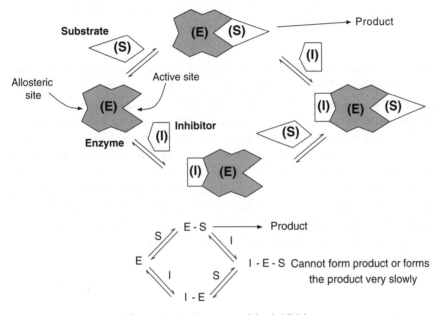

Figure 9.19 Non-competitive inhibition

In both types of non-competitive inhibition the binding of the inhibitor to the enzyme ultimately results in the formation of an enzyme–substrate–inhibitor (I–E–S) complex, which will not yield the normal products for the enzyme-catalysed process. However, some normal products will be formed since not all the E–S complex formed will be converted to the I–E–S complex (Fig. 9.19). Furthermore, as non-competitive inhibition is reversible the E–S complex can be reformed by the loss of the inhibitor. Consequently, effective inhibition will depend on maintaining a relatively high concentration of the inhibitor in contact with the enzyme in order to force the equilibria to favour the formation of I–E and I–E–S.

Where single substrate enzyme processes exhibit Michaelis–Menten kinetics both types of non-competitive inhibition exhibit characteristic changes to their Lineweaver–Burk plots (Fig 9.20). Both types of non-competitive inhibitors decrease the value of V_{max}. Furthermore, K_m remains the same for pure non-competitive inhibition but changes for mixed non-competitive inhibition.

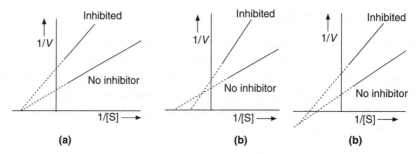

Figure 9.20 The Lineweaver–Burk plots for **(a)** pure and **(b)** the two general cases of mixed non-competitive inhibition

The fact that in non-competitive inhibition the inhibitor does not bind to the active site of the enzyme means that the structure of the substrate cannot be used as the basis for designing new drugs that act in this manner to inhibit enzyme action.

Non-competitive inhibitors have been used to treat aquired immune defficiency disease (AIDS). AIDS is now known to be caused by the HIV virus (see section 10.14.2). This is a retrovirus that uses an enzyme reverse transcriptase (RT) to reproduce. One approach to controlling human immunodeficiency virus (HIV) and hence AIDS is to inhibit RT, which would prevent virial replication. Random screening of chemical data bases yielded a number of so-called non-nucleoside reverse transcriptase inhibitors (NNRTIs). Nevirapine, delaviridine and efavirenz are non-competitive NNRTIs and as such are used in combination with other AIDS drugs to control HIV levels in AIDS cases (see section 10.14.3).

Uncompetitive inhibition

Uncompetitive inhibitors are believed to form a complex with the enzyme–substrate complex.

$$E + S \rightleftharpoons ES + I \rightleftharpoons IES$$

The substrate residue in this IES complex is unable to react and form its normal product. An uncompetitive inhibitor does not change the slope of its Lineweaver–Burk plot of $1/V$

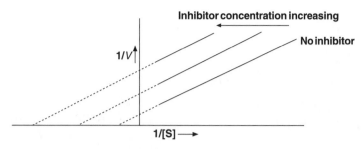

Figure 9.21 The changes in the Lineweaver–Burk plot for increasing amounts of an uncompetitive inhibitor

against [S] but moves it towards the left-hand side of the plot as the concentration of the inhibitor increases (Fig. 9.21).

The substrate is often used as the lead compound when designing new uncompetitive inhibitors

Uncompetitive inhibitors active against 5α-reductase

Two isoforms of 5α-reductase are known, type I found mainly in the liver and type II that occurs mainly in the prostate gland and testes. Type I is concerned with the metabolism of testosterone and related compounds while type II catalyses the reduction of testosterone to dihydrotestosterone (DHT), which is responsible for a number of male characteristics. DHT is also involved in benign prostate hyperplasia (BPH), an enlargement of the prostate that affects many older men. Examination of people deficient in 5α-reductase led researchers to suggest that inhibition of 5α-reductase type II might be used to treat BPH. Investigations by medicinal chemists resulted in the development of the 5α-reductase inhibitors episteride (SmithKline-Beecham) and finasteride (Merck Research Laboratories). However, episteride was not developed beyond its Phase I trials, probably because finasteride was more potent.

Testosterone	Episteride	Finasteride

Finasteride is a relatively selective uncompetitive inhibitor of 5α-reductase type II. The mechanism of its action is believed to involve it bonding to the NADH cofactor of the enzyme (Fig 9.22) to form a NADPH-finasteride complex. This complex has a half life of about 30 days and so, as a result, the 5α-reductase is almost irreversibly inhibited by the finasteride. Finasteride is also used to treat male pattern baldness.

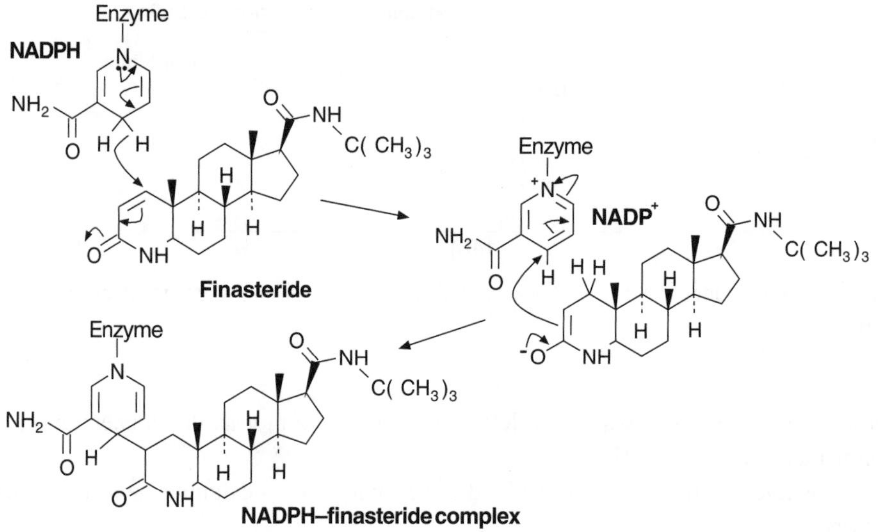

Figure 9.22 A simplification of the mechanism of the action of finasteride. A–H is an active hydrogen atom source

It is believed that 5α-reductase type I also plays a part in the progression of hormone-dependent prostate cancer. As a result, dutasteride, a drug that inhibits both types of 5α-reductase, was developed. This drug has been approved for use in treating BPH.

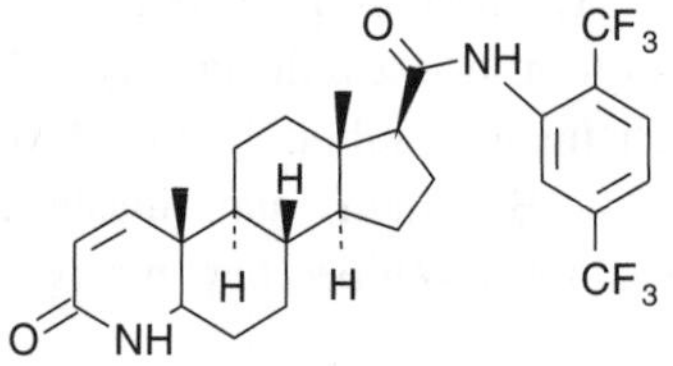

Dutasteride

9.9.2   Irreversible inhibition

Irreversible inhibitors bind to the enzyme by either strong non-covalent or strong covalent bonds. Compounds bound by strong non-covalent bonds will slowly dissociate, releasing the enzyme to carry out its normal function. However, whatever the type of binding the enzyme will resume its normal function once the organism has synthesised a sufficient number of additional enzyme molecules to overcome the effect of the inhibitor.

Irreversible inhibitors may be classified for convenience as active site-directed inhibitors and suicide or irreversible mechanism-based inhibitors (IMBI).

Active site-directed inhibitors

Active site-directed inhibitors are compounds that bind at or near to the active site of the enzyme. These inhibitors usually form strong covalent bonds with the functional groups that are found at the active site or close to that site (Fig. 9.23). The inhibitor effectively reduces the concentration of the active enzyme, which in turn reduces the rate of formation of the normal products of the enzyme process.

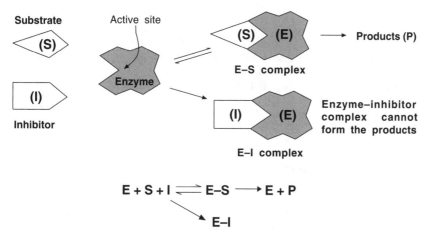

Figure 9.23 Irreversible inhibition

The action of many irreversible inhibitors depends on the inhibitor possessing a functional group that can react with a functional group at the active site of the enzyme. Consequently, as the functional groups found at the active sites of enzymes are usually nucleophiles, the incorporation of strongly electrophilic groups in the structure of a substrate can be used to develop new inhibitors (Table 9.3). Furthermore, this approach can also be used to enhance the action of a known inhibitor. In all cases the product of the reaction of the inhibitor with the enzyme must be relatively stable if inhibition is to be effective.

Many of the inhibitors developed in this way are too reactive and as a result too toxic to be used clinically. However, they have been used to identify the amino acid residues forming the active site of an enzyme. The inhibitor, when it irreversibly binds to a functional group at the active site, effectively acts as a chemical label for the amino acid residue containing that group. Identification of the labelled residue indicates which residues are involved in forming the active site of the enzyme.

Most of the active site-directed irreversible inhibitors in clinical use were not developed from a substrate. They were obtained or developed by other routes and only later was their mode of action discovered. For example, aspirin, first used clinically at the end of the nineteenth century, was a direct development from the use of salicylic acid as an antipyretic by Carl Buss in 1875.

Table 9.3 Examples of the electrophilic groups used to produce active site-directed inhibitors

Nucleophilic group of enzyme (E)		Electrophilic group		Product
E-NH$_2$	Anhydrides	RCOOCOR	Amides	RCONH-E
	Ketones	>C=O	Imines	>C=N-E
	Arenesulphonyl halides	RSO$_2$X	Arenesulphonamides	RSO$_2$NH-E
E-COOH	Epoxides	R$-\overset{O}{\triangle}$	Hydroxyesters	RCH(OH)CH$_2$CO$_2$-E
	α-Haloacetates	X-CH$_2$CO$_2$$^-$	Half esters	O$_2$CCHCO$_2$-E
E-OH	Phosphoryl halides	(RO)$_2$PO(X)	Phosphates	(RO)$_2$OPO-E
	Carbamates	RNHCOOR	Carbamates	RNHCOO-E
E-SH	α-Haloesters	X-CH$_2$CO$_2$R	Sulphides	ROCOCH$_2$-S-E
	α-Haloacetates	X-CH$_2$CO$_2$	Sulphides	ROCOCH$_2$-S-E
	Epoxides	R$-\overset{O}{\triangle}$	Hydroxy sulphides	RCH(OH)CH$_2$-S-E

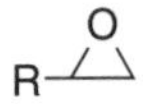

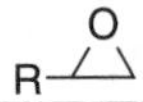

Anti-inflammatory drugs: a case study

Aspirin is believed to irreversibly inhibit prostaglandin synthase. This enzyme, which is also known as cyclooxygenase (COX), exists as two isozymes, COX-1 and COX-2. Both of these forms catalyse the metabolism of arachidonic acid to prostaglandin G$_2$ (PGG$_2$), which is further oxidised to prostaglandin H$_2$ (PGH$_2$) and a number of other prostaglandins (Fig.9.24). Many of these metabolic products are biologically active. It is believed that excessive concentrations of these biologically active compounds are associated with inflammation, pain, swelling and fevers. Arachidonic acid is a component of the phospholipid matrix of most cell membranes and as such is not normally available for metabolism. It is released from the cell matrix in response to a traumatic event such as a wound, hormonal stimulation or a toxin. The subsequent oxidation followed by the action of the prostaglandin metabolites is thought to be responsible for inflammation, pain, swelling and fever.

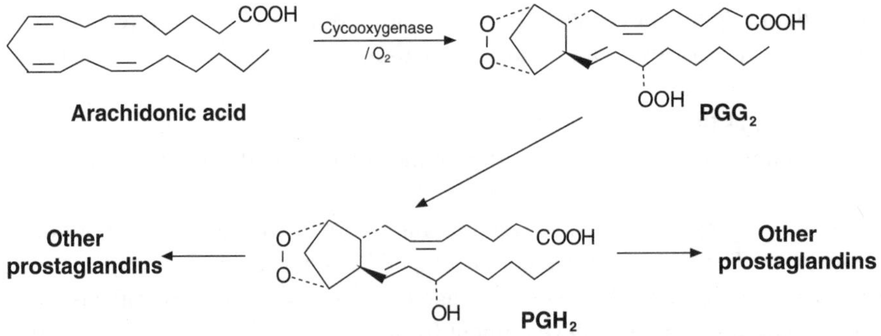

Figure 9.24 The oxidation of arachidonic acid by COX -1 and COX-2

Experimental evidence suggests that aspirin irreversibly inhibits both COX-1 and COX-2 by acetylating serine hydroxy groups at the enzyme's active site, probably by a transesterification mechanism.

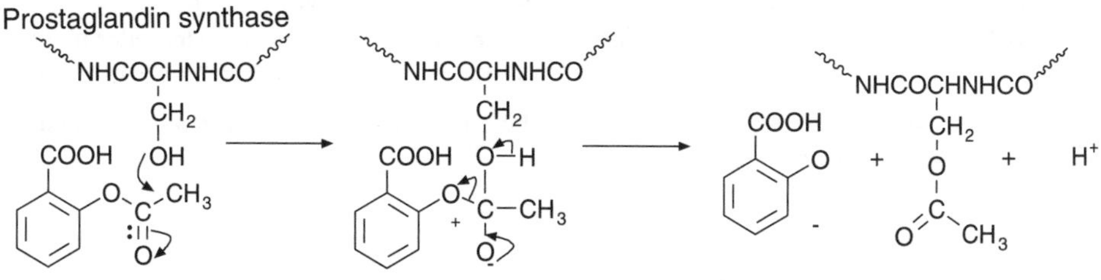

Aspirin, ibuprofen and other non-steroidal anti-inflammatory drugs (NSAIDs) inhibit both forms of COX. Their long-term use caused ulceration of both the kidney and the GI tract in significant numbers of patients. However in 1991 D. Simmons *et al.* discovered that COX-2 was responsible for the production of prostaglandins that led to inflammation and other unwanted side effects, while COX-1 produced prostaglandins that were necessary for normal body processes. This discovery prompted the pharmaceutical companies to carry out research to discover selective COX-2 inhibitors that would not have these unwanted side effects. With this objective in mind, G. D. Searle tested over 2500 compounds from the company's agrochemical library for selective COX-2 inhibition. Seven of these compounds were found to be suitable as COX-2 inhibitors but one also exhibited anti-inflammatory activity and was marketed in 1999 as celeoxib. Other companies followed suit and this ultimately led to the discovery and development of rofecoxib (VIOXX) and valdecoxib. Rofecoxib has now been withdrawn because its long-term use led of an increased risk of heart attacks in patients with a history of ischaemic heart disease.

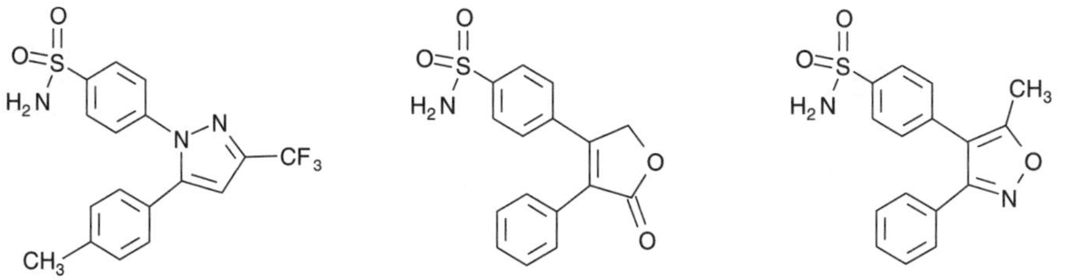

Celeoxib   Rofecoxib   Valdecoxib

Suicide inhibitors

Suicide inhibitors, alternatively known as K_{cat} or irreversible mechanism-based inhibitors (IMBI), are irreversible inhibitors that are often analogues of the normal substrate of the enzyme. The inhibitor binds to the active site where it is modified by the enzyme to produce a reactive group that reacts irreversibly to form a stable inhibitor–enzyme complex

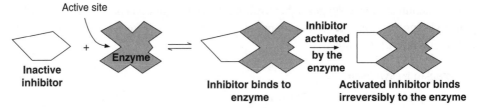

Figure 9.25 A schematic representation of suicide inhibition in which the enzyme activates the inhibitor to form a strong covalent bond with the enzyme

(Fig.9.25). This subsequent reaction may or may not involve functional groups at the active site. These mechanisms mean that suicide inhibitors are likely to be reasonably specific in their action since they can only be activated by a particular enzyme. Consequently, this specificity means that drugs designed as suicide inhibitors could exhibit a lower degree of toxicity.

The action of suicide inhibitors usually involves either the prosthetic group of the enzyme or its coenzyme. For example, 7-aminomethylisatoic anhydride is a suicide inhibitor of serine protease thrombin, which catalyses the hydrolysis of peptides and proteins. This catalysis involves the rapid hydrolysis of acyl-enzyme intermediates. The inhibitor reacts with the hydroxy group of a serine residue of thrombin to form an acyl derivative, which rapidly decarboxylates to form a hydrolysis-resistant complex stable enough to inhibit the enzyme. It is possible that the resistance to hydrolysis may be due to hydrogen bonding between the *ortho*-amino group and the oxygen atom bound to the enzyme (Enz).

The electrophilic groups formed by many suicide inhibitors often take the form of α,β-unsaturated carbonyl compounds and imines. These electrophiles react by a type of Michael addition with nucleophilic groups (Nu), such as the OH of serine residues, the SH of cysteine residue and the ω-NH_2 of lysine residues frequently found at the active sites of enzymes.

The formation of the α,β-unsaturated carbonyl compound or imine often requires the reverse of a Michael addition at the active site of the enzyme.

X = O, S or NR
L = a stable leaving group

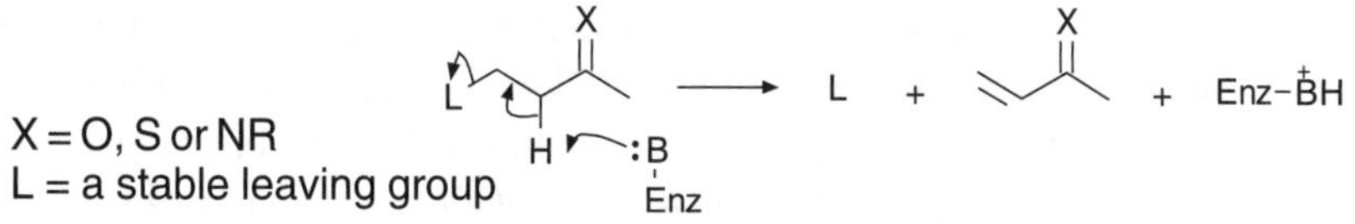

A wide variety of structures have been found to act as sources of the electrophilic groups of suicide inhibitors (Fig. 9.26). However, these structures will only give rise to an electrophilic group if the compound containing the structure can act as a substrate for the enzyme.

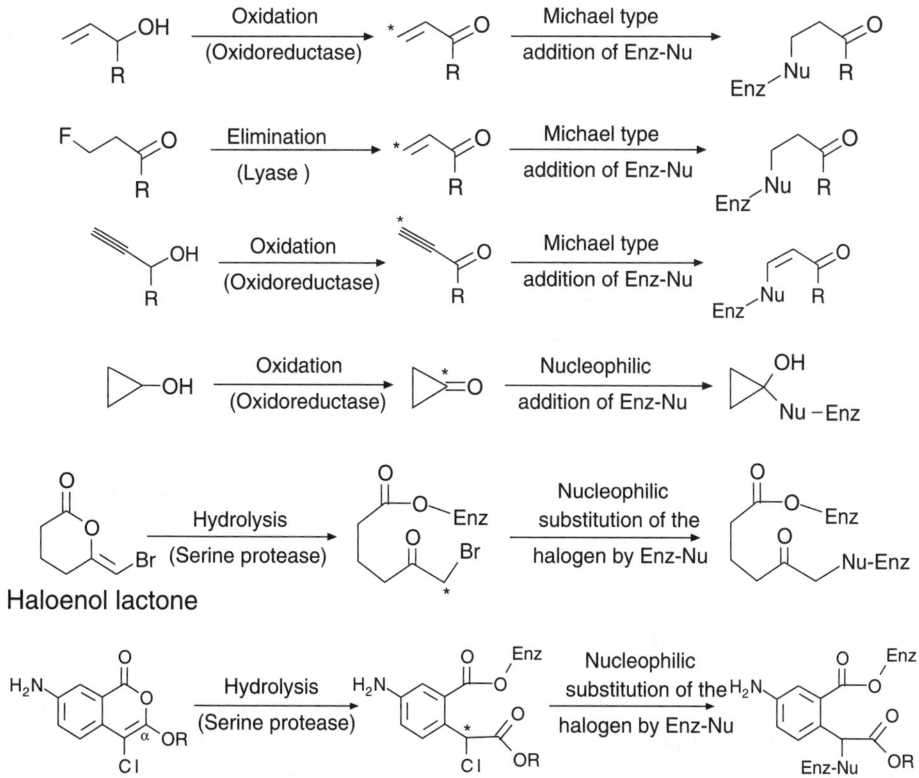

Figure 9.26 Examples of the reactions to enhance or form the electrophilic centres (*) of suicide inhibitors and their subsequent reaction with a nucleophile at the active site of the enzyme. The general structures used for the enzyme–inhibitor complexes are to illustrate the reactions; the enzyme ester groups may or may not be present in the final complex

The effect of suicide inhibitors increases as the concentration of the inhibitor increases relative to the concentration of the enzyme. Consequently, for processes involving single substrates that follow simple Michaelis–Menten kinetics the Lineweaver–Burk plot for different concentrations of the inhibitor is a series of straight lines (Fig.9.21).

Tienilic acid, a suicide inhibitor

The diuretic tienilic acid acts as a suicide inhibitor of cytochrome P-450 (CYP-450). Tienilic acid is oxidised by CYP-450 to its corresponding sulphoxide derivative (Fig. 9.27), which acts as the suicide inhibitor. The combined electron withdrawing effects of the keto carbonyl and the sulphoxide groups of this sulphoxide derivative make carbon-5 electrophilic enough to undergo a Michael reaction with a thiol group in the enzyme's active site. This bonds the tienilic acid to the enzyme by a strong sulphide link, which blocks the active site and irreversibly inhibits the enzyme. However, as CYP-450 is an important enzyme in many essential metabolic pathways (see section 12.4.1), tienilic acid is no longer used as a diuretic.

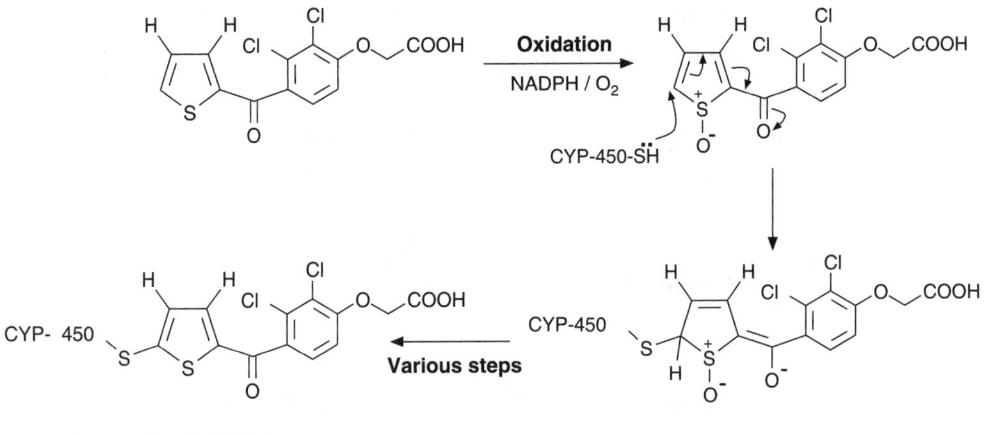

Figure 9.27 An outline of the mechanism of the inhibition of cytochrome P-450 by tienilic acid

9.10 Transition state inhibitors

The substrate in an enzyme-catalysed reaction is converted to the product through a series of transition state structures (Fig. 9.11). Although these transition state structures are transient, they bind to the active site of the enzyme and therefore must have structures that are compatible with the structure of the active site. Consequently, it has been proposed that *stable compounds* with structures similar to those of these transition state structures could bind to the active site of an enzyme and act as inhibitors for that enzyme. Compounds that fulfil this requirement are known as *transition state inhibitors*.

A number of transition state inhibitors have been identified. For example, the antibiotic coformycin, which inhibits adenosine deaminase, has a structure similar to that of the transition state of adenosine in the process for the conversion of adenosine to inosine which is catalysed by adenosine deaminase.

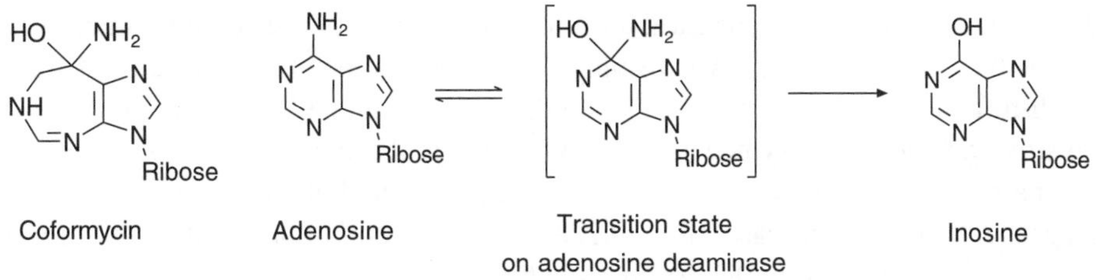

Coformycin   Adenosine   Transition state   Inosine
                                   on adenosine deaminase

Transition state inhibitors can act in a reversible or irreversible manner. Consequently, a consideration of the transition state structure offers a possible method of approach for discovering new drugs. Structures that might act as transition state inhibitors may be deduced using a knowledge of the mechanism of action of the target enzyme, classical chemistry and mechanistic theory. If a molecular model of the active site is available, docking of potential structures may assist in selecting the most appropriate structure for further investigation. This method of approach was used to design the experimental anticancer drug sodium N-phosphonoacetyl-L-aspartate (PALA, Fig. 9.28). In the early

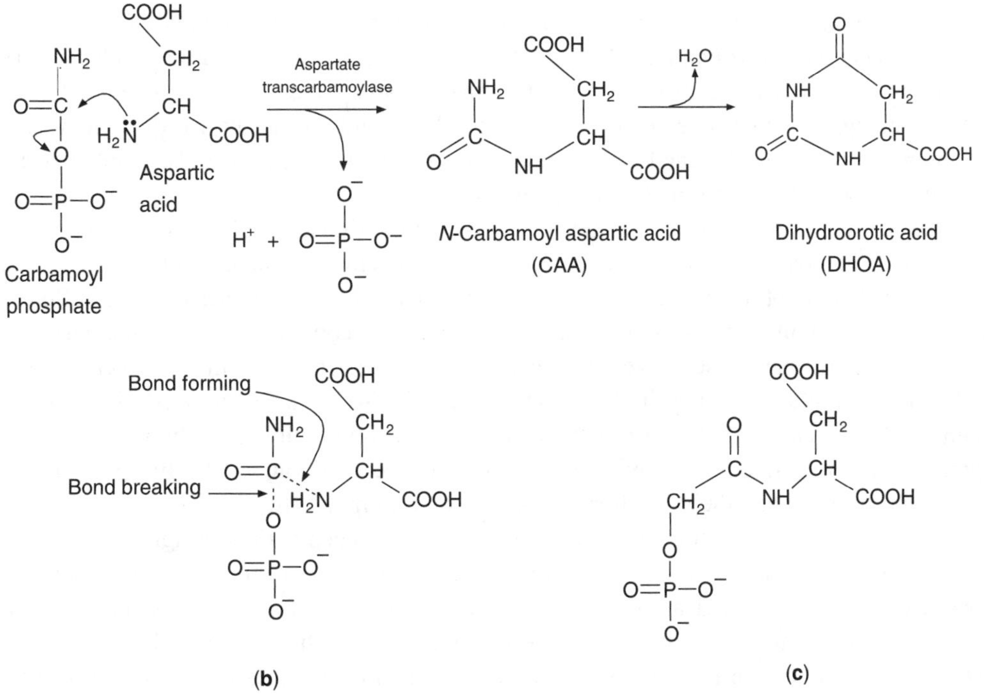

Figure 9.28 **(a)** The first steps in the biosynthesis of uracil. **(b)** The proposed transition state for the carbamoyl phosphate/aspartic acid stage in pyrimidine synthesis. **(c)** Sodium N-phosphonoacetyl-L-aspartate (PALA)

1950s, it was observed that some rat liver tumours appeared to utilise more uracil in DNA formation than healthy liver. As a result of this observation it was suggested that one approach to treating cancers was to develop drugs that inhibited the formation of uracil.

The first step in the biosynthesis of pyrimidines is the condensation of aspartic acid with carbamoyl phosphate to form *N*-carbamoyl aspartic acid, the reaction being catalysed by aspartate transcarbamoylase (Fig. 9.28a). It has been proposed that the transition state for this conversion (Fig. 9.28b) involves the simultaneous loss of phosphate with the attack of the nucleophilic amino group of the aspartic acid on the carbonyl group of the carbamoyl phosphate. Consequently, PALA (Fig. 9.28c) was designed to have a similar structure to this transition state but without the amino group necessary for the next stage in the synthesis, which is the conversion of *N*-carbamoyl aspartic acid to dihydroorotic acid.

It was found that PALA bound 10^3 times more tightly to the enzyme than the normal substrate. Furthermore, the drug was found to be effective against some cancers in rats.

9.11 Enzymes and drug design: some general considerations

Enzymes are obvious targets in drug design. Inhibition offers a method of either preventing or regulating cell growth. However, the design team must first decide whether this strategy will achieve the desired therapeutic result. This decision requires a detailed knowledge of both the chemistry and pharmacology of the condition for which the drug is intended, as well as commercial considerations. In many cases, the detailed chemical and pharmacological knowledge required is not available and so the discovery of new lead compounds relies on the random screening of collections of compounds, combinatorial chemistry (see Chapter 5) or intuitive guesswork.

One advantage of targeting enzymes is their diversity. Consequently, an enzyme process that occurs in a pathogen may not occur in humans. This means that an inhibitor active in a pathogen should not inhibit the same process in humans. Many enzymes occur in different forms in different species. Consequently an enzyme common to both humans and microorganisms may be selectively inhibited in one species but virtually unaffected in the other. For example, trimethoprim inhibits bacterial dihydrofolate reductase (DHFR) but not human DHFR. However, the use of these selective enzyme inhibitors in humans does not preclude the occurrence of unwanted side effects (see section 9.9.2). In spite of this potential disadvantage they do offer a possible route to more effective drugs.

Non-competitive inhibition offers the least rational approach to drug design as it does not affect the active site. Consequently, the structure of the substrate and its chemical behaviour cannot be used as the basis for the design of a new drug. Conversely, with inhibition involving the active site of the enzyme, the structure of the substrate and its chemical reactivity can be used as the basis of rational approaches to the design of drug substances. A problem with reversible inhibitors is that the inhibition can be reversed by increasing the concentration of the substrate, which automatically starts to occur naturally as soon as the enzyme is inhibited. The reversible nature of the inhibition means that as the substrate's concentration increases it is better able to compete with the inhibitor for the active site. In addition, it is more likely to displace the inhibitor from the active site. These problems do not occur with irreversible inhibitors. However, high concentrations of irreversible inhibitors are usually required if they are to be effective, which also increases the risk of unwanted side effects. A further complication in designing inhibitors is that isoenzymes often respond differently to inhibitors. In extreme cases an inhibitor will inhibit one enzyme but not all of its isoenzymes. Consequently, an inhibitor may not have the desired therapeutic effect.

9.12 Examples of drugs used as enzyme inhibitors

Enzyme inhibitors are used for a wide range of medical conditions (Table 9.4). This section briefly describes the discovery and development of two classes of inhibitors used in this capacity as being representative of the drugs in current clinical use. Other clinically used inhibitors are described as they arise naturally in the text.

Table 9.4 Examples of drugs used as enzyme inhibitors

Enzyme	Inhibitor	Condition or action
Dihydropteroate synthetase	Sulphamethoxazole	Bacteriostatic
Dihydrofolate reductase	Methotrexate	Cancer
Thymidylate synthase	Fluorouracil	Cancer
Angiotensin converting enzyme	Captopril	Hypertension
β-Lactamase	Penicillins	Bacteriocide
HIV reverse transcriptase	Zidovudine	AIDS
Cyclooxygenase	Aspirin	Analgesic
Xanthine oxidase	Allopurinol	Gout

9.12.1 Sulphonamides

In 1935 the German bacteriologist Gerhard Domagk, who had been evaluating the effect of a variety of substances on streptococci strains, successfully treated his daughter for a virulent streptococcal infection using the azo dye prontosil (Fig. 9.29). Subsequent investigation showed that this drug was reduced *in vivo* to the active agent *p*-aminobenzenesulphonamide (sulphanilamide). Prontosil was in fact, acting as a prodrug (see section 12.9). The discovery of this lead compound led, in the next decade, to the synthesis and testing of large numbers of sulphonamides for antibacterial action.

Sulphonamides are competitive inhibitors of folic acid synthesis. They are bacteriostatics preventing the replication of bacteria. This control of the spread of the bacteria enables the body's natural defences to gain strength and destroy the microorganism. Sulphonamides act by inhibiting dihydropteroate synthetase, which catalyses the reaction of 2-amino-4-hydroxy-6-hydroxymethyl-7,8-dihydropteridine diphosphate with *p*-aminobenzoic acid

Reduction
Prontosil
p-Aminobenzenesulphonamide

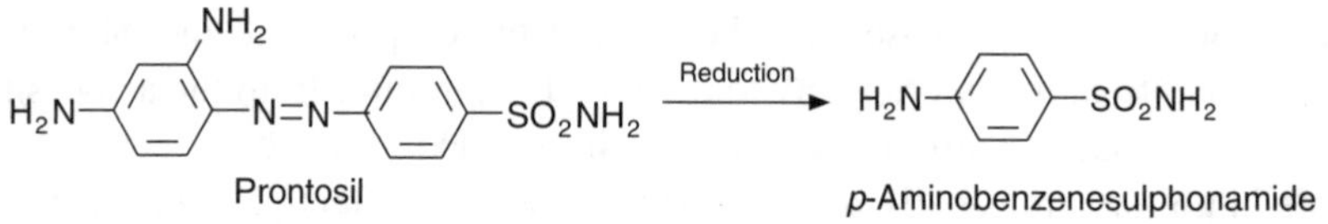

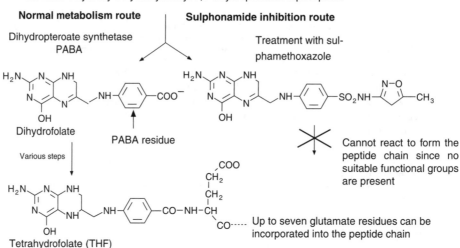

Figure 9.29 The conversion of prontosil into the active form of the drug: sulphanilamide

(PABA) to form dihydrofolate (Fig.9.30). The sulphonamide binds to the active site and reacts with the 2-amino-4-hydroxy-6-hydroxymethyl-7,8-dihydropteridine diphosphate to form a product that cannot be converted into folic acid. As a result, the bacteria cannot produce enough tetrahydrofolate (THF), which is used as a coenzyme in the production of purines required for DNA synthesis and subsequent cell reproduction.

2-Amino-4-hydroxy-6-hydroxymethyl-7,8-dihydropteridine diphosphate
Normal metabolism route
Sulphonamide inhibition route
Dihydropteroate synthetase
PABA
Treatment with sulphamethoxazole
Dihydrofolate
PABA residue
Cannot react to form the peptide chain since no suitable functional groups are present
Various steps
Tetrahydrofolate (THF)
Up to seven glutamate residues can be incorporated into the peptide chain

Figure 9.30 An outline of the chemistry of the inhibition of bacterial folic acid synthesis by sulphamethoxazole

Sulphonamides have a similar structure to PABA. Consequently they are believed to act by competing for the same active site as PABA.

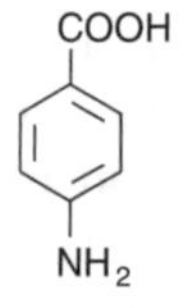

Para-aminobenzoic acid (PABA)

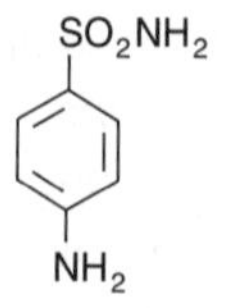

p-Aminobenzenesulphonamide

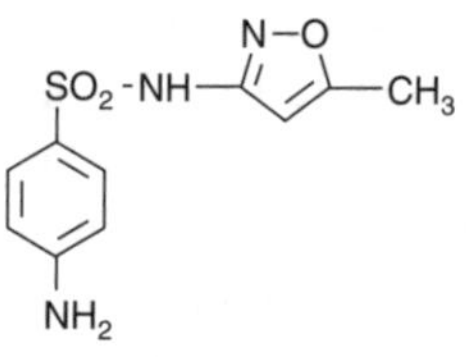

Sulphamethoxazole

It is interesting to note that humans are not able to synthesise folic acid and have to obtain it from their diet. Consequently, dihydropteroate inhibitors should have no effect on humans. However, because of the diversity of the chemical processes occurring in the human body these inhibitors may still give rise to unwanted side effects as they are able to bind to active sites and receptor sites controlling other processes.

9.12.2 Captopril and related drugs

The angiotensins are peptide hormones that have an important role in the control of blood pressure and electrolyte balance. Angiotensin I is inactive but angiotensin II is a potent vasoconstrictor. However, angiotensin III is a less potent vasoconstrictor but is involved in controlling the release of sodium ions from the kidney. They are produced from angiotensinogen (an α-globulin produced by the liver) by a series of hydrolyses catalysed by renin, angiotensin-converting enzymes (ACE) and an aminopeptidase (Fig. 9.31). Overproduction of angiotensins is thought to be one of the causes of hypertension (high blood pressure). Angiotensin-converting enzymes are also involved in a number of other processes that can result in hypertension.

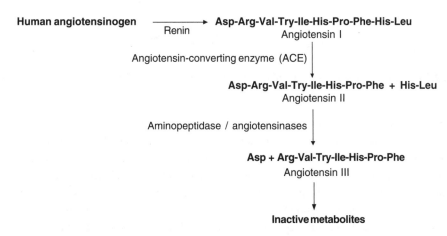

Figure 9.31 The conversion of angiotensinogen to the angiotensins I, II and III

Initial experimental work showed that ACE could be inhibited by small peptides, the most effective of these peptides having a proline residue at the C-terminal, an alanine at the penultimate position and an aromatic amino acid residue at the antepenultimate position (Fig. 9.32).

These peptides were not suitable for oral administration as drugs since they underwent extensive hydrolysis in the stomach and GI tract. Further investigation of the structure of ACE showed that it contained one zinc ion per molecule of the enzyme,

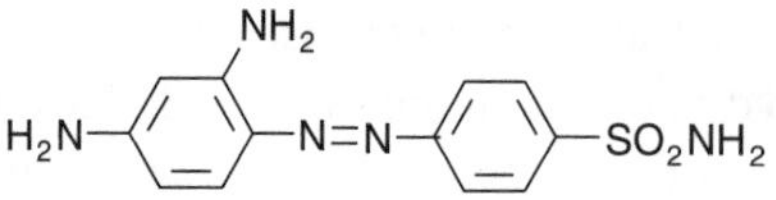

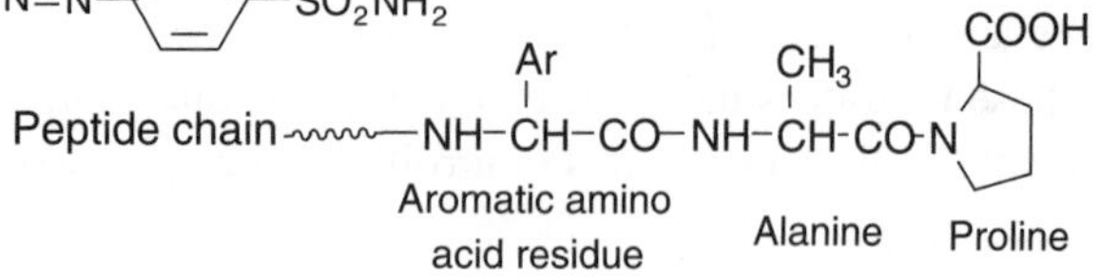

Figure 9.32 The general structure of the C-terminal of the most effective peptide inhibitors of ACE

which was thought to coordinate with the carbonyl of the peptide bond (Fig. 9.33). This coordination is believed to increase the polarisation of the carbonyl group, facilitating the nucleophilic attack of a water molecule bound to the enzyme to hydrolyse the peptide bond.

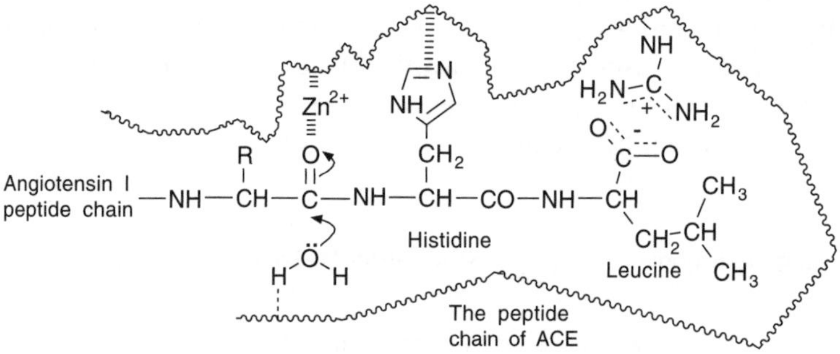

Figure 9.33 A schematic representation of the binding of the C-terminal of angiotensin I to the active site of ACE

Experimental investigations also indicated that the structure of ACE was similar to carboxypeptidase A, a zinc containing pancreatic digestive enzyme whose structure had been determined by X-ray crystallography. This enzyme was found to be strongly inhibited by 2R-benzylsuccinic acid (Fig. 9.34a). Consequently, using all this information and computer simulation of the active site of carboxypeptidase A, a series of carboxyacylproline derivatives (Fig. 9.34b) were synthesised and tested. All of the compounds synthesised were found to be weak inhibitors of ACE but subsequent replacement of the terminal carboxylate group by a

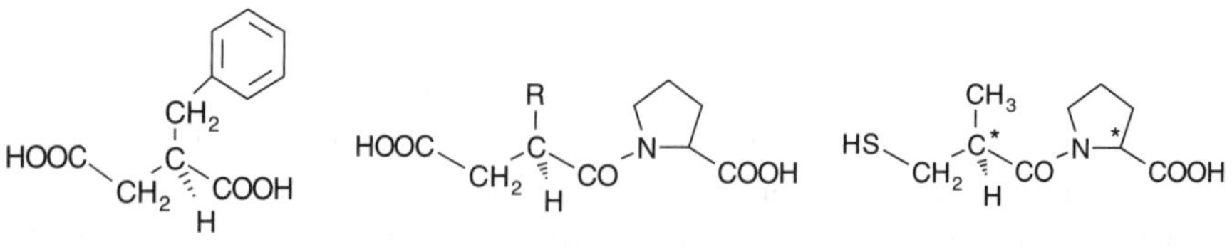

(a)   (b)   (c)

Figure 9.34 **(a)** 2R-Benzylsuccinic acid (carboxypeptidase A inhibitor). **(b)** Carboxyacylproline derivatives (ACE inhibitors); R is a variety of structures. **(c)** Captopril (ACE inhibitor); its chiral centres are marked with an asterisk

thiol group resulted in compounds that were strong ACE inhibitors. Thiols are better ligands than carboxylate ions for coordinating with zinc(II) ions, which probably accounts for the increase in potency. The most potent of these thiol derivatives was captopril (Fig. 9.34c), which is now an important drug in the treatment of hypertension.

Captopril is a competitive ACE inhibitor. It has two chiral centres and as a result four epimers, the *S,S*-epimer being the most active. Captopril has a number of unwanted side effects, including skin rashes, dry coughs, palpitations and renal impairment. This, coupled with the fact that thiols are readily oxidised *in vivo* to disulphides, has led to the development of a number of other ACE inhibitors with structures similar to captopril (Fig. 9.35). The approach adopted was to increase the potency of the carboxyacylproline derivatives by increasing their degree of binding to the enzyme by making more like the products of ACE hydrolysis (Fig. 9.32). The methylene group of the chain was replaced by a secondary amine (-NH-) to make the structure look more like a peptide. Since the third residue of angiotensin I was phenylalanine (Phe), an aromatic residue was incorporated to increase the binding to the active site. It is interesting to note that in the most effective peptide inhibitors of ACE the third amino acid was an aromatic residue (Fig. 9.32). The result was the eventual production of enalaprilat and the clinically used compounds enalapril (synthesis, see Fig. 15.11) and lisinopril, all of which have structures related to that of captopril (Fig. 9.35). However, they were all stronger inhibitors of ACE than captopril. Although enalaprilat is more effective than captopril it is poorly absorbed when administered orally because of its strong polar nature. As a result, it can only be given by intravenous injection. Reduction of its polar nature by converting the carboxyl group to its ethyl ester produced, enalapril which is more easily absorbed and is also more potent than enalaprilat. Unfortunately both enalapril

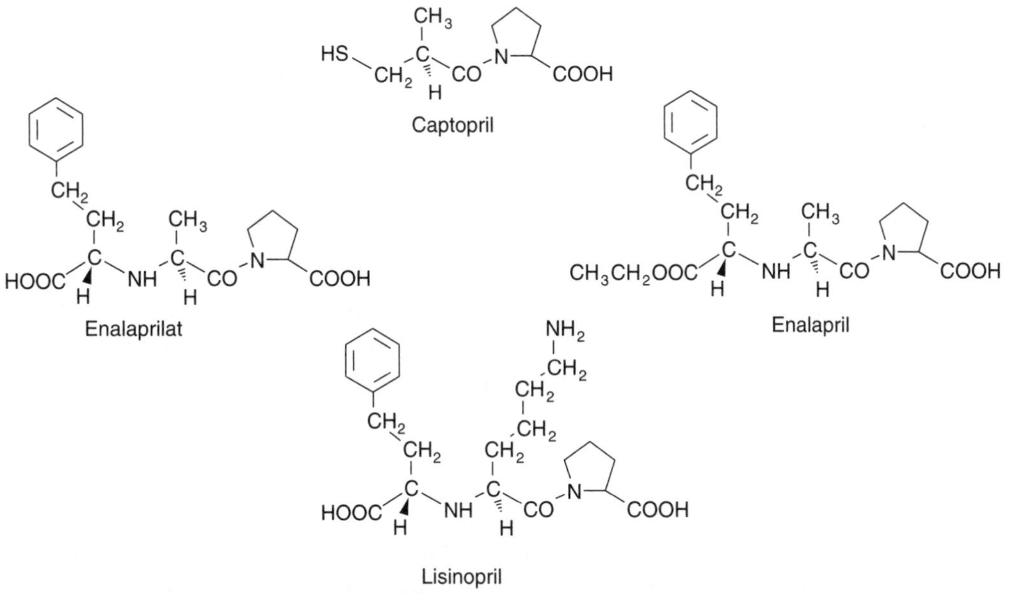

Figure 9.35   ACE inhibitors

and lisinopril still suffer from the same side effects as captopril but they do have the advantage that lower doses of the drug are required, which reduces the possibility of side effects.

A number of new ACE inhibitors have been developed (Fig. 9.36b). Like enalapril they all have a 2S-aminophenylbutanoic acid ethyl ester group except fosinopril, which has a phosphinyl group. Like enalapril they are prodrugs (see section 12.9), depending on hydrolysis of their ester groups to release the diacidic active form of the inhibitor (Fig. 9.36a). This hydrolysis occurs in the liver and intestine. Unfortunately, they all exhibit unwanted side effects to some extent.

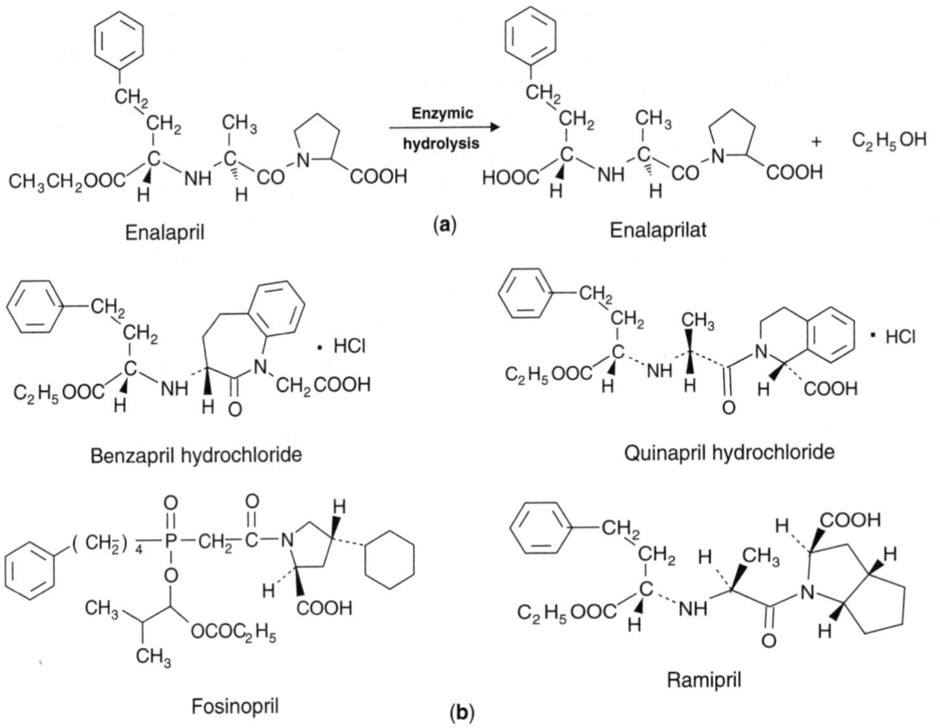

Figure 9.36 (a) The hydrolysis of enalapril to its active form enalaprilat. (b) ACE prodrug inhibitors

ACE inhibitors are not the only drugs used to treat hypertension. Losartan (see section 4.6.1) and a number of other drugs (Fig. 9.37) act as competitive antagonists (see section 8.5.3) blocking the AT_1 receptor by which angiotensin II produces its vasoconstriction effect.

9.12.3 Statins

Heart attacks and strokes are responsible for over a quarter of a million deaths in Britain every year. Raised cholesterol levels are believed to be a future indication of the possibility of such attacks. Cholesterol is a water-insoluble steroid that is an essential component of

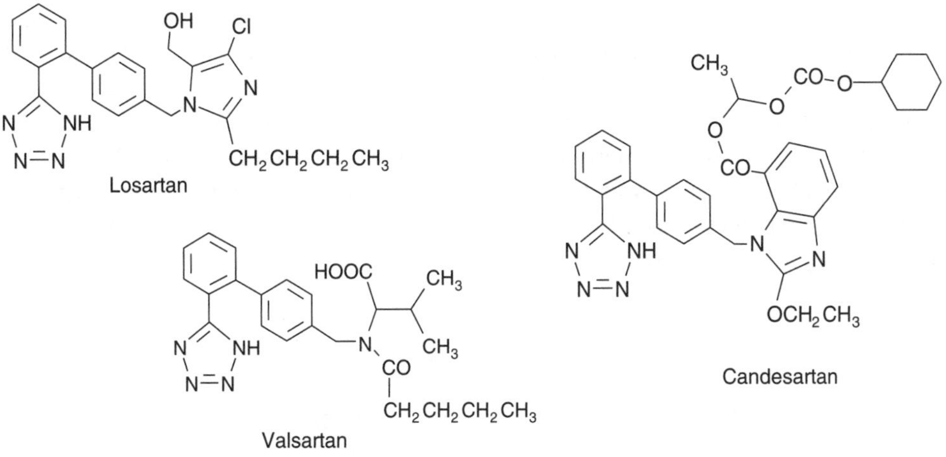

Losartan

Valsartan

Candesartan

Figure 9.37 Examples of the competitive antagonists used to treat hypertension

cell membranes (see sections 7.2 and 7.2.1) and the precursor of steroidal hormones and bile acids. It enters the body through the small intestine and is transported through the circulatory system by lipoproteins. These lipoproteins are macromolecules that consist of a spherical shell of a monolayer of lipoproteins and carbohydrates in which are embedded a few cholesterol molecules. This shell surrounds a non-polar core containing cholesterol, triglycerides and other lipid substances. The polar groups in the shell are orientated into the plasma, which ensures that the structure remains soluble, and the complete structure is held together by non-covalent bonds. A number of different types of lipoprotein have been identified (Table 9.5).

Table 9.5 Lipoprotein classes and composition

Type	Composition (figures are approximate)/Notes
Chylomicrons	~90% Triglycerides Originate from the diet and have the lowest density
Very-low-density lipoprotein (VLDL)	60% Triglycerides, 12% cholesterol, 18% phospholipides Originates in the liver, normally rapidly converted to IDL
Intermediate-density lipoprotein (IDL)	Normally rapidly converted into LDL
Low-density lipoprotein (LDL)	10% Triglycerides, 50% cholesterol Accounts for 65% of plasma cholesterol in most humans
High-density lipoprotein (HDL)	25% Cholesterol, 50% protein Accounts for 17% of the plasma cholesterol in most humans

Under normal conditions these lipoproteins transport cholesterol to where it is needed. The level of cholesterol in the plasma does not result in an excessive deposit of cholesterol in arterial walls. However, excessive cholesterol in the diet and/or changes in the body's chemistry can result in excessive deposition of cholesterol in the arteries with a subsequent increase in arterial blockage, heart attacks and strokes. In this case LDL-cholesterol is the major problem. Consequently, in recent years considerable work has been carried out in investigating compounds that would reduce excessive plasma cholesterol levels.

Humans either obtain cholesterol from their diet or by biosynthesis from ethanoate by a long and complicated pathway first outlined by Konrad Bloch. The rate limiting step in this pathway is the reduction of β-hydroxy-β-methylglutaryl-coenzyme A (HMG-CoA) to mevalonate by HMG-CoA reductase.

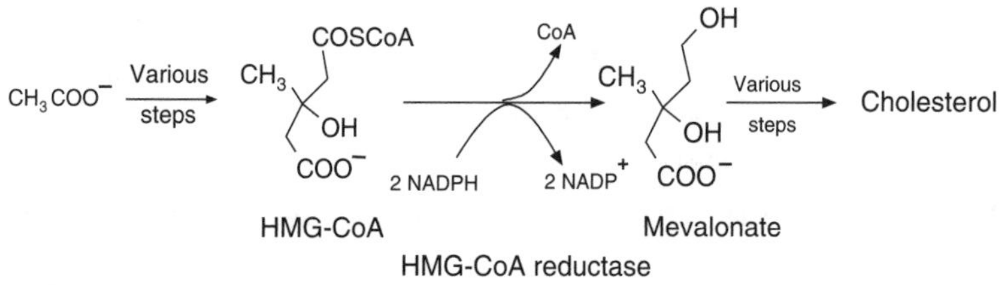

This point in the biosynthesis was selected for the development of an inhibitor for HMG-CoA in an attempt to control the level of cholesterol in the plasma and hence its arterial deposition. This approach led to the discovery of lovastatin and mevastatin (Fig. 9.38). Lovastatin was originally isolated from a number of different fungi by several different companies. For example, Merck obtained it from *Aspergillus terreus*. Mevastatin was isolated from *Penicillium cillium citrum*, but although it inhibited HMG-CoA reductase it failed an initial toxicity test and was withdrawn from clinical trials. Further

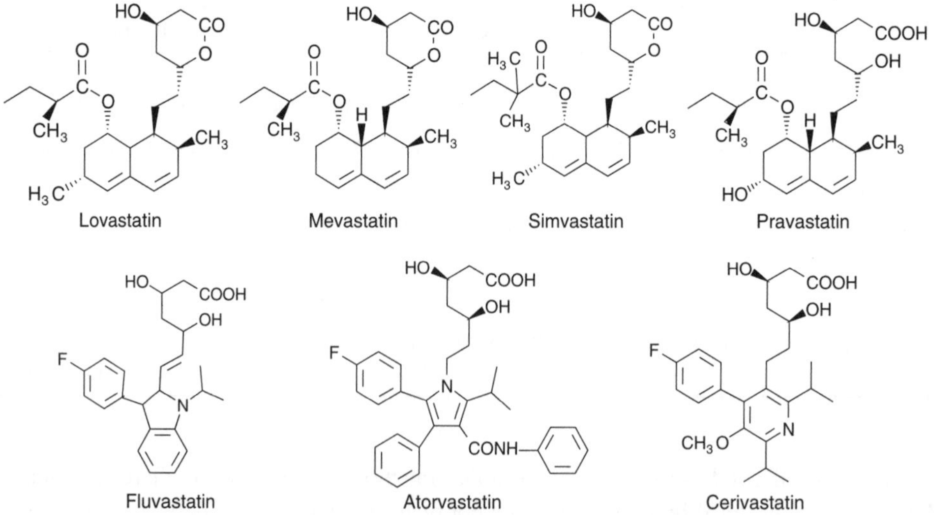

Figure 9.38 Examples of the statins used to reduce cholesterol levels in humans

SAR investigations yielded a number of other compounds, collectively referred to as statins, that are now in clinical use. They are reasonably effective in reducing the levels of LDL-cholesterol circulating in the plasma.

Lovastatin and Simvastatin are prodrugs (see section 12.9). They are administered orally and pass unchanged into the liver where the lactone ring is hydrolysed to the active hydroxy-heptanoic acid form of the drug.

Lovastrain (inactive)   Active form of the drug

Pravastatin, fluvastatin, atorvastatin and cerivastatin have hydroxy-heptanoic acid side chains that correspond to the lactone containing side chains of lovastatin and simvastatin (Fig. 9.38). However, cerivastatin was withdrawn in 2001 as it was associated with a higher incidence of rhabomyolysis than other statins. Rhabomyolysis is a condition where muscle fibres degenerate. This together with muscular pains is one of the unwanted side effects of the use of statins. These muscular problems are believed to arise because all statins inhibit coenzyme Q_{10} (CoQ_{10}) as well as HMG-CoA reductase. The hydroxy acid side chains of pravastatin, fluvastatin and atorvastatin are more hydrophilic than the lactone rings of lovastatin and simvastatin and consequently are more polar and less able to cross the blood–brain barrier (see section 7.2.9) than lovastatin and simvastatin and so have fewer CNS side effects than these drugs.

It is emphasised that statins reduce the level of LDL-cholesterol in the plasma. Many but not all of the other types of lipoproteins (see Table 9.5) may be reduced by changes in diet and the use of other suitable drugs.

9.13 Enzymes and drug resistance

Drug resistance occurs when a drug no longer has the desired clinical effect. This may be due to either a natural inbuilt resistance in some individuals and organisms or may arise naturally in the course of treatment. The former is probably due to differences in the genetic code of individuals within a species whilst the latter arises because of natural selection. In natural selection the drug kills the susceptible strains of an organism but does not affect other strains of the same organism. Consequently, these immune strains multiply and become the common strain of the organism, which subsequently results in ineffective drug treatment.

Resistance occurs on an individual basis and so is not usually detected until a wide sample of the population has been treated with or indirectly exposed to the drug. Its detection necessitates the discovery of new drugs to treat the condition. Its emergence is probably due to the widespread and poorly controlled use of a drug. For example, the generous use of antibiotics in farming is strongly suspected to be the reason for an increase in antibiotic-resistant strains of bacteria in humans. The response of medicinal chemists to resistance is either to devise new drugs or to modify existing drugs. This approach suffers from the high probability of being unsuccessful as well as time consuming and expensive. In the light of human experience it would be better if, in future, we reduced the possibility of resistance by using the effective existing drugs more intelligently.

Drug resistance can be linked to a change in either the permeability of the membranes of the organism or an enzyme system(s) of the organism. Enzymes may be involved in drug resistance in a number of ways (see section 7.4.2) but in many cases resistance may be due to several different processes occurring at approximately the same time.

9.13.1 Changes in enzyme concentration

A significant increase or decrease from the normal concentration of an enzyme can result in resistance to a drug. The overproduction of an enzyme can have two effects:

1. The target process catalysed by the enzyme will not be inhibited because excess enzyme is produced. For example, the resistance of malarial parasites is believed to be caused by overproduction of dihydrofolate reductase due to the drug stimulating the parasite's RNA.

2. The increased production of enzymes that inactivate the drug, for example β-lactamases inactivate most penicillins and cephalosporins by hydrolysing their β-lactam rings (see section 7.4.2). A number of enzymes deactivate inhibitors by incorporating (conjugation) phosphate by phosphorylation of hydroxyl groups, adenine by adenylation of hydroxyl groups or acetyl by acetylation of amino groups in the inhibitor's structure (Fig. 9.39). ATP is believed to be the usual provider of phosphate and adenylic acid, whilst acetyl coenzyme A is thought to be the normal source of acetyl groups. For

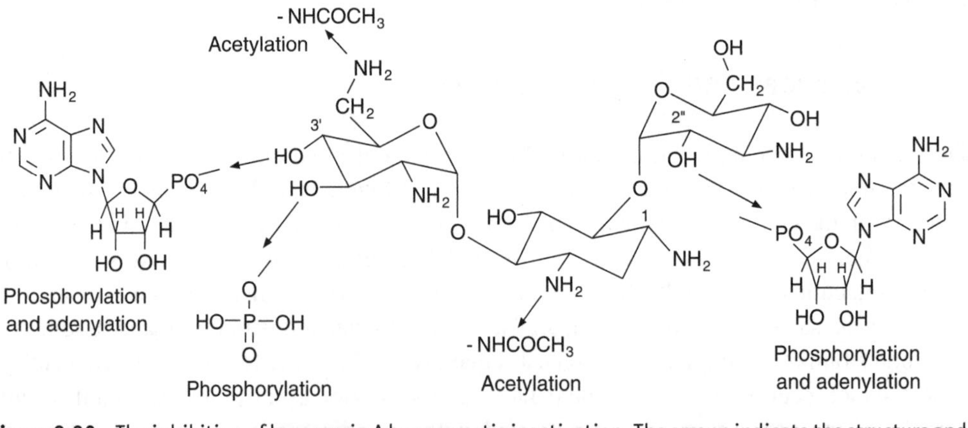

Figure 9.39 The inhibition of kanamycin A by enzymatic inactivation. The arrows indicate the structure and position of the result of an enzyme reaction

example, resistance to the antibiotic kanamycin A can occur by all three routes (Fig. 9.39) although kanamycin A-resistant bacteria do not normally use all three routes.

Many aminoglycoside antibiotics are susceptible to this type of enzyme inhibition. However, amikacin (AK) where the C_1-NH_2 of kanamycin A has been acylated by S-α-hydroxy-γ-aminobutanoic acid is not susceptible to 3′-phosphorylation and 2″-adenylation.

The underproduction of an enzyme could result in insufficient enzyme being present to produce the active form of a drug from a prodrug. For example the resistance to the antileukaemia drug 6-mercaptopurine is caused by a reduced production of hypoxanthine–guanine phosphoribosyltransferase, the enzyme required to convert the prodrug to its active ribosyl 5′-monophosphate derivative.

These changes in the production of the enzyme are believed to be due to genetic changes in the organism.

9.13.2 An increase in the production of the substrate

Increased production of the substrate can prevent competitive reversible inhibitors from binding to the active site in sufficient quantities to be effective (see section 9.9.1). The high concentration of the substrate moves the position of equilibrium to favour the formation of the E–S complex.

$$I\text{–}E \rightleftharpoons I + E + S \rightleftharpoons E''''S$$

For example, the inhibition of dihydropteroate synthetase by sulphonamides (see section 9.12.1) results in a build-up of p-aminobenzoic acid. This increase in substrate concentration prevents sulphonamides from inhibiting dihydropteroate synthetase, which is a key enzyme in the production of the RNA necessary for bacterial reproduction. Similarly, a build-up of angiotensin I will overcome the effect of ACE inhibitors (see section 9.12.2). It is thought to be the reason for the concentration of plasma angiotensin II returning to normal in some cases where there has been a chronic administration of ACE inhibitors.

9.13.3 Changes in the structure of the enzyme

Changes in the structure of the target enzyme result in a structure that is not significantly inhibited by the drug. However, the modified enzyme is still able to produce the normal product of the reaction, which allows the unwanted metabolic pathway to continue to function. For example, resistance to the antibiotic trimethoprim is believed to be due to a plasmid (see section 10.14.1)-directed change in the structure of dihydrofolate reductase in the bacteria.

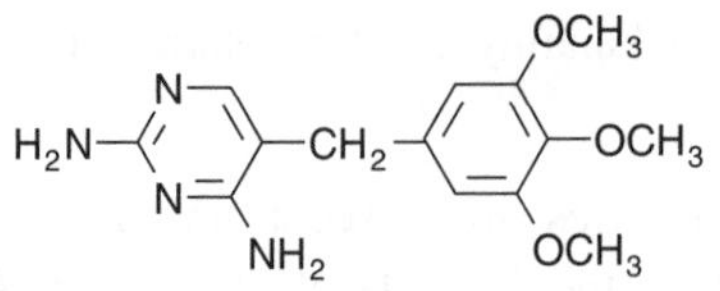

Trimethoprim

Similarly, the resistance of the *E. coli* strains to sulphonamides has been shown to be due to their containing a sulphonamide-resistant dihydropteroate synthase.

9.13.4   The use of an alternative metabolic pathway

The blocking of a metabolic pathway by a drug can result in the opening of a new pathway controlled by a different enzyme that is not inhibited by the same drug.

9.14   Ribozymes

A number of biological reactions in which certain RNA molecules act as catalysts have been discovered. These catalytic RNAs exhibit many of the same general properties as protein-based enzymes. For example, they are substrate specific, increase reaction rate and reappear unchanged at the end of the reaction. However, in a number of cases their action appears to be significantly enhanced by the presence of protein subunits. These subunits do not act as catalysts for the reaction in the absence of the ribozyme.

9.15   Questions

1   Distinguish between each of the following terms: (a) isoenzyme and isofunctional enzyme; (b) enzyme and proenzyme; (c) active and allosteric site; (d) modulator and activator; and (e) single and double displacement reactions in the context of enzyme reactions.

2   What information can be obtained from the systematic name of an enzyme?

3   A Ca^{2+} ion channel blocker is administered to a patient. What would be the expected effect of this treatment on NO synthase.

4   Outline, in general terms, the part played by the active site of an enzyme when it catalyses a reaction.

5   Explain how the Lineweaver–Burk plots can be used to indicate the general type of mechanism employed by an enzyme inhibitor. What major assumptions is your explanation based on?

6   Describe the main features of competitive, non-competitive and irreversible inhibition of enzymes.

7 Compounds A and B are inhibitors of a monosubstrate (S) enzyme process that exhibits simple Michaelis–Menten kinetics. Use the data in Table 9.6 to determine the general type of enzyme inhibition processes exhibited by A and B. Decide, on the basis of this information, whether A or B could be considered a suitable candidate for further investigation into its suitability as a lead compound for the development of a new drug. Give an explanation for your decision.

Table 9.6 Experimental results (V in $\mu mol\ s^{-1}$ and [S] in mmol dm^{-3})

| No inhibitor | | [A] mmol dm^{-3} | | | | [B] mmol dm^{-3} | | | |
| | | 0.102 | | 0.217 | | 0.137 | | 0.315 | |
[S]	V	[S]	V	[S]	V	[S]	V	[S]	V
0.083	0.122	0.125	0.063	0.333	0.062	0.091	0.091	0.087	0.060
0.125	0.155	0.153	0.072	0.500	0.075	0.118	0.101	0.143	0.079
0.200	0.194	0.200	0.095	1.000	0.094	0.222	0.122	0.250	0.089
0.500	0.259	0.500	0.139	20.00	0.137	0.666	0.144	0.500	0.095

8 (a) Explain the meaning of the term 'suicide inhibitor'.

(b) Compound A acts as a suicide inhibitor of a serine protease. Outline a possible mechanism for its action and suggest modifications to its structure that might improve its action.

9 (a) What is the natural substrate of lactate dehydrogenase?

(b) How could this substrate be modified to produce a possible active site-directed inhibitor for lactate dehydrogenase? Outline the reasoning behind your modification.

10 Explain how the transition state of a reaction can be used to design an enzyme inhibitor. The transition state of an enzyme-catalysed reaction involves pyrrole binding to the active site of the enzyme. Suggest modifications to the structure of pyrrole that would (i) increase the binding and (ii) decrease the binding of the modified molecule to the active site. Explain the electronic basis of your choices.

11 What is drug resistance and how does it arise? Outline the ways by which enzymes are linked to drug resistance. How can drug resistance be minimised?

12 You are required to design analogues of lovastatin. (a) What structural features would be expected to produce active compounds? (b) How, in general terms, could these structural features be changed?

10
Nucleic acids

10.1 Introduction

The nucleic acids are the compounds that are responsible for the storage and transmission of the genetic information that controls the growth, function and reproduction of all types of cells. They are classified into two general types: the *deoxyribonucleic acids* (*DNA*) whose structures contain the sugar residue β-D-deoxyribose; and the *ribonucleic acids* (*RNA*) whose structures contain the sugar residue β-D-ribose (Fig. 10.1).

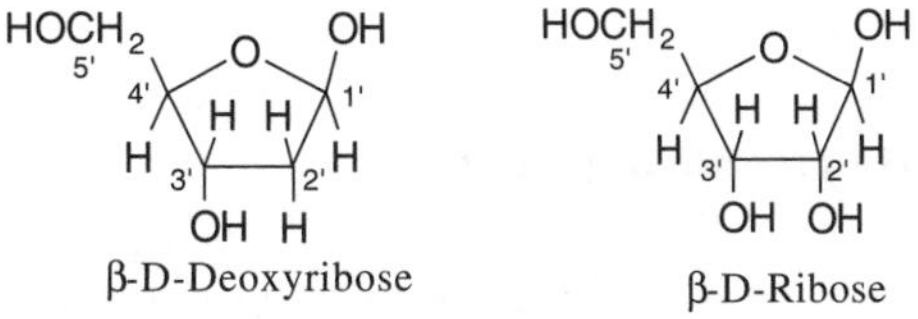

Figure 10.1   The structures of β-D-deoxyribose and β-D-ribose

Both types of nucleic acids are polymers based on a repeating structural unit known as a *nucleotide* (Fig.10.2). These nucleotides form long single-chain polymer molecules in both DNA and RNA.

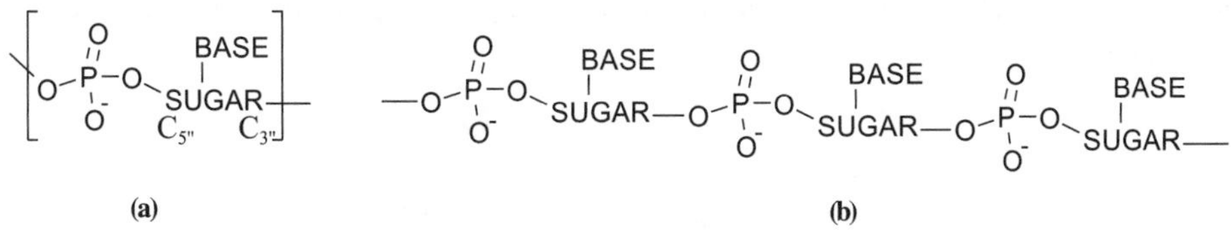

Figure 10.2   The general structures of (**a**) nucleotides and (**b**) a schematic representation of a section of a nucleic acid chain

Medicinal Chemistry, Second Edition   Gareth Thomas
© 2007 John Wiley & Sons, Ltd

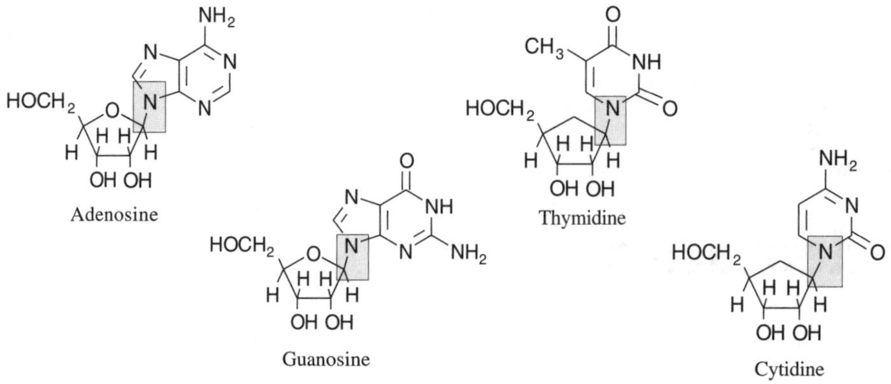

NH₂

HOCH₂

Adenosine

CH₃

HOCH₂

Thymidine

HOCH₂

Guanosine

NH₂

HOCH₂

Cytidine

Figure 10.3   Examples of the structures of some of the nucleosides found in RNA. The β-*N*-glycosidic link is shaded. The corresponding nucleosides in DNA are based on deoxyribose and use the same name but with the prefix deoxy

Each nucleotide consists of a purine or pyrimidine base bonded to the 1′-carbon of a sugar residue by a β-*N*-glycosidic link (Fig. 10.3). These base–sugar subunits, which are known as *nucleosides*, are linked through the 3′- and 5′-carbons of their sugar residues by phosphate units to form the polymer chain.

10.2   Deoxyribonucleic acid (DNA)

DNA occurs in the nuclei of cells in the form of a very compact DNA–protein complex called *chromatin*. The protein in chromatin consists mainly of *histones*, a family of relatively small positively charged proteins. The DNA is coiled twice around an octomer of histone molecules with a ninth histone molecule attached to the exterior of these minicoils to form a structure like a row of beads spaced along a string (Fig.10.4). This 'string of beads' is coiled and twisted into compact structures known as miniband units, which form the basis of the structures of *chromosomes*. Chromosomes are the structures that form duplicates during cell division in order to transfer the genetic information of the old cell to the two new cells.

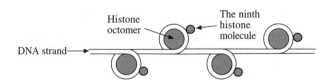

Figure 10.4   The 'string of beads' structure of chromatin. The DNA strand is wound twice around each histone octomer. A ninth histone molecule is bound to the exterior surface of the coil

10.2.1 Structure

DNA molecules are large with relative molecular masses up to one trillion (10^{12}). The principal bases found in their structures are adenine (A), thymine (T), guanine (G) and cytosine (C), although derivatives of these bases are found in some DNA molecules (Fig. 10.5). Those bases with an oxygen function have been shown to exist in their keto form.

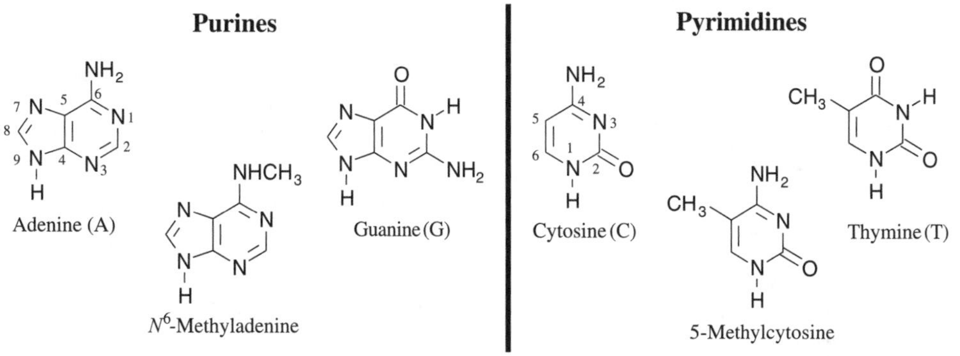

Purines   **Pyrimidines**

Adenine (A) N^6-Methyladenine   Guanine (G)   Cytosine (C)   5-Methylcytosine   Thymine (T)

Figure 10.5 The purine and pyrimidine bases found in DNA. The numbering is the same for each type of ring system

Chargaff showed that the molar ratios of adenine to thymine and guanine to cytosine are always approximately 1:1 in any DNA structure, although the ratio of adenine to guanine varies according to the species from which the DNA is obtained. This and other experimental observations led Crick and Watson in 1953 to propose that the three-dimensional structure of DNA consisted of two single-molecule polymer chains held together in the form of a double helix by hydrogen bonding between the same pairs of bases, namely: adenine to thymine (A–T) and cytosine to guanine (C–G) (Fig. 10.6). These pairs of bases, which are referred to as *complementary base pairs*, form the internal structure of the helix. They are hydrogen bonded in such a manner that their flat structures lie parallel to one another across the inside of the helix. The two polymer chains forming the helix are aligned in opposite directions. In other words, at the ends of the structure one chain has a free 3'-OH group whilst the other chain has a free 5'-OH group. X-ray diffraction studies have since confirmed this as the basic three-dimensional shape of the polymer chains of B-DNA, the natural form of DNA. This form of DNA has about ten bases per turn of the helix. Its outer surface has two grooves known as the minor and major grooves, respectively, which act as the binding sites for many ligands. Two other forms of DNA, the A and Z forms, have also been identified but it is not certain whether these forms occur naturally in living cells.

Electron microscopy has shown that the double helical chain of DNA is folded, twisted and coiled into quite compact shapes. A number of DNA structures are cyclic and these

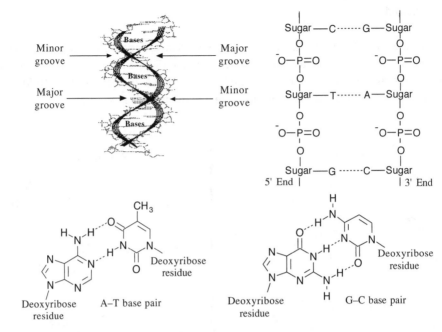

Figure 10.6 The double helical structure of B-DNA. Interchanging of either the bases of a base pair and/or base pair with base pair does not affect the geometry of this structure. Reproduced from G. Thomas, *Chemistry for Pharmacists and the Life Sciences*, 1996, by permission of Prentice Hall, a Pearson Education Company

compounds are also coiled and twisted into specific shapes. These shapes are referred to as supercoils, supertwists and superhelices, as appropriate.

10.3   The general functions of DNA

The DNA found in the nuclei of cells has three functions:

1. To act as a repository for the genetic information required by a cell to reproduce that cell.

2. To reproduce itself in order to maintain the genetic pool when cells divide.

3. To supply the information that the cell requires to manufacture specific proteins.

Genetic information is stored in a form known as *genes* by the DNA found in the nucleus of a cell (see section 10.4).

The duplication of DNA is known as *replication*, It results in the formation of two identical DNA molecules, which carry the same genetic information from the original cell to the two new cells that are formed when a cell divides (see section 10.5).

The function of DNA in protein synthesis is to act as a template for the production of the various RNA molecules necessary to produce a specific protein (see section 10.6).

10.4 Genes

Each species has its own internal and external characteristics. These characteristics are determined by the information stored and supplied by the DNA in the nuclei of its cells. This information is carried in the form of a code based on the consecutive sequences of bases found in sections of the DNA structure (see section 10.7). This code controls the production of the peptides and proteins required by the body. The sequence of bases that act as the code for the production of one specific peptide or protein molecule is known as a gene.

Genes can normally contain from several hundred to 2000 bases. Changing the sequence of the bases in a gene by adding, subtracting or changing one or more bases may cause a change in the structure of the protein whose production is controlled by the gene. This may have a subsequent knock-on effect on the external or internal characteristics of an individual. For example, an individual may have brown instead of blue eyes' or their insulin production may be inhibited, which could result in that individual suffering from diabetes. An increasing number of medical conditions have been attributed to either the absence of a gene or the presence of a degenerate or faulty gene in which one or more of the bases in the sequence have been changed. For example it is now known that a faulty gene is the cause of cystic fibrosis. Once the gene responsible for a disease condition is known, it will be possible to use the information to treat and develop new specific drugs for that disease (see section 10.15.2).

In simple organisms, such as bacteria, genetic information is usually stored in a continuous sequence of DNA bases. However, in higher organisms the bases forming a particular gene may occur in a number of separate sections known as *exons* separated by sections of DNA that do not appear to be a code for any process. These non-coding sections are referred to as *introns*. For example, the gene responsible for the β-subunit of haemoglobin consists of 990 bases. These bases occur as three exons separated by two introns (Fig. 10.7).

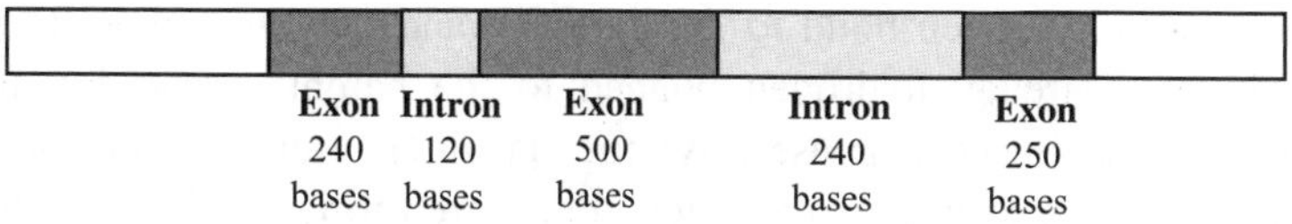

Exon	Intron	Exon	Intron	Exon
240	120	500	240	250
bases	bases	bases	bases	bases

Figure 10.7 A schematic representation of the gene for the β-subunit of haemoglobin

The complete set of genes that contain all the hereditary information of a particular species is called a *genome*. The Human Genome Project, initiated in 1990 and largely finished in 2000, sets out to identify all the genes that occur in human chromosomes and also the sequence of bases in these genes. It has created a gene index held in public data bases that can be used to locate the genes responsible for particular biological processes and medical conditions.

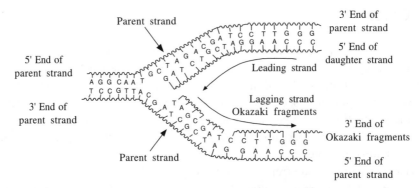

Figure 10.8 A schematic representation of the replication of DNA. The arrows show the direction of growth of the leading and lagging strands. Reproduced from G. Thomas, *Chemistry for Pharmacists and the Life Sciences*, 1996, by permission of Prentice Hall, a Pearson Education Company

10.5   Replication

The reproduction of DNA is known as *replication*. It is believed to start with the unwinding of a section of the double helix (Fig. 10.8). Unwinding may start at the end or more commonly in a central section of the DNA helix. It is initiated by the binding of the DNA to specific receptor proteins that have been activated by the appropriate first messenger (see section 8.4). The separated strands of the DNA act as templates for the formation of a new daughter strand. Individual nucleotides, which are synthesised in the cell by a complex route, bind by hydrogen bonding between the bases to the complementary parent nucleotides. This hydrogen bonding is specific in that only the complementary base pairs can hydrogen bond. In other words, the hydrogen bonding can only be between either thymine and adenine or cytosine and guanine. This means that the new daughter strand is an exact replica of the original DNA strand bound to the parent strand. Consequently, replication will produce two identical DNA molecules.

As the nucleotides hydrogen bond to the parent strand they are linked to the adjacent nucleotide, which is already hydrogen bonded to the parent strand, by the action of enzymes known as DNA polymerases. As the daughter strands grow the DNA helix continues to unwind. However, *both* daughter strands are formed at the same time in the $5'$ to $3'$ direction. This means that the growth of the daughter strand that starts at the $3'$ end of the parent strand can continue smoothly as the DNA helix continues to unwind. This strand is known as the *leading strand*. However, this smooth growth is not possible for the daughter strand that started from the $5'$ end of the parent strand. This strand, known as the *lagging strand*, is formed in a series of sections, each of which still grows in the $5'$ to $3'$ direction. These sections, which are known as Okazaki fragments after their discoverer, are joined together by the enzyme DNA ligase to form the second daughter strand.

Replication, which starts at the end of a DNA helix, continues until the entire structure has been duplicated. The same result is obtained when replication starts at the centre of a

DNA helix. In this case unwinding continues in both directions until the complete molecule is duplicated. This latter situation is more common.

DNA replication occurs when cell division is imminent. At the same time new histones are synthesised. This results in a thickening of the chromatin filaments into chromosomes (see section 10.2). These rod-like structures can be stained and are large enough to be seen under a microscope.

10.6   Ribonucleic acid (RNA)

Ribonucleic acid is found in both the nucleus and the cytoplasm. In the cytoplasm RNA is located mainly in small spherical organelles known as *ribosomes*. These consist of about 65 per cent RNA and 35 per cent protein. Ribonucleic acids are classified according to their general role in protein synthesis: messenger RNA (mRNA); transfer RNA (tRNA); and ribosomal RNA (rRNA). Messenger RNA informs the ribosome as to what amino acids are required and their order in the protein, that is, they carry the genetic information necessary to produce a specific protein. This type of RNA is synthesised as required and once its message has been delivered it is decomposed. Transfer RNA transports the required amino acids in the correct order to the ribosome, where ribosomal RNA (rRNA) controls the synthesis of the required protein (see section 10.8).

The structures of RNA molecules consist of a single polymer chain of nucleotides with the same bases as DNA with the exception of thymine, which is replaced by uracil (Fig. 10.9). Uracil, where appropriate, forms a complementary base pair with adenine. These chains often contain single-stranded loops separated by short sections of a distorted double helix. These structures are known as *hairpin loops*.

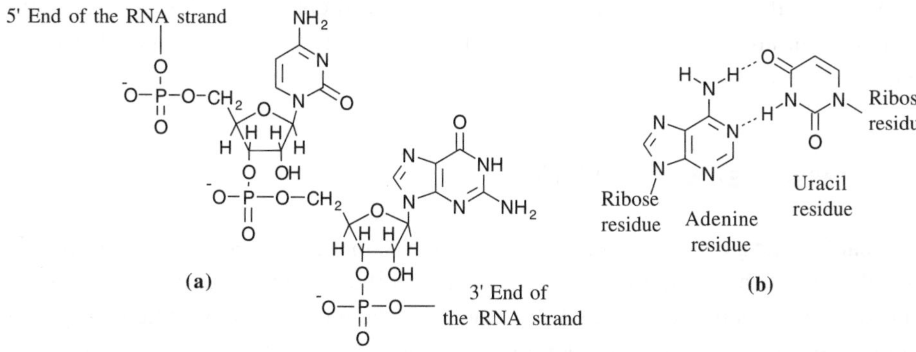

Figure 10.9   (a) The general structure of a section of an RNA polymer chain. (b) The hydrogen bonding between uracil and adenine. Reproduced from G. Thomas, *Chemistry for Pharmacists and the Life Sciences*, 1996, by permission of Prentice Hall, a Pearson Education Company

All types of RNA are formed from DNA by a process known as *transcription*, which occurs in the nucleus. It is thought that the DNA unwinds and the RNA molecule is formed

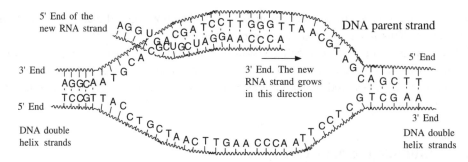

Figure 10.10 A schematic representation of a transcription process. Reproduced from G. Thomas, *Chemistry for Pharmacists and the Life Sciences*, 1996, by permission of Prentice Hall, a Pearson Education Company

in the 5′ to 3′ direction. It proceeds smoothly with the 3′ end of the new strand bonding to the 5′ end of the next nucleotide (Fig. 10.10). This bonding is catalysed by enzymes known as RNA polymerases. Since only complementary base pairs can hydrogen bond, the order of bases in the new RNA strand is determined by the sequence of bases in the parent DNA strand. In this way DNA controls the genetic information being transcribed into the RNA molecule. The strands of DNA also contain start and stop signals, which control the size of the RNA molecule produced. These signals are in the form of specific sequences of bases. It is believed that the enzyme *rho factor* could be involved in the termination of the synthesis and the release of some RNA molecules from the parent DNA strand. However, in many cases there is no evidence that this enzyme is involved in the release of the RNA molecule.

The RNA produced within the nucleus by transcription is known as *heterogeneous nuclear RNA* (*hnRNA*), *premessenger RNA* (*pre-mRNA*) or *primary transcript RNA* (*ptRNA*). Since the DNA gene from which it is produced contains both exons and introns, the hnRNA will also contain its genetic information in the form of a series of exons and introns complementary to those of its parent gene.

10.7   Messenger RNA (mRNA)

mRNA carries the genetic message from the DNA in the nucleus to a ribosome. This message instructs the ribosome to synthesise a specific protein. mRNA is believed to be produced in the nucleus from hnRNA by removal of the introns and the splicing together of the remaining exons into a continuous genetic message, the process being catalysed by specialised enzymes. The net result is a smaller mRNA molecule with a continuous sequence of bases that are complementary to the gene's exons. This mRNA now leaves the nucleus and carries its message in the form of a code to a ribosome.

The code carried by mRNA was broken in the 1960s by Nirenberg and other workers. These workers demonstrated that each naturally occurring amino acid had a DNA code that consisted of a sequence of three consecutive bases known as a *codon* and that an amino acid

Table 10.1 The genetic code. Some codons act as start and stop signals in protein synthesis (see sections 10.10 and 10.11). Codons are written left to right, 5' to 3'

Code	Amino acid	Code	Amino acid	Code	Amino acid	Code	Amino acid
UUU	Phe	CUU	Leu	AUU	Ile	GUU	Val
UUC	Phe	CUC	Leu	AUC	Ile	GUC	Val
UUA	Leu	CUA	Leu	AUA	Ile	GUA	Val
UUG	Leu	CUG	Leu	AUG	Met	GUG	Val
UCU	Ser	CCU	Pro	ACU	Thr	GCU	Ala
UCC	Ser	CCC	Pro	ACC	Thr	GCC	Ala
UCA	Ser	CCA	Pro	ACA	Thr	GCA	Ala
UCG	Ser	CCG	Pro	ACG	Thr	GCG	Ala
UAU	Tyr	CAU	His	AAU	Asn	GAU	Asp
UAC	Tyr	CAC	His	AAC	Asn	GAC	Asp
UAA	**Stop**	CAA	Gln	AAA	Lys	GAA	Glu
UAG	**Stop**	CAG	Gln	AAG	Lys	GAG	Glu
UGU	Cys	CGU	Arg	AGU	Ser	GGU	Gly
UGC	Cys	CGC	Arg	AGC	Ser	GGC	Gly
UGA	**Stop**	CGA	Arg	AGA	Arg	GGA	Gly
UGG	Trp	CGG	Arg	AGG	Arg	GGG	Gly

could have several different codons (Table 10.1). In addition, three of the codons are stop signals, which instruct the ribosome to stop protein synthesis. Furthermore, the codon that initiates the synthesis is always AUG, which is also the codon for methionine. Consequently, all protein synthesis starts with methionine. However, few completed proteins have a terminal methionine since this residue is normally removed before the peptide chain is complete. Moreover, methionine can still be incorporated in a peptide chain since there are two different tRNAs that transfer methionine to the ribosome (see section 10.8): one is specific for the transfer of the initial methionine whilst the other will only deliver methionine to the developing peptide chain. By convention, the three letters of a codon's triplets are normally written with their 5' ends on the left and their 3' ends on the right.

The mRNA's codon code is known as the *genetic code*. Its use is universal, with all living matter using the same genetic code for protein synthesis. This suggests that all living matter must have originated from the same source and is strong evidence for Darwin's theory of evolution.

10.8 Transfer RNA (tRNA)

tRNA is also believed to be formed in the nucleus from hnRNA. tRNA molecules are relatively small and usually contain 73–94 nucleotides in a single strand. Some of these nucleotides may contain derivatives of the principal bases, such as 2'-O-methylguanosine (OMG) and inosine (I). The strand of tRNA is usually folded into a three-dimensional L shape. This structure, which consists of several loops, is held in this shape by hydrogen bonding between complementary base pairs in the stem sections of these loops and also by hydrogen bonding between bases in different loops. This results in the formation of sections of double helical structures. However, the structures of most tRNAs are represented in two dimensions as a *cloverleaf* (Fig. 10.11).

2'-*O*-Methylguanosine   Inosine

tRNA molecules carry amino acid residues from the cell's amino acid pool to the mRNA attached to the ribosome. The amino acid residue is attached through an ester linkage to the ribose residue at the 3' terminal of the tRNA strand, which almost invariably has the sequence CCA. This sequence plus a fourth nucleotide projects beyond the double helix of the stem. Each type of amino acid can only be transported by its own specific tRNA molecule: in other words, a tRNA that carries serine residues will not transport alanine residues. However, some amino acids can be carried by several different tRNA molecules.

The tRNA recognises the point on the mRNA where it has to deliver its amino acid through the use of a group of three bases known as an *anticodon*. This anticodon is a sequence of three bases found on one of the loops of the tRNA (Fig. 10.11). The anticodon can only form base pairs with the complementary codon in the mRNA. Consequently, the tRNA will only

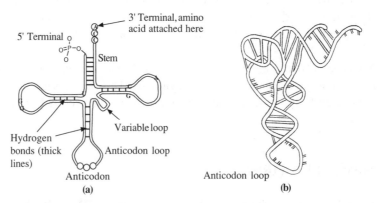

Figure 10.11 The general structures of tRNA. (**a**) The two-dimensional cloverleaf representation showing some of the invariable nucleotides that occur in the same positions in most tRNA molecules. (**b**) The three-dimensional L shape. From *Chemistry*, by Linus Pauling and Peter Pauling. Copyright © 1975 by Linus Pauling and Peter Pauling. Used with permission of W. H. Freeman and Company

hydrogen bond to the region of the mRNA that has the correct codon, which means its amino acid can only be delivered to a specific point on the mRNA. For example, a tRNA molecule with the anticodon CGA will only transport its alanine residue to a GCU codon on the mRNA. This mechanism controls the order in which amino acid residues are added to the growing protein, which always occurs from the N-terminal of the protein (see section 10.10).

10.9 Ribosomal RNA (rRNA)

Ribosomes contains about 35 per cent protein and 65 per cent rRNA. Their structures are complex and have not yet been fully elucidated. However, they have been found to consist of two sections, which are referred to as the *large* and *small* subunits. Each of these subunits contains protein and rRNA. In *E. coli* the small subunit has been shown to contain a 1542-nucleotide rRNA molecule whilst the large contains two rRNA molecules of 120 and 2094 nucleotides, respectively.

Experimental evidence suggests that rRNA molecules have structures that consist of a single strand of nucleotides whose sequence varies considerably from species to species. The strand is folded and twisted to form a series of single-stranded loops separated by sections of distorted double helix. The double helical segments are believed to be formed by hydrogen bonding between complementary base pairs. The general pattern of loops and helixes is very similar between species even though the sequence of nucleotides is different. However, little is known about the three-dimensional structures of rRNA molecules and their interactions with the proteins found in the ribosome.

10.10 Protein synthesis

Protein synthesis starts from the N-terminal of the protein. It proceeds in the $5'$ to $3'$ direction along the mRNA and may be divided into four major stages: activation; initiation; elongation; and termination. Activation is the formation of the tRNA–amino acid complex. Initiation is the binding of mRNA to the ribosome and activation of the ribosome. Elongation is the formation of the protein, whilst termination is the ending of the protein synthesis and its release from the ribosome. All these processes normally require the participation of protein catalysts known as *factors*, as well as other proteins whose function is not always known. GTP and sometimes ATP act as sources of energy for the processes.

10.10.1 Activation

It is believed that an amino acid (AA) from the cellular pool reacts with adenosine triphosphate (ATP) to form a reactive amino acid–adenosine monophosphate (AA–AMP) complex. This complex reacts with the tRNA specific for the amino acid to form an aminoacyl–tRNA complex, the reaction being catalysed by a synthase that is specific for that amino acid.

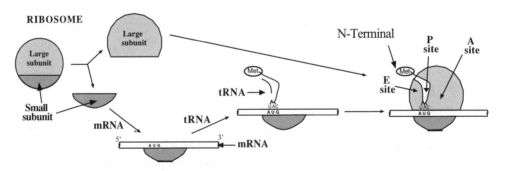

10.10.2 Initiation

The mechanism of initiation is more complex than the outline given in this text. It is well documented but the finer details are still not known. Initiation is thought to start with the two subunits of the ribosome separating and the binding of the mRNA to the smaller subunit. Protein synthesis then starts by the attachment of a methionine–tRNA complex to the mRNA, so that it forms the N-terminal of the new protein. Methionine is always the first amino acid in all protein synthesis because its tRNA anticodon is also the signal for the ribosome system to start protein synthesis. Since the anticodon for methionine–tRNA is UAC, this synthesis will start at the AUG codon of the mRNA. This codon is usually found within the first 30 nucleotides of the mRNA. However, few proteins have an N-terminal methionine because once protein synthesis has started the methionine is usually removed by hydrolysis. As soon as the methionine–tRNA has bound to the mRNA the larger ribosomal subunit is believed to bind to the smaller subunit so that the mRNA is sandwiched between the two subunits (Fig. 10.12). This large subunit is believed to have three binding sites called the *P* (*peptidyl*), *A* (*acceptor*) and *E* (*exit*) sites. It attaches itself to the smaller subunit so that its P site is aligned with the methionine–tRNA complex bound to the mRNA. This P site is where the growing protein will be bound to the ribosome. The A site, which is thought to be

Figure 10.12 A schematic representation of the initiation of protein synthesis

adjacent to the P site, is where the next amino acid–tRNA complex binds to the ribosome so that its amino acid can be attached to the peptide chain. The A site is where the discharged tRNA is transiently bound before it leaves the ribosome.

10.10.3 Elongation

Elongation is the formation of the peptide chain of the protein by a stepwise repetitive process. A great deal is known about the nature of this process but its exact mechanism is still not fully understood.

The process of elongation is best explained by the use of a hypothetical example. Suppose the sequence of codons of the mRNA including the start codon AUG is AUGUUGGCUGGA...etc. The elongation process starts with the methionine–tRNA complex binding to the AUG codon of the mRNA (Fig. 10.13). Since the second codon is UUG the second amino acid in the polypeptide chain will be leucine. This amino acid is transported by a tRNA molecule with the anticodon AAC because this is the only anticodon that matches the UUG codon on the mRNA strand. The leucine–tRNA complex 'docks' on the UUG codon of the mRNA and binds to the A site. This docking and binding is believed to involve ribosome proteins, referred to as *elongation factors*, and energy supplied by the hydrolysis of guanosine triphosphate (GTP) to guanosine diphosphate (GDP). Once the leucine–tRNA has occupied the A site the methionine is linked to the leucine by means of a

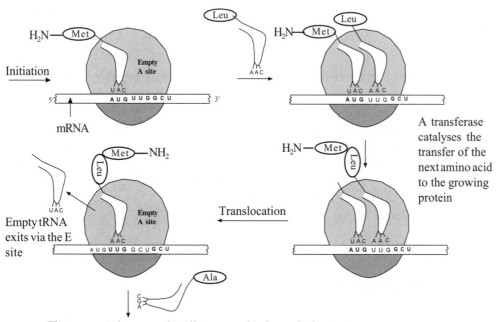

The sequence is repeated until a stop codon is reached

Figure 10.13 A diagrammatic representation of the process of elongation in protein synthesis

peptide link whose carbonyl group originates from the methionine. This reaction is catalysed by the appropriate transferase. It leaves the tRNA on the P site empty and produces an (NH_2)-Met-Leu–tRNA complex at the A site. The empty tRNA is discharged through the E site and at the same time the whole ribosome moves along the mRNA in the $3'$ to $5'$ direction so that the dipeptide–tRNA complex moves from the A site to the P site. This process is known as *translocation*. It is poorly understood but it leaves the A site empty and able to receive the next amino acid–tRNA complex. The whole process is then repeated in order to add the next amino acid residue to the peptide chain. Since the next mRNA codon in our hypothetical example is GCU this amino acid will be alanine (see section 10.7, Table 10.1). Subsequent amino acids are added in a similar way, the sequence of amino acid residues in the chain being controlled by the order of the codons in the mRNA.

10.10.4 Termination

The elongation process continues until a stop codon is reached. This codon cannot accept an amino acid–tRNA complex and so the synthesis stops. At this point the peptide–tRNA chain occupies a P site and the A site is empty. The stop codon of the mRNA is recognised by proteins known as *release factors*, which promote the release of the protein from the ribosome. The mechanism by which this happens is not fully understood but they are believed to convert the transferase responsible for peptide synthesis into a hydrolase that catalyses hydrolysis of the ester group linking the polypeptide to its tRNA. Once released the protein is folded into its characteristic shape, often under the direction of molecular *chaperone* proteins.

10.11 Protein synthesis in prokaryotic and eukaryotic cells

The general sequence of events in protein synthesis is similar for both eukaryotic and prokaryotic cells. In both cases the hydrolysis of GTP to GDP is the source of energy for many of the processes involved. However, the structures of prokaryotic and eukaryotic ribosomes are different. For example, the ribosomes of prokaryotic cells of bacteria are made up of 50S (where S is a Svedberg unit) and 30S rRNA subunits whilst the ribosomes of mammalian eukaryotic cells consist of 60S and 40S rRNA subunits. The differences between the prokaryotic and eukaryotic ribosomes are believed to be the basis of the selective action of some antibiotics (see section 10.12).

10.11.1 Prokaryotic cells

The first step in protein synthesis is the correct alignment of mRNA on the small subunit of the ribosome. In prokaryotic cells this alignment is believed to be due to binding by base pairing between bases at the $3'$ end of the rRNA of the ribosome and bases at the $5'$

φX174 phage A protein	-AAU-CUU-**GGA-GG**C-UUU-UUU-**AUG**-GUU-CGU-
Ribosomal protein S12	-AAA-ACC-**AGG-AG**C-UAU-UUA-**AUG**-GCA-ACA-
trpL. leader	-GUA-AA**A-AGG-G**UA-UCG-ACA-**AUG**-AAA-GCA-
araB	-UUU-GGA-U**GG-AG**U-GAA-ACG-**AUG**-GCG-AUU-

Figure 10.14 Examples of Shine–Dalgarno sequences (bold larger type) of mRNA recognised by *E. coli* ribosomes. These sequences usually lie between 4 to 10 nucleotides upstream of the AUG start codon for the specified protein in the left-hand column

end of the mRNA. This ensures the correct alignment of the AUG anticodon of the mRNA with the P site of the ribosome. The mRNA sequence of bases responsible for this binding occurs as part of the 'upstream' (5′ terminal end) section of the strand before the start codon. This sequence is often known as the *Shine–Dalgarno* sequence after its discoverers. Shine–Dalgarno sequences vary in length and base sequence (Fig. 10.14). The initiating tRNA in prokaryotic cells is a specific methionine–tRNA known as $tRNA_f^{Met}$, which is able to read the start codon AUG but not when AUG is part of the elongation sequence. $tRNA_f^{Met}$ is unique in that the methionine it carries is usually in the form of its *N*-formyl derivative.

$$N\text{-Formyl-methionine-tRNA}_f{}^{Met}$$
$$(\,fMet\text{-}tRNA_f{}^{Met}\,)$$

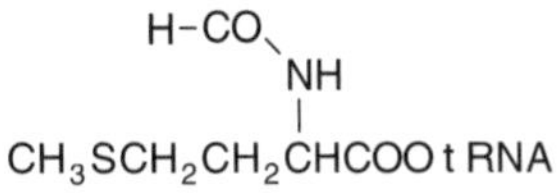

$$\begin{array}{c} H\text{-}CO \diagdown \\ NH \\ | \\ CH_3SCH_2CH_2CHCOO\,t\,RNA \end{array}$$

When AUG is part of the elongation sequence methionine is added to the growing protein by a different transfer RNA known as $tRNA_m^{Met}$, which also has the anticodon UAC. However, $tRNA_m^{Met}$ cannot initiate protein synthesis. Elongation follows the general mechanism for protein synthesis (see section 10.9). It requires a group of proteins known as elongation factors and energy supplied by the hydrolysis of GTP to GDP. Termination normally involves three release factors.

Experimental work has shown that an mRNA strand actively synthesising proteins will have several ribosomes attached to it at different places along its length. These multiple ribosome structures are referred to as *polyribosomes* or *polysomes*. The polysomes of prokaryotic cells can contain up to 10 ribosomes at any one time. Each of these ribosomes will be simultaneously producing the same polypeptide or protein: the further the ribosome has moved along the mRNA, the longer the polypeptide chain. The process resembles the assembly line in a factory. Each mRNA strand can, in its lifetime, produce up to 300 protein molecules.

In prokaryotic but not eukaryotic cells (see section 7.1), ribosomes are found in association with DNA. This is believed to be due to the ribosome binding to the mRNA as it is produced by transcription from the DNA. Furthermore, these ribosomes have been shown to start producing the polypeptide chain of their designated protein before transcription is complete. This means that in bacteria protein synthesis can be very rapid and in some cases faster than transcription. It has been reported that in some bacteria an average of 10 amino acid residues are added to the peptide chain every second.

10.11.2   Eukaryotic cells

The initiation of protein synthesis in eukaryotic cells follows a different route from that found in prokaryotic cells although it still uses a methionine–tRNA to start the process. Eukaryotic mRNAs have no Shine–Dalgarno sequences but are characterised by a 7-methyl-GTP unit at the 5′ end of the mRNA strand and a poly-adenosine nucleotide tail at the 3′ end of the strand (Fig. 10.15).

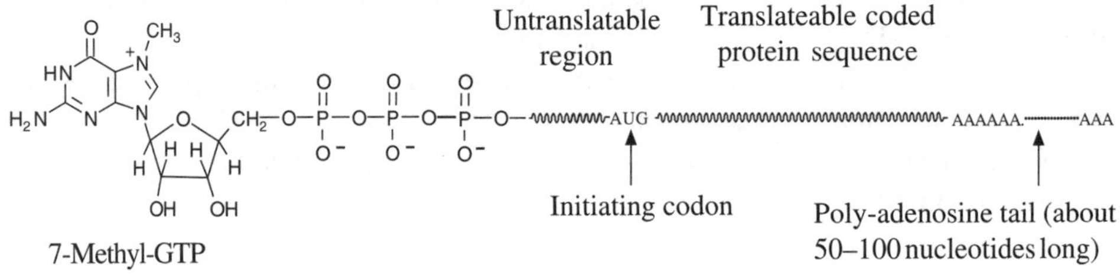

Figure 10.15 The general structure of eukaryotic mRNA molecules, where ∿∿∿∿ represents the mRNA strand

In eukaryotic cells, the initiating tRNA is a unique form of the activated methionine–tRNA (tRNA$_i^{Met}$). However, unlike in the case of prokaryotic cells, the methionine residue it carries is not formylated. The initiating process is started by this tRNA$_i^{Met}$ binding to the 40S subunit of the ribosome to form the so-called *preinitiation complex*, the process requiring the formation of a complex between tRNA$_i^{Met}$, various eukaryotic initiation factors (eIFs) and GTP.

The absence of the Shine–Dalgarno sequence means that an alternative mechanism must be available to align the first AUG codon of the mRNA with the P site of the ribosome. This mechanism is believed to direct the preinitiation complex to the first AUG codon of the mRNA.

Elongation in eukaryotic ribosomes follows the general mechanism for protein synthesis (see section 10.10.3) but involves different factors and proteins from those utilised by prokaryotic ribosomes. Termination only requires one release factor, unlike in prokaryotic ribosomes where three release factors are usually required.

10.12   Bacterial protein synthesis inhibitors (antimicrobials)

Many protein inhibitors inhibit protein synthesis in both prokaryotic and eukaryotic cells (Table 10.2). This inhibition can take place at any stage in protein synthesis. However,

Table 10.2 Examples of drugs that inhibit protein synthesis

Drug	Action
Chloramphenicol (see section 10.12.2)	Blocks the enzymes that catalyse the transfer of the new amino acid residue to the peptide chain, that is, prevents elongation in *prokaryotic* cells.
Cyclohexidine	Inhibits translocation of mRNA in *eukaryotic* ribosomes
Erythromycin (see section 10.12.4)	Blocks the enzymes that catalyse the transfer of the new amino acid residue to the peptide chain, that is, prevents elongation in *prokaryotic* cells
Fusidic acid	Inhibits dissociation of the protein from the ribosome in both *prokaryotic* and *eukaryotic* cells
Streptomycin (see this section)	Inhibits initiation of prokaryotic protein synthesis. It also causes the misreading of codons, which results in the insertion of incorrect amino acid residues into the structure of the protein formed in *prokaryotic* cells
Tetracycline (see section 10.12.3)	Inhibits the binding of aminoacyl–tRNA to the A site of *prokaryotic* ribosomes

some inhibitors have a specific action in that they inhibit protein synthesis in prokaryotic cells but not in eukaryotic cells, or vice versa. Consequently, a number of useful drugs have been discovered that will inhibit protein synthesis in bacteria but either have no effect or a very much reduced effect on protein synthesis in mammals.

The structures and activities of the drugs that inhibit protein synthesis are very diverse. Consequently, only a few of the more commonly used drugs and structurally related compounds will be discussed in greater detail in this section.

10.12.1 Aminoglycosides

Aminoglycosides are a group of compounds that have structures in which aminosugar residues in the form of mono- or polysaccharides are attached to a substituted 1,3-diaminocyclohexane ring by modified glycosidic-type linkages (Fig.10.16). This ring is either streptidine (streptomycin) or deoxystreptamine (amikacin, kanamycin, netilmicin, neomycin, paromycin, gentamicin and tobramycin).

Streptomycin was the first aminoglycoside discovered (Schatz *et al.* 1944) It was isolated from cultures of the soil Actinomycete *Streptomyces griseus*. It acts by interfering with the initiation of protein synthesis in bacteria. The binding of streptomycin to the 30S ribosome inhibits initiation and also causes some amino acid–tRNA complexes to misread the mRNA codons. This results in the insertion of incorrect amino acid residues into the protein chain, which usually leads to the death of the bacteria. The mode of action of the other aminoglycosides has been assumed to follow the same pattern even though most of the

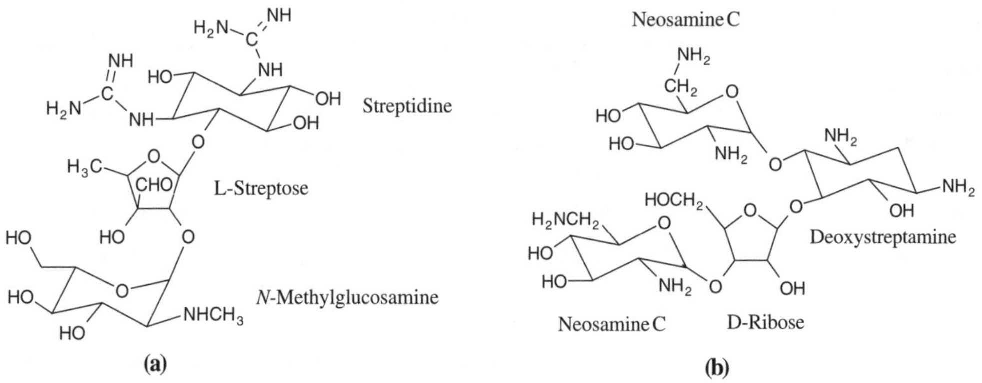

Figure 10.16 The structures of (**a**) streptomycin and (**b**) neomycin C

investigations into the mechanism of the antibacterial action of the aminoglycosides has been carried out on streptomycin.

The majority of the aminoglycosides have been isolated from various microorganisms (Table 10.3). However, several of these drugs, such as kanamycin (Fig. 10.17) and neomycin (Fig. 10.16b), are obtained as mixtures of closely related compounds. Amikacin and netilmicin are obtained by semisynthetic methods from kanamiycin A and sisomicin, respectively (Fig. 10.18).

The clinically used aminoglycosides have structures closely related to that of streptomycin. They are essentially broad spectrum antibiotics although they are normally used to treat serious Gram-negative bacterial infections (see section 7.2.5). Aminoglycosidic drugs are very water soluble. They are usually administered as their water-soluble inorganic salts but their polar nature means that they are poorly absorbed when administered orally. Once in the body they are easily distributed into most body fluids. However, their polar nature means that they do not easily penetrate the CNS, bone, fatty and connective tissue. Moreover, aminoglycosides tend to concentrate in the kidney where they are excreted by glomerular filtration.

The activity of the aminoglycosides is related to the nature of their ring substituents. Consequently, it is convenient to discuss this activity in relation to the changes in the substituents of individual rings but in view of the diversity of the structures of aminoglycosides it is difficult to identify common trends. As a result, this discussion will be largely limited to kanamycin (Fig. 10.17). However, the same trends are often true for other aminoglycosides whose structures consist of three rings including a central deoxystreptamine residue.

Changing the nature of the amino substituents at positions 2′ and 6′ of ring I has the greatest effect on activity. For example, kanamycin A, which has a hydroxy group at position 2′, and kanamycin C, which has a hydroxy group at position 6′, are both less active than kanamycin B, which has amino groups at the 2′ and 6′ positions. However, the removal of one or both of the hydroxy groups at positions 3′ and 4′ does not have any effect

Table 10.3 Clinically used aminoglycosides and their sources

Drug	Source	Discoverer (date)	Notes
Streptomycin	*S. grisius*	Schatz *et al.* (1944)	*S grisius* also produces other antibiotic substances such as cycloheximide
Amikacin	Semisynthetic synthesis	Kawaguchi *et al.* (1972)	Resistant to most inactivating enzymes of most bacteria
Kanamycin	*S. kanamyceticus*	Umezawa *et al.* (1957)	Occurs as three closely related compounds (Fig. 10.17). Kanamycin A is the least toxic and the main component of the commercial drug. An increasing number of resistant strains of bacteria (see section 9.13.1)
Netilmicin	Semisynthetic synthetic	Weinstein *et al.* (1975).	Inactivated by many bacterial enzymes that acetylate aminoglycosides
Neomycin	*S. fradiae*	Wakeman and Lechevalier (1949)	Little development of bacterial-resistant strains
Paromycin	*S. rimosus* forma *paromomycinus*	Frohardt *et al.* (1959)	
Gentamicin	*Microspora purpurea*	Researchers at the Schering Corporation, Bloomfield, New Jersey (1963)	Occurs as three closely related compounds (Fig. 10.19). Resistant strains of bacteria are increasing. They act by acetylation and adenylation of the drug (see section 9.13.1). Strongly active against *P. aeruginosa*, which is highly resistant to penicillins, cephalosporins, tetracyclines, quinolones and chloramphenicol
Tobramycin	*S. tenebrarius*	Unidentified (1976)	Active against strains of *P. aeruginosa*

Key:
Kanamycin A: $R_1 = NH_2$; $R_2 = OH$
Kanamycin B: $R_1 = NH_2$; $R_2 = NH_2$
Kanamycin C: $R_1 = OH$; $R_2 = OH$

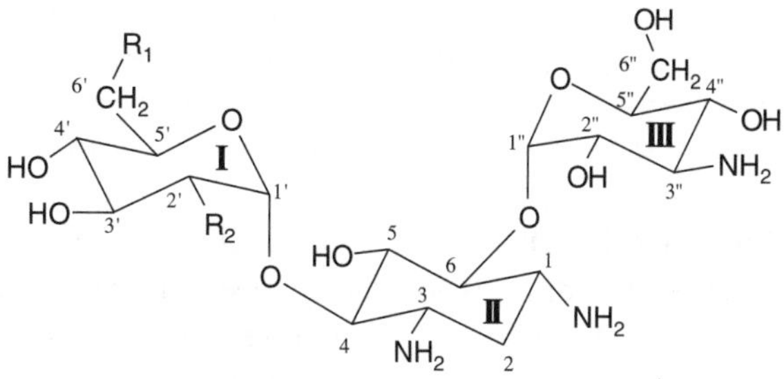

Figure 10.17 The kanamycins

on the potency of the kanamycins. Resistance to bacterial enzyme acetylation is increased by methylation at either the 6′-carbon or 6′-amino group without a significant loss in activity.

Most modifications to ring II (the deoxystreptamine ring) greatly reduce the potency of the kanamycins. However, N-acylation and alkylation of the amino group at position 1 can give compounds with some activity. For example, acylation of kanamycin A gives amikacin (Fig. 10.18), which has a potency of about 50 per cent of that of kanamycin A (Fig. 10.17). In spite of this, amikacin is a useful drug for treating some strains of Gram-negative bacteria because it is resistant to deactivation by bacterial enzymes. However, netilmicin (Fig. 10.18), formed by alkylation of sisomicin, is as potent as its parent aminoglycoside.

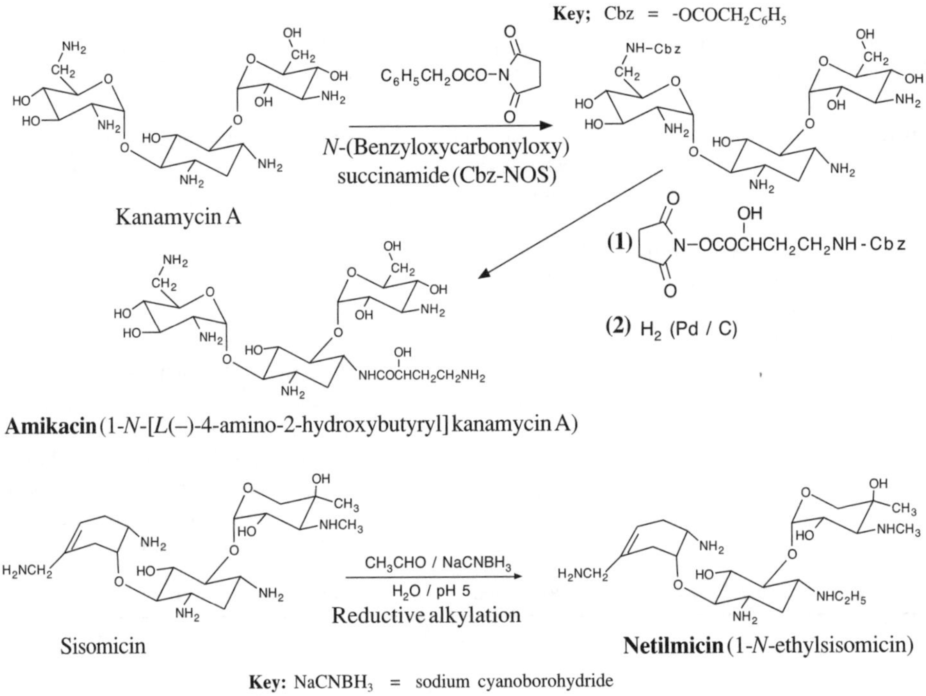

Figure 10.18 An outline of the chemistry involved in the synthesis of the antibiotics amikacin and Netilmicin. Cbz is N-(benzyloxycarbonyloxy)succinamide. This reagent is frequently used as a protecting group for amines as it is easily removed by hydrogenation

Changing the substituents of ring III does not usually have such a great effect on the potency of the drug as similar changes in rings I and II. For example, removal of the 2″-hydroxy group of gentamicin results in a significant drop in activity. However, replacement of the 2″-hydroxy group of gentamicin (Fig. 10.19) by amino groups gives the highly active seldomycins.

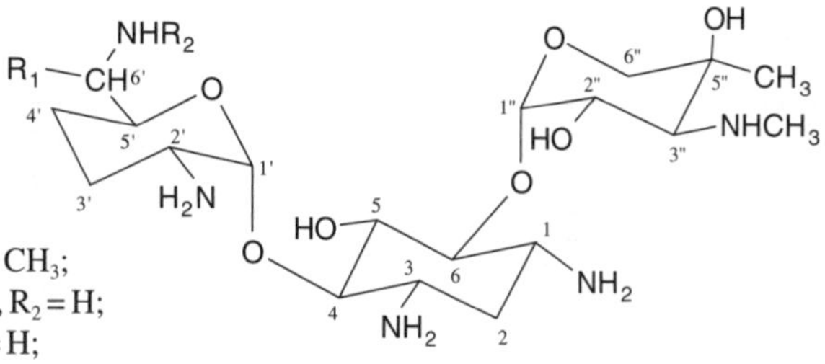

Key:
Gentamicin C_1; $R_1 = R_2 = CH_3$;
Gentamicin C_2; $R_1 = CH_3, R_2 = H$;
Gentamicin C_{1a}; $R_1 = R_2 = H$;

Figure 10.19 The structures of gentamicin

The highly polar nature of aminoglycosidic drugs means that they will not normally be easily absorbed from the GI tract. Consequently, they are usually administered intravenously. A high proportion of the administered dose is excreted unchanged in the first 24 hours after administration by glomerular filtration.

Aminoglycoside drug-resistant strains of bacteria are now recognised as a serious medical problem. They arise because dominant bacteria strains have emerged that possess enzymes that effectively inactivate the drug. These enzymes act by catalysing the acylation, phosphorylation and adenylation of the drug (see section 9.13.1), which results in the formation of inactive drug derivatives.

10.12.2 Chloramphenicol

Chloramphenicol was first isolated by Ehrlich *et al.* in 1947 from the microorganism *Streptomyces venezuelae*, which was found in a soil sample from Venezuela. It is a broad spectrum antibiotic whose structure contains two asymmetric centres. However, only the D-(–)-*threo* form is active. Its use can cause serious side effects and so it is recommended that chloramphenicol is only used for specific infections. It is often administered as its palmitate in order to mask its bitter taste. The free drug is liberated from this ester by esterase-catalysed hydrolysis in the duodenum. Chloramphenicol has a poor water solubility (2.5 g dm^{-3}) and so it is sometimes administered in the form of its more soluble sodium hemisuccinate salt (see section 2.9.4), which acts as a prodrug.

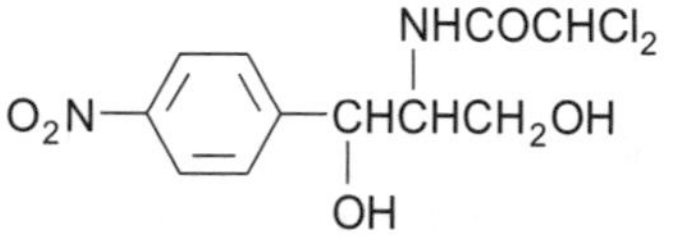

D-(–)-*Threo*-chloramphenicol

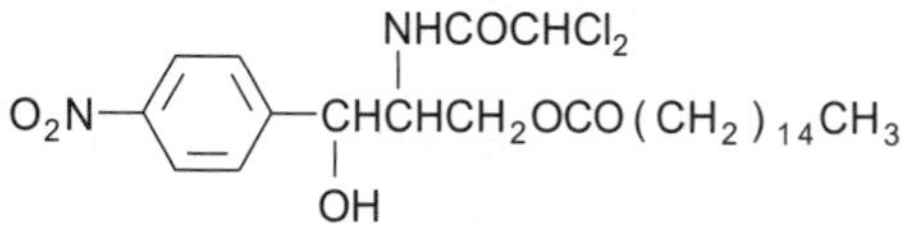

D-(–)-*Threo*-chloramphenicol palmitate

Chloramphenicol is believed to act by inhibiting the elongation stage in protein synthesis in prokaryotic cells. It binds reversibly to the 50S ribosome subunit and is thought to prevent the binding of the aminoacyl–tRNA complex to the ribosome. However, its precise mode of action is not understood.

Table 10.4 The activity against *E. coli* of some analogues of chloramphenicol relative to chloramphenicol

Analogue	Chloramphenicol activity against *E. coli*
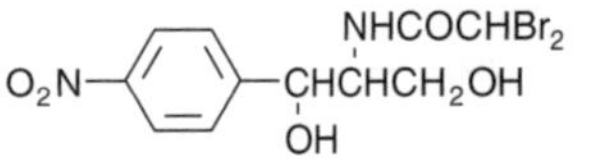 O₂N—⟨benzene⟩—CHCHCH₂OH with NHCOCHBr₂ and OH	About 0.8
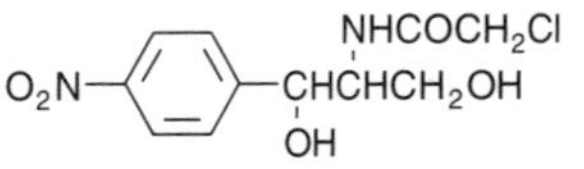 O₂N—⟨benzene⟩—CHCHCH₂OH with NHCOCH₂Cl and OH	About 0.4
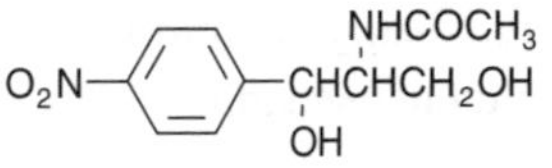 O₂N—⟨benzene⟩—CHCHCH₂OH with NHCOCH₃ and OH	Almost inactive
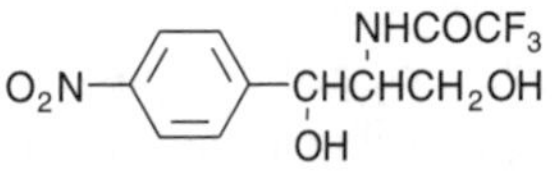 O₂N—⟨benzene⟩—CHCHCH₂OH with NHCOCF₃ and OH	1.7

Investigation of the activity of analogues of chloramphenicol showed that activity requires a *para*-electron withdrawing group. However, substituting the nitro group with other electron withdrawing groups gave compounds with a reduced activity. Furthermore, modification of the side chain, with the exception of the difluoro derivative, gave compounds that had a lower activity than chloramphenicol (Table 10.4). These observations suggest that D-(–)-*threo*-chloramphenicol has the optimum structure of those tested for activity.

The synthesis of chloramphenicol was first reported by Controulis J *et al.* in 1949 (Fig. 10.20). Numerous synthetic routes have since been devised for the synthesis of chloramphenicol, the commercial routes usually starting with 4-nitroacetophenone. Chloramphenicol is now manufactured by both totally synthetic and microbiological routes.

10.12.3 Tetracyclines

Tetracyclines are a family of natural and semisynthetic antibiotics isolated from various *Streptomyces* species. The first member of the group, chlortetracycline, was obtained in 1948 by Duggar from *Streptomyces aureofaciens*. A number of highly active semisynthetic analogues have also been prepared from naturally occurring compounds (Table 10.5). The tetracyclines are a broad spectrum group of antibiotics active against many Gram-positive and Gram-negative bacteria, rickettsiae, mycoplasmas, chlamydiae and some protozoa that cause malaria. A number of the natural and semisynthetic compounds are in current medical use.

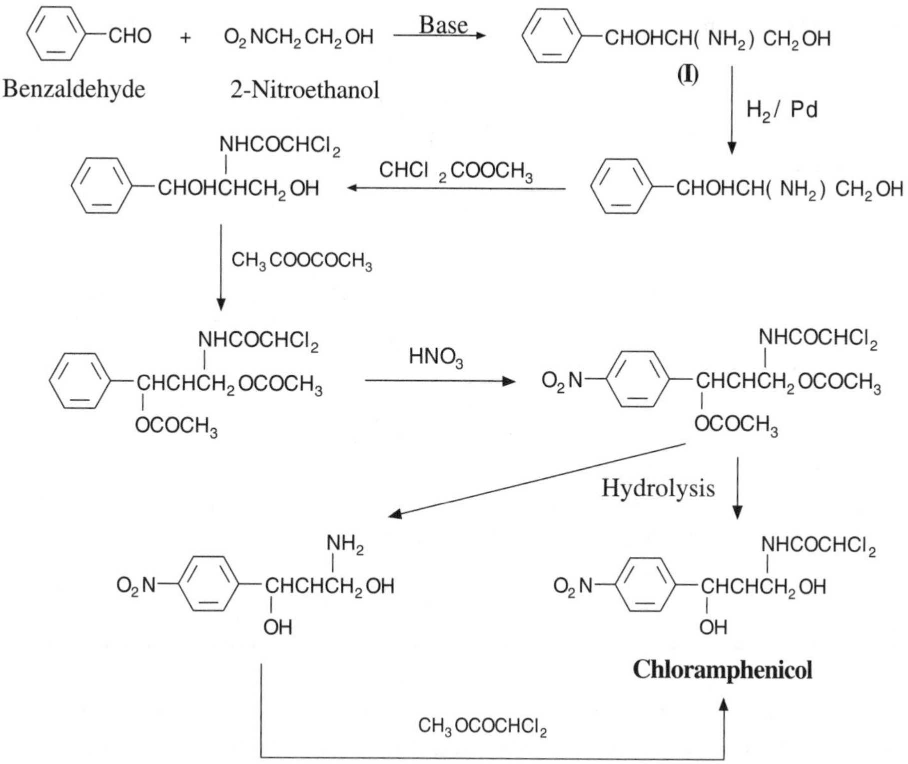

Benzaldehyde —CHO + $O_2NCH_2CH_2OH$ 2-Nitroethanol —Base→ —CHOHCH(NH$_2$)CH$_2$OH (I)

(I) | H_2/ Pd

NHCOCHCl$_2$
—CHOHCHCH$_2$OH ←CHCl$_2$COOCH$_3$— —CHOHCH(NH$_2$)CH$_2$OH

| CH$_3$COOCOCH$_3$

NHCOCHCl$_2$
—CHCHCH$_2$OCOCH$_3$
OCOCH$_3$
—HNO$_3$→ O_2N— NHCOCHCl$_2$ —CHCHCH$_2$OCOCH$_3$ OCOCH$_3$

Hydrolysis

O_2N— NH$_2$ —CHCHCH$_2$OH OH

O_2N— NHCOCHCl$_2$ —CHCHCH$_2$OH OH

Chloramphenicol

$CH_3OCOCHCl_2$

Figure 10.20 An outline of the synthesis of chloramphenicol by J. Controulis *et al.* Both benzaldehyde and 2-nitroethanol are readily available from commercial suppliers. Two products were obtained from the reduction of compound I: a crystalline fraction A and a gum B. Fraction A was found to give rise to chloramphenicol with the inactive *erythro* configuration but fraction B gave rise to chloramphenicol with the active *threo* configuration

The structures of the tetracyclines are based on an in-line fused four-ring system. Their structures are complicated by the presence of *up to* six chiral carbons in the fused ring system. These normally occur at positions 4, 4a, 5, 5a, 6 and 12a, depending on the symmetry of the structure. The configurations of these centres in the active compounds have been determined by X-ray crystallography (Table 10.5). This technique has also confirmed that C_1 to C_3 and C_{11} to C_{12} were conjugated structures.

Tetracyclines are amphoteric, forming salts with acids and bases. They normally exhibit three pK_a values in the regions 2.8–3.4 (pK_{a1}), 7.2–7.8 (pK_{a2}) and 9.1–9.7 (pK_{a3}), the last being the values for the corresponding ammonium salts. These values have been assigned by Leeson *et al.* to the structures shown in Table 10.5 and have been supported by the work of Rigler *et al.* However, the assignments for pK_{a2} and pK_{a3} are opposite to those suggested by Stephens *et al.* Tetracyclines also have a strong affinity for metal ions, forming stable chelates with calcium, magnesium and iron ions. These chelates are usually insoluble in water, which accounts for the poor absorption of tetracyclines in the presence of drugs and

Table 10.5 The structures of the tetracyclines

R$_1$ R$_2$ R$_3$ R$_4$ N(CH$_3$)$_2$ pK$_{a3}$ (of the ammonium salt)

pK$_{a2}$ R$_5$ pK$_{a1}$

Tetracycline	R$_1$	R$_2$	R$_3$	R$_4$	R$_5$
Chlortetracycline	Cl	CH$_3$	OH	H	CONH$_2$
Demeclocycline	Cl	H	OH	H	CONH$_2$
Doxycycline	H	CH$_3$	OH	CONH$_2$	
Melocycline	Cl	=CH2	–	OH	CONH$_2$
Methacycline	H	CH$_2$		OH	CONH$_2$
Minocycline	N(CH$_3$)$_2$	H	CONH$_2$		
Oxytetracycline	H	CH$_3$	OH	CONH$_2$	
Rolitetracycline	H	CH$_3$	OH	H	CONHCH$_2$-N
Tetracycline	H	CH$_3$	OH	H	CONH$_2$

foods that contain these metal ions. However, this affinity for metals appears to play an essential role in the action of tetracyclines.

Tetracyclines are transported into the bacterial cell by passive diffusion and active transport. Active transport requires the presence of Mg^{2+} ions and possibly ATP as an energy source. Once in the bacteria, tetracyclines act by preventing protein elongation by inhibiting the binding of the aminoacyl–tRNA to the 30S subunit of the prokaryotic ribosome. This binding has also been shown to require magnesium ions.

Tetracyclines also penetrate mammalian cells and bind to eukaryotic ribosomes. However, their affinity for eukaryotic ribosomes is lower than that for prokaryotic ribosomes and so they do not achieve a high enough concentration to disrupt eukaryotic protein synthesis. Unfortunately, bacterial resistance to tetracyclines is common. It is believed to involve three distinct mechanisms, namely, active transport of the drug out of the bacteria by membrane spanning proteins, enzymic oxidation of the drug and ribosome protection by chromosomal protein determinants.

The structure–activity relationships of tetracyclines have been extensively investigated and reported. Consequently, the following paragraphs give only a synopsis of these relationships. This synopsis only considers general changes to both the general structure of the tetracyclines (Fig. 10.21) and the substitution patterns of their individual rings.

Activity in the tetracyclines requires four rings with a *cis* A/B ring fusion. Derivatives with three rings are usually inactive or almost inactive. In general, modification to any of the substituent groups in the positions C-10, C-11, C-11a, C-12, C-12a, C-1, C-2, C-3

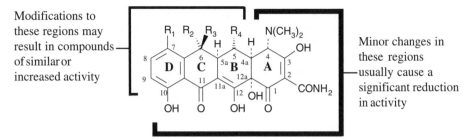

Modifications to these regions may result in compounds of similar or increased activity

Minor changes in these regions usually cause a significant reduction in activity

Figure 10.21 General structure–activity relationships in the tetracyclines

and C-4 results in a significant loss in activity. For example, replacement of the 2-carboxamide group by aldehyde and nitrile groups results in a significant loss in activity. N-Alkylation of the 2-amido group usually reduces activity, the reduction in activity increasing with increasing size of the group. Changes to the 4-dimethylamino group also usually reduce activity. This group must have an α-configuration and partial conversion of this group to its β-epimer under acidic conditions at room temperature significantly reduces activity.

α-Epimer of the 4-diethylamino group

β-Epimer of the 4-diethylamino group

In addition, either removal of the α-dimethylamino group at position 4 or replacement of one or more of its methyl groups by larger alkyl groups also reduces activity. Ester formation at C-12a gives inactive esters, with the exception of the formyl ester that hydrolyses in aqueous solution to the parent tetracycline. Alkylation of C-11a also gives rise to a loss of activity.

Modification of the substituents at positions 5, 5a, 6, 7, 8 and 9 may lead to similar or increased activity. Minor changes to the substituents at these positions tend to change pharmacokinetic properties rather than activity (Table 10.6). A number of active derivatives have been synthesised by electrophilic substitution of C-7 and C-9 but the effect of introducing substituents at C-8 has not been studied since this position is difficult to substitute.

10.12.4 Macrolides

Macrolides are a group of compounds that act as antibiotics. They have a similar spectrum of activity to the penicillins and are active against bacteria that are resistant to the penicillins.

Table 10.6 The pharmacokinetic properties of tetracycline and some of its analogues. The values given are representative values only as variations between individuals can be quite large

Tetracycline	R_1	R_2	R_3	R_4	$P_{o/w}$ at pH 7.5	V_D (dm^3kg^{-1})	Cl_T (dm^3h^{-1})	$t^{1/2}$ (h)	Protein binding
Chlortetracycline	Cl	CH$_3$	OH	H	0.12	1.2		6	55
Demeclocycline	Cl	H	OH	H	0.05	1.8		12	70
Doxycycline	H	CH$_3$	OH	CONH$_2$	0.63	0.7	1.7	16	90
Methacycline	H	CH$_2$	CONH$_2$	OH	0.40				
Minocycline	N(CH$_3$)$_2$	H	CONH$_2$	1.5	5	15	70		
Oxytetracycline	H	CH$_3$	OH	CONH$_2$	0.03	1.5		9	30
Tetracycline	H	CH$_3$	OH	H	0.04	2	15	6	45

The original compounds were isolated from the actinomycetes. They are characterised by structures that include a large lactone ring, an aminosugar and a ketone group. The lactone ring is often unsaturated and usually contains 12, 14 or 16 carbon atoms. A number of macrolides are in clinical use: azithromycin, clarithromycin and erythromycin (Fig. 10.22). Azithromycin and clarithromycin are both produced by semisynthetic methods.

Little is known about the mechanism of action of the macrolides. However, it is known that erythromycin binds to the 50S ribosomal subunit of bacterial ribosomes but not mammalian ribosomes. As a result, erythromycin inhibits the translocation step in bacterial protein synthesis. Resistance to erythromycin by some Gram-negative bacteria is believed to be due to the drug being unable to pass through the cell walls of these bacteria (see section 7.2.5). A more specific form of resistance is exhibited by certain strains of *S. aureus*. These strains methylate a particular adenine residue at the erythromycin binding site, which prevents erythromycin binding to the bacterial 50S ribosomal subunit. However, it does not prevent the methylated ribosome from producing the designated bacterial protein.

10.12.5 Lincomycins

Lincomycins are a group of antibiotics containing a sulphur functional group. They were originally isolated from *Streptomyces lincolnenis* but some, such as clindamycin, are now

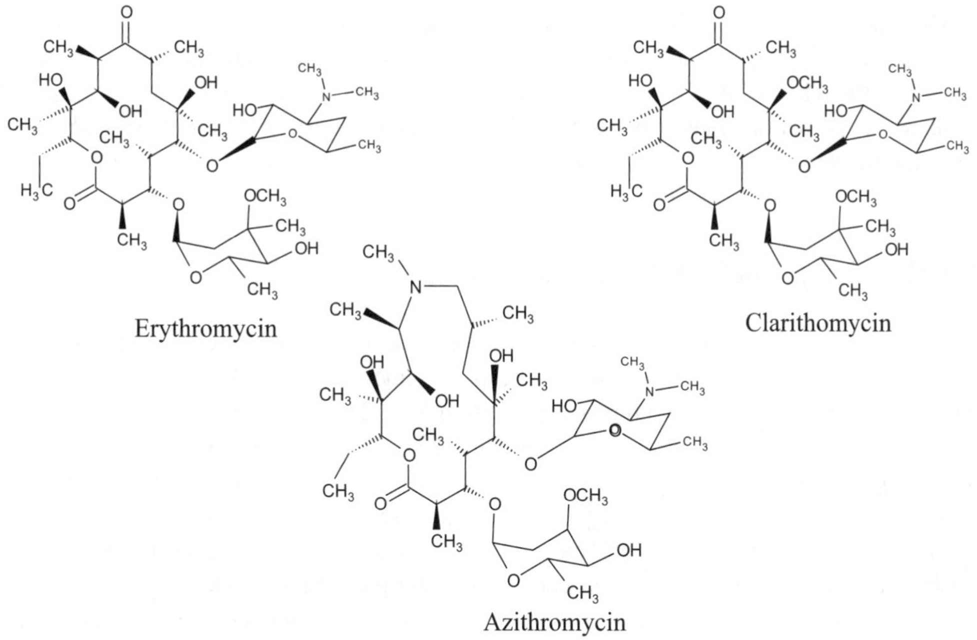

Figure 10.22 Examples of some of the macrolides in clinical use

synthesised by semisynthetic methods. The most active members of the group are lincomycin and clindamycin (Fig. 10.23). Clindamycin was synthesised from lincomycin by Magerlain *et al.* in 1967 by treatment of lincomycin with chlorine. This process caused inversion of the configuration of carbon-7. Clindamycin is less toxic than lincomycin and is absorbed more easily.

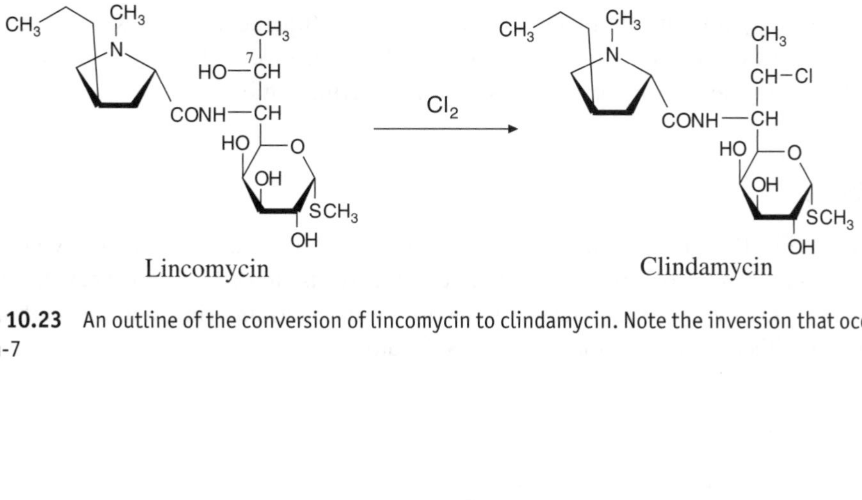

Figure 10.23 An outline of the conversion of lincomycin to clindamycin. Note the inversion that occurs at carbon-7

The lincomycins are mainly active against Gram-positive bacteria (see section 7.2.5), some strains of *Plasmodium*, actinomycetes, mycoplasma and certain anaerobic bacteria. They are thought to inhibit protein synthesis by binding to the 50S bacterial ribosomal subunit. This action may be bactericidal or bacteriostatic. Bacterial resistance to lincomycins appears to occur by mechanisms similar to those found with macrolides.

10.13   Drugs that target nucleic acids

Drugs that target DNA and RNA either inhibit their synthesis or act on existing nucleic acid molecules. Those that inhibit the synthesis of nucleic acids usually act as either antimetabolites or enzyme inhibitors (see section 9.9). Drugs that target existing nucleic acid molecules can, for convenience, be broadly classified into intercalating agents, alkylating agents and chain cleaving agents. However, it should be realised that these classifications are not rigid and drugs may act by more than one mechanism. Those drugs acting on existing DNA usually inhibit transcription whilst those acting on RNA normally inhibit translation. In both cases the net result is the prevention or slowing down of cell growth and division. Consequently, the discovery of new drugs that target existing DNA and RNA is a major consideration when developing new drugs for the treatment of cancer and bacterial and other infections due to microorganisms.

10.13.1   Antimetabolites

Antimetabolites are compounds that block the normal metabolic pathways operating in cells. They act by either replacing an endogenous compound in the pathway by a compound whose incorporation into the system results in a product that can no longer play any further part in the pathway, or inhibiting an enzyme in the metabolic pathway in the cell. Both of these types of intervention inhibit the targeted metabolic pathway to a level that hopefully has a significant effect on the health of the patient.

The structures of antimetabolites are usually very similar to those of the normal metabolites used by the cell. Those used to prevent the formation of DNA may be classified as antifolates, pyrimidine antimetabolites and purine antimetabolites. However, because of the difficulty of classifying biologically active substances (see section 1.8), antimetabolites that inhibit enzyme action are also classified as enzyme inhibitors.

Antifolates

Folic acid (Fig.10.24) is usually regarded as the parent of a family of naturally occurring compounds known as folates. These folates are widely distributed in food. They differ from folic acid in such ways as the state of reduction of the pteridine ring, and one carbon unit attached to either or both of the N-5 and N-10 atoms.

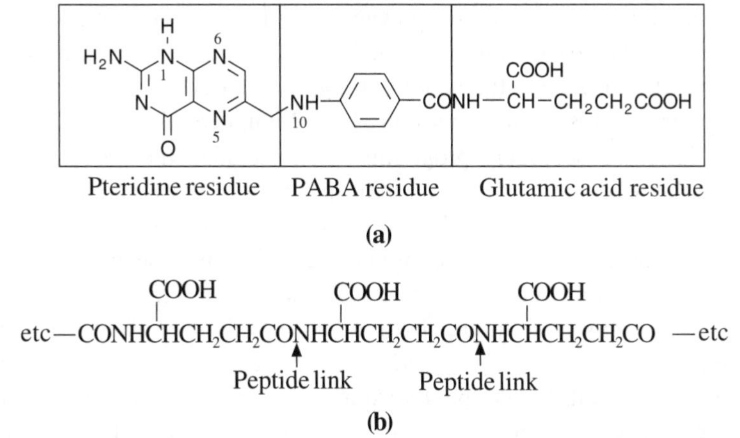

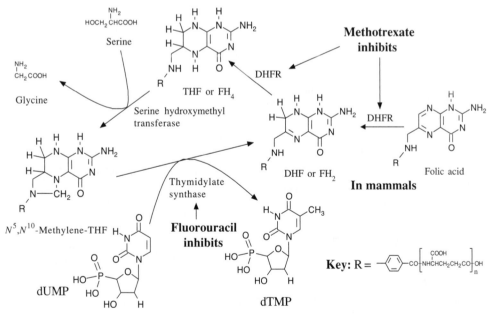

Pteridine residue   PABA residue   Glutamic acid residue

(a)

(b)

Figure 10.24 **(a)** The structure of folic acid. In blood folic acids usually have one glutamate residue. However, in the cell they are converted to polyglutamates. **(b)** A fragment of a polyglutamate chain

In mammals, folates are converted by a two-step process into tetrahydrofolates (THF or FH4) by the action of the enzyme dihydrofolate reductase (DHFR, Fig. 10.25). THF is converted to N^5, N^{10}-methylene-THF, which transfers a methyl group to 2′-deoxyuridylate in the biosynthesis of purines and thymine, which are essential for the biosynthesis of

Figure 10.25 An outline of the synthesis of deoxythymidylate monophosphate (dTMP) from 2-deoxyuridylate monophosphate (dUMP). N^5, N^{10}-Methylene-THF is the source of the methyl group in the conversion. DHFR is dihydrofolate reductase

DNA. Inhibition of DHFR will either prevent or reduce the formation of purines and thymine essential for the production of nucleic acids. This will ultimately lead to cell death.

The discovery of antifolate antimetabolite drugs was based on work carried out in the late 1930s and early 1940s by R. Lewisohn at the Mount Sinai Hospital, New York. He carried out an intensive screening programme that included the recently discovered B vitamins. Lewisohn used yeast extracts as a cheap source of these vitamins and found that crude extracts from these sources could cause regression of tumours in mice. He also found that barley extract could cause the same tumour regression in these animals. As a result of this work Lewisohn speculated that the active substance in the active extracts might be folic acid. Subsequent investigations in collaboration with Lederle showed that the folic acid (pteroylglutamic acid) supplied by Lederle was ineffective against spontaneous breast cancer tumours in mice but that the pteroyltriglutamic acid, also supplied by Lederle, was highly effective in causing regression in these tumours in mice. This information was followed up by Lederle who synthesised diopterin and teropterin as potential antitumour drugs (Fig. 10.26). Both of these compounds underwent a Phase I trial in 1947 and a slight degree of success was observed against a range of cancers. It was also observed that in cases of acute leukaemia these folates stimulated the growth of leukaemia cells. In the same year, researchers at Lederle reported that a crude methylated derivative of folic acid had a depressant action on the formation of the red and white cells of rats and mice. This observation stimulated the team at Lederle to produce pteroylaspartic acid (Fig. 10.26),

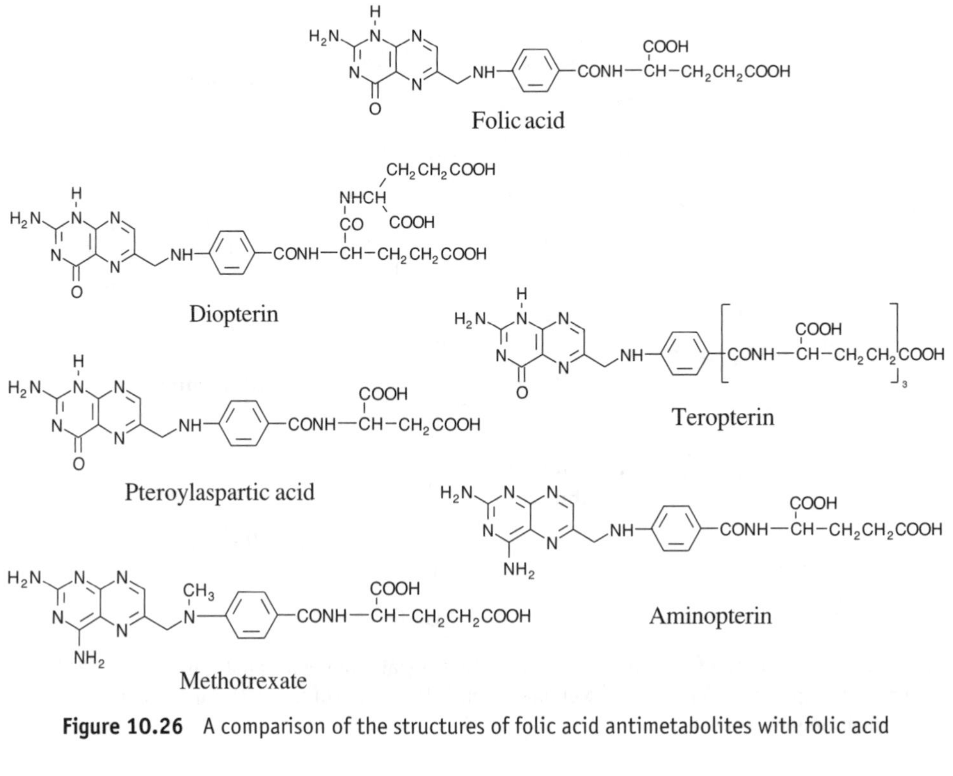

Figure 10.26 A comparison of the structures of folic acid antimetabolites with folic acid

which was found to be active against acute myeloid leukaemia. Pteroylaspartic acid was replaced by aminopterin (Fig. 10.26), which was more potent but suffered from a number of serious toxic side effects. This was soon replaced by the less toxic methotrexate (Figs. 10.25 and 10.26).

Methotrexate is the only folate antimetabolite in clinical use. It is distributed to most body fluids but has a low lipid solubility, which means that it does not readily cross the blood–brain barrier. It is transported into cells by the folate transport system and at high blood levels an additional second transport mechanism comes into operation. Once in the cell it is metabolised to the polyglutamate, which is retained in the cell for considerable periods of time, probably due to the polar nature of the polymer. Methotrexate is used to treat a variety of cancers, including head and neck tumours, and in low doses rheumatoid arthritis. Its use has apparently caused permanent remission in some cases of choriocarcinoma. However, methotrexate can cause vomiting, nausea, oral and gastric ulceration and depression of bone marrow, as well as other unwanted serious side effects.

Purine antimetabolites

The discovery of purine antimetabolites is based on the work, started in 1942, of G. Hitchings at the Wellcome Research Laboratories at Tuckahoe, New York. USA. He recognised that all cells require pyrimidine bases to form nucleic acids and divide. Consequently, he speculated that preventing the formation of these bases would inhibit cell proliferation and so it might be possible to block the reproduction of malignant cancer cells.

Hitchings' initial approach was to screen analogues of thymine for activity against *Lactobacillus casei*. This did not yield any compounds that were suitable for clinical trials. He continued his investigation using analogues of adenine and, in 1948, G. Elion synthesised diaminopurine, which was found to be active against acute leukaemia in children (Fig. 10.27). However, it was not as potent as aminopterin and methotrexate (Fig. 10.26). Elion went on to synthesise 6-mercaptopurine in 1950, which proved to be very effective against leukaemia.

Further investigations of purine analogues by Hitchings and Elion resulted in the synthesis in 1955 of 6-thioguanine (Fig. 10.27d), which was also found to be active against leukaemia. However, they did not find any other purine antimetabolites with higher

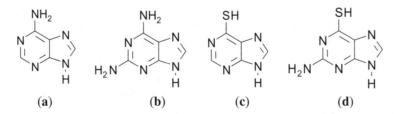

Figure 10.27   (a) Adenine, (b) aminopurine, (c) 6-mercaptopurine and (d) 6-thioguanine

potencies than 6-mercaptopurine. Both 6-mercaptopurine and 6-thioguanine are used clinically in the treatment of leukaemias.

As in the case of many drug discovery investigations, spin off from the work of Hitching and Elion led to the discovery of allopurinol and azathioprine (Fig. 10.28). The former is used to treat gout while the latter is used as an immunosuppressant in human organ transplants.

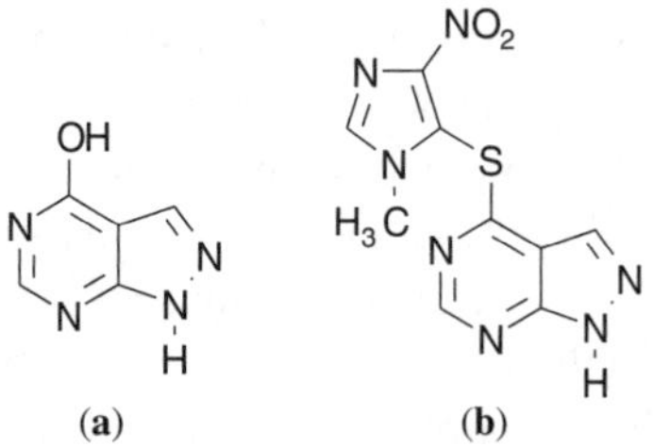

Figure 10.28   (a) Allopurinol and (b) azathioprine

Allopurinol was the result of attempts to increase the potency of 6-mercaptopurine by finding an inhibitor for xanthine oxidase, the enzyme that rapidly metabolises a large part of the administered dose of 6-mercaptopurine to inactive thiouric acid (Fig. 10.29), thereby reducing the drug's effectiveness.

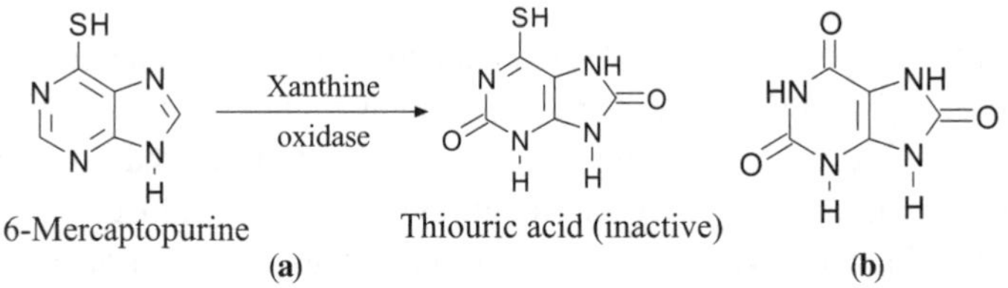

Figure 10.29   (a) The conversion of 6-mercaptopurine to thiouric acid. (b) Uric acid

Although allopurinol inhibited the enzyme, its use did not result in any therapeutic benefit. However, W. Rundles recognised its importance as an inhibitor of xanthine oxidase. This enzyme is important in the biosynthesis of uric acid, which crystallises in the joints of patients suffering from gout causing them considerable discomfort. The use of allopurinol to treat gout has considerably reduced the pain suffered by patients.

Azathioprine was one of a number of analogues synthesised by Hitchings and Elion in an attempt to produce a prodrug (see section 12.9.1) for 6-mercaptopurine in order to improve its effectiveness. It was found to be active but gave disappointing results. However, the discovery by R. Schwartz that 6-mercaptopurine was a very effective immunosuppressant resulted in azathioprine being screened by Hitchings for immunosuppressant activity. Azathioprine was found to be highly effective in this role. Today it is also used in the treatment of inflammatory bowel disease, myasthenia gravis and rheumatic disease.

Purine antimetabolites have been found to inhibit the synthesis of DNA and in some cases RNA by a number of different mechanisms. For example, 6-mercaptopurine is metabolised to the ribonucleotide 6-thioguanosine-5′-phosphate. This exogenous nucleotide inhibits several pathways for the biosynthesis of endogenous purine nucleotides. In contrast, 6-thioguanine is converted in the cell to the ribonucleotide 6-thioinosine-5′-phosphate. This ribonucleotide disrupts DNA synthesis by being incorporated into the structure of DNA as a false nucleic acid. Resistance to these two drugs arises because of a loss of the 5-phosphoribosyl transferase required for the formation of their ribonucleotides.

Pyrimidine antimetabolites

These are antimetabolites whose structures closely resemble those of the endogenous pyrimidine bases (Fig. 10.30a). The first pyrimidine antimetabolite was discovered in the 1950s by Robert Duschinsky and Robert Schnitzer working at Hoffmann-La Roche in Nutley, New Jersey. They had been requested to synthesise fluorouracil and other fluorine pyrimidine analogues by Charles Heidelberger at the McArdle Laboratory for cancer research at the University of Winsconsin. He had noted that Abraham Cantarow and Karl Paschkis had found that radioactive uracil had been more rapidly absorbed into rat liver tumours than into normal liver cells. Heidleberger decided, on the basis of his previous research work into the inhibition of enzymes by fluoroethanoic acid, to test the effect of incorporating fluorine into the structure of uracil and other pyrimidines. Duschinsky and Schnitzer synthesised a number of fluorinated pyrimidines and found that fluorouracil was active against tumours in rats and mice. It is now used to treat some solid tumours, including breast and GI tract cancers.

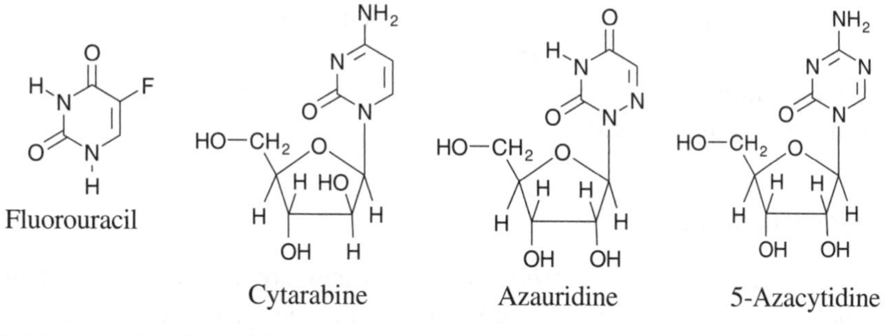

Figure 10.30 Examples of pyrimidines that act as antimetabolites. It should be noted that cytarabine only differs from cytidine by the stereochemistry of the 2′-carbon

Pyrimidine antimetabolites usually act by inhibiting one or more of the enzymes that are required for DNA synthesis. For example, fluorouracil is metabolised by the same metabolic pathway as uracil to 5-fluoro-2′-deoxyuridylic acid (FUdRP). FUdRP inhibits the enzyme thymidylate synthetase, which in its normal role is responsible for the transfer of a methyl group from the coenzyme methylenetetrahydrofolic acid ($MeTH_4$) to the C5

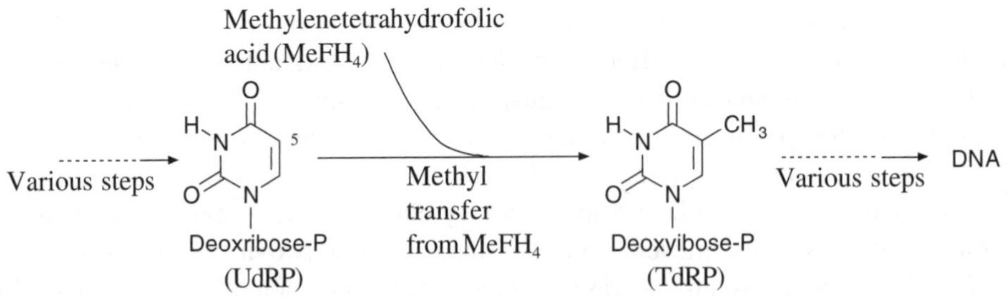

Figure 10.31 An outline of the mechanism of the intervention of fluorouracil in pyrimidine biosynthesis. Fluorouracil follows the same pathway but its unreactive C–F bond prevents methylation

atom of deoxyuridylic acid (UdRP, Fig. 10.31). The presence of the unreactive C5–F bond in UFdRP blocks this methylation, which prevents the formation of deoxythymidylic acid (TdRP) and its subsequent incorporation into DNA (Fig. 10.31). The fluorine is a similar size to the hydrogen atom (atomic radii: F, 0.13 nm; H, 0.12 nm) and it was thought that this similarity in size would give a drug that would cause little steric disturbance to the biosynthetic pathway. Analogues containing larger halogen atoms do not have any appreciable activity.

Pyrimidine antimetabolites are mainly used in the treatment of cancers (Table 10.7) and as antiviral agents (see section 10.14.3). Antiviral antimetabolites usually inhibit the production of a compound necessary for the production of viral nucleic acids.

10.13.2 Enzyme inhibitors

Enzyme inhibitors may be classified for convenience as those that either inhibit the enzymes directly responsible for the formation of nucleic acids or the variety of enzymes that catalyse the various stages in the formation of the pyrimidine and purine bases required for the formation of nucleic acids.

Topoisomerase inhibitors

Topoisomerases, or *nicking and closing enzymes* as they are also known, are families of enzymes that are responsible for the supercoiling, the cleavage and rejoining of DNA. There are two types of enzyme, namely type I and type II topoisomerases.

Type I topoisomerases are also known as DNA gyrases. They act by breaking (*nicking*) one of the strands of DNA. This allows the other strand to pass through the gap and as a result the double strand either unwinds by one turn or forms a new turn of the double helix of the DNA. After unwinding or winding the break is automatically resealed. This type of behaviour occurs in replication and transcription (see sections 10.5 and 10.6).

Type II topoisomerases act by breaking both strands of the DNA. This allows the broken strands to move apart and provide a gap through which a second part of the same

Table 10.7 Some examples of pyrimidine antimetabolites in clinical use

Antimetabolite	Used to treat	Active form	Brief outline of mechanism
Cytarabine (Ara-C) 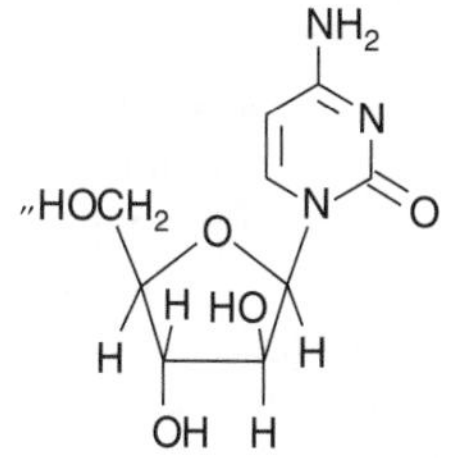	Acute Leukaemias	Cytarabine triphosphate (Ara-CTP) 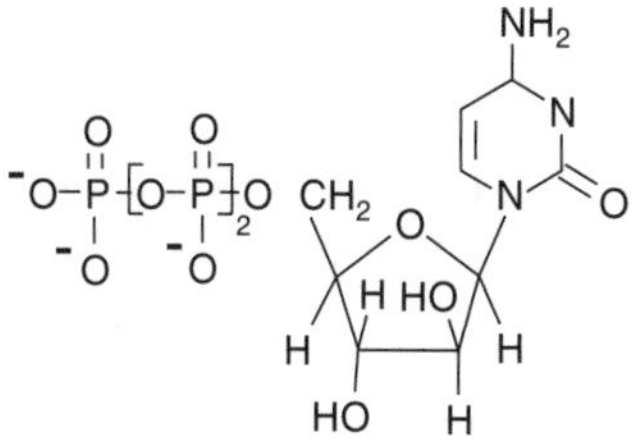	The Ara-CTP inhibits the conversion of cytidylic acid to 2′-deoxycytidylic acid. It also inhibits DNA polymerase and is incorporated into new DNA. The latter results in mis-coding when the DNA is replicated or transcribed. Consequently, cell growth is inhibited
Gemcitabine 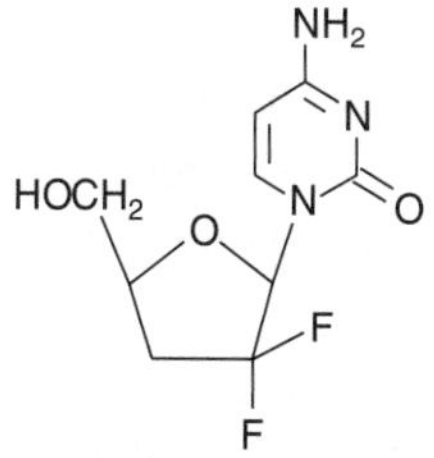	Local or metastatic non-small-cell lung, pancreatic and bladder cancers	Gemcitabine triphosphate 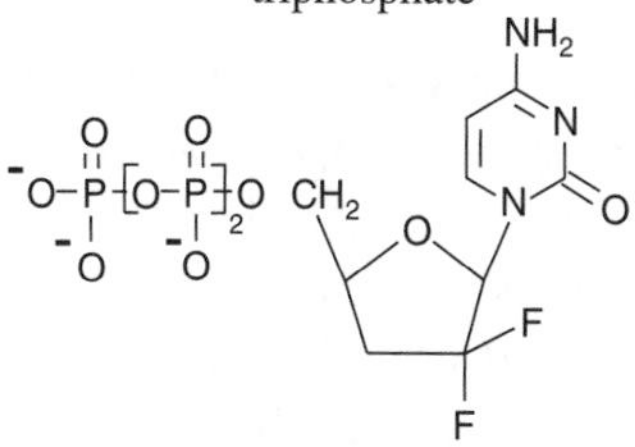	The triphosphate inhibits ribonucleic reductase. It is also incorporated into DNA instead of 2′-deoxycytidine triphosphate. As a result of these interventions, the cell dies
Capecitobine (a prodrug) 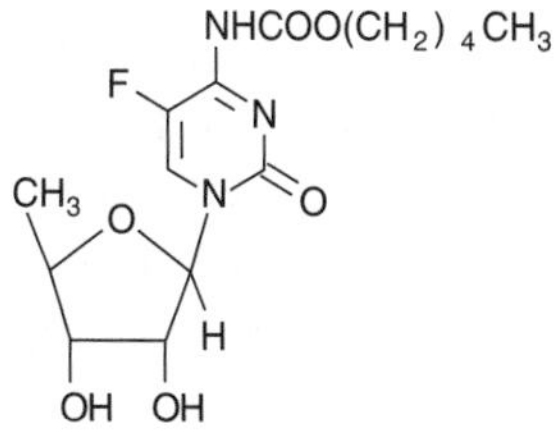	Metastatic colorectal cancer	Fluorouracil 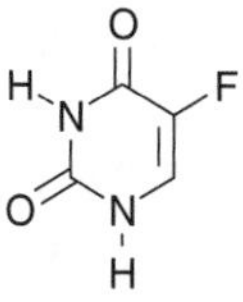	Capecitobine is converted by cytidine deaminase in tumour tissue into 5′-deoxy-5-fluorocytidine. Cytidine deaminase occurs in higher concentrations in tumour tissue than in normal tissue. The 5′-deoxy-5-fluorocytidine is further metabolised by thymine phosphorylase to fluorouracil
Flucytosine (a prodrug) 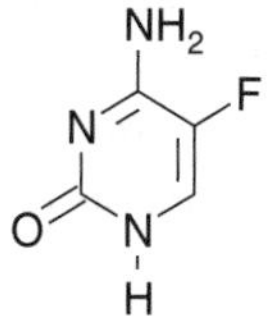	Systemic yeast infections	Fluorouracil 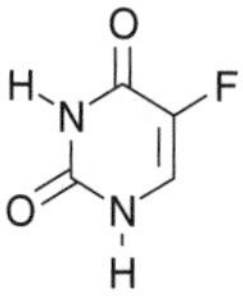	The fluorouracil is liberated by the action of a cytosine deaminase found only in fungi and not in humans. The action of fluorouracil is outlined earlier in this section

Table 10.7   (*Continued*)

Antimetabolite	Used to treat	Active form	Brief outline of mechanism
Idoxuridine (a prodrug)	Topical infections of herpes simplex	Idoxuridine 5′-triphosphate	The drug is converted to the 5′-triphosphate in virus-infected cells mainly by viral thymidine kinase. Idoxuridine binds to the viral form in preference to the corresponding human form of the enzyme. The triphosphorylated form of the drug inhibits viral DNA polymerase and also incorporation of the drug into viral DNA

double-stranded DNA strand can pass in order to alter the overall shape of the molecule. Once the strand has passed through the gap, *in effect to the other side of the DNA chain*, the broken ends of the strand are moved together by the enzyme and are automatically reconnected before the enzyme disengages.

Both types of enzyme are found in prokaryotes and eukaryotes. Their mechanisms of action appear to be very similar. They are believed to break (*nick*) the DNA strand by forming a complex with the DNA. Once the nick is formed the appropriate number of strands move through the gap and the topographical rearrangement of the molecule occurs. The enzyme remains attached to the single or double strand of the DNA until closure of the gap in the strand or strands has taken place.

The inhibition of topoisomerases has the effect of preventing DNA replication and transcription. Some of the compounds in clinical use that interfere with the action of topoisomerases are shown in Figure 10.32. Their mechanisms of action are varied and not fully elucidated. They range from enzyme inhibition to intercalation and DNA–topoisomerase complex stabilisation. A number of the compounds are believed to act by more than one mechanism.

Inhibition of ribonucleotide reductase inhibitors

Ribonucleotide reductase (RNR) is found in all living cells. It catalyses the reduction of all types of ribonucleotide diphosphates to deoxyribonucleotide diphosphates (Fig. 10.33a). The holoenzyme is thought to consist, of two separate proteins known as R1 and R2, iron II centres and, unusually, a tyrosine free radical group. X-ray crystallographic studies on *E. coli* RNR have shown that the iron and tyrosyl radical are situated in a hydrophobic pocket in the interior of the R2 protein. This is different from mammalian R2 where they are more accessible. Inhibition of RNR prevents DNA synthesis as there appears to be no other route for the formation of deoxyribonucleotides. Consequently, RNR is an important target for discovering new anticancer and antiviral drugs.

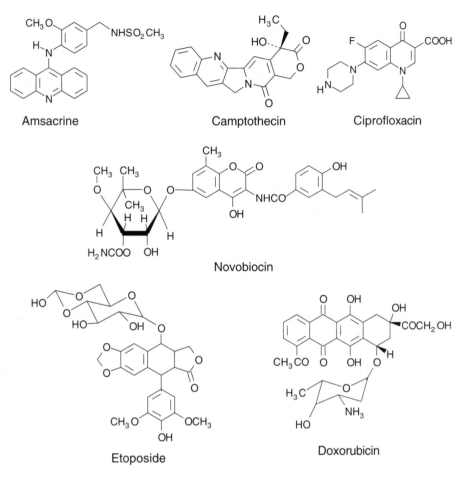

Figure 10.32 Examples of topoisomerase inhibitors. Amsacrine is used to treat ovarian carcinomas, lymphomas and myelogenous leukaemias. It is believed to act by intercalation followed by stabilisation of the DNA–topoisomerase complex. Both ciprofloxacin and novobiocin act as antibiotics because they inhibit DNA gyrase but not eukaryotic type II topoisomerases. However, novobiocin is not used as it has a number of adverse side effects and also bacteria rapidly develop resistance to this drug. Etoposide and doxorubicin inhibit eukaryotic type II topoisomerases, which is why they are used in the treatment of some cancers. Etoposide is thought to act by stabilising the DNA–topoisomerase II complex after cleavage of the DNA. DNA cleavage is also believed to occur because of free radical action involving free radicals produced by the oxidation of the 4′-hydroxyphenolic group. Doxorubicin is believed to prevent the action of topoisomerase II by intercalation (see section 10.13.3) followed by stabilisation of the drug–enzyme complex at the point where the DNA is cleaved. Camptothecin is an antitumour agent. It is thought to act by forming a stable drug–enzyme–DNA complex with type I topoisomerases

Hydroxyurea (Fig. 10.33b) is the only drug in clinical use at the present time. It is believed to act by destroying the tyrosine free radical of the R2 protein by reducing it to tyrosine. It is also thought that it may inhibit the enzyme by chelating to the iron centre of R2. Hydroxyurea is used to treat chronic myelogenous leukaemia and head, neck and metastatic ovarian carcinoma cancers. It has also been used as an antiviral to treat HIV-I.

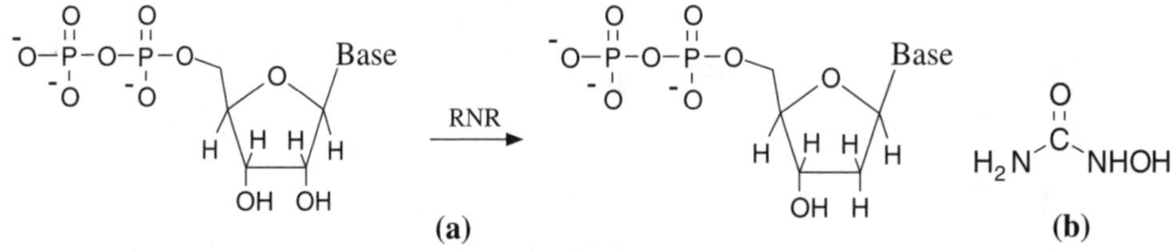

Figure 10.33 (a) The catalytic action of RNR. (b) Hydroxyurea (hydroxycarbamide)

Enzyme inhibitors for pyrimidine and purine precursor systems

A wide range of compounds are active against a number of the enzyme systems that are involved in the biosynthesis of purines and pyrimidines in bacteria. For example, sulphonamides inhibit dihydropteroate synthetase (see section 9.12.1), which prevents the formation of folic acid, whilst trimethoprim inhibits dihydrofolate reductase, which prevents the conversion of folic acid to tetrahydrofolate (see section 10.13.1). In both of these examples the overall effect is the inhibition of purine and pyrimidine synthesis, which results in the inhibition of the synthesis of DNA. This restricts the growth of the bacteria and ultimately prevents it replicating, which gives the body's natural defences time to destroy the bacteria. Since sulphonamides and trimethoprim inhibit different stages in the same metabolic pathway they are often used in conjunction (Fig. 10.34). This allows the clinician to use lower and therefore safer doses.

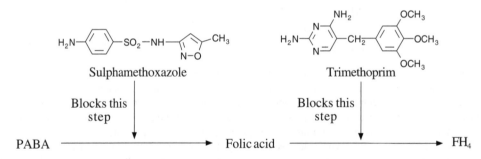

Figure 10.34 Sequential blocking using sulphamethoxazole and trimethoprim

10.13.3 Intercalating agents

Intercalating agents are compounds that insert themselves between the bases of the DNA helix (Fig. 10.35). This insertion causes the DNA helix to partially unwind at the site of the intercalated molecule. This inhibits transcription, which blocks the replication process of the cell containing the DNA. However, it is not known how the partial unwinding prevents transcription but some workers think it inhibits topoisomerases (see section 10.12.2). Inhibition of cell replication can lead to cell death, which reduces the size of a tumour, the

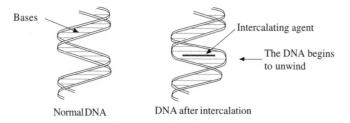

Figure 10.35 A schematic representation of the distortion of the DNA helix by intercalating agents. The horizontal lines represent the hydrogen-bonded bases. The rings of these bases and intercalating agent are edge-on to the reader

number of *'free'* cancer cells or the degree of infection, all of which will contribute to improving the health of the patient (see section 8.2).

The insertion of an intercalating agent appears to occur via either the minor or major grooves of DNA. Compounds that act as intercalating agents must have structures that contain a flat fused aromatic or heteroaromatic ring section that can fit between the flat structures of the bases of the DNA. It is believed that these aromatic structures are held in place by hydrogen bonds, van der Waals' forces and charge transfer bonds (see section 8.2).

Drugs whose mode of action includes intercalation are the antimalarials quinine and chloroquine, the anticancer agents mitoxantrone and doxorubicin, and the antibiotic proflavine (Fig. 10.36). In each of these compounds it is the flat aromatic ring system that is responsible for the intercalation. However, other groups in the structures may also

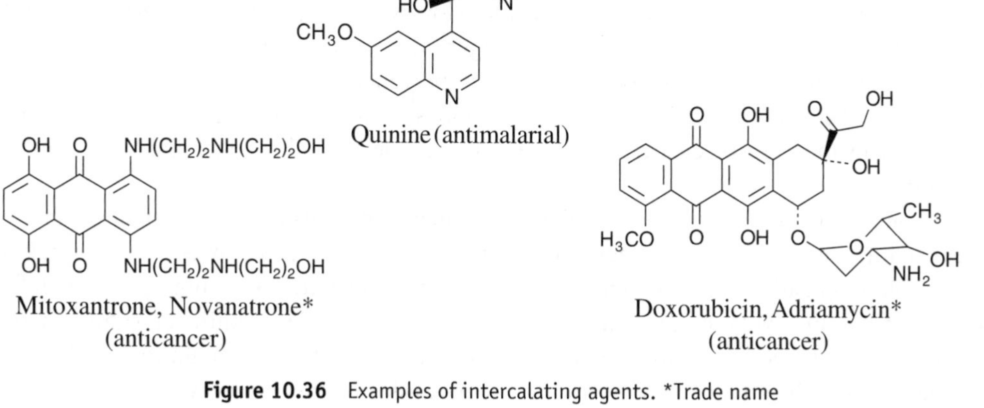

Proflavine (3,6-diaminoacridine)
(antibiotic)

Chloroquine (antimalarial)

Quinine (antimalarial)

Mitoxantrone, Novanatrone*
(anticancer)

Doxorubicin, Adriamycin*
(anticancer)

Figure 10.36 Examples of intercalating agents. *Trade name

contribute to the binding of a drug to the DNA. For example, the amino group of the sugar residue of doxorubicin forms an ionic bond with the negatively charged oxygens of the phosphate groups of the DNA chain, which effectively lock the drug into place. A number of other drugs appear to have groups that act in a similar manner.

Some intercalating agents exhibit a preference for certain combinations of bases in DNA. For example, mitoxantrone appears to prefer to intercalate with cytosine–guanosine-rich sequences. This type of behaviour does open out the possibility of selective action in some cases.

10.13.4 Alkylating agents

Alkylating agents are believed to bond to the nucleic acid chains in either the major or minor grooves. In DNA the alkylating agent frequently forms either *intrastrand* or *interstrand* cross-links. *Intrastrand* cross-linking agents form a bridge between two parts of the same chain (Fig. 10.37). This has the effect of distorting the strand, which inhibits transcription.

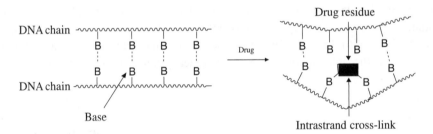

Figure 10.37 A schematic representation of the intrastrand cross-linking

Interstrand cross-links are formed between the two separate chains of the DNA, which has the effect of locking them together (Fig. 10.38). This also inhibits transcription. In RNA only intrastrand cross-links are possible. However, irrespective of whether or not it forms a bridge, the bonding of an alkylating agent to a nucleic acid inhibits replication of that nucleic acid. In the case of bacteria this prevents an increase in the size of the infection and so buys the body time for its immune system to destroy the existing bacteria. However, in the case of cancer it may lead to cell death and a beneficial reduction in tumour size.

The nucleophilic nature of the nucleic acids means that alkylating agents are usually electrophiles or give rise to electrophiles. For example, it is believed that a weakly electrophilic β-carbon atom of an aliphatic nitrogen mustard alkylating agent, such as mechlorethamine (Mustine), is converted to the more highly electrophilic aziridine ion by an internal nucleophilic substitution of a β-chlorine atom. This is thought to be followed by the nucleophilic attack of the N7 of a guanine residue on this ion by what appears to be

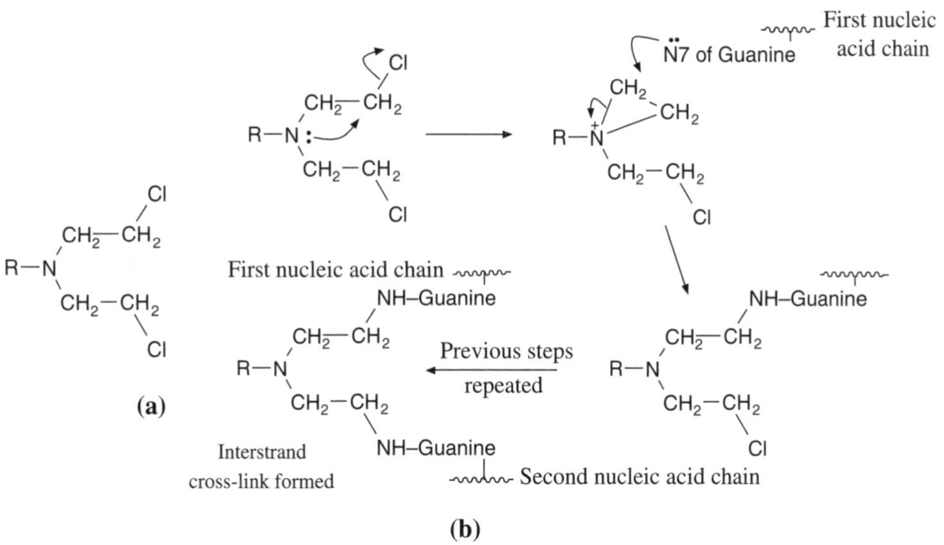

Figure 10.38 (a) The general structure of nitrogen mustards. (b) The proposed mechanism for forming interstrand cross-links by the action of aliphatic nitrogen mustards

an S_N2 type of mechanism. Since these drugs have two hydrocarbon chains with β-chloro groups, each of these chloro groups is believed to react with a guanine residue in a different chain of the DNA strand to form a cross-link between the two nucleic acid chains (Fig. 10.38).

The electrophilic nature of alkylating agents means that they can also react with a wide variety of other nucleophilic biomacromolecules. This accounts for many of the unwanted toxic effects that are frequently observed with the use of these drugs. In the case of the nitrogen mustards, attempts to reduce these side effects have centred on reducing their reactivity by discouraging the formation of the aziridine ion before the drug reaches its site of action. The approach adopted has been to reduce the nucleophilic character of the nitrogen atom by attaching it to an electron withdrawing aromatic ring. This produced analogues that would only react with strong nucleophiles and resulted in the development of chlorambucil (Fig. 10.39a). This drug is one of the least toxic nitrogen mustards, being active against malignant lymphomas, carcinomas of the breast and ovary and lymphocytic leukaemia. It has been suggested that because of the reduction in the nucleophilicity of the nitrogen atom these aromatic nitrogen mustards do not form an aziridine ion. Instead they react by direct substitution of the β-chlorine atoms by guanine, which is a strong nucleophile, by an S_N1 type of mechanism (Fig. 10.39b).

Further attempts to reduce the toxicity of nitrogen mustards were based on making the drug more selective. Two approaches have yielded useful drugs. The first was based on the fact that the rapid synthesis of proteins that occurs in tumour cells requires a large supply of amino acid raw material from outside the cell. Consequently, it was thought that the presence of an amino acid residue in the structure of a nitrogen mustard might lead to an

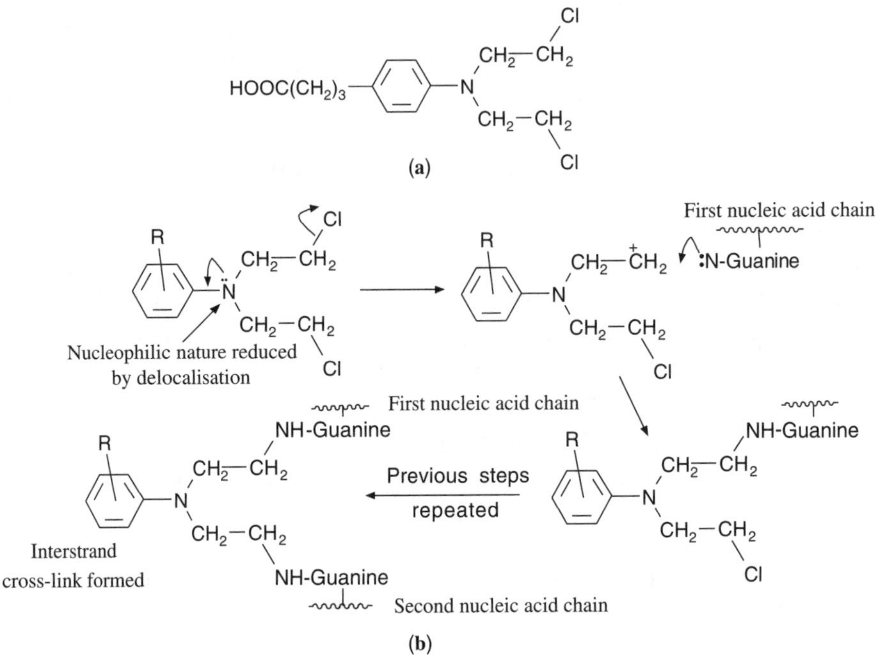

Figure 10.39   (a) The structure of chlorambucil and (b) a proposed mode of action for some aromatic nitrogen mustards

increased uptake of that compound. This approach resulted in the synthesis of the phenylalanine mustard melphalan. The L form of this drug is more active than the D form and so it has been suggested that the L form may be transported into the cell by means of an L-phenylalanine active transport system.

The second was based on the fact that some tumours were thought to contain a high concentration of phosphoramidases. This resulted in the synthesis of nitrogen mustard analogues whose structures contained phosphorus functional groups that could be attacked by this enzyme. It led to the development of the cyclophosphamide (Fig.10.40), which has a wide spectrum of activity. However, the action of this prodrug has now been shown to be due to phosphoramide mustard formed by oxidation by microsomal enzymes in the liver rather than hydrolysis by tumour phosphoramidases. The acrolein produced in this process is believed to be the source of myelosuppression and haemorrhagic cystitis associated with the use of cyclophosphamide. However, co-administration of the drug with sodium 2-mercaptoethanesulphonate (MESNA) can relieve some of these symptoms (see next section).

Some alkylating agents act by decomposing to produce electrophiles that bond to a nucleophilic group of a base in the nucleic acid. For example, the imidazotetrazinone temozolomide enters the major groove of DNA where it reacts with water to from nitrogen, carbon dioxide, an aminoimidazole and a methyl carbonium ion ($^+CH_3$). This methyl carbonium ion then methylates the strongly nucleophilic N7 of the guanine

Temozolomide

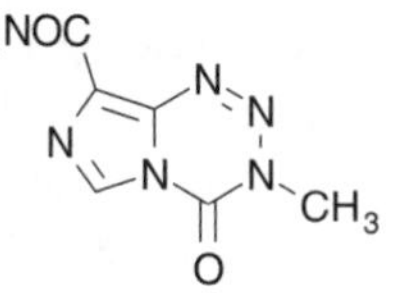

A range of different classes of compound can act as nucleic acid alkylating agents. Within these classes, which include triazeneimidazoles, alkyldimethanesulphonates, nitrosoureas, platinum complexes and carbinolamines, a number of compounds have been found to be useful drugs. In many cases their effectiveness is improved by the use of combinations of drugs. Their modes of action are usually not fully understood but a large amount of information is available concerning their structure–activity relationships.

Sodium 2-mercaptoethanesulphonate (MESNA)

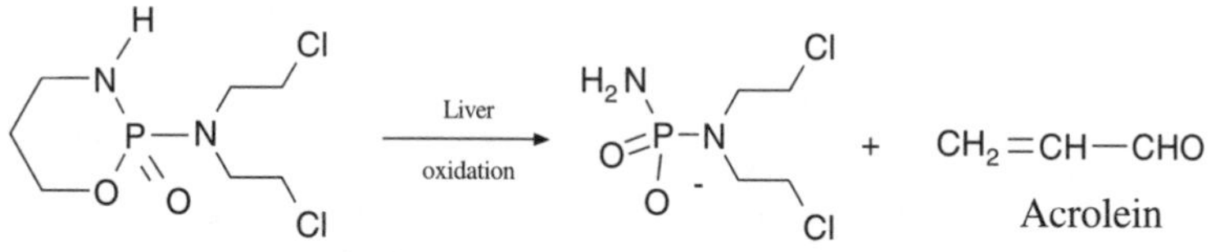

Cyclophosphamide   Phosphoramide mustard

Figure 10.40 Cyclophosphamide and the formation of phosphoramide mustard, the active form of this drug

MESNA is used to reduce the unwanted side effects of some cytotoxic drugs, such as cyclophosphamide (Fig. 10.40). It may be administered orally or by infusion. After absorption or parental administration, it is oxidised to the disulphide. This is reduced in the kidney to the thiol where it reacts with acrolein and other toxic metabolites to form compounds that are excreted in the urine.

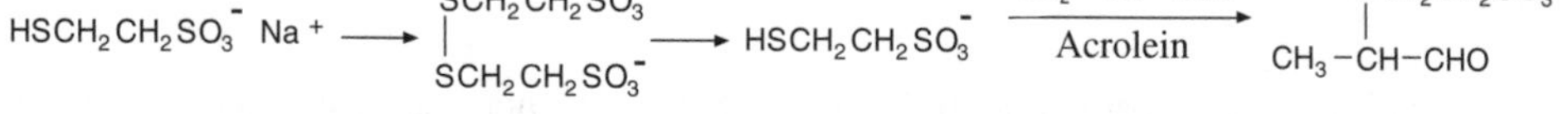

MESNA   Disulphide   Excreted

10.13.5 Antisense drugs

The concept of antisense compounds or sequence-defined oligonucleotides offers a new specific approach to designing drugs that target nucleic acids. The idea underlying this approach is that the antisense compound contains the sequence of complementary bases to those found in a short section of the target RNA that contains part of the genetic message being carried by a mRNA molecule. On reaching the mRNA, the antisense compound binds to this section by hydrogen bonding between the complementary base pairs. This inhibits translation of the message carried by the mRNA, which inhibits the production of a specific protein responsible for a disease state in a patient. Furthermore, the binding of an antisense molecule to mRNA may also result in cleavage of the RNA strand of the RNA–antisense drug complex by the enzyme RNase H.

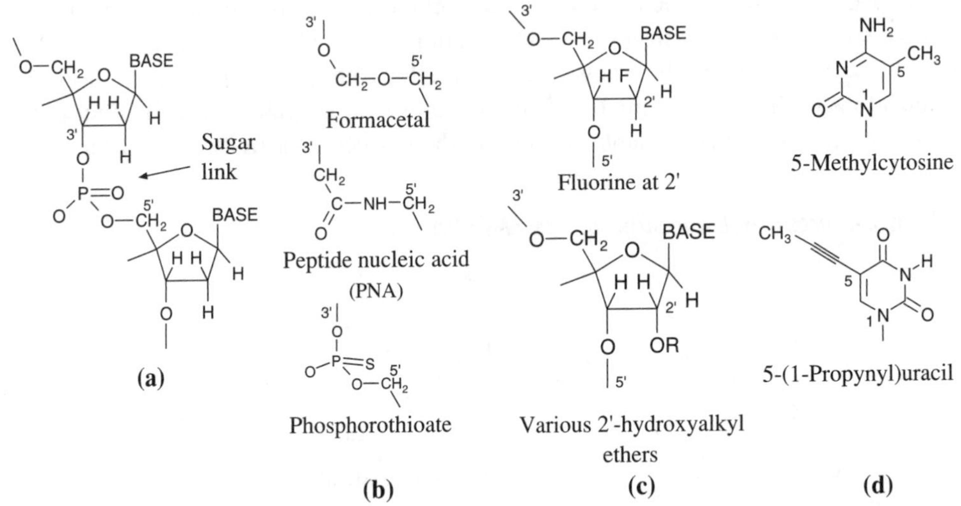

Figure 10.41 Development routes for antisense drugs. Examples of: (**a**) a section of the backbone of a deoxyribonucleic acid; (**b**) backbone modifications; (**c**) sugar residue modifications; and (**d**) base modifications

Antisense compounds were originally short lengths of nucleic acid chains that had base sequences that were complementary to those found in their target RNA. These short lengths of nucleic acid antisense compounds were found to be unsuitable as drugs because of poor binding to the target site and short half-lives due to enzyme action. However, they provided lead compounds for further development. Development is currently taking three basic routes:

1. Modification of the backbone linking the bases to increase resistance to enzymic hydrolysis (Fig. 10.41).

2. Changing the nature of the sugar residue by either replacing some of the free hydroxy groups by other substituents or forming derivatives of these groups (Fig. 10.41).

3. Modifying the nature of the substituent groups of the bases (Fig. 10.41).

Statistical analysis shows that by using the four different bases usually found in nucleic acids antisense oligonucleotides with about 15–25 bases should be specific for one sequence of bases in only one type of mRNA molecule. This prediction opens out the

Table 10.8 Examples of active antisense compounds

Name	Structure (outline only)	Action	Status
Oblimersen	18 Oligonucleotides linked by phosphorothioate links	Binds to the initiation codon of the mRNA for the protein Bcl-2. This protein suppresses apoptosis (cell death)	Anticancer agent in Phase III trials
Fomivirsen (Vitraene)	21 Oligonucleotides linked by phosphorothioate links	Blocks the translation of viral RNA	In clinical use for the treatment of inflammation of the eye caused by cytomegalovirus
Trecovirsen	25 Oligonucleotides linked by phosphorothioate links	Prevents the formation of viral proteins by blocking viral mRNA	Withdrawn from clinical trials as too toxic, blocking the action

possibility of the discovery of highly specific drugs. Using this approach a number of potential drug candidates with about 10–25 bases have been discovered (Table 10.8).

Antisense compounds are also able to bind to DNA. In this case they are designed to bind to a defective gene in the DNA and so are known as *antigene* agents. Antigene binding results in the formation of a local triple helix that prevents the action of the defective gene.

10.13.6 Chain cleaving agents

The interaction of chain cleaving agents with DNA results in the breaking of the nucleic acid into fragments. Currently, the main cleaving agents are the bleomycins (Fig. 10.42), discovered in 1966, and their analogues.

The bleomycins are a group of naturally occurring glycoproteins that exhibit antitumour activity. When administered to patients they tend to accumulate in the squamous cells and so are useful for treating cancers of the head, neck and genitalia. However, the bleomycins cause pain and ulceration of areas of skin that contain a high concentration of keratin, as well as other unwanted side effects.

The action of the bleomycins is not fully understood. It is believed that the bithiazole moiety (domain X in Fig. 10.42) intercalates with the DNA. In bleomycin A2 the resulting adduct is thought to be stabilised by ionic bonding between the phosphate units of the DNA and the sulphonium ion of the side chain. The intercalation is believed to be followed by domain Y forming a complex with Fe(II). This complex reacts with oxygen to form free radicals such as the highly reactive hydroxyl and superoxide free radicals. These reactive radicals are believed to be formed so close to the DNA chain that they readily react with it to cleave the phosphodiester bonds. The chemical nature of this cleavage is such that the nucleic acid sections cannot be rejoined by DNA ligases.

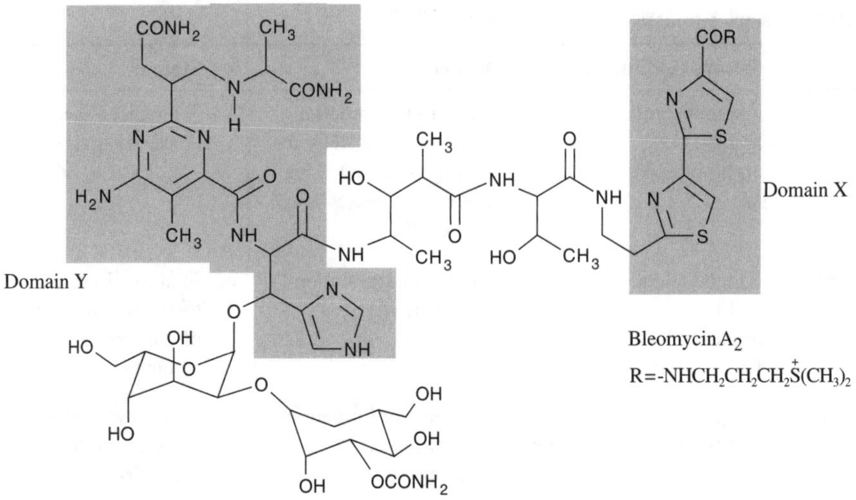

Figure 10.42 The bleomycins. The drug bleomycin sulphate is a mixture of a number of bleomycins

A number of other compounds that act by DNA strand cleavage are known but have not been developed because they are too toxic. An exception to this is calicheamicin γ^1, which is too toxic to be used in its own right as an anticancer agent. However, its toxic properties are modified when it is used as a conjugated-humanised monoclonal antibody adduct (see section 10.15.2). This adduct, known as gemtuzumab, is now licenced to treat acute myeloid leukaemia in older patients.

10.14   Viruses

Viruses are infective agents that are considerably smaller than bacteria. They are essentially packages, known as *virions*, of chemicals that invade host cells. However, viruses are not independent and can only penetrate a host cell that can satisfy the specific needs of that virus. The mode of penetration varies considerably from virus to virus. Once inside the host cell viruses take over the metabolic machinery of the host and use it to produce more viruses. Replication is often lethal to the host cell, which may undergo lysis to release the progeny of the virus. However, in some cases the virus may integrate into the host chromosome and become dormant. The ability of viruses to reproduce means that they can be regarded as being on the borderline of being living organisms.

10.14.1   Structure and replication

Viruses consist of a core of either DNA or, as in the majority of cases, RNA fully or partially covered by a protein coating known as the *capsid*. The capsid consists of a number

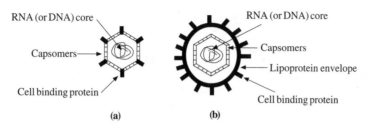

Figure 10.43 (a) Schematic representations of the structure of a virus (a) without a lipoprotein envelope (naked virus) and (b) with a lipoprotein envelope

of polypeptide molecules known as *capsomers* (Fig.10.43). The capsid that surrounds most viruses consists of a number of different capsomers although some viruses will have capsids that only contain one type of capsomer. It is the arrangement of the capsomers around the nucleic acid that determines the overall shape of the virion. In the majority of viruses, the capsomers form a layer or several layers that completely surround the nucleic acids. However, there are some viruses in which the capsomers form an open-ended tube that holds the nucleic acids.

In many viruses the capsid is coated with a protein-containing lipid bilayer membrane. These are known as *enveloped viruses*. Their lipid bilayers are often derived from the plasma membrane of the host cell and are formed when the virus leaves the host cell by a process known as *budding*. Budding is a mechanism by which a virus leaves a host cell without killing that cell. It provides the virus with a membrane whose lipid components are identical to those of the host (Fig. 10.43). This allows the virus to penetrate new host cells without activating the host's, immune systems.

Viruses bind to host cells at specific receptor sites on the host's cell envelope. The binding sites on the virus are polypeptides in its capsid or lipoprotein envelope. Once the virus has bound to the receptor of the host cell the virus–receptor complex is transported into the cell by receptor-mediated endocytosis (see section 7.3.6). In the course of this process the protein capsid and any lipoprotein envelopes may be removed. Once it has entered the host cell the viral nucleic acid is able to use the host's cellular machinery to synthesise the nucleic acids and proteins required to replicate a number of new viruses (Fig. 10.44). A great deal of information is available concerning the details of the mechanism of virus replication but this text will only outline the main points. For greater detail the reader is referred to specialist texts on virology.

10.14.2 Classification

RNA-viruses can be broadly classified into two general types, namely: *RNA-viruses* and *RNA-retroviruses*.

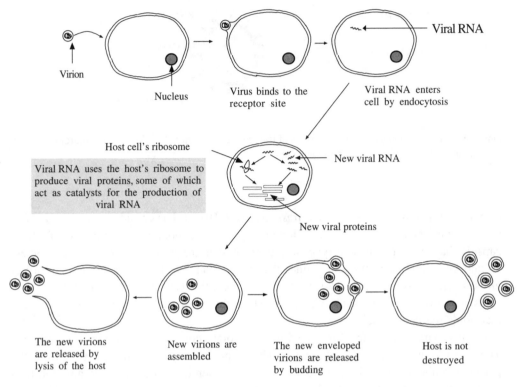

Figure 10.44 A schematic representation of the replication of RNA-viruses

RNA-viruses

RNA-virus replication usually occurs entirely in the cytoplasm. The viral mRNA either forms part of the RNA carried by the virion or is synthesised by an enzyme already present in the virion. This viral mRNA is used to produce the necessary viral proteins by translation using the host cell's ribosomes and enzyme systems. Some of the viral proteins are enzymes that are used to catalyse the reproduction of more viral mRNA. The new viral RNA and viral proteins are assembled into a number of new virions that are ultimately released from the host cell by either lysis or budding (see section 10.14.1).

Retroviruses

Retroviruses synthesise viral DNA using their viral RNA as a template. This process is catalysed by enzyme systems known as *reverse transcriptases* that form part of the virion. The viral DNA is incorporated into the host genome to form a so-called *provirus*. Transcription of the provirus produces new 'genomic' viral RNA and viral mRNA. The viral mRNA is used to produce viral proteins, which together with the 'genomic' viral RNA are assembled into new virions. These virions are released by budding (see section 10.14.1), which in many cases does not kill the host cell. Retroviruses are responsible for some forms of cancer and AIDS.

DNA-viruses

Most *DNA-viruses* enter the host cell's nucleus where formation of viral mRNA by transcription from the viral DNA is brought about by the host cell's polymerases. This viral mRNA is used to produce viral proteins by translation using the host cell's ribosomes and enzyme systems. Some of these proteins will be enzymes that can catalyse the synthesis of more viral DNA. This DNA and the viral proteins synthesised in the host cell are assembled into a number of new virions that are ultimately released from the host by either cell lysis or budding (see section 10.14.1).

10.14.3   Viral diseases

Viral infection of host cells is a common occurrence. Most of the time this infection does not result in illness as the body's immune system can usually deal with such viral invasion. When illness occurs it is often short lived and leads to long-term immunity. However, a number of viral infections can lead to serious medical conditions (Table 10.9). Some viruses like HIV, the aetiological agent of AIDS, are able to remain dormant in the host for a number of years before becoming active, whilst others such as herpes zoster (shingles) can give rise to recurrent bouts of the illness. Both chemotherapy and preventative vaccination are used to treat patients. The latter is the main clinical approach since it has been difficult to design drugs that only target the virus. However, a number of antiviral drugs have been developed and are in clinical use.

Table 10.9   Examples of some of the groups of viruses that cause disease

Virus group	Disease	Characteristics	
		RNA/DNA	**Envelope/naked**
Parvovirus	Gastroenteritis	DNA	Naked
Herpes	Cold sores	DNA	Enveloped
Picornavirus	Polio and hepatitis A	RNA	Naked
Retrovirus	AIDS and leukaemia	RNA	Enveloped
Paramyxovirus	Measles, mumps and para-influenza	RNA	Enveloped
Rhabdovirus	Rabies	RNA	Enveloped

AIDS

AIDS is a disease that progressively destroys the human immune system. It is caused by the human immunodeficiency virus (HIV), which is a retrovirus. This virus enters and destroys human T4 lymphocyte cells. These cells are a vital part of the human immune system. Their destruction reduces the body's resistance to other infectious diseases, such as pneumonia, and some rare forms of cancer.

The entry of the virus into the body usually causes an initial period of acute ill health with the patient suffering from headaches, fevers and rashes, amongst other symptoms. This is followed by a period of relatively good healthy where the virus replicates in the lymph nodes. This relatively healthy period normally lasts a number of years before full-blown AIDS appears. Full-blown AIDS is characterised by a wide variety of diseases such as bacterial infections, neurological diseases and cancers. Treatment is more effective when a mixture of antiviral agents is used (see next section).

10.14.4   Antiviral drugs

It has been found that viruses utilise a number of virus-specific enzymes during replication. These enzymes and the processes they control are significantly different from those of the host cell to make them a useful target for medicinal chemists. Consequently, antiviral drugs normally act by inhibiting viral nucleic acid synthesis, inhibiting attachment to and penetration of the host cell or inhibiting viral protein synthesis.

Nucleic acid synthesis inhibitors

Nucleic acid synthesis inhibitors usually act by inhibiting the polymerases or reverse transcriptases required for nucleic acid chain formation. However, because they are usually analogues of the purine and pyrimidine bases found in the viral nucleic acids, they are often incorporated into the growing nucleic acid chain. In this case their general mode of action frequently involves conversion to the corresponding 5'-triphosphate by the host cell's cellular kinases. This conversion may also involve specific viral enzymes in the initial monophosphorylation step. These triphosphate drug derivatives are incorporated into the nucleic acid chain where they terminate its formation. Termination occurs because the drug residues do not have the 3'-hydroxy group necessary for the phosphate ester formation required for further growth of the nucleic acid chain. This effectively inhibits the polymerases and transcriptases that catalyse the growth of the nucleic acid (Fig. 10.45).

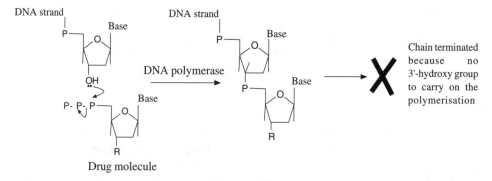

Figure 10.45 An outline of the mechanism for DNA chain termination used by some antiviral drugs. The growing chain terminates because R cannot react with the next triphosphate nucleotide (for DNA replication, see section 10.5). Key: P = phosphate residue and R = hydrogen and various other groups (see Table 10.12)

It is not possible to list all the known antiviral agents in this text so only a representative selection are discussed.

Aciclovir Aciclovir was the first effective antiviral drug. It is effective against a number of herpes viruses, notably simplex, varicella-zoster (shingles), varicella (chickenpox) and Epstein–Barr virus (glandular fever). It may be administered orally and by intravenous injection as well as topically. Orally administered doses have a low bioavailability.

Aciclovir

The action of aciclovir is more effective in virus-infected host cells because the viral thymidine kinase is a more efficient catalyst for the monophosphorylation of aciclovir than the thymidine kinases of the host cell. This leads to an increase in the concentration of the aciclovir triphosphate, which has 100-fold greater affinity for viral DNA polymerase than human DNA polymerase. As a result, it preferentially competitively inhibits viral DNA polymerase and so prevents the virus from replicating. However, resistance has been reported due to changes in the viral mRNA responsible for the production of the viral thymidine kinase. Aciclovir also acts by terminating chain formation. The aciclovir–DNA complex formed by the drug also irreversibly inhibits DNA polymerase.

Vidarabine Vidarabine is active against herpes simplex and herpes varicella-zoster. However, the drug does give rise to nausea, vomiting, tremors, dizziness and seizures. In addition it has been reported to be mutagenic, teratogenic and carcinogenic in animal studies. Vidarabine is administered by intravenous infusion and topical application. It has a half-life of about one hour, the drug being rapidly deaminated to arabinofuranosyl hypoxanthine (ara-HX) by adenosine deaminase. This enzyme is found in the serum and red blood cells. Ara-HX, which also exhibits a weak antiviral action, has a half-life of about 3.5 hours.

Vidarabine   Arabinofuranosyl hypoxanthine (ara-HX)

Ribavirin Ribavirin is effectively a guanosine analogue. It is active against a wide variety of DNA and RNA viruses but the mechanism by which it acts is not understood. It is mainly used in aerosol form to treat influenza and other respiratory viral infections. Intravenous administration in the first 6 days of onset has been effective in reducing deaths from Lassa fever to 9 per cent. Ribavirin has also been shown to delay the onset of full-blown AIDS in

patients with early symptoms of HIV infection (see section 10.14.3). However, administration of the drug has been reported to give rise to nausea, vomiting, diarrhoea, deterioration
of respiratory function, anaemia, headaches and abdominal pain. The mechanism by which it
acts may differ from one virus to another.

Ribavirin (Tribavirin)

Zidovudine (AZT)   Zidovudine was originally synthesised in 1964 as an analogue of
thymine by J. Horwitz as a potential antileukaemia drug. It was found to be unsuitable for use
in this role and for 20 years was ignored, even though in 1974 W. Osterag *et al.* reported that
it was active against Friend leukaemia virus, a retrovirus. However, the identification in 1983
of the retrovirus HIV as the source of AIDS resulted in the virologist M. St Clair setting up a
screening programme for drugs that could attack HIV. Fourteen compounds were selected
and screened against Friend leukaemia virus and a second retrovirus called Harvey sarcoma
virus. This screen led to the discovery of zidovudine (AZT), which was rapidly developed
into clinical use on selected patients in 1986.

Zidovudine (AZT)

AZT is converted by the action of cellular thymidine kinase to the $5'$-triphosphate. This
inhibits the enzyme reverse transcriptase in the retrovirus, which effectively prevents it
from forming the viral DNA necessary for viral replication. The incorporation of AZT into
the nucleic acid chain also results in chain termination because the presence of the $3'$-azide
group prevents the reaction of the chain with the $5'$-triphosphate of the next nucleotide
waiting to join the chain (Fig. 10.45). AZT is also active against mammalian DNA
polymerase and although its affinity for this enzyme is about 100-fold less this action is
thought to be the cause of some of its unwanted side effects.

Zidovudine is active against the retroviruses (see section 10.14.2) that cause AIDS
(HIV virus) and certain types of leukaemia. It also inhibits cellular α-DNA polymerase but
only at concentrations in excess of 100-fold greater than those needed to treat the viral
infection. The drug may be administered orally or by intravenous infusion. The
bioavailability from oral administration is good, the drug being distributed into most body
fluids and tissues. However, when used to treat AIDS it has given rise to gastrointestinal

disorders, skin rashes, insomnia, anaemia, fever, headaches, depression and other unwanted effects. Resistance increases with time. This is known to be due to the virus developing mutations' which result in changes in the amino acid sequences in the reverse transcriptase.

Didanosine   Didanosine is used to treat some AZT-resistant strains of HIV. It is also used in combination with AZT to treat HIV. Didanosine is administered orally in dosage forms that contain antacid buffers to prevent conversion by the stomach acids to hypoxanthine. However, in spite of the use of buffers the bioavailability from oral administration is low. The drug can cause nausea, abdominal pain and peripheral neuropathy, amongst other symptoms. Drug resistance occurs after prolonged use.

Didanosine is converted by viral and cellular kinases to the monophosphate and then to the triphosphate. In this form it inhibits reverse transcriptase and in addition its incorporation into the DNA chain terminates the chain because the drug has no 3'-hydroxy group (Fig. 10.45).

Didanosine

Host cell penetration inhibitors

The principal drugs that act in this manner are amantadine and rimantadine (Fig. 10.46). Both amantadine and rimantadine are also used to treat Parkinson's disease. However, their mode of action in this disease is different from their action as antiviral agents.

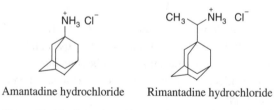

Amantadine hydrochloride   Rimantadine hydrochloride

Figure 10.46 Examples of host cell penetration inhibitors

Amantadine hydrochloride   Amantadine hydrochloride is effective against influenza A virus but is not effective against the influenza B virus. When used as a prophylactic, it is believed to give up to 80 per cent protection against influenza A virus infections. The drug acts by blocking an ion channel in the virus membrane formed by the viral protein M_2. This is believed to inhibit the disassembly of the core of the virion and its penetration of the host (see section 10.14.1).

Amantadine hydrochloride has a good bioavailability on oral administration, being readily absorbed and distributed to most body fluids and tissues. Its elimination time is

12–18 hours. However, its use can result in depression, dizziness, insomnia and gastrointestinal disturbances, amongst other unwanted side effects.

Rimantadine hydrochloride Rimantadine hydrochloride is an analogue of amantadine hydrochloride. It is more effective against influenza A virus than amantadine. Its mode of action is probably similar to that of amantadine. The drug is readily absorbed when administered orally but undergoes extensive first-pass metabolism. However, in spite of this, its elimination half-life is double that of amantadine. Furthermore, CNS side effects are significantly reduced.

Inhibitors of viral protein synthesis

The principal compounds that act as inhibitors of protein synthesis are the *interferons*. These compounds are members of a naturally occurring family of glycoprotein hormones (RMM 20 000–160 000), which are produced by nearly all types of eukaryotic cell. Three general classes of interferons are known to occur naturally in mammals, namely: the α-interferons produced by leucocytes, β-interferons produced by fibroblasts and γ-interferons produced by T lymphocytes. At least twenty α-, two β- and two γ-interferons have been identified.

Interferons form part of the human immune system. It is believed that the presence of virions, bacteria and other antigens in the body switches on the mRNA that controls the production and release of interferon. This release stimulates other cells to produce and release more interferon. Interferons are thought to act by initiating the production in the cell of proteins that protect the cells from viral attack. The main action of these proteins takes the form of inhibiting the synthesis of viral mRNA and viral protein synthesis. α-Interferons also enhance the activity of killer T cells associated with the immune system. (see section 14.5.5).

A number of α-interferons have been manufactured (see Table 10.10 on page 394) and proven to be reasonably effective against a number of viruses and cancers. Interferons are usually given by intravenous, intramuscular or subcutaneous injection. However, their administration can cause adverse effects, such as headaches, fevers and bone marrow depression, that are dose related.

The formation and release of interferon by viral and other pathological stimulation has resulted in a search for chemical inducers of endogenous interferon. Administration of a wide range of compounds has resulted in the induction of interferon production. However, no clinically useful compounds have been found for humans' although tilorone is effective in inducing interferon in mice.

Tilorone

10.15   Recombinant DNA technology (genetic engineering)

The body requires a constant supply of certain peptides and proteins if it is to remain healthy and function normally. Many of these peptides and proteins are only produced in very small quantities. They will only be produced if the correct genes are present in the cell (see section 10.10). Consequently, if a gene is missing or defective an essential protein will not be produced, which can lead to a diseased state. For example, cystic fibrosis is caused by a defective gene. This faulty gene produces a defective membrane protein, cystic fibrosis transmembrane regulator (CFTR), which will not allow the free passage of chloride ions through the membrane. The passage of chloride ions through a normal membrane into the lungs is usually accompanied by a flow of water molecules in the same direction. In membranes that contain CFTR the transport of water through the membrane into the lungs is reduced. As a result, the mucus in the lung thickens. This viscous mucus clogs the lungs and makes breathing difficult, a classic symptom of cystic fibrosis. It also provides a breeding ground for bacteria that cause pneumonia and other illnesses.

Several thousand hereditary diseases found in humans are known to be caused by faulty genes. Recombinant DNA (rDNA) technology (genetic engineering) offers a new way of combating these hereditary diseases by either replacing the faulty genes or producing the missing peptides and proteins so that they can be given as a medicine (see section 10.15.2).

The first step in any use of recombinant DNA technology is to isolate or copy the required gene. There are three sources of the genes required for cloning. The two most important are genomic and copy or complementary DNA (cDNA) libraries. In the first case the library consists of DNA fragments obtained from a cell's genome, whilst in the second case the library consists of DNA fragments synthesised by using the mRNA for the protein of interest The third is by the automated synthesis of DNA, which is only feasible if the required base sequence is known. This may be deduced from the amino acid sequence of the required protein if it is known. Once the gene has been obtained it is inserted into a carrier (*vector*) that can enter a host cell and be replicated, propagated and transcribed into mRNA by the cellular biochemistry of that cell. This process is often referred to as *gene cloning*. The mRNA produced by the cloned DNA is used by the cell ribosomes (see section 10.10) to produce the protein encoded by the cloned DNA. In theory, gene cloning makes it possible to produce any protein provided that it is possible to obtain a copy of the corresponding gene. Products produced using recombinant DNA usually have recombinant, r or rDNA in their names.

10.15.1   Gene cloning

Bacteria are frequently used as host cells for gene cloning. This is because they normally use the same genetic code as humans to make peptides and proteins. However, in bacteria the mechanism for peptide and protein formation is somewhat different. It is not restricted to the chromosomes but can also occur in extranuclear particles called *plasmids*. Plasmids are large circular supercoiled DNA molecules whose structure contains at least one gene and a start site for replication. However, the number of genes found in a plasmid is fairly limited, although bacteria will contain a number of identical copies of the same plasmid.

It is possible to isolate the plasmids of bacterial cells. The isolated DNA molecules can be broken open by cleaving the phosphate bonds between specific pairs of bases by the action enzymes known as *restriction enzymes* or *endonucleases*. Each of these enzymes, of which over 500 are known, will only cleave the bonds between specific nucleosides. For example, EcoR I cleaves the phosphate link between guanosine and adenosine whilst Xho I cuts the chain between cytidine and thymine nucleosides. Cutting the strand can result in either *blunt ends*, where the endonuclease cuts across both chains of the DNA at the same points, or *cohesive ends* (sticky ends), where the cut is staggered from one chain to the other (Fig. 10.47). The new non-cyclic structure of the plasmid is known as *linearised* DNA in order to distinguish it from the new *insert* or *foreign* DNA. This foreign DNA must contain *the required gene, a second gene system that confers resistance to a specific antibiotic and any other necessary information*. It should be remembered that a eukaryotic gene is made up of exons separated by introns, which are sequences that have no apparent use.

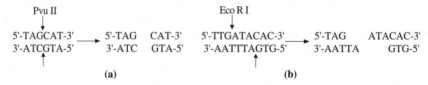

Figure 10.47   (a) Blunt and (b) cohesive cuts with compatible adhesive cuts

Mixing the foreign DNA and the linearised DNA in a suitable medium results in the formation of extended plasmid loops when their ends come into contact (Fig. 10.48). This contact is converted into a permanent bond by the catalytic action of an enzyme called DNA ligase. When the chains are cohesive the exposed single chains of new DNA must contain a complementary base sequence to the exposed ends of the linearised DNA. The hydrogen bonding between these complementary base pairs tends to bind the chains together prior to the action of the DNA ligase, hence the name "sticky ends". The new DNA of the modified plasmid is known as *recombinant DNA* (*rDNA*). However, the random nature of the techniques used to form the modified plasmids means that some of the linearised DNA reforms the plasmid without incorporating the foreign DNA, that is, a mixture of both types of plasmid is formed. The modified plasmids are separated from the unmodified plasmids when they are reinserted into a bacterial cell.

The new plasmids are reinserted into the bacteria by a process known as *transformation*. Bacteria are mixed with the new plasmids in a medium containing calcium chloride. This medium makes the bacterial membrane permeable to the plasmid. However, not all bacteria will take up the modified plasmids. Such bacteria can easily be destroyed by specific antibiotic action since they do not contain plasmids with the appropriate protecting gene. This makes isolation of the bacteria with the modified plasmids relatively simple. These modified bacteria are allowed to replicate and, in doing so, produce many copies of the modified plasmid. Under favourable conditions one modified bacterial cell can produce over 200 copies of the new

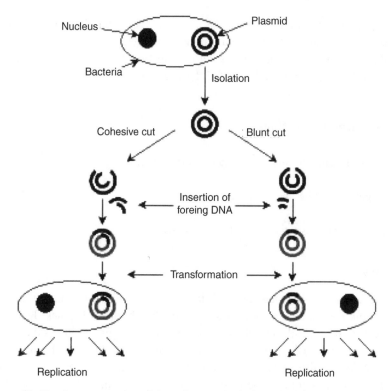

Figure 10.48. A representation of the main steps in the insertion of a gene into a plasmid

plasmid. The gene in these modified plasmids will use the bacteria's internal machinery to automatically produce the appropriate peptide or protein. Since many bacteria replicate at a very rapid rate this technique offers a relatively quick way of producing large quantities of essential naturally occurring compounds that cannot be produced by other means.

Plasmids are not the only vectors that can be used to transport DNA into a bacterial host cell. Foreign DNA can also be inserted into *bacteriophages* and *cosmids* by similar techniques. Bacteriophages (phage) are viruses that specifically infect bacteria whilst a cosmid is a hybrid between a phage and a plasmid that has been especially synthesised for use in gene cloning. Plasmids can be used to insert fragments containing up to 10 kilobase-pairs (kbp), phages up to 20 kbp and cosmids 50 or more kbp.

It is not always necessary to use a vector to place the recombinant DNA in a cell. If the cell is large enough, the recombinant DNA may be placed in the cell by using a micropipette whose overall tip diameter is less than 1 μm. Only a small amount of the recombinant DNA inserted in this fashion is taken up by the cell's chromosomes. However, this small fraction will increase to a significant level as the cell replicates (Fig. 10.48).

Host cells for all methods of cloning are usually either bacterial or mammalian in origin. For example, bacterial cells often used are *E. coli* and eukaryotic yeast while mammalian cell lines include Chinese hamster ovary (CHO), baby hamster kidney (BHK) and African green monkey kidney (VERO). In all cases small-scale cultures of the host cell plus vector are grown to find the culture containing the host with the required gene that gives the best yield of the desired protein. Once this culture has been determined the process is scaled up via a suitable pilot plant to production level (see section 16.6). The mammalian cell line cultures normally give poorer yields of the desired protein.

10.15.2   Medical applications

The main uses of gene cloning in the medical field are:

● to correct genetic faults and absences;

● to manufacture rare essential natural compounds.

Gene therapy

A wide range of undesirable medical conditions are due to either the presence of defective genes that contain an incorrect base sequence or the absence of genes. For example, a defective gene is responsible for the substitution of a glutamic acid residue by a valine residue in haemoglobin (Hb). This results in the formation of HbS, which is responsible for sickle cell anaemia, a disease commonly found in central and west Africa. The absence of genes producing a growth hormone leads to stunted growth in children.

The use of gene cloning in medicine is known as *gene therapy*. Gene therapy may involve either the replacement of a faulty gene with a normal gene or the addition of a new extra gene that will produce proteins that could fight a diseased state such as cancer or viral infections. There are two fundamental approaches to these treatments: *germline* and *somatic*. The germline approach is concerned with preventing inherited diseases by transferring a cloned gene to a germ cell (sperm or egg cells), while the somatic approach is concerned with treating existing diseases by transplanting the cloned gene to other body cells (somatic cells).

In germline therapy, for example, a fertilised egg may be removed from the mother and the required gene inserted by cloning techniques. If this egg is replanted in the mother and if the process has been successful the normal replacement gene will be present in all the cells of the new individual formed from the egg. This procedure could, in theory, be used to prevent all inherited disorders, such as cystic fibrosis, haemophilia and insulin-dependent diabetes.

The somatic approach uses cells that are not involved in the reproduction of the organism and so any alterations will not be transmitted to any future generations of cells and people. The technique involves removing cells from the body, infection with the required gene and the return of the cells to the body. This technique appears to offer a solution to inherited genetic disorders such as haemophilia and thalassaemia. It has been used to successfully treat children with adenosine deaminase (ADA) deficiency. This is a single gene deficiency disease that leads to an almost total lack of white blood cells and, as a result, almost no natural immunity.

Manufacture of pharmaceuticals

The body produces large numbers of peptides and proteins, often in extremely small quantities, which are essential for its well being. The absence of the genes responsible for the production of these peptides and proteins means that the body does not produce these essential compounds, resulting in a deficiency disease that is usually fatal. Treatment by supplying the patient with sufficient amounts of the missing compounds is normally successful. However, extraction from natural sources is usually difficult and yields are often low. For example, it takes half a million sheep brains to produce 5 mg of somatostatin, a growth hormone that inhibits the secretion of the pituitary growth hormone. Furthermore, unless the source of the required product is donated blood, there is a limit to the number of cadavers available for the extraction of compounds suitable for use in humans. Moreover, there is also the danger that compounds obtained from human sources may be contaminated by viruses such as AIDS, hepatitis, Creutzfeld–Jakob disease (mad cow disease) and others that are difficult to detect. Animal sources have been used but only a few human protein deficiency disorders can be treated with animal proteins.

Gene cloning is used to obtain a wide range of human recombinant proteins. However, some proteins will also need post-translational modification such as glycosylation and/or the modification of amino acid sequences. These modifications may require forming different sections of the peptide chain in the culture medium and chemically combining these sections *in vitro*. The genes required for these processes are synthesised using the required peptide as a blueprint. For example, human recombinant insulin prior to 1986 was produced in this manner (Fig. 10.49). The genes for the A and B chains of insulin were synthesised separately. They were cloned separately, using suitable plasmids, into two different bacterial strains. One of these strains is used to produce the A chain whilst the other is used to produce the B strain. The chains are isolated and attached to each other by *in vitro* disulphide bond formation. This last step is inefficient and human recombinant insulin is now made by forming recombinant proinsulin by gene cloning. The proinsulin is converted to recombinant insulin by proteolytic cleavage.

Recombinant peptides and proteins are normally produced by fermentation of the rDNA-infected host cells (see section 10.15.1). Both their isolation from the fermentation broth and subsequent purification are difficult because at each step of these procedures the peptide/protein must remain intact and biologically active. Isolation usually starts by using a suitable filtration technique to remove particulate contaminants and/or heating to

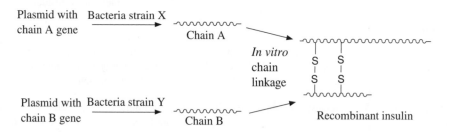

Figure 10.49 An outline of the original pre-1986 method of producing recombinant human insulin

inactivate viruses provided that the rDNA-peptide/protein is sufficiently stable to heat. After filtration, if necessary, the volume of the liquid is reduced prior to precipitating the rDNA-peptide/protein from solution by either salting out (see section 6.7.1) or the addition of water-miscible organic solvents such as polyethylene glycol and trifluoroethanoic acid. The precipitated rDNA-peptide/protein is often purified by dialysis to remove salts before being crudely separated into its component proteins using ion exchange or gel filtration. The crude products are further purified by high-resolution chromatography and the products collected from the column and assayed for activity. At the end of the procedure the product may be sterilised before formulation. The names of products obtained using recombinant technology are usually prefixed by either *recombinant* or *r*, while *h* is used as an abbreviation for human. A wide range of drug products are now produced by recombinant gene technology (Table 10.10).

Table 10.10 Examples of recombinant gene technology products in clinical use

Product	Used to treat	Source(s) / Notes
Hormones		
Recombinant human insulin	Diabetes type 1	Recombinant *E coli.* (Humulin) Recombinant *S. cerevisiae* (Novolin) Indistinguishable from human insulin
Recombinant human growth hormone (rhGH)	Growth hormone deficiency Turner's syndrome	Mammalian (mouse) cell culture Has an identical amino acid sequence to that found in the endogenous compound
Cytokines		
Recombinant human erythropoietin-α	Anaemia of chronic renal failure in cancer and HIV patients	Mammalian cell line The rDNA protein has the same amino acid sequence to that found in the endogenous compound
Filgrastim, granulocyte colony-stimulating factor (G-CSF)	Neutropenia in patients undergoing cancer chemotherapy	The rDNA protein has the same amino acid sequence to that found in the endogenous compound except for an N-terminal methionine residue
Interferons		
Recombinant α_{2a}-interferon	Hairy cell leukaemia, chronic hepatitis C and AIDS-related Kaposi's sarcoma	Recombinant *E coli. system* and is purified by high-affinity mouse monoclonal antibody chromatography
Recombinant β_{1a}-interferon	Relapsed multiple sclerosis	Recombinant CHO cells Recombinant interferon-α_{1a} is equivalent to that secreted by human fibroblasts

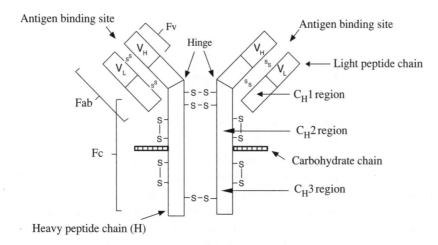

Figure 10.50 A plan of the structure of an antibody showing some of the nomenclature used in connection with this molecule. The nomenclature for both heavy chains is identical. Similarly, that for both light chains is also identical. The Fab and Fc regions meet at the hinge in both chains

Monoclonal antibodies (Mabs)

Antibodies or immunoglobins (Ig) are glycoproteins that constitute an important part of the body's immune system. The simplest antibodies are Y-shaped molecules that consist of two identical so-called ~23 kD light (L) and two identical so-called 53–75 kD heavy (H) peptide chains held together by S–S links and non-covalent interactions (Fig. 10.50). There are five known major classes of Ig, namely IgM, IgG, IgA, IgD and IgE, respectively. Immunoglobins of the same class have similar heavy chains except for the variable regions although subtle differences result in a class being subdivided into sub-types known as *isotypes*. Both the heavy and light chains contain the variable amino acid sequence regions. These regions occur at the N-terminals of the light and heavy chains and are known as the V_L and V_H regions or domains, respectively. Parts of these variable regions have a greater variability when compared with the variable domains of other antibodies. These regions are known as *hypervariable regions* or *complementarity-determining regions (CDRs)*. CDRs are brought into close contact in the antibody's native conformation and are believed to be the antigen binding sites. This variation confers specificity to the antibody. The sugar moiety is usually attached to the C_H2 domain of the heavy chain. Removal of the sugar has no effect on antigen binding but does change other properties of the antibody, such as its half-life in human serum.

Antibodies act by recognising macromolecules, for example proteins known as *antigens*, on foreign invading bodies such as cells, proteins and carbohydrates. They bind to specific chemical groupings known as *epitopes* found on these antigens (Fig. 10.51). Epitopes on the surface of a cell usually consist of 5–7 amino acid residues. Natural antigens usually have hundreds of different epitopes and so a number of different antibodies can bind to the same foreign body. These antibodies are produced by a type of cell known as a B lymphocyte (B cell). Each antibody is produced by its own particular B lymphocyte. When

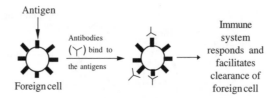

Figure 10.51   An outline of the mode of action of an antibody

a particular B lymphocyte meets its antigen it undergoes proliferation to produce more cells that produce that antibody. These antibodies bind to the antigen, which activates the immune system to facilitate the clearance of the foreign body. All individuals posses a wide repertoire of antibodies with different variable regions that are able to recognise a diverse range of foreign antigens. A number of mechanisms operate to prevent the development of antibodies to the body's own antigens but unfortunately these occasionally break down and autoimmune diseases develop.

The ability of antibodies to recognise foreign bodies provides a possible vehicle for the delivery of drugs and diagnostic aids to specific areas of the body. The use of B lymphocytes to produce antibodies is not practical as they have too short a life in the laboratory. However, in the 1970s Kohler and Milstein immortalised B cell lines by fusing them with malignant myeloma cells to produce hybrid cells (*hybridomas*), which have a longer life *in vitro*. These hybridomas retained the antibody producing properties of the B lymphocytes but could be grown indefinitely in a suitable culture medium. This made it possible to produce large quantities of pure antibodies that could recognise a single epitope. Antibodies that only recognise a single epitope are known as *monoclonal antibodies* (*Mabs*).

The production of a specific Mab usually starts with the inoculation of a rodent with the target antigen. The animal is sacrificed after a few weeks when it will have generated the antibodies to that antigen. Its spleen cells are collected because they will contain a high level of B lymphocyte cells and it is probable that some will be responsible for the production of the antibody for the target antigen. These B lymphocytes are fused with myeloma cells in a medium that only allows the fused spleen–myeloma cells to survive. The individual hybridomas are separated and grown in individual cultures. The cultures are screened to find the hybridoma that secretes the desired antibody. Once this hybridoma has been identified the clone can be grown on a larger scale to produce large numbers of cells that secrete the same monoclonal antibody.

The early monoclonal antibodies were produced from mouse B lymphocytes. They are known as *murines*. However, their use in humans resulted in the formation of human anti-mouse antibodies (HAMA) after repeated doses of the antibody. HAMA can cause severe allergic reactions, which limits their clinical use. In addition, murine monoclonals have a short half-life. These properties severely limit the therapeutic use of murine monoclonal antibodies. To overcome this problem *chimeric* and *humanised* antibodies have been produced by genetic engineering. Chimeric antibodies have the V_L and V_H domains of the murine grafted onto a human antibody by transferring the nucleotide sequence for those domains into the human antibody gene. Humanised antibodies have the CDR of the murine grafted onto a human antibody by transferring the nucleotide sequence for the mouse CDR to that human antibody gene. The presence of the human monoclonal antibody reduces the possibility of the patient's immune system generating antibodies against the introduced monoclonal antibody. Furthermore, the half-lives of humanised monoclonals are longer than those of chimeric monoclonals.

The clinical use of monoclonal antibodies depends on identifying an antigen that is unique to the target site. Once identified the relevant antibody can be produced by the method described previously. Since the introduction of monoclonal antibody technology, over 20 products worldwide have been accepted for clinical use. The majority are either chimeric or humanised antibodies (Table 10.11). Unfortunately the use of antibodies has not been as effective as was first thought. This is believed to be due to inhibition of the antibody by other secretions from the target cell. However, in many patients the response to this therapy has been good.

The fact that an antibody will bind to a specific antigen means that it can be employed as a vehicle for the delivery of both drugs and diagnostic aids to specific sites in the body. In the former case a drug residue is usually attached to the antibody by means of a linker to form an antibody–drug conjugate (Fig. 10.52a). The antibody delivers the drug to the target site where, after passing into the target cell, it is released to attack and destroy the cell. An example of one such antibody–drug conjugate in clinical use is Gemtuzumab (Table 10.11), which is used to treat acute myeloid leukaemia. This is a humanised monoclonal antibody conjugate of calicheamicin (Fig. 10.52b). Calicheamicin is an anticancer agent obtained as a mixture of seven related compounds from cultures of *Micromonospora echinospora* ssp. *Calicheansis*. It is too toxic to be used on its own as a drug. To be effective antibody–drug conjugates require the drug to be more cytotoxic than the normal cytotoxic drugs, the antibody to be selective for an antigen that is mainly found on the target cells, capable of cell penetration by receptor-mediated endocytosis and able to release the drug once the conjugate has reached the target. A novel approach to drug release is found in antibody-directed enzyme therapy (see page 473)

Antibodies also have a use in radiotherapy. It is possible using antibodies to deliver a radioactive isotope to specific parts of the body. For example, an antibody–yttrium-90 (Y-90) conjugate is currently being evaluated for the destruction of bone marrow cells prior to autologous stem cell transplantation in patients with multiple myeloma at Southampton General Hospital. The medium energy β-emissions of Y-90 are able to penetrate several layers of cells. The rational is that the antigen only delivers the isotope to the target cancer cells where selective binding of the antigen–isotope conjugate to a cancer cell results in the β-emissions killing the cancer cell (Fig. 10.53). It also allows lower total doses of radiation to be used and so it is hoped that the emissions do less damage to the healthy organs and blood vessels that the conjugate passes through on route to its bone marrow target than whole body irradiation, the current treatment method. In addition, it is also hoped that the directed destruction of myeloma bone marrow cells will also result in a more efficient destruction of these cells. Other medium energy β-emitters being evaluated for use as anticancer agents include iodine-125 and- 131 and rhenium-186 and- 188. Ibritumomab, a

Table 10.11 Examples of monoclonal antibodies and antibody fragments approved for clinical use. Reproduced from G. Walsh, *Biopharmaceuticals, Biochemistry and Biotechnology*, 2nd edn, 2004, by permission of John Wiley and Sons Ltd

Product	Company	Type / target	Use
ANTIBODIES			
Rituxan (Rituximab)	Genetech / IDEC Pharmaceuticals	Chimeric Mab directed against CD20 B-lymphocyte surface antigen	Non-Hodgkin's lymphoma
Heceptin (Trastuzumab)	Genetech (USA) Roche Registration (EU)	Humanised antibody directed against HER2	Metastatic breast cancer
Abciximab (ReoPro)	Centor	Fab fragments from a chimeric Mab directed against the platelet surface receptor GBII$_b$ / III$_a$ α-chain of the IL-2 receptor	Prevention of blood clots
Remicade (Infliximab)	Centocor	Chimeric Mab directed against TNF-α	Treatment of Crohn's disease
CONJUGATES			
Ibritumomab (Zevalin)	IDEC Pharmaceuticals	Murine antibody conjugate with ^{90}Y directed against CD20 antigen on the surface of B-lymphocytes	Non-Hodgkin's lymphoma
Mylotarg (Gemtuzumab)	Wyeth Ayerst	Humanised anticancer conjugate targeted against the CD33 antigen found on leukaemic blast cells	Acute myeloid leukaemia
DIAGNOSTIC AIDS			
Tecnemab KI	Sorin	Murine Mab fragments directed against high molecular weight melanoma-associated antigen	Diagnosis of cutaneous melanoma lesions
LeukoScan (Sulesomab)	Immunomedics	Murine Mab fragments, directed against NCA90, a surface granulocyte non-specific cross-reacting antigen	Diagnosis of infection and inflammation in the bone of osteomyelitis patients

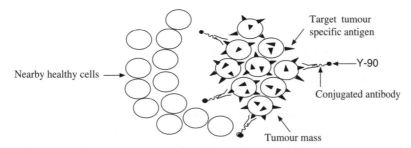

Figure 10.52 (**a**) A schematic representation of the antibody–drug conjugate. (**b**) Calicheamicin γ^1, the most abundant of the calicheamicins

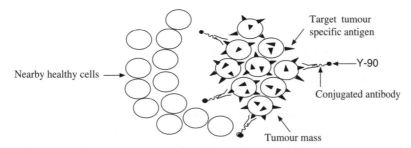

Figure 10.53 A representation of the action of an antibody–radioactive isotope complex in destroying cancer cells. The β-emissions from the Y-90 isotope will kill several layers of both the tumour and healthy cells

Y-90 conjugate (Table 10.11), has been accepted for clinical use to treat non-Hodgkin's lymphoma and may be more therapeutic than unconjugated anti-CD20 antibodies.

The use of antibodies to transport drugs to their site of action is limited by the ability of the conjugate to penetrate the tumour mass or target cell. This is believed to be due to the physical size of the intact antibody. Consequently, fragments of antibodies have been used as the basis of conjugates (Table 10.11). These fragments are generated by means of recombinant DNA technology. However, their use results in fragment-conjugates that have a much lower half-life in human serum than whole antibodies.

A number of antibody-conjugates are also used as diagnostic aids (Table 10.11). For example, the γ-emitter indium-111 Oncoscint (Fig. 10.54) is a conjugate used to image colorectal and ovarian carcinomas. It is prepared by oxidising part of the carbohydrate moieties with periodate. The aldehydes produced are reacted with the amino groups of glycyl-tyrosyl-lysine-*N*-diethylene triaminopentaacetic acid to form the appropriate Schiff's bases. These imine groups were reduced by sodium cyanoborohydride to the more stable secondary amines. The product of this reaction chelates with the indium-111 to form indium-111 Oncoscint. The accumulation of indium-111 Oncoscint in tumours is visualised by the use of a whole body gamma camera.

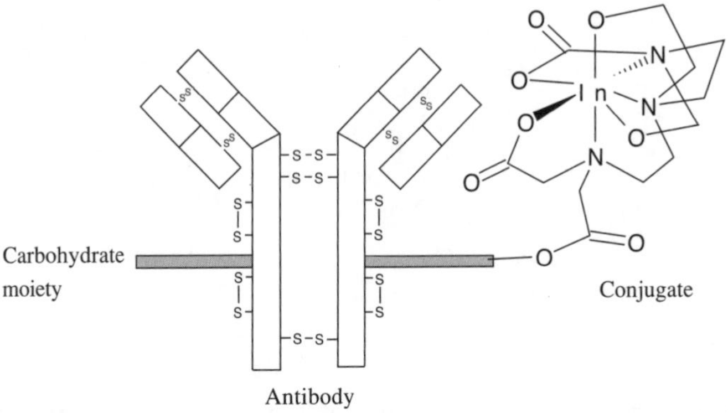

Figure 10.54 An outline of the structure of indium-111 Oncoscint (Satumomabpendetide)

Genetically modified crops and animals as sources of drugs

The advent of genetic engineering has made it possible to produce recombinant protein drugs by using plants and animals as bioreactors. This source of medicinal compounds from genetically modified plants and animals offers the hope of producing compounds with the same activity as the native compounds they are supplementing. Its success depends on the ability of the surrogate plant or animal to produce an active form of the recombinant protein in sufficient quantity to make it a commercially viable proposition and on the safety of the product in medical trials.

The discovery process for producing a recombinant protein is dependent on first determining the nature of the gene responsible for producing the compound required. Obtaining that gene can be by isolation from a natural source, from a gene library, by DNA synthesised from the mRNA for the gene or synthesised from the nucleotides corresponding to the amino acid sequence of the protein by automated synthesis. This last is only used for the production of small recombinant proteins. The required DNA sequence is inserted into the plants by either the use of a suitable vector such as an *Agrobacterium*-based vector or by chemical, electrical and physical methods such as microprojectile bombardment. Animals are usually cloned using techniques based on directly microinjecting exogenous DNA into an egg cell, fertilising the egg and breeding the animal using a surrogate mother. The resultant genetically modified plants and animals are screened for a high expression level of the desired recombinant protein and suitable specimens are selected for reproduction. Large-scale reproduction of the selected plant or animal is carried out and the recombinant protein is isolated (see Chapter 6). The isolated recombinant protein is tested for biological activity. If this is successful the recombinant is submitted for further development and commercial scale production (see Chapter 16). A wide variety of recombinant proteins have been produced from plants and animals (Table. 10.12) and a limited number have or are undergoing clinical trials, although progress towards large scale use is slow.

Table 10.12 Examples of some of the recombinant proteins that have been obtained from plants and animals using recombinant gene engineering

Product	Source	Possible use	Yield
β-Interferon	Tobacco plants	Treatment of relapsed multiple sclerosis	0.003 of total soluble plant protein
GM-CSF	Tobacco plants	Stimulation of human neutrophile formation	250 ng cm^{-3} extract
Fibrinogen	Sheep's milk	Clotting agent	5 g l^{-1}
α$_1$-Antitrypsin	Goat's milk	Infections of the lung of patients with cystic fibrosis	20 g l^{-1}
Antithrombin-III	Goat's milk	Anticlotting agent	14 g l^{-1}
Hepatitis B surface antigen	Tobacco plants		0.007 of soluble leaf protein
Human protein C	Pig's milk	Anticoagulant agent	1 g l^{-1}

10.16 Questions

1 Distinguish carefully between the members of the following pairs of terms:

(a) nucleotide and nucleoside;

(b) introns and exons;

(c) codons and anticodons.

2 Explain how cytosine and guanine form a complementary base pair.

3 The sequence AATCCGTAGC appears on a DNA strand. What would be the sequence on (a) the complementary chain of this DNA and (b) a transcribed RNA chain?

4 What are the two main functions of DNA?

5 How does RNA differ from DNA? Outline the functions of the three principal types of RNA.

6 What is the genetic code?

7 Why is thymine never found in a human codon?

8 Explain in outline the significance of the P, A and E sites of ribosomes.

9 What is the sequence of amino acid residues in the peptide formed from the mRNA:

UUCGUUACUUAGAUGCCCAGUGGUGGGUACUAAUGGCUCGAG

10 How does protein synthesis in prokaryotic cells differ from that in eukaryotic cells?

11 Explain why the structure of chloramphenicol is probably the optimum one for antibiotic activity.

12 Outline a general strategy for discovering a more active antibiotic than tetracycline using tetracycline as the lead compound.

13 Draw the general structure of nitrogen mustards. How do these drugs inhibit the transcription of DNA?

14 Explain why the incorporation of didanosine into a strand of RNA stops the synthesis of the RNA.

15 Describe the way in which the antibiotic proflavine disrupts the transcription of DNA. What part do the amino groups of proflavine play in the mode of action of this drug?

16 What are antisense drugs? How do they inhibit mRNA?

17 Explain the meaning of the term antimetabolite. Outline a strategy for designing a new antimetabolite for a biological process.

18 Describe the essential differences between RNA-viruses, retroviruses and DNA-viruses.

19 Outline, using suitable examples, the general mode of action of antiviral drugs.

20 Explain how recombinant DNA is produced using plasmids.

21 Describe the difference between germline and somatic gene therapy.

22 Suggest a structural explanation for the activity of doxorubicin. How could this structural feature be used in the general design of analogues of this drug with possible anticancer activity in mind? Give a reason for your answer.

23 Outline some of the essential structural features of a monoclonal antibody. Explain the meaning of the terms (a) murine and (b) humanised antibody.

24 What in general are the requirements for an effective monoclonal antibody–drug conjugate?

11

Pharmacokinetics

11.1 Introduction

The action of a drug is initially dependent on it reaching its site of action in sufficient concentration for a long enough period of time for a significant pharmacological response to occur. This build-up in drug concentration and the maintenance of this concentration over a period of time depend on the route of drug administration, the efficiency of drug absorption, the rate at which the drug is transported to its site of action, the rate of drug metabolism on route and, at the site of action, the rate of drug excretion as well as the age, gender and physiological state of the patient. Pharmacokinetics is the study of the relationships between drug response in the patient and these factors.

Pharmacokinetics is based on the hypothesis that the magnitude of the responses to a drug, both therapeutic and toxic, is a function of its concentration at its site of action. The relationship of this concentration to the dose administered is not simple. Once a drug is absorbed into the body it must find its way to its site of action. In the course of this transportation some of the drug will be metabolised (Chapter 12) and some will be irreversibly excreted by the liver and/or kidneys and/or lungs. The *irreversible* processes by which a drug is prevented from reaching its site of action are collectively referred to as *elimination*. Uptake into the tissues is not regarded as an elimination process since it is usually reversible, the drug returning to the general circulation system (systemic circulation) in the course of time.

The process of elimination means that the concentration of the drug reaching the desired site of action may not be high enough to provide the required therapeutic effect. Consequently, it is important to have a method of monitoring the concentration of a drug in contact with its site of action. However, since the precise site of action is often unknown this is not usually possible. As a result, the pharmacokinetic behaviour of a drug is usually monitored by following the concentration of the drug in the plasma and other suitable body fluids (Fig.11.1). These measurements are statistically correlated with

Medicinal Chemistry, Second Edition   Gareth Thomas
© 2007 John Wiley & Sons, Ltd

the effects of the drug on patients. However, it is often difficult to obtain data using humans, so many investigations are carried out using animals. The results of these experiments have been extrapolated to humans (see section 11.7) with varying degrees of success. Consequently, it is necessary to carry out trials on humans before the drug is released for clinical use.

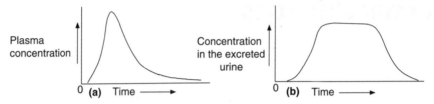

Figure 11.1 Typical variations in the concentration of a drug with time in samples of (**a**) plasma and (**b**) urine after the administration of a single oral dose of the drug at time = 0. In both cases the precise shape of the graph will depend on the drug being studied

A drug is therapeutically successful when its plasma concentration lies within its therapeutic window (see section 1.6 and Fig. 11.2). However, it should be realised that the therapeutic window of a drug can vary considerably from patient to patient. Consequently, the clinically acceptable values used for a dosage form can still cause unacceptable toxic effects in some individuals.

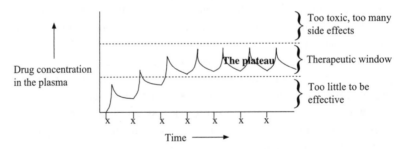

Figure 11.2 A schematic representation of a therapeutic window. Successive orally administered doses are given at the points x. Too high a dose results in the plateau being above the therapeutic window whilst too low a dose gives a plateau below the therapeutic window

When a drug is given orally to a patient, its concentration in the plasma gradually increases to a maximum as it is absorbed from the GI tract. Once in the plasma the processes of elimination start to reduce the concentration of the drug. However, in most cases absorption is faster than elimination. Consequently, a succession of doses at regular time intervals will usually result in a build-up of the drug concentration to the desired level in the plasma and by inference the correct concentration at the site of action for therapeutic success. Since, in the normal course of events, the concentration of the drug never reaches zero before the next dose is administered one would expect a steady rise in the drug's concentration in the plasma. However, there is a limit to the amount of a drug the plasma can contain and this, coupled with the elimination processes, results in the drug concentration reaching a *plateau*. If the dose is correct, this plateau will lie within the therapeutic window of the drug (see Fig 11.2 and section 11.5.3).

11.1.1 General classification of pharmacokinetic properties

The pharmacokinetic behaviour of a drug after administration (Fig. 11.3) is broadly classified into the general regions of absorption, distribution, metabolism and excretion (ADME). The parameters associated with each of these general regions are used to specify the pharmacokinetic properties of drugs in the body. The methods used to calculate these parameters are independent of the method of administration although the values obtained will depend on the administrative route. For example, intravascular routes will not normally give values for absorption parameters. However, intravascular routes do give higher concentrations of the drug in the general circulatory system (*systemic circulatory system*).

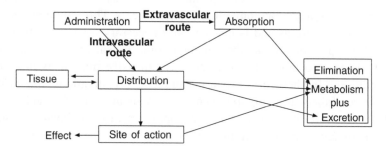

Figure 11.3   The general stages and their relationships in the life cycle of a drug after administration

Each of the general regions of pharmacokinetics and their associated parameters will be discussed later in this chapter under the most appropriate method of administration. However, it is emphasised that the parameters apply across the board and are not confined to the method of administration under which they are introduced.

11.1.2 Drug regimens

A drug regimen is the way in which a drug is administered to a patient. It normally includes the method of administration, the dose, its frequency of administration and the

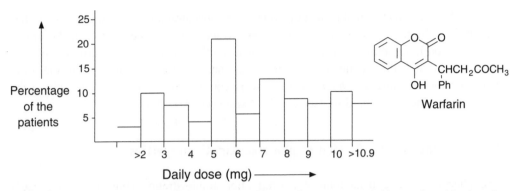

Figure 11.4 The variation of the daily dose of warfarin needed to produce similar prothrombin times in adult patients. The prothrombin time is the time taken for citrated plasma treated with calcium and standardised reference thromboplastin to clot. (Reprinted with permission from J. Koch-Wesser, The serum level approach to individualisation of drug dosage. *Eur.J.Clin.Pharmacol.* **9**, 1–8 (1975), © 1975, Springer-Verlag

duration of the treatment. The response of individual patients to the same doses of a drug can be very varied (Fig. 11.4). It will depend on the age and weight of the subject as well as the severity of the disease. Consequently, to use any drug effectively, the drug regimen should be tailored to an individual's requirements. This would require extensive and expensive investigations and is not normally necessary for most patients suffering from medical complaints such as the common cold, thrush, diarrhoea and bronchitis. However, it can lead to more effective and less expensive treatments in more critical medical conditions. In these cases, pharmacokinetic investigations aim to determine the most successful course of treatment by balancing the desirable therapeutic effects against undesirable side effects. The optimum drug regimen for any drug is the one that maintains the concentration of the drug within the therapeutic window for the patient.

11.1.3 The importance of pharmacokinetics in drug discovery

Pharmacokinetics is important in a number of aspects of drug design. It can prevent the discarding of what could be an important drug. For example, quinacrine, developed during the Second World War as an alternative to quinine, was found to be too toxic when administered in high enough doses to be effective. However, a study of its pharmacokinetic properties showed that it had a slow elimination rate and so rapidly accumulated to toxic levels in the plasma. Once this elimination pattern was discovered it was a simple matter to use high initial doses and to follow this with smaller doses just sufficient to maintain the drug's concentration within its therapeutic window.

Quinacrine

Quinine

A study of the pharmacokinetic properties of a compound indicates which properties need to be modified in order to produce a more effective analogue. Consider, for example, a drug that is not suitable for development because it has too short a duration of action. The logical way forward is to determine which analogues have a slower elimination rate by testing on suitable animal models. The analogues selected will depend on what is believed to be the structural cause of this rapid elimination. If, for example, the drug is rapidly metabolised by esterases in the plasma, compounds that are more stable to these enzymes would be tested. Alternatively, the drug may be very water soluble and as a result poorly absorbed, in which case the approach is to produce and test less-water-soluble analogues.

A study of pharmacokinetics forms the basis of the design of the dosage forms. Drugs with a narrow therapeutic window require smaller and more frequent doses or a change in the method of administration. If a potential drug is administered incorrectly it will be ineffective in trials and consequently the drug will be discarded even though it could have been of considerable therapeutic value if it had been administered correctly. For example, the uterine stimulant oxytocin could have been overlooked if it had not been administered correctly in trials. Investigations of its pharmacokinetic properties showed that it has a narrow therapeutic window and is eliminated within minutes of entering the systemic circulation. In addition, it is also metabolised by enzymes in the GI tract. These characteristics make it impossible to administer the drug orally. Consequently, the only way to accurately maintain the oxytocin concentration in the plasma is to administer it by carefully monitored continuous intravenous infusion.

Cys-Tyr-Ile-Gln-Asn-Cys-Pro-Leu-Gly(NH$_2$)

Oxytocin

11.2 Drug concentration analysis and its therapeutic significance

The determination of drug concentration in the body requires taking samples of biological fluids from patients and test animals. Samples are taken by either *invasive* or *non-invasive* methods. Invasive methods include the removal of blood, spinal fluid and tissue (biopsy) samples whilst non-invasive methods include collecting urine, faeces, expired air and saliva samples. The information obtained from these samples will depend on their source. For example, changes in the drug concentration in plasma are usually good indications of the changes occurring in the tissue whilst the concentration of a drug in the faeces may either indicate the degree of biliary excretion of the drug or, when compared with the orally administered dose, show the degree of absorption of the drug. However, most analytical methods are designed for plasma analysis and so plasma concentrations are the most commonly reported measurement.

The cells in tissue are in contact with the extracellular fluid, which has a similar composition to that of blood plasma. Consequently, the concentration of a drug in the blood plasma is found in many cases to be a good measure of that drug's concentration at its receptor sites, which are usually found in tissue cells. This means that monitoring the level of a drug in a patient's plasma is often a good method of checking that the dose level is correct for effective therapeutic action. However, for some drugs, such as those used in cancer chemotherapy, the concentration of the drug in the plasma is not a good indicator of pharmacological response. Consequently, clinical decisions should not be based solely on blood plasma levels of a drug.

The assessment of the pharmacokinetic behaviour of a drug in a patient requires an accurate method for assaying that drug. Many drugs are substances that are racemates or mixtures of compounds with similar structures. The components of these substances can have similar or completely different pharmacological actions (see Table 1.1). Drug licencing now normally requires that all the components of a mixture are tested separately even though the drug may be administered as a racemate or mixture. Consequently, it is necessary to separate the components and develop individual assays for each pure component as well as the individual components in the mixture or racemate. However, separation of components is not always possible and can be very expensive. Consequently, medicinal chemists try to produce drugs that are not optically active or mixtures of compounds.

To obtain meaningful pharmacokinetic data it is necessary to analyse for the metabolites of the drug as well as for the drug itself. This is particularly important if the metabolites are either active or toxic. For example, the activity of a drug may be due to a large increase in concentration of an active metabolite rather than the drug itself (see section 12.2). Similarly, the unrecognised increase in concentration of a toxic metabolite could lead to the unnecessary abandonment of what could have been a useful drug. This type of problem has been reduced in a number of cases by the use of a suitable dosage form. For example, sodium 2-mercaptoethanesulphonate (MESNA) reduces the toxic effects of the antineoplastic agent cyclophosphamide (Fig. 10.40) by increasing the rate of elimination of the toxic metabolites of this drug (see section 10.13.4).

Identification of the metabolites is also of importance when assessing the plasma concentration of a drug using methods based on the use of radioactive isotopes such as ^{14}C and ^{3}H. The use of these isotopes makes it possible to rapidly assay compounds in many areas of the body and so follow the route taken by the radioactive drug. However, metabolism of the drug can result in all the labelled atoms being transferred to a metabolite. Consequently, it is important that the chemical identity of the substance containing the tracer is known if the data are to be accurately interpreted. This is particularly important if the potential drug is acting as a prodrug. The rate of change of plasma concentration with time is normally recorded as a plot of concentration against time (Fig. 11.5). In practice the shape of the graph usually indicates that the process exhibits either zero- or first-order kinetics. This means that the data can be expressed mathematically using the general equations (11.1) and (11.2), where k is the rate constant and C is the concentration of the drug normally associated with these kinetic processes. For first-order processes, a plot of $\log_e C$ against t will be a straight line with a slope of k.

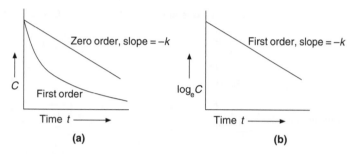

Figure 11.5 (a) Zero- and first-order concentration–time and (b) first-order $\log_e$ concentration–time plots

$$\text{Zero order} \quad \text{Rate} = k \quad \text{and} \quad C = C_0 - kt \qquad (11.1)$$

$$\text{First order} \quad \text{Rate} = k\,C \quad \text{and} \quad C = C_0 e^{-kt} \qquad (11.2)$$

11.3 Pharmacokinetic models

The accurate assessment of the results of a pharmacokinetic investigation requires the use of mathematical methods. In order to apply these methods to the behaviour of a drug in what is a complex biological system, it is necessary to use so-called *model* systems. These models simulate the rate relationships between drug absorption, distribution, response and elimination in the various sections of the biological system. The accuracy of all pharmacokinetic models in describing the drug concentration changes and relating these changes to pharmacological and toxic responses depends on the accurate assay of drug concentrations in the plasma and tissues. Since it is often impossible to obtain the required samples from human subjects, pharmacokinetic models are often developed from data obtained from animals.

Pharmacokinetic models enable the medicinal chemist to use mathematical equations to describe the relationships between the concentrations of a drug in different tissues and, as a result, predict the concentrations of a drug in a tissue for any drug regimen. This information has a wide variety of uses, such as correlating drug doses with pharmacological and toxic responses and determining an optimum dose level for an individual. However, as these models are based on a simplification of what is a very complex system, it is necessary to treat the results of these analyses with some degree of caution until the model has been rigorously tested.

A commonly used pharmacokinetic model is the *compartmental model*. These models visualise the biological system as a series of interconnected compartments (Fig. 11.6), which allows the biological relationships between these compartments to be expressed in the form of mathematical equations. In all compartmental models, a compartment is defined as a group of tissues that have a similar blood flow and drug affinity. It is not

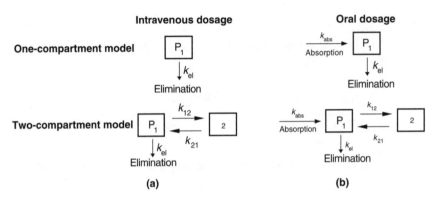

Figure 11.6 Compartmental models in which a drug is administered by (a) intravenous injection and (b) orally. **Key: P_1** is the plasma and highly perfused tissue compartment, which is the central compartment in all compartmental models. It is the first destination of a drug. Other compartments are indicated by 2. k_{el} is the rate constant for the elimination of the drug from P_1, k_{12} is the rate constant for the transfer of the drug from P_1 to compartment 2 and k_{21} is the rate constant for the transfer of the drug from compartment 2 to P_1

necessarily a defined anatomical region in the body, however all compartmental model systems assume that:

• the compartments communicate with each other by reversible processes;

• rate constants are used as a measure of the rate of entry and exit of a drug from a compartment;

• the distribution of a drug within a compartment is rapid and homogeneous;

• each drug molecule has an equal chance of leaving a compartment.

Compartmental models are normally based on a central compartment that represents the plasma and the highly perfused tissues (Fig.11.6). Elimination of a drug is assumed to occur only from this compartment since the processes associated with elimination occur mainly in the plasma and the highly perfused tissues of the liver and kidney. Other compartments are connected to the central compartment as required by the nature of the investigation. The simplest compartmental model is the one-compartmental model in which the compartment represents the circulatory system and all the tissues perfused by the drug (Fig. 11.6). This is the model used in this text.

An alternative model that is also widely used is the *flow* or *perfusion model*. This model uses the compartment concept but the compartments represent the anatomical regions of the body (Fig.11.7). The perfusion model uses blood flow to assess the distribution of the drug to the various organs in the body and the degree of tissue binding as a measure of the uptake of the drug into an organ. This means that the drug concentration in an organ is determined using the blood flow, the size of the organ and the partition of the drug between the blood and the organ. However, these factors may be affected by the physiological state of the subject and this must be taken into account when drawing general conclusions from

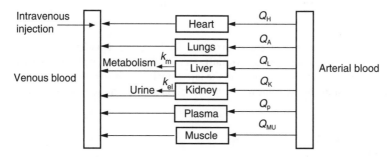

Figure 11.7 A simple drug perfusion model for an intravenous injection of a drug. Each box represents either an organ or a tissue type. Tissues and organs that are impervious to the drug are not included. Tissues with similar degrees of blood perfusion are grouped together. **Key:** k-subscript represents the rate constants for removal of the drug from the appropriate compartment and Q-subscript represents the rate of blood perfusion of the tissue

investigations using perfusion models. An advantage of the perfusion model is that the deductions from animal studies can in some cases be accurately extrapolated to humans. For further details of the perfusion model and its uses the reader is referred to more specialised texts, as this introductory text will only deal with the one-compartment model.

11.4 Intravascular administration

The main methods of intravascular administration are intravenous (IV) injection and infusion. When a single dose of a drug is administered to a patient by intravenous injection, the dose is usually referred to as an *IV bolus*. Intravascular administration places the drug directly in the patient's circulatory system, which bypasses the body's natural barriers to drug absorption. Once it enters the circulatory system the drug is rapidly distributed to most tissues since a dynamic equilibrium is speedily reached between the drug in the blood and the tissue. This means that a fast IV bolus injection will almost immediately give a high initial concentration of the drug in the circulatory system but this will immediately start to fall because of elimination processes (Fig. 11.8a). However, the plasma concentration of a drug administered by intravenous infusion will increase with time until the rate of infusion is equal to the rate of elimination (Fig. 11.8b). At this point the drug plasma concentration remains constant until infusion is stopped, whereupon it falls. Administration by intravenous infusion can be used to accurately maintain the required concentration of a drug in the blood and tissues.

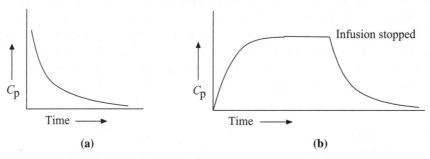

Figure 11.8 The variation of the concentration of a drug in the plasma (C_p) with time when administered by **(a)** a rapid single intravenous injection and **(b)** intravenous infusion. With rapid intravenous injections the graph does not show the time taken to carry out the injection; it is normally taken as being spontaneous. In these cases the curve starts at the point where the first plasma concentration measurements were taken

11.4.1   Distribution

Once a drug is absorbed it is distributed to all the accessible tissues of the body. Consequently, the use of the term *distribution* in pharmacokinetics refers to the transfer of the drug from its site of absorption to its site of action. The main drug distribution route is the circulatory system. The rate and extent of drug distribution will depend on the chemical structure of the drug, the rate of blood flow (Q), the ease of transport of the drug through membranes (see section 7.3), the binding of the drug to the many proteins found in the blood (see next section), the metabolism (see section 1.7.1) and the excretion processes (see section 1.7.1) that occur on its route. Only a small proportion of the administered dose will reach the site of action, and the remainder will either undergo elimination or be absorbed into the tissues it meets en route. The former processes are irreversible but the latter are reversible and so will gradually release the drug back to the general circulatory system and its site of action.

The binding of drugs to plasma proteins

A proportion of the drug molecules that enter the general circulatory system bind to serum proteins by electrostatic interactions, hydrogen bonding, dipole–dipole interactions and other types of van der Waals' forces. The binding is reversible, with the bound and free drug forming a dynamic equilibrium mixture in the blood.

$$\text{Free drug} \rightleftharpoons \text{Bound drug}$$

The degree of binding is defined as:

$$\text{Percentage of protein binding} = \frac{\text{Concentration of bound drug} \times 100}{\text{Total concentration of bound and free drug}} \quad (11.3)$$

The most important protein with regard to binding is albumin, but drugs also bind to the α_1-acid glycoproteins and lipoproteins. Weakly acidic drugs tend to bind to albumin whilst weakly basic drugs prefer to bind to the α_1-acid glycoproteins. A drug may be bound to a plasma protein for a considerable time during which it will not be available to act at its target site.

The reversible binding of drugs to plasma proteins has a significant effect on a number of pharmacokinetic parameters that are dependent on the concentration of the free drug in the plasma. The bound drug has no effect. For example, drugs with a low percentage of protein binding will have a higher plasma concentration of the free drug and so will be more readily available for metabolism and excretion and to exert either a therapeutic or toxic effect than drugs with a high percentage plasma protein binding. However, in the latter case the high percentage of protein binding could result in a longer duration of action as the drug is released from the protein over a longer period of time.

The degree of binding of a drug to plasma proteins may be considerably affected by pathological states, such as renal failure and inflammation, or a change in physiological status due to pregnancy, fasting and malnutrition. In addition, competition from other drugs can also affect binding. This may have clinical implications. For example, a number of drugs can displace warfarin (Fig. 11.4) from its albumin binding sites, which increases the concentration of warfarin in the blood. This increased concentration can lead to an increase in clot formation time (prothrombin) and a subsequent increase in the possibility of haemorrhage.

Intravenous injection (IV bolus)

The administration of a drug by a rapid intravenous injection places the drug in the circulatory system where it is distributed to all accessible body compartments and tissues. The one-compartment model (Fig. 11.6a) of drug distribution assumes that the administration and distribution of the drug in the plasma and associated tissues are instantaneous. This does not happen in practice and is one of the possible sources of error when using this model to analyse experimental pharmacokinetic data.

Once in the circulatory system the concentration of drug begins to decline (Fig. 11.9a). This decline, which is due to elimination processes (see section 1.7.1), is normally reported by plotting a graph of the plasma concentration of the drug against time. These graphs are usually close to being exponential curves (Fig. 11.8a), that is, the elimination of the drug normally exhibits first-order kinetics. This observation enables medicinal chemists to describe these elimination processes to an acceptable degree of accuracy by the mathematical equations used for first-order kinetics, that is:

$$\text{Rate of elimination} = k_{el}C_p \qquad (11.4)$$

and

$$C_p = C_o e^{-k_{el}t} \qquad (11.5)$$

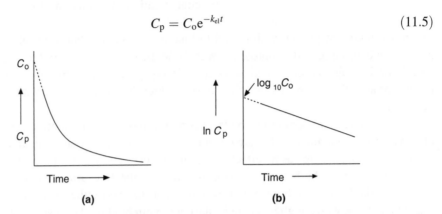

Figure 11.9 Extrapolation of plots of **(a)** the concentration (C_p) and **(b)** $\ln C_p$ (log to base e) against time for the changes in plasma concentration of a drug with time

where:
  C_o is the plasma concentration of the drug in the body at a time $t = 0$, that is, the concentration immediately following bolus injection;
  C_p is the plasma concentration of the drug in the body at a time t;
  t is the lapsed time between the administration and the measurement; and
  k_{el} is the rate constant for the irreversible elimination of the drug.

The plasma concentration C_p of a drug in the system is related to the total amount of the drug (D) in the system by the relationship:

$$C_p = D/V_d \qquad (11.6)$$

where V_d is the apparent volume of distribution. The value of V_d is a measure of the total volume of the body perfused by the drug. It is an apparent volume because it is the volume of plasma that is equivalent to the total volume of the body readily perfused by the drug. It is not the volume of the tissue and the circulatory system actually perfused by the drug. This means that the values of V_d for a drug can be considerably higher than the volume of the blood in the circulatory system, which is usually about 5 litres for a 70 kg person. Values of V_d are usually recorded in terms of litres per kilogram (Table 11.1), which gives a value of 0.071 1 kg^{-1} for a 70 kg person. A value of less than 0.071 for V_d indicates that the drug is probably distributed mainly within the circulatory system, whilst values greater than 0.071 indicate that the drug is distributed in both the circulatory system and specific tissues.

Table 11.1 The values of the pharmacokinetic parameters of some common drugs recorded in the literature (various sources)

Drug	V_d (1 kg^{-1})	$t_{1/2}$ (hours)	% Plasma binding
Aspirin	0.1–0.2	0.28	90% below 100 µg and 50% above 4 µg
Ampicillin	0.3	1–2	About 20%
Cimetidine	2.1	1–3	13–26 %
Chlorpromazine	20	Mean 15–30	95–98%
Diazepam	1.0	Mean 48	98–99 %
Ibuprofen	0.14	2	99 %
Propranolol	3.9	2–6	90 %
Morphine	3.0–5.0	2–3	20–35 %
Warfarin	0.15	Mean 42	97–99 %

The value of V_d may be calculated by substituting the values for the dose D_o administered and the plasma concentration at time $t = 0$ in equation (11.6). However, the time taken to administer a drug and for that drug to achieve a homogeneous distribution throughout the system means that it is not possible to measure the total amount of drug present in the plasma at time $t = 0$. Consequently, a theoretical value for C_p at time $t = 0$, obtained by extrapolating a plot of $\ln C$ (log to base e) against t to $t = 0$ (Fig. 11.9b), is normally used to calculate V_d.

Elimination and elimination half-life

Elimination is the term used to represent the *irreversible* processes that remove a drug from the body during its journey to its site of action and after its action. The processes involved are metabolism and excretion. The main centre for metabolism is the liver. Excretion occurs mainly through the kidney and liver but some also occurs *via* the lungs and sweat. Excretion through the kidney occurs by glomerular filtration and tubular secretion. Tubular secretion occurs by active transport (see section 7.3.5) and so will remove protein-bound drugs as well as free drugs. However, in tubular secretion the drug has to dissociate from the protein before it is transported and ultimately excreted in the urine. However, tubular reabsorption often returns a considerable proportion of a drug and other substances to the systemic circulation. The degree of reabsorption will depend on the drug's ability to cross membranes. Consequently, drugs that are polar or highly ionised at the pH of the urine are less likely to be reabsorbed. For example, the polar β-lactams are not readily reabsorbed. Furthermore, acidic urine will enhance the ionisation of basic drugs and so reduce their reabsorption. Similarly, basic urine will have the same effect on acidic drugs. The reabsorption of drugs can be enhanced by changing the pH of a patient's urine using oral doses of sodium hydrogen carbonate or potassium citrate to make the urine more basic and ammonium chloride to make the urine more acidic. Biliary excretion involving the liver is not fully understood. The process is specific, only occurring for compounds with relative molecular masses greater than about 500.

The rate of elimination is an important characteristic of a drug. Too rapid an elimination necessitates frequent repeated administration of the drug if its concentration is to reach its therapeutic window. Conversely, too slow an elimination could result in accumulation of the drug in the patient, which might give an increased risk of toxic effects. Most drug eliminations follow first-order kinetics (equation 11.2) but there are some notable exceptions, such as ethanol that exhibits zero-order kinetics (equation 11.1). This allows the calculation of blood alcohol levels at any time after drinking the alcohol even though the blood sample was taken some time after the alcohol was drunk. The fact that most drug elimination follows first-order kinetics means that it is not usually possible to determine the time at which a dose of a drug would be completely cleared from the system as the plasma concentration–time curves are exponential (Fig. 11.8a). Consequently, both biological half-life and the eliminationrate constants (k_{el}) are used as indicators of the rate of elimination of a drug from the system. Half-lives are normally quoted in the literature and k_{el} values are calculated as required.

Half-life ($t_{1/2}$):   This is the time taken for the concentration of a drug to fall (be eliminated) to half of its original value. Consequently, for $t_{1/2}$:

$$0.5 = \frac{C_{p2}}{C_{p1}} \tag{11.7}$$

where C_{p1} is the initial plasma concentration and C_{p2} is the plasma concentration after a lapsed time equal to the half-life $t_{1/2}$. Therefore, for *first-order* elimination processes substituting equation (11.7) in equation (11.2):

$$C_{p2} = C_{p1}e^{-k_{el}t_{1/2}}$$

where k_{el} is the elimination rate constant,
and

$$\frac{C_{p2}}{C_{p1}} = 0.5 = e^{-k_{el}t_{1/2}}$$

Taking log to base e:

$$\ln 0.5 = -k_{el}t_{1/2}$$

that is:

$$-0.693 = -k_{el}t_{1/2}$$

therefore

$$t_{1/2} = \frac{0.693}{k_{el}} \tag{11.8}$$

Consequently, for first-order reactions the half-life is a constant. However, for other orders of reaction it is not constant but dependent on the concentration (C_t) of the drug at time $t = t$. For example, for a zero-order reaction:

$$t_{1/2} = C_t/2k \tag{11.9}$$

Elimination rate constant: For *first-order* elimination processes the value of the elimination rate constant may be calculated by interpreting the experimental data using the logarithmic forms of equation (11.2).

for log to base e : $\quad \ln C_p = \ln C_o - k_{el}t \tag{11.10}$

for log to base 10 : $\quad \log_{10} C_p = \log_{10} C_o - \dfrac{k_{el}t}{2.303} \tag{11.11}$

Both of these equations give straight line plots (Fig. 11.10) and so it is possible to obtain a value of k_{el} by measuring the slope of the graph provided that the experimental data give a reasonable straight line. Half-life values may also be calculated from these graphs. It is advisable to take an average of several measurements of $t_{1/2}$ made from different initial values of C_p in order to obtain an accurate value for the half-life. Alternatively $t_{1/2}$ may be calculated by substituting the value of k_{el} in equation (11.8).

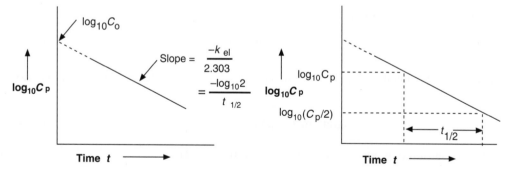

Figure 11.10 Determination of the values of $t_{1/2}$ and k_{el} from logarithmic plots of plasma concentration against time. The logarithm plot for log to base 10 is shown but a natural logarithmic plot would be similar except the slope would be equal to k_{el}

The measurement of the half-life of a potential drug may give the medicinal chemist a reference point when developing drugs from a lead compound. It might enable him to compare the pharmacological effect of a lead with its analogues on a numerical basis and, as a result, provide an indication of the best course of action to take for the successful development of a useful drug. For example, if a lead has a short duration of action, analogues with larger $t_{1/2}$ and smaller k_{el} values than those of the lead are more likely to give the required pharmacological effect. Similarly, if the lead is too toxic, analogues with smaller $t_{1/2}$ and larger k_{el} values need to be developed. It is emphasised that $t_{1/2}$ and k_{el} data are not infallible and should not be considered in isolation. The more and wider the range of information one has, the more likely it is that a successful course of action will be pursued.

Clearance and its significance

Clearance (Cl) is the *volume* of blood in a defined region of the body that is cleared of a drug in *unit time*. For example, total clearance (Cl_T) is the volume of blood in the whole body cleared of the drug in unit time whilst hepatic clearance (Cl_H) is the volume of blood passing through the liver that is cleared in unit time. Clearance is an artificial concept in that it is not possible for a drug to be removed from only one part of the total volume of the blood in the body or organ. However, since the clearance of a drug from a specific region of the body is measured as the volume cleared per unit time and the plasma concentration of a drug is the mass of drug per unit volume, the mathematical product C_pCl will be the mass of the drug cleared from the specified region in unit time, that is, the rate of elimination of the drug from that region of the body. In other words, clearance is the parameter that relates the rate of elimination of a drug from a defined region of the body to the plasma concentration of that drug. For example, the rate of elimination of a drug from the whole of the body is given by the equation:

$$\text{Rate of elimination of a drug from the whole body} = Cl_T Cp \qquad (11.12)$$

The rate of elimination of a drug is usually first order, therefore making this assumption and substituting in equation (11.2):

$$\text{Rate of elimination} = k_{el}D \qquad (11.13)$$

where D is the amount of drug in the body at time t. Substituting for D from equation (11.6) gives:

$$\text{Rate of elimination} = k_{el}V_dC_p \qquad (11.14)$$

therefore, substituting equation (11.14) in equation (11.12):

$$Cl_T\ C_p = k_{el}V_dC_p$$

and simplifying

$$Cl_T = k_{el}V_d \qquad (11.15)$$

hence, substituting equation (11.8) in equation (11.15):

$$Cl_T = V_d \frac{0.693}{t_{1/2}} \tag{11.16}$$

For first-order elimination clearance is a constant since both $t_{1/2}$ and k_{el} are constant. However, should the order of the elimination change due to a change in the biological situation, such as the drug concentration increasing to the point where it saturates the metabolic elimination pathways, then clearance may not be constant.

The clearance (Cl) of a drug from a region of the body is the sum of all the clearances of all the contributing processes in that region. For example, hepatic clearance (Cl_H) is the sum of the clearances due to metabolism (Cl_M) and excretion (Cl_{Bile}) in the liver, that is:

$$Cl_H = Cl_M + Cl_{Bile} \tag{11.17}$$

For an IV bolus that places the drug directly in the circulatory system, total clearance (Cl_T) of the drug from the body can also be determined from blood plasma measurements. The area under the curve (AUC) of a C_p against t plot represents the total amount of the drug that reaches the circulatory system in the time t. It can be used to calculate total clearance since it is related to the dose administered by the relationship:

$$\text{Dose} = Cl_T \cdot \text{AUC} \tag{11.18}$$

which may be derived from equation (11.12) as follows:

$$\text{Rate of elimination} = Cl_T \cdot C_p \tag{11.12}$$

that is:

$$dD/dt = Cl_T \cdot C_p$$

and so

$$dD = ClT \cdot C_p dt \tag{11.19}$$

Integrating equation (11.19) between the limits D and 0 and t and 0:

$$D = Cl_T \cdot C_p t \tag{11.20}$$

For intravenous doses the amount of drug in the body at $t = 0$ is the dose administered. Therefore, as $C_p t$ is the area under the plasma concentration–time curve for the drug:

$$\text{Dose} = Cl_T \cdot \text{AUC} \tag{11.18}$$

This relationship can also be used to calculate the clearance that occurs in a specific time by simply measuring the AUC for that time. Furthermore, analysis of the total amount of a drug in the urine can be used to estimate renal clearance (Cl_R) because the total amount of unchanged drug found in the urine (U) will be related to its plasma concentration by a similar mathematical expression to equation (11.18)

$$U = Cl_R \cdot AUC \tag{11.21}$$

The relationship (11.18) holds true regardless of the way in which a single dose of the drug is administered. However for enteral routes the dose is the amount absorbed (see section 11.5.1), not the dose administered.

Clearance will vary with body weight and so for comparison purposes values are normally quoted per kilogram of body weight (Table 11.2). It also varies with the degree of protein binding. A large proportion of a drug with a high degree of protein binding will not be available to take part in the metabolic and excretion processes. In other words it will not be so readily available for elimination as a drug with a low degree of protein binding.

Table 11.2 Clearance values of some drugs

Drug	Clearance (cm^3 min^{-1} kg^{-1})	Drug	Clearance (cm^3 min^{-1} kg^{-1})
Atropine	8	Bumetamide	3
Bupivacaine	8	Caffeine	1–2
Disopyramide	0.5–2 (dose dependent)	Ethambutol	9
Mepivacaine	5	Pentobarbitone	0.3–0.5
Ranitidine	About 10	Vancomycin	About 1

Clearance is a more useful concept in pharmacokinetics than either $t_{1/2}$ or k_{el}. It enables blood flow rate (Q), which controls the rate at which a drug is delivered to a specific region of the body, to be taken into account when assessing the pharmacokinetic behaviour of the drug in that region of the body (see section 11.5). Clearance values enable the medicinal chemist to compare the effect of structural changes on drug behaviour and as a result decide which analogues might yield drugs with the desired pharmacokinetic properties.

All drugs administered by IV bolus injection are carried to their site of action by the blood. In order to reach its destination, a drug may have to pass through several organs where some of the drug may be lost by elimination. This loss is known as *extraction*. The effect of extraction on the distribution of the drug may be rationalised by the use of the concept of clearance. Consider, for example, a closed system in which a pump is pumping a fluid from a reservoir through an organ before being returned to the reservoir (Fig. 11.11). This system may be regarded as being analogous to the heart pumping the blood through an organ. If the organ eliminates some of the drug from the blood by excretion and metabolism, the proportion of the drug removed by a single transit of the total dose of the drug through the organ is defined as:

$$E = \frac{C_{in} - C_{out}}{C_{in}} \tag{11.22}$$

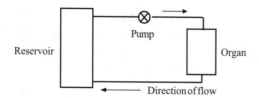

Figure 11.11 A simple extraction system

where E is known as the extraction ratio and C_{in} and C_{out} are the concentrations of the drug at the entry and exit points from the organ. Since the liver is a major site of metabolism and excretion the hepatic extraction (E_H) values of many drugs have been determined (Table 11.3).

Table 11.3 The hepatic extraction values of some drugs

Drug E_H value < 0.3 (low)	Drug E_H value 0.3–0.7	Drug E_H value > 0.7 (high)
Antripyrine	Aspirin	Cocaine
Diazepam	Codeine	Lignocaine
Nitrazepam	Nifedipine	Nicotine
Warfarin	Nortriptyline	Propranolol

The extraction ratio has no units. Its values range from 0 to 1. A value of 0.4 means that 40 per cent of the drug is irreversibly removed as it passes through the organ. Consequently, as a drug is delivered to an organ via the circulatory system the clearance of the organ is related to the blood flow rate by the relationship:

$$Cl = Q \cdot E \qquad (11.23)$$

where Q (volume per unit time) is the rate of blood flow.

Hepatic extraction ratios are important in the design of dosage forms because the liver plays a major part in the extraction of drugs from the circulatory system. For example, propranolol has an E value for hepatic extraction of about 0.7. If this drug is administered by IV bolus injection most of the drug would be distributed throughout the general circulatory system before reaching the liver. However, if the drug is administered orally 70 per cent of the dose would never reach the general circulatory system. This is because orally administered drugs that are absorbed from the GI tract pass through the liver before they enter the general circulatory system (Fig. 11.12). Consequently, the hepatic extraction ratio of a drug is useful in determining its dose level and how it is to be administered. For example, if a drug has a high hepatic extraction ratio a much higher dose of the drug must be used if it is to be given orally than if it were given by intravenous injection. This would increase the risk of toxic side effects.

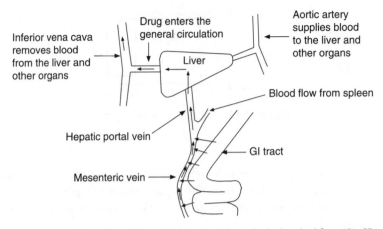

Figure 11.12 A schematic outline of the route of drugs (←) absorbed from the GI tract

It is difficult to obtain drug concentration data by sampling the human hepatic portal vein and so investigations are normally carried out on animals. Unfortunately, animals can behave in quite a different fashion to humans and so the conclusions drawn from this source can only be used as a rough guide to drug behaviour. However, extraction ratios are another piece of measurable information that can be used to link the required desirable characteristics of a potential new drug with the chemical structures of the analogues of a lead. Furthermore, it can lead to the avoidance of the loss of a potential drug by indicating the most effective dosage form.

Intravenous infusion

In intravenous infusion, the drug is infused into the vein at a steady rate. Initially the plasma concentration of the drug increases as the amount of the infused drug exceeds the amount of the drug being eliminated (Fig 11.13). However, as the concentration of the drug in the plasma increases, the rate of elimination also increases until the rate of infusion equals the rate of elimination, at which point the concentration of the drug in the plasma remains constant. As long as the infusion rate is kept constant, the drug plasma concentration will remain at this steady state level (C_{ss}). When infusion is stopped the drug plasma concentration will fall,

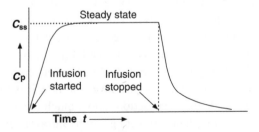

Figure 11.13 Plasma concentration changes with time in intravenous infusion

usually in an exponential curve because the biological situation is now the same as if a dose of the drug had been given at that time by IV bolus injection (see section 11.4).

The concentration of a drug is normally maintained at a constant value during intravenous infusion. This means that the infusion process is zero order (see equation 11.1). Since elimination can be assumed to be first order (see section 11.4.1), the rate of change of the plasma concentration of a drug can be described by the mathematical relationship:

Rate of change of plasma concentration = Rate of infusion − Rate of elimination

that is:

$$\frac{\mathrm{d}C_p}{\mathrm{d}t} = k_o - k_{el}C_p \qquad (11.24)$$

where k_o is the rate constant for the infusion. The value of the rate of elimination $(k_{el}C_p)$ increases as the plasma concentration increases until it equals the rate of infusion and there is no change in the drug plasma concentration, that is: $\mathrm{d}C_p/\mathrm{d}t = 0$. At this point the plasma concentration has reached the steady state and so:

Rate of elimination = Rate of infusion

that is:

$$k_{el}C_p = k_o \qquad (11.25)$$

but at the steady state

$$C_{ss} = C_p \qquad (11.26)$$

and so:

$$C_{ss} = \frac{k_o}{k_{el}} \qquad (11.27)$$

Since k_{el} is related to the total clearance (see section 11.4.1) by:

$$Cl_T = k_{el}V_d \qquad (11.28)$$

Substituting for k_{el} in equation (11.27) gives:

$$C_{ss} = \frac{k_o V_d}{Cl_T} = \frac{k_o^*}{Cl_T} \qquad (11.29)$$

where k_o^* is the amount of drug infused per unit time. Equations (11.27) and (11.29) can be used to calculate the rate of infusion required to achieve a specific steady state plasma concentration. Furthermore, since these equations are independent of time an increase in the rate of infusion and a subsequent increase in the value of k_o will not result in a reduction of the time taken to reach a specific value of C_{ss}. It will simply increase the value of C_{ss} as the rate of elimination and hence k_{el} will remain constant. Consequently, too high a rate of infusion could increase the steady state plasma concentration of the drug to a value above the top limit of its therapeutic window, which in turn would increase the chances of a toxic response from the patient.

A time-dependent relationship can be determined for the initial part of the infusion before the plasma concentration reaches the steady state. Integration of equation (11.24) gives the relationship:

$$t = \frac{-1}{k_{el}}\ln(1 - C_p/C_{ss}) \tag{11.30}$$

where t is the time taken to reach the drug plasma concentration C_p. The initial stage of the intravenous infusion normally follows first-order kinetics. Consequently, substituting equation (11.8) in equation (11.30) leads to the relationship:

$$t = \frac{-t_{1/2}}{0.693}\ln(1 - C_p/C_{ss}) \tag{11.31}$$

that is:

$$t = -t_{1/2}\ 1.44\ln(1 - C_p/C_{ss}) \tag{11.32}$$

Equation (11.32) allows the calculation of the time taken to reach the effective therapeutic plasma concentration, which is normally taken as being 90 per cent of the C_{ss} value. This time is dependent on the half-life value: the shorter the half-life, the sooner the C_{ss} plateau is reached. For example, when infused at a suitable rate, the antibiotic penicillin G has a half-life of about 30 minutes and reaches its effective therapeutic value in 100 minutes, whilst procainamide with a half-life of 2.8 hours requires 9 hours to reach its effective therapeutic value. Calculations, based on equation (11.32), can be used to devise the best dosage form for potential and existing drugs.

To reduce the time required to obtain an effective therapeutic plasma concentration, a single IV bolus injection may be given in conjunction with an intravenous infusion. As a result, the C_{ss} drug plasma concentration is reached almost immediately. However, once the drug enters the blood stream it undergoes elimination no matter what its original source, IV bolus or infusion. This elimination is compensated for by a build-up in the drug concentration from the intravenous infusion (Fig. 11.14). At any time t the total concentration of the drug in the system is the sum of the drug from the IV bolus and the infusion. Its value is equal to the steady state concentration C_{ss} of the drug. The net result is that the patient almost immediately receives an almost constant effective therapeutic dose of the drug. The pharmacokinetics of biological situations, such as the inhalation of a

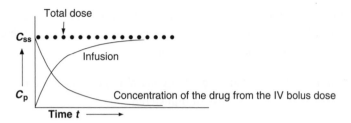

Figure 11.14 The effect of a single IV bolus on the plasma concentration of a drug administered by intravenous infusion

general anaesthetic gas and the slow release of a drug from an implant, in which a drug is introduced into the blood at a constant rate, may also be described using a similar approach to that described for intravenous diffusion.

11.5 Extravascular administration

The most common form of enteral dosage form is oral administration. Consequently, this section will mainly discuss the pharmacokinetics of orally administered drugs and will largely ignore other enteral routes. Most orally administered drugs are absorbed from the GI tract. This absorption may be considered to take place in two stages, namely, dissolution of the dosage form (see section 11.5.1) and transfer of the drug through the GI tract lining into the circulatory system (see section 11.5.2). Dissolution is the rate at which the dosage form passes into the aqueous fluid of the GI tract and diffuses to the lining of the GI tract, while absorption is the actual transfer of the drug through the tissues surrounding the GI tract into the circulatory system. The rate of dissolution is not usually a significant contributor to the rate of absorption (see section, 11.5.2) *unless* the release of the drug from the dosage form is slow or the drug is either a liquid or solid with a poor water solubility (Kaplan, less than 10 mg cm^{-3} at pH 7).

Drugs absorbed from the GI tract must pass through the liver in order to reach the general circulation system (Fig. 11.12). As a result, a fraction of the drug is lost by metabolism and excretion as it passes through the GI tract membrane, liver and other organs before it reaches the systemic circulation. These losses are referred to as the *first-pass effect* or *first-pass metabolism*. First-pass metabolism is effectively the elimination of the drug before it enters the general circulatory system. In this circuit the main areas of excretion and metabolism are the enzyme-rich liver and lungs and so the term *first-pass metabolism* is usually taken to refer to the elimination of a drug by these two organs. However, since the liver is the first organ the drug passes through after absorption from the GI tract and it is also the principal area of metabolism, the effect of the lungs is often ignored and the term first-pass metabolism is frequently used as though it involves only the liver.

The physiology of drug absorption from the GI tract has a direct effect on the *bioavailability (F)* of a drug. Bioavailability is defined as the fraction of the dose of a drug that enters the general circulatory system, that is:

$$F = \frac{\text{Amount of drug that enters the general circulatory system}}{\text{Dose administered}} \qquad (11.33)$$

Since the area under the plasma concentration–time curve (AUC) for a drug is a measure of the total amount of a drug reaching the general circulatory system in a time t, the bioavailability of a drug may also be defined in terms of the AUC as:

$$F = \text{AUC}/\text{Dose} \qquad (11.34)$$

If all the dose of a drug reached the circulatory system, the bioavailability as defined by equation (11.33) would have a value of unity. Therefore if E is the extraction for the first pass:

$$F = 1 - E \qquad (11.35)$$

For orally administered drugs, equation (11.35) approximates to:

$$F = 1 - E_H \qquad (11.36)$$

where E_H is the hepatic extraction ratio. Therefore, drugs with high hepatic extraction values $(E_H \sim 1)$ will seldom reach the general circulatory system in sufficient quantity to be therapeutically effective if administered orally.

Furthermore, for these drugs, since $E \sim 1$, equation (11.23) becomes:

$$Q_H = Cl_H \qquad (11.37)$$

that is, the hepatic clearance (Cl_H) of substances that undergo high first-pass clearence approaches the rate of blood flow to the liver (Q_H). Consequently, the determination of the hepatic clearance of a potential drug after intravascular administration using animal experiments is used to determine whether a potential drug will have a high first-pass metabolism effect. Potential drugs that undergo a high first-pass metabolism will need further structural modification or the use of a different delivery system.

Bioavailability studies are used to compare the efficiency of the delivery of the dosage forms of a drug to the general circulatory system as well as the efficiency of the route of administration for both licensed drugs and new drugs under development. Two useful measurements are *relative* and *absolute bioavailability*.

Relative availability   Relative bioavailability may be used to compare the relative absorptions of the different dosage forms of the same drug and also the relative availabilities of two different drugs with the same action when delivered using the same type of dosage form. It is defined for equal doses as:

$$\text{Relative bioavailability} = \frac{\text{AUC for drug A (or dosage form A)}}{\text{AUC for drug B (or dosage form B)}} \tag{11.38}$$

Percentage relative bioavailability figures may be obtained by multiplying equation (11.38) by 100. A correction must be made if different drug doses are used. In which case, equation (11.39) becomes:

$$\text{Relative bioavailability} = \frac{(\text{AUC for drug A or dosage form A})/\text{Dose A}}{(\text{AUC for drug B or dosage form B})/\text{Dose B}} \tag{11.39}$$

Example 5.1. *The variation of the plasma concentration with time for a single dose of a drug administered orally using either one 100 mg tablet or one 5 cm³ dose of a linctus containing 100 mg of the drug was determined for a number of healthy volunteers. The results of the study of the variation of plasma concentration with time showed that the tablet curve had an AUC of 43.7 μg h cm⁻³ while the linctus curve had an AUC of 42.5 μg h cm⁻³. Since the doses of the drug are the same, substituting these experimental figures in equation (11.38) gives:*

$$\textit{Relative bioavailability} = \frac{\textit{AUC tablet}}{\textit{AUC linctus}} = \frac{43.7}{42.5} = 1.028$$

This means that there was almost no difference between the two dosage forms as far as absorption (bioavailability) is concerned but drug absorption was slightly better from the tablet than the linctus. A value significantly higher than 1 would have suggested that the bioavailability of the drug from the tablet is much better than from the linctus whilst a value significantly less than 1 would have indicated the reverse was true.

This type of calculation is useful in drug design as it ensures that the dosage forms used in trials are effective in delivering the drug to the general circulation. It is also used by licensing authorities as a check on the efficacy of products when manufacturers change the dosage form of a drug in clinical use.

Absolute bioavailability   Absolute bioavailability is used as a measure of the efficiency of the absorption of the drug. It is defined in terms of the total dose of the drug the body would receive if the drug was placed directly in the general circulation by an IV bolus injection, that is:

$$\text{Absolute bioavailability(F)} = \frac{\text{AUC for oral dosage form/oral dose}}{\text{AUC for IV dosage form/IV dose}} \tag{11.40}$$

Example 5.1. *The change in plasma concentration with time for a series of the analogues of a lead compound was followed using a number of healthy volunteers and the data obtained were recorded in Table 11.4. Predict from these data which analogues would be the most profitable to develop.*
For analogue A, substituting in equation (11.39):

$$\text{Absolute bioavailability } (F) \text{ of } A = \frac{67.3/100}{47.8/50} = 0.71$$

Similarly the absolute bioavailabilities of analogues B and C are 0.53 and 0.97, respectively. This means that analogue C would have the best absorption and on this basis would give the best chance of a successful outcome. However, it should be realised that the final decision as to which analogue to develop would not be based solely on its bioavailability. It would be based on a consideration of all the pharmacokinetic and pharmacodynamic data obtained for the three analogues.

Table 11.4 Experimental results of plasma concentration–time studies for three analogues

Analogue	Oral dose (mg)	AUC (μg h cm^{-3})	IV dose (mg)	AUC (μg h cm^{-3})
A	100	67.3	50	47.8
B	100	49.2	50	46.5
C	100	91.2	50	46.7

11.5.1 Dissolution

Dissolution is the rate at which the dosage form passes into the aqueous fluid of the GI tract and diffuses to the lining of the GI. It depends on the chemical structure and the crystal structures of the drug, the pH of the medium containing the drug (see section 2.11), the lipid aqueous medium partition coefficient of the drug (see sections 2.12 and 3.7.2) and the surface area of the absorbing region of the GI tract and the nature of the dosage form.

The process of dissolution may be considered to take place in two steps, namely, the formation of a layer of a saturated solution of the drug around each of the separate particles of the dosage form (Fig. 11.15) and the diffusion of the drug from that saturated layer to the surface of the lining of the GI tract. Its rate is given by the Noyes–Whitney relationship:

$$\text{Rate of dissolution } (dm/dt) = kA(C_s - C) \tag{11.41}$$

where m is the mass of solid that dissolves in the GI tract aqueous medium in a time t, A is the surface area of the dissolving particles, C_s is the solubility of the drug in the saturated solution layer surrounding the particle and C is the concentration of the drug in the diffusing medium. It follows from equation (11.41) that an increase in the surface area of the dosage form of a drug with a poor water solubility should increase the rate of

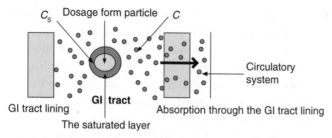

Figure 11.15 A representation of the process of dissolution

dissolution of that drug. Consequently, the reduction in size of the particles of the dose of such a drug will increase its rate of dissolution with a subsequent increase in its bioavailability and a reduction in the dose level required to achieve a specified level of activity. For example, the antifungal drug griseofulvin has a poor water solubility and thus a slow rate of dissolution in the GI tract. Micronisation of this drug results in a more rapid dissolution, which subsequently increases its plasma levels to the extent that a lower dose of the drug may be given. However, for drugs that have a good water solubility, reduction in particle size does not significantly improve bioavailability. It may reduce it because of an increased chance of chemical and/or enzymic degradation in the GI tract.

A considerable number of drugs exist in a number of different crystalline forms (*polymorphic* or *allotropic forms*) and/or a non-crystalline form (*amorphous form*). However, only one polymorph is stable at a given temperature and pressure, but others may be metastable under the same conditions of temperature and pressure. The different polymorphs (allotropic forms) of a drug will dissolve at different rates and so will have different dissolution rates. Metastable polymorphs will usually have a higher solubility than the more stable forms of the drug. Therefore, it is possible to use the metastable form of the drug in a drug delivery system provided that it reverts to the stable form at a reasonable rate. The amorphous form of a drug is normally more soluble than the crystalline form. Consequently, it is possible to use the amorphous form when the crystalline form has a slow dissolution rate. For example, the amorphous forms of the antibiotics novobiocin and chloramphenicol dissolve faster in water than their crystalline forms.

11.5.2 Absorption

Absorption is the process by which the drug passes through the tissues surrounding the GI tract into the circulatory system (see section 1.7.1). The small surface area and the short time a drug normally takes to pass through the stomach means that drug absorption is often less in this region of the GI tract than in the small intestine, which has a far larger surface area. However, absorption in the stomach is better if the drug is taken after a meal, as the presence of food in the stomach slows down the passage of the drug through the stomach.

The process of drug absorption from the GI tract occurs by either *transcellular* or *paracellular absorption*. In trascellular absorption the drug either passes through the relevant cell membranes and the body of the cell or around the interior of the cell membrane to reach the other side of the cell. Paracellular absorption occurs by the drug diffusing through pores between the cells. Transcellular absorption is believed to be the faster since as cell membranes have a larger surface area than the pores used in paracellular absorption. For example, the surface area of the pores in the small intestine is about 0.01 per cent of the whole surface. Transcellular absorption can occur along the whole length of the GI tract. In contrast paracellular absorption is restricted by pore size and molecular mass and so it occurs largely in the small intestine, which has the largest pores (ca 0.06–0.08 nm). Since drugs only spend about six hours in this area of the GI tract it means that compounds that are absorbed by the paracellular route are usually incompletely absorbed.

A guide to the extent and route of absorption of small simple molecules may be obtained from their octanol $\log D_{7.4}$ values (see section 3.4.1). Log $D_{7.4}$ negative values are believed to indicate that the compound is likely to follow a paracellular route while positive values are thought to indicate a transcellular absorption route. The situation is more complicated with larger more complex molecules, as more physicochemical properties need to be considered (see sections 1.4 and 1.7.1).

11.5.3   Single oral dose

When a single dose of a drug is administered orally its plasma concentration increases to a maximum value (C_{max}) at time (t_{max}) before falling with time (Fig. 11.16). The increase in plasma concentration occurs as the drug is absorbed. It is accompanied by elimination, which starts from the instant the drug is absorbed. The rate of elimination increases as the concentration of the drug in the plasma increases to the maximum absorbed dose. At this point, the rate of absorption equals the rate of elimination. Once absorption ceases, elimination becomes the dominant pharmacokinetic factor and plasma concentration falls.

The change in plasma concentration–time curve for a single oral dose is useful in a number of ways. It shows the time taken for the drug to reach its therapeutic window concentration (Fig. 11.16) and the period of time the plasma concentration lies within the therapeutic window. The latter is an important consideration when selecting the time interval (t_{di}) for administering repeat doses in order to maintain the plasma concentration within the therapeutic window. It also shows whether the drug reaches toxic values above the upper limit of the therapeutic window. Moreover, since the time (t_{max}) taken to reach the maximum plasma concentration (C_{max}) is an approximate measure of the rate of absorption of the drug, the value of t_{max} can be used to compare the absorption rates of that drug from different dosage forms. However, in view of the difficulty of taking serum samples at exactly the right time both t_{max} and C_{max} are normally determined by calculation (see section 11.5.4). Both of these factors are important in the selection of drugs for further development and the design of their dosage forms.

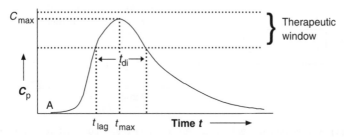

Figure 11.16 The change in plasma concentration of a drug with time due to a single oral dose of the drug. A small time lapse A occurs before the drug reaches the blood. This is mainly the time taken for the drug to reach its site of absorption from the mouth

The absorption of a drug into the general circulatory system is a complex process. Drugs given orally dissolve in the GI tract fluids before being absorbed through the GI tract membrane. The rate of absorption depends on both the drug's chemical nature and the physical conditions at the site of absorption. However, most drugs exhibit approximately first-order absorption kinetics except when there is a high local concentration that saturates the absorptive mechanism, in which case zero-order characteristics are often found. Elimination also normally follows first-order kinetics but can change if the elimination processes are saturated by a high drug concentration due to the use of a high dose.

The rate of change of the amount (A) of an orally administered drug in the body with time t will depend on the relative rates of absorption and elimination, that is:

$$\frac{\mathrm{d}A}{\mathrm{d}t} = \text{Rate of absorption} - \text{Rate of elimination} \tag{11.42}$$

This equation may be used to calculate the changes in a drug's plasma concentration with time. The nature of the calculation, which is beyond the scope of this text, will depend on the order of the absorption and elimination processes and the type of compartmental model used. For example, for a one-compartment model in which the drug exhibits first-order absorption and elimination (Fig. 11.17), using equation (11.42) it is possible to show that:

$$C_p = \frac{FD_0}{V_d} \cdot \frac{k_{pab}}{(k_{ab} - k_{el})} \left(e^{-k_{el}t} - e^{-k_{ab}t}\right) \tag{11.43}$$

where k_{ab} and k_{el} are the absorption and elimination rate constants, respectively, and D_0 is the dose administered.

The elimination rate constant, like other pharmacokinetic parameters, is in theory independent of the method of administration. Therefore, provided that the dose is below the saturation concentration of the elimination processes, the value of k_{el} may be determined by an independent experiment using a single IV bolus of the drug (see section 11.4.1). Substituting this value of k_{el} together with the values of F (see equation 11.33), V_d (see section 11.6) D_0, C_p

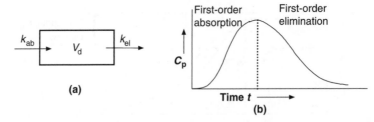

Figure 11.17 **(a)** A one-compartment model for a single orally administered dose. **(b)** The plasma concentration-time curve for a drug that exhibits first-order kinetics for both its absorption and elimination

and t in equation (11.43) will give a value for k_{ab}. It is necessary to make several calculations using different values of C_p in order to obtain an accurate figure. The absorption rate constant k_{ab} may be used to compare the relative rates of absorption of drugs with the same action as well as the relative rates of absorption of the same drug from different dosage forms.

The elimination rate constant may also be calculated from the elimination section of the drug plasma concentration–time curve for the oral dose of the drug (Fig.11.17b). In this section of the curve the rate of absorption is zero. Consequently the absorption term in equation (11.42) is zero and, as a result, equation (11.43) simplifies to:

$$C_p = \frac{FD_0}{V_d} \cdot \frac{k_{ab}}{(k_{ab} - k_{el})} (e^{-k_{el}t}) \qquad (11.44)$$

taking log to base e:

$$\ln C_p = \ln\left(\frac{FD_0}{V_d} \cdot \frac{k_{ab}}{(k_{ab} - k_{el})}\right) - k_{el}t \qquad (11.45)$$

Expressing equation (11.45) in terms of log to base 10:

$$\log C_p = \log\left(\frac{FD_0}{V_d} \cdot \frac{k_{ab}}{(k_{ab} - k_{el})}\right) - \frac{k_{el}t}{2.303} \qquad (11.46)$$

Hence a plot of $\log C$ against t will be a straight line (Fig.11.18) with a slope of $-k_{el}/2.303$ and an intercept on the y axis equal to:

$$\log\left(\frac{FD_0}{V_d} \cdot \frac{k_{ab}}{(k_{ab} - k_{el})}\right) \qquad (11.47)$$

The rate constant k_{el} for the elimination is calculated from the slope of the graph. Its value is used to calculate the rate constant k_{ab} for the absorption of the drug by substituting it in either equation (11.44) or the expression for the intercept, equation (11.47).

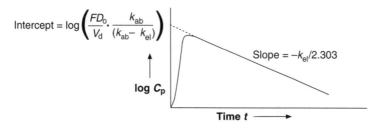

Figure 11.18 The log C_p against t plot for the elimination stage of a first-order absorption and elimination of a drug

The mathematical approach outlined in this section may be generally applied to drugs that do not exhibit first-order absorption and elimination processes. Consider, for example, a drug that exhibits zero-order absorption and first-order elimination kinetics. The rate of change of the drug in the body will be given by substituting the relevant rate expressions in equation (11.42), which gives:

$$\frac{dA}{dt} = k_o - k_{el}A \tag{11.48}$$

where k_0 is the zero rate constant for the absorption process. However, it should be realised that some drug absorption and elimination processes do not exhibit zero- or first-order kinetics and so these processes cannot always be so easily quantified.

11.5.4 The calculation of t_{max} and C_{max}

The calculation of t_{max} and C_{max} for a drug normally starts from the relevant equation for the change in plasma concentration of the drug with time derived from equation (11.42). For example, for drugs that exhibit first-order absorption and elimination kinetics, equation (11.41) (see section 11.5.1) shows how C_p changes with time. Differentiation of equation (11.41) with time t gives:

$$\frac{dC_p}{dt} = \frac{FD_o}{V_d} \cdot \frac{k_{ab}}{(k_{ab} - k_{el})} \left(-k_{el}e^{-k_{el}t_{max}} + k_{ab}e^{-k_{ab}t_{max}} \right) \tag{11.49}$$

but at t_{max} $dC_p/dt = 0$ and so:

$$\frac{FD_o}{V_d} \cdot \frac{k_{ab}}{(k_{ab} - k_{el})} \left(-k_{el}e^{-k_{el}t_{max}} + k_{ab}e^{-k_{ab}t_{max}} \right) = 0 \tag{11.50}$$

simplifying:

$$-k_{el}e^{-k_{el}t_{max}} + k_{ab}e^{-k_{ab}t_{max}} = 0 \tag{11.51}$$

and:

$$k_{ab}e^{-k_{ab}t_{max}} = k_{el}e^{-k_{el}t_{max}} \tag{11.52}$$

taking log to base e:

$$\ln k_{ab} - k_{ab}t_{max} = \ln k_{el} - k_{el}t_{max} \tag{11.53}$$

and so:

$$t_{max} = \frac{\ln k_{el} - \ln k_{ab}}{k_{el} - k_{ab}} \tag{11.54}$$

Once t_{max} has been calculated C_{max} can be found by substitution of the values of F (see equation 11.33), V_d (see section 11.4.1), D_o, k_{ab} and k_{el} for the drug in equation (11.43). Since F and V_d are both constants, C_{max} is proportional to the dose administered: the larger the dose, the greater C_{max}.

11.5.5 Repeated oral doses

For a drug to be therapeutically effective its plasma concentration must be maintained within its therapeutic window for a long enough period of time to obtain the desired therapeutic effect. This can only be achieved by the use of repeat doses at regular time intervals. Initially, for each dose, the rate of absorption will exceed the rate of elimination and so the plasma concentration of the drug will steadily increase as the number of doses increases (Fig. 11.19). However, as the plasma concentration increases so does the rate of elimination and so, eventually, the plasma concentration will reach a plateau. The plateau plasma concentration will vary between a maximum and minimum value; the time interval between these values depends on the time interval between the doses. The values of k_{ab} and k_{el} may be used to calculate the plateau maximum and minimum drug plasma concentration values. This is useful in designing multiple dosage form regimens.

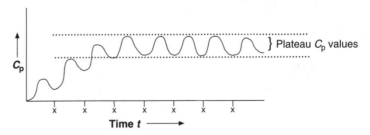

Figure 11.19 The general changes in plasma concentration with time for repeated oral doses. Repeat doses were administered at regular times intervals x

The time taken to achieve a plateau concentration may be reduced by using a larger than usual initial dose. This *loading dose*, as it is known, gives a relatively high initial plasma concentration that acts as an elevated starting point for the succeeding normal doses. This reduces the time taken to reach the plateau concentration, that is, the therapeutic window, and so can be particularly useful in cases of serious illness.

11.6 The use of pharmacokinetics in drug design

Pharmacokinetic data are used to differentiate between active substances with good and poor pharmacokinetic characteristics. For example, substances with poor absorption, high first-pass metabolism and an unsuitable half-life (too long or too short) will normally be discarded in favour of substances with more appropriate pharmacokinetic properties.

Pharmacokinetics is used in all the development stages of a drug from preclinical to Phase IV trials. Legislation normally demands that the absorption, elimination, distribution, clearance, bioavailability, $t_{1/2}$ and V_d of all existing and new drugs must be defined using preclinical trials. In theory all these parameters can be scaled up from animal experiments to predict the behaviour of the substance in humans but the correlation is usually only approximate (see section 11.7). The same parameters are required for Phase I trials. Phase I results for bioavailability and $t_{1/2}$ are used by the pharmaceutical industry as part of the evidence for deciding whether further investigation of a potential drug would be justified. For example, if the first-pass elimination of the potential drug is high (low bioavailability) the drug is not likely to be therapeutically effective using an oral dosage form. This poses questions such as: will the community allow IV bolus and infusion methods; and is the action of the drug unique enough to justify the expense of these dosage forms. A positive answer to these and other questions means that further development could occur. However, a negative answer means that the development is stopped and a new analogue would be selected for investigation. Phase I trials are also used to predict appropriate dose levels for the expected patient community.

Once the drug has been found to be safe and effective in the Phase I trials the evaluation of pharmacokinetic parameters is required for the diseased state, age (young and old) and gender in the Phase II trials. It is especially important to determine the effect of reduced liver and kidney function on the elimination of the new drug in order to avoid toxic effects due to the use of too high a dose level in patients with these conditions. All this information is used to develop effective safe dosage forms for new drugs and to check new dosage forms for existing drugs.

11.7 Extrapolation of animal experiments to humans

The extrapolation of the results of animal experiments to humans can be made using the relationship:

$$P = x \cdot m^n \qquad (11.55)$$

where P is the parameter being extrapolated, such as Cl, V_d or $t_{1/2}$, m is the body mass of the animal used to determine P and x and n are constants characteristic of the parameter and compound being investigated. This relationship may also be expressed in the form:

$$\log P = \log x + n \log m \qquad (11.56)$$

Since equation (11.56) corresponds to that for a straight line $(y = mx + c)$, the values of n and x may be determined by plotting a graph of $\log P$ against $\log m$ for a number of animals with different body weights. The value of n is the slope of the graph while $\log x$ is the

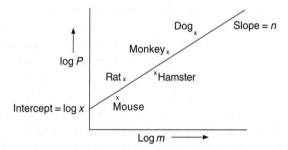

Figure 11.20 A simulation of the logarithmic change of a pharmacokinetic parameter with body mass for various animals

intercept on the y axis. For example, animal experiments (Fig.11.20) have shown that the plasma clearance $(P = Cl_p)$ for cyclophosphamide is related to body weight by the equation:

$$Cl_p = 16.7\, m^{0.754} \qquad (11.57)$$

In general, experimental work has shown that the values of n tend to be of the order of 0.75 for clearance, 0.25 for half-life and 1 for volume of distribution.

11.8   Questions

1   Explain the meaning of each of the following terms: (a) therapeutic window; (b) drug regimen; (c) IV bolus; (d) clearance; and (e) first-pass effect.

2   Discuss the general importance of quantitative analysis in pharmacokinetic studies. Include in the discussion a brief description of the problems that can give rise to inaccurate interpretation of kinetic data.

3

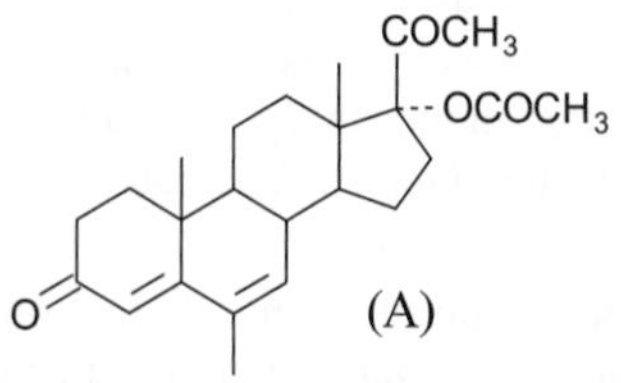

(A)

Megestrol acetate (A) is an oral contraceptive. What pharmacokinetic parameters should be determined for other potential oral contraceptives in order to compare their actions with compound A? Give reasons for choosing your selected parameters.

4 Digoxin and cyclosporin have narrow therapeutic windows. The rate of elimination of cyclosporin is much faster than that of digoxin. If the doses required for therapeutic success are similar, how does this information affect their drug regimens?

5 A patient was given a single dose of 30 mg of a drug by an IV bolus injection. The drug plasma concentration was determined at set time intervals, the data obtained being recorded in Table 11.5.

Calculate:

(a) the value of the elimination constant of the drug;

(b) the apparent volume of distribution of the drug;

(c) the clearance of the drug.

What fundamental assumption has to be made in order to calculate these values?

Table 11.5

Time (hours)	Plasma concentration ($\mu g \; cm^{-3}$)
1	5.9
2	4.7
3	3.7
4	3.0
5	2.4
6	1.9

6 The clearance of a drug from a body compartment is $5 \; cm^3 \; min^{-1}$. If the compartment originally contained 50 mg of this drug, calculate the amount of drug remaining in the system after 1, 2, 3, 4, 5, 6, 7, 8, 9 and 10 minutes if the volume of the compartment was $50 \; cm^3$. Plot a graph of concentration against time and determine the values of $t_{1/2}$ and k_{el} for the compartment and the drug.

7 A dose of 50 mg of the drug used in question 5 was administered orally to the same patient who received the IV bolus in question 5. Plasma samples were taken at regular time intervals and a graph of plasma concentration against time was plotted. If the area under this curve was 5.01, calculate the absolute bioavailability of the drug in the patient, assuming first-order absorption and elimination. Comment on the value of the figure obtained. You may use the IV data recorded in question 5.

8 The data in Table 11.6. are based on plasma concentration–time curves for a number of analogues of a lead compound. Calculate the relevant pharmacokinetic parameter(s) and indicate the best analogue for further investigation if a drug with a reasonable duration of action is required. Assume that the absorption and elimination of the drug follow first-order kinetics.

Table 11.6

Analogue	Elimination rate constant (m^{-1})	AUC IV bolus (30 mg dose) $\mu g\ m\ cm^{-3}$	AUC single oral dose (30 mg dose) $\mu g\ m\ cm^{-3}$
A	0.1386	30.4	31.5
B	0.0277	31.2	47.6
C	0.0462	100.3	81.4
D	0.0173	69.7	81.9

9 A patient is being treated with morphine by intravenous infusion. The steady state plasma concentration of the drug is to be maintained at 0.04 $\mu g\ cm^{-3}$. Calculate the rate of infusion necessary, assuming a first-order elimination process (for morphine V_d is 4.0 dm^3 and $t_{1/2}$ is 2.5 h).

12

Drug metabolism

12.1   Introduction

Drug metabolism or biotransformations are the chemical reactions that are responsible for the conversion of drugs into other products (*metabolites*) within the body before and after they have reached their sites of action. It usually occurs by more than one route (Fig. 12.1, *R*-(+)-warfarin). These routes normally consist of a series of enzyme-controlled reactions. Their end products are normally pharmacologically inert compounds that are more easily excreted than the original drug. The reactions involved in these routes are classified for convenience as *Phase I* (see section 12.4) and *Phase II* (see section 12.6) reactions. Phase I reactions either introduce or unmask functional groups that are believed to act as a centre for Phase II reactions. The products of Phase I reactions are often more water soluble and so more readily excreted than the parent drug. Phase II reactions produce compounds that are often very water soluble and usually form the bulk of the inactive excreted products of drug metabolism.

The rate of drug metabolism controls the duration and intensity of the action of many drugs by controlling the amount of the drug reaching its target site. In addition, the metabolites produced may be pharmacologically active (see section 12.2). Consequently, it is important in the development of a new drug to document the behaviour of the metabolic products of a drug as well as that of their parent drug in the body. Furthermore, in the case of prodrugs, metabolism is also responsible for liberating the active form of the drug.

12.1.1   The stereochemistry of drug metabolism

The body contains a number of non-specific enzymes that form part of its defence against unwanted xenobiotics. Drugs are metabolised both by these enzymes and the more specific enzymes that are found in the body. The latter enzymes usually catalyse the metabolism of drugs that have structures related to those of the normal substrates of the enzyme and so are to a certain

Medicinal Chemistry, Second Edition   Gareth Thomas
© 2007 John Wiley & Sons, Ltd

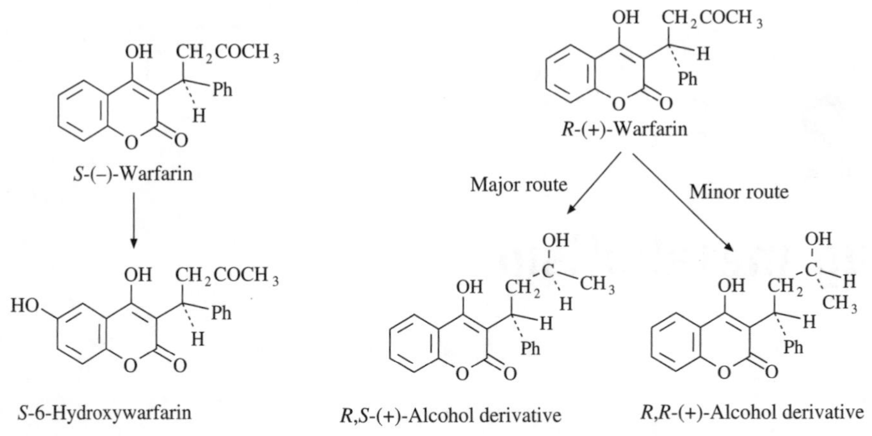

Figure 12.1   The different metabolic routes of S-($-$)-warfarin and R-($+$)-warfarin in humans

extent stereospecific. The stereospecific nature of some enzymes means that enantiomers may be metabolised by different routes, in which case they could produce different metabolites (Fig. 12.1).

A direct consequence of the stereospecific nature of many metabolic processes is that racemic modifications must be treated as though they contained two different drugs, each with its own pharmacokinetic and pharmacodynamic properties. Investigation of these properties must include an investigation of the metabolites of each of the enantiomers of the drug. Furthermore, if a drug is going to be administered in the form of a racemic modification, the metabolism of the racemic modification must also be determined since this could be different from that observed when the pure enantiomers are administered separately.

12.1.2   Biological factors affecting metabolism

The metabolic differences exhibited by a species are believed to be due to variations in *age*, *gender* and *genetics*. *Diseases* can also affect drug metabolism. In particular, diseases such as cirrhosis and hepatoma that affect the liver, which is the major site for metabolism, will seriously impair the metabolism of a number of drugs. Diseases of organs such as the kidneys and lungs, which are less important centres for metabolism, will also effect the excretion of the resulting metabolic products. Consequently, when testing new drugs, it is essential to design trials to cover all these aspects of metabolism.

Age   The ability to metabolise drugs is lower in the very young (under 5) and the elderly (over 60). However, it is emphasised that the quoted ages are approximate and the actual changes will vary according to the individual and their life style.

In the foetus and the very young (*neonates*) many metabolic routes are not fully developed. This is because the enzymes required by metabolic processes are not produced in sufficient quantities until several months after birth. For example, when chloramphenicol was used to treat bacterial infections in premature babies it was found to have a high mortality rate, which fell considerably when the babies were 30 days old. This high mortality rate was attributed to the premature babies having too little glucuronate transferase, the enzyme that catalyses the conversion of the drug to its readily excreted water-soluble glucuronate. Consequently, the concentration of the drug built up to fatal levels in the babies. It is now known that the body synthesises glucuronate transferase over the first 30 days of life and as a result the mortality rate has fallen.

Chloramphenicol   Chloramphenicol glucuronide derivative

Some xenobiotics are able to cross the placenta from the mother to the foetus and so any drugs used by the mother are also likely to pass into the foetus. Since a number of the enzyme systems that are required by the body for Phase II drug metabolism reactions have been found to be present in negligible to low concentrations in the foetus, these drugs can affect biological processes in the developing foetus, resulting in teratogenic and other undesirable effects. Furthermore, where drug metabolism occurs the more water-soluble metabolites of the drug are likely to accumulate on the foetal side of the placenta. Obviously, if these metabolic products are pharmacologically active they could also be detrimental to the development of the foetus. However, not all drugs cause damage to the foetus: administration of betamethasone to the mother several days before birth has been shown to aid the development of the surfactants in the lungs of the foetus.

Betamethasone

Children (above 5) and teenagers usually have the same metabolic routes as adults. However, their smaller body volume means that smaller doses are required to achieve the desired therapeutic effect. It is well documented that a person's ability to metabolise drugs decreases with age. This decline in a person's ability to metabolise drugs has been attributed to the physiological changes that accompany aging. It does not pose too many problems between the approximate ages of 20 and 60 years but can become more significant in the elderly. In general the body gradually loses its capacity to metabolise and eliminate the drug and its metabolites (Fig. 12.2). This leads to higher blood concentrations of the drug and the possibility of an increase in the adverse effects of the drug. This is normally countered by giving reduced doses of relevant drugs to the patient.

Gender The metabolic pathway followed by a drug is normally the same for both males and females. However, some gender-related differences in the metabolism of anxiolytics,

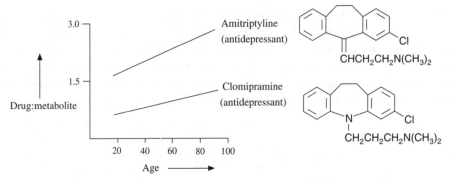

Figure 12.2 The variation of the blood : metabolite ratio of some drugs with age

hypnotics and a number of other drugs have been observed. For example, Wilson has found that diazepam has an average half-life of 41.9 hours in females but only 32.5 hours in males. Such differences have been attributed in some instances to significant differences in enzyme concentrations. For example, women have a significantly lower concentration of alcohol dehydrogenase and so do not metabolise alcohol so rapidly as men.

Pethidine   Chlorpromazine   Diazepam

Pregnant women will also exhibit changes in the rate of metabolism of some drugs. For example, the metabolism of the analgesic pethidine and the antipsychotic chlorpromazine is reduced during pregnancy.

Genetic variations Variations in the genetic codes of individuals can result in the absence of enzymes, low concentrations of enzymes or the formation of enzymes with reduced activity. These differences in enzyme concentration and activity result in individuals exhibiting different metabolic rates and in some cases different pharmacological responses for the same drug. For example, the antituberculous drug isoniazid is metabolised by a Phase II reaction: acetylation. In some patients this acetylation is fast and in others it is slow. Slow acetylation is found in 75 per cent of Caucasians and Negroes but in only 10 per cent of Japanese and Eskimos. Patients are often classified as fast or slow acetylators.

Isoniazid

An individual's inability to metabolise a drug could result in that drug accumulating in the body. This could give rise to unwanted effects.

12.1.3   Environmental factors affecting metabolism

The metabolism of a drug is also affected by life style. Poor diet, drinking, smoking, and drug abuse may all have an influence on the rate of metabolism. The use of over-the-counter self-medicaments may also affect the rate of metabolism of an endogenous ligand or a prescribed drug. Since the use of over-the-counter medicaments is widespread, it can be difficult to assess the results of some large-scale clinical trials.

12.1.4   Species and metabolism

Different species often respond differently to a drug. For example, a dose of 50 mg kg^{-1} of body mass of hexobarbitone will anaesthetise humans for several hours but the same dose will only anaesthetise mice for a few minutes.

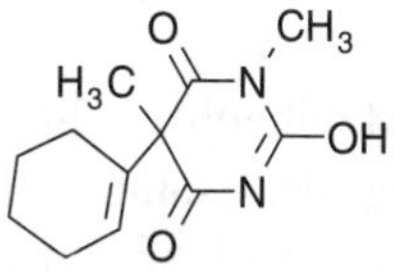

Hexobarbitone

The main reason for the different reponses to a drug by members of different species is believed to be due to differences in their metabolism. These metabolic differences may take either the form of different metabolic pathways for the same compound or different rates of metabolism when the pathway is the same. Both deviations are though to be due to enzyme variations, deficiencies and sufficiencies.

12.1.5   Enzymes and metabolism

The activity of enzymes may be enhanced or inhibited by the presence of metabolic products. This results in either a reduction or an increase in drug concentration in the body. In the former case a reduction will result in a shorter and less effective drug action. Conversely, an accumulation will lead to prolonged drug action and an increase in the possibility of adverse drug effects. These effects are not limited to drug metabolites but can be brought about by xenobiotics in general. For example, eating grapefruit can increase the bioavailability of the anxiolytic diazepam and the anticonvulsant carbamazepine.

12.2   Secondary pharmacological implications of metabolism

Metabolites may be either pharmacologically inactive or active. Active metabolites may exhibit a similar activity to the drug, a different activity or be toxic.

12.2.1 Inactive metabolites

Routes that result in inactive metabolites are classified as *detoxification* processes. For example, the detoxification of phenol results in the formation of phenyl hydrogen sulphate, which is pharmacologically inactive. This compound is very water soluble and so is readily excreted through the kidney.

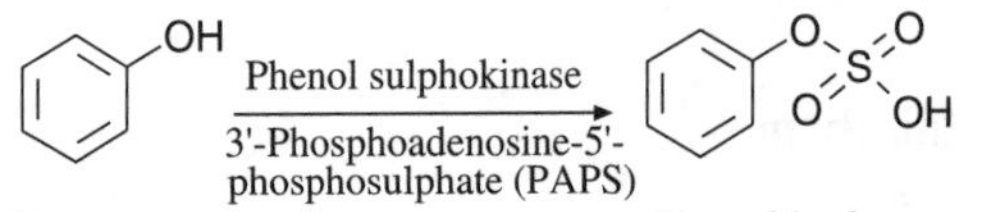

Phenol

Phenol sulphokinase
3'-Phosphoadenosine-5'-phosphosulphate (PAPS)

Phenyl hydrogen sulphate

12.2.2 Metabolites with a similar activity to the drug

In this situation the metabolite can exhibit either a different potency or duration of action or both with respect to the original drug. For example, the anxiolytic diazepam, which has a sustained action, is metabolised to the anxiolytic temazepam, which has a short duration of action. This in turn is further metabolised by demethylation to the anxiolytic oxazepam, which also has a short duration of action.

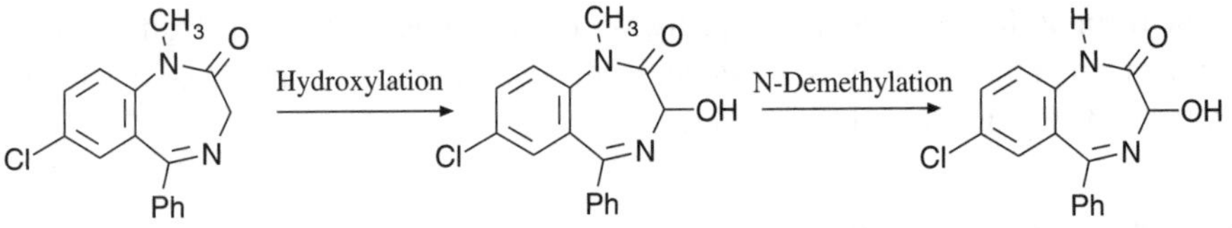

Diazepam   Temazepam   Oxazepam

12.2.3 Metabolites with a dissimilar activity to the drug

In these cases the activity of a metabolite has no relationship to that of its parent drug. For example, the antidepressant iproniazid is metabolised by dealkylation to the antituberculous drug isoniazid.

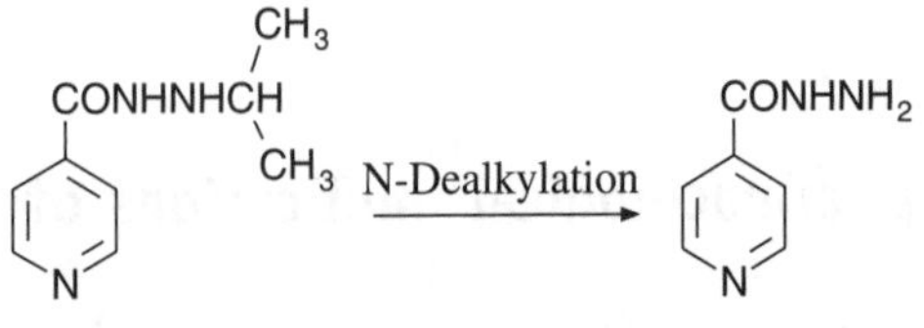

Iproniazid   Isoniazid

12.2.4 Toxic metabolites

The toxic action usually arises because the metabolite either activates an alternative receptor or acts as a precursor for other toxic compounds. For example, deacylation of the analgesic phenacetin yields *p*-phenetidine, which is believed to act as the precursor of substances that cause the condition methaemoglobinaemia. This condition, which causes headaches, shortness of breath, cyanosis, sickness and fatigue, is caused by the presence of methaemoglobin (a modification of haemoglobin) in the blood (Fig. 12.3). Phenacetin is also metabolised via its *N*-hydroxy derivative, which is believed to cause liver damage.

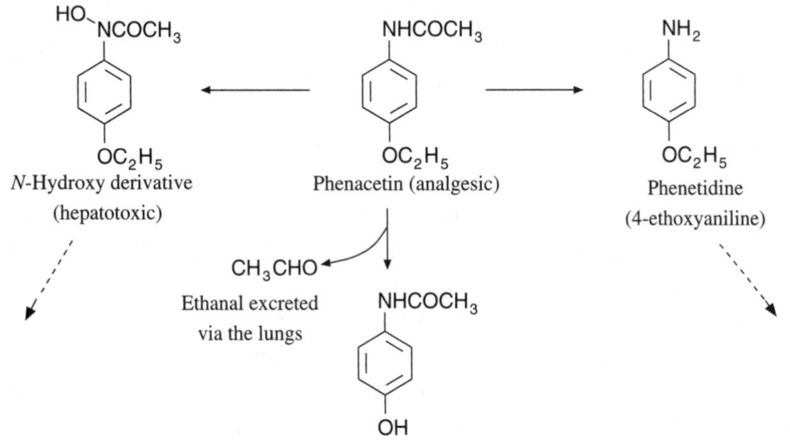

Figure 12.3 An outline of part of the metabolism of phenacetin

12.3 Sites of action

Drug metabolism can occur in all tissues and most biological fluids. However, the widest range of metabolic reactions occurs in the liver. A more substrate-selective range of metabolic processes takes place in the kidney, lungs, brain, placenta and other tissues.

Orally administered drugs may be metabolised as soon as they are ingested. However, the first region where a significant degree of drug metabolism occurs is usually in the GI tract and within the intestinal wall. Once absorbed from the GI tract, many potential and existing drugs are extensively metabolised by first-pass metabolism (see section 11.5). For example, the first-pass metabolism of some drugs like lignocaine is so complete that they cannot be administered orally. The bioavailability of other drugs, such as nitroglycerine (vasodilator), propranolol (antihypertensive) and pethidine (narcotic analgesic), is significantly reduced by their first-pass metabolism.

Lignocaine   Nitroglycerine   Propranolol   Pethidine

12.4 Phase I metabolic reactions

The main Phase I metabolic reactions are oxidation, reduction and hydrolysis. Within each of these areas the reactions undergone will largely depend on the nature of the available enzymes, the hydrocarbon skeleton of the drug and the functional groups present. A knowledge of the structural features of a molecule makes it possible to predict its most likely metabolic reactions and products. However, the complex nature of biological systems makes a comprehensive prediction difficult. Consequently, in practice the identification of the metabolites of a drug and their significance is normally carried out during its preclinical and Phase I trials. However, prediction of the possible products can be of some help in these identifications, although it should not be allowed to obscure the possible existence of unpredictable metabolites.

The identity of metabolites is obtained by a variety of investigative methods. A commonly used method is to incorporate radioactive tracers, such as C-14 and tritium (H-3 or T) into the drug. After ingestion by the test animal any radioactive compounds are isolated from the relevant organs, urine or faeces and identified by an appropriate analytical method. C-14 is preferred to tritium since the latter is prone to exchange reactions and the C–T bond can be broken in the course of the enzyme-catalysed reaction. Consequently, when using tritium, it must be incorporated into the structure of the drug in such a way that it cannot be replaced by other non-radioactive hydrogen isotopes or exchange with the active hydrogen groups of compounds that occur naturally in the test animal.

12.4.1 Oxidation

Oxidation is by far the most important Phase I metabolic reaction. One of the main enzyme systems involved in the oxidation of xenobiotics appears to be the so-called *mixed function oxidases* or *monooxygenases*, which are found mainly in the smooth endoplasmic reticulum of the liver but also occur, to a lesser extent, in other tissues. The mechanism by which these mixed function oxidases operate may involve a series of steps, each step being controlled by an appropriate enzyme system. These steps can be either oxidative or reductive in nature. For example, oxidation of aliphatic C–H bonds involving the cytochrome P-450 family is believed to take place by a series of interrelated steps. These steps are catalysed by cytochrome P-450, cytochrome P-450 reductase and cytochrome b_5 reductase, with either NADPH or NADH as coenzyme (Fig. 12.4).

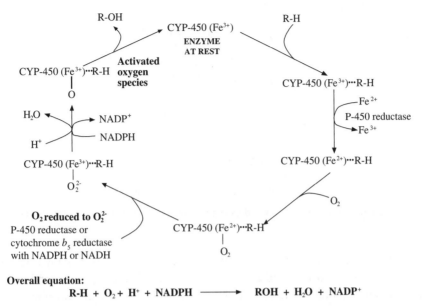

Figure 12.4 The proposed mechanism of action of cytochrome P-450. The abbreviation CYP-450 is used for this system because the P-450(Fe^{2+})$\cdots$R-H complex forms a derivative with carbon monoxide, which has a λ_{max} at 450 nm

Cytochrome P-450 (CYP-450) is a haem-protein molecule in which the haem residue is protoporphyrin IX containing coordinated iron. When the system is at rest the iron is in the fully oxidised Fe^{3+} state. It has been proposed that the substrate initially binds to the active site. This is followed by a reduction of the Fe^{3+} of cytochrome P-450 to the Fe^{2+} state by a one-electron transfer catalysed by P-450 reductase. At this point molecular oxygen binds to the Fe^{2+} and is converted by a series of steps to an activated oxygen species. This species reacts to form the appropriate product.

The complete system is non-specific, catalysing the oxidation of a wide variety of substrates. This is probably due to both the non-selectivity of the enzyme and also the existence of numerous isozymes. Furthermore, lipid-soluble xenobiotics are good substrates for the cytochrome P-450 monooxygenases because high concentrations of the enzyme system are found in lipoidal tissue.

Flavin monooxygenases (FMO) are also an important family of non-selective mixed function oxidases. These enzymes, which have a FAD (flavin adenine dinucleotide) prosthetic group, require either NADH or NADPH as coenzyme. They catalyse the oxidation of nucleophilic groups such as aromatic rings, amines, thiols and sulphides but will not metabolise substrates with anionic groups. Like cytochrome P-450, the mechanism of FMO-catalysed oxidation is thought to involve a number of oxidative and reductive steps. The flavin residue is believed to play a major part in the oxidation, which is thought to occur by the attack of the nucleophile on the terminal oxygen of the hydroxyperoxide residue formed on this residue (Fig. 12.5).

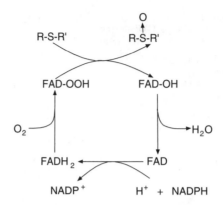

Figure 12.5 A simplification of the proposed mechanism of action of FMO in the metabolism of a sulphide

A number of other enzymes such as monoamine oxidase, alcohol dehydrogenase and xanthine oxidase are also involved in drug metabolism. These enzymes tend to be more specific, oxidising xenobiotics related to the normal substrate for the enzyme.

12.4.2   Reduction

Reduction is an important reaction for the metabolism of compounds that contain reducible groups, such as aldehydes, ketones, alkenes, nitro-compounds, azo-compounds and sulphoxides. The products of many of these reductions are functional groups that subsequently could undergo further Phase II reactions (see section 12.6) to form derivatives that are more water soluble than the original drug. Reduction of some functional groups results in the formation of stereoisomers. Although this suggests that two metabolic routes may be necessary to deal with the products of the reduction, only one product usually predominates. For example, $R(+)$-warfarin is reduced to a mixture of the corresponding $R,S(+)$ and $R,R(+)$ diastereoisomers, the $R,S(+)$ isomer being the major product (Fig. 12.1).

The enzymes used to catalyse metabolic reductions are usually specific in their action. For example, aldehydes and ketones are reduced by soluble aldo–keto reductases, nitro groups by soluble nitro reductases and azo groups by multicomponent microsomal reductases. All these enzymes are found mainly in the liver but also occur in the kidneys and other tissues. Many of these enzymes require NADPH as a coenzyme.

12.4.3   Hydrolysis

Hydrolysis is an important metabolic reaction for drugs whose structures contain ester and amide groups. All types of esters and amides can be metabolised by this route. Ester hydrolysis is often catalysed by non-specific esterases in the liver, kidney and other tissues, as well as pseudocholinesterases in the plasma. Amide hydrolysis is also catalysed by non-specific esterases as well as by non-specific amidases, carboxypeptidases, amino peptidases and decyclases. These enzymes are found in various tissues in the body. More specific enzyme systems are able to hydrolyse sulphate and glucuronate conjugates as well as hydrate epoxides, glycosides and other moieties. In all these various reactions the products formed may be subsequently converted by Phase II reactions into inactive and/or more water-soluble conjugates that are more easily excreted.

The hydrolysis of esters is usually rapid whilst that of amides if often much slower. This makes esters suitable as prodrugs (see section 12.9) and amides a potential source for slow-release drugs.

12.4.4 Hydration

Hydration, in the context of metabolism, is the addition of water to a structure. Epoxides are readily hydrated to diols (see epoxides, Table 12.1), the reaction being catalysed by the enzyme epoxide hydrolase. These epoxides are often formed as a result of a previous metabolic reaction (see alkenes, Table 12.1).

12.4.5 Other Phase I reactions

The reactions involved in Phase I metabolism are not limited to those discussed in the previous sections. In theory, any suitable organic reaction could be utilised in a metabolic route. For example, the initial stage in the metabolism of L-dopa is decarboxylation.

Levodopa   Dopamine

12.5 Examples of Phase I metabolic reactions

The reactions in this section (Table 12.1) are classified according to functional group. Consequently, it is emphasised that the reactions used in this section to illustrate a general process represent only one of the possible reactions by which the drug being used as the example could be metabolised since each of the functional groups in the drug is a potential starting point for a metabolic pathway. Moreover, it is again emphasised that most drugs are metabolised by several routes, each of which normally involves a series of reactions. In other words, the metabolite of one reaction becomes the starting point of a succeeding metabolic reaction, this process being repeated until the drug has been converted to a compound that can either be excreted or is pharmacologically inert. Since the rate at which compounds are metabolised varies, some metabolites may accumulate in the body (see section 12.7). In

Table 12.1　Examples of the more common metabolic reactions undergone by drugs

Functional group	General reaction	Notes
Alkanes	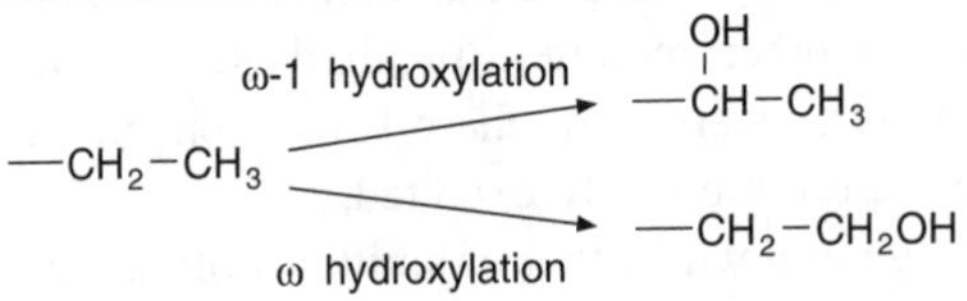 $-CH_2-CH_3$ $\xrightarrow{\omega\text{-1 hydroxylation}}$ $-\overset{OH}{\underset{}{CH}}-CH_3$; $\xrightarrow{\omega \text{ hydroxylation}}$ $-CH_2-CH_2OH$	Hydroxylation of an α-C–H next to an electron withdrawing group is preferred to the ω or ω-1 hydroxylation
Alkenes	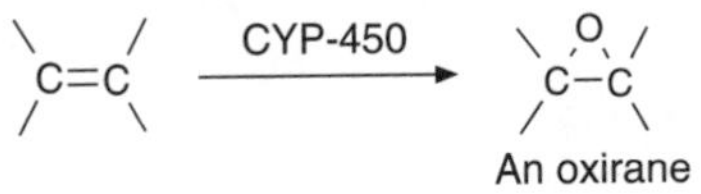 $\underset{/}{\overset{\backslash}{C}}=\underset{\backslash}{\overset{/}{C}}$ $\xrightarrow{\text{CYP-450}}$ an oxirane	Oxiranes are relatively stable but are thought to undergo further reaction with nucleophiles. Reduction of alkenes is rare
Alkynes	 $-C{\equiv}C-$ $\longrightarrow$ An oxirene	Most alkyne groups appear to be stable to metabolism but some are oxidised to oxirenes
Aromatic C–H	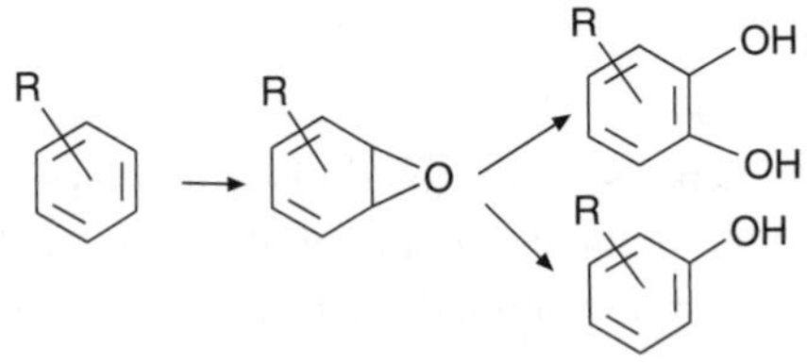 (benzene with R) → (arene oxide with R) → R-catechol (diol OH, OH) and R-phenol (OH)	Hydroxylation, often in the *para* position, is believed to proceed by an epoxide intermediate known as an arene oxide
Heterocyclic rings	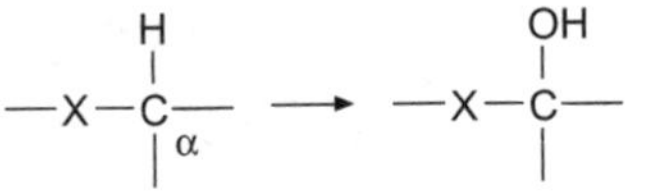 $-X-\overset{H}{\underset{\alpha}{C}}- \longrightarrow -X-\overset{OH}{\underset{}{C}}-$ X = N, S,O α-Hydroxylation. 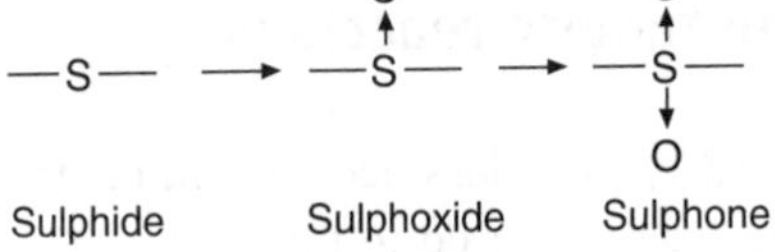 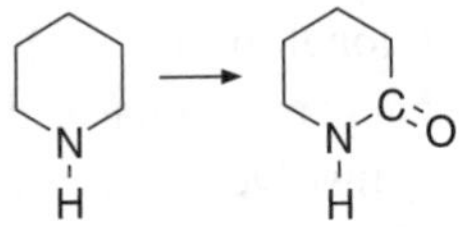 $-S- \longrightarrow -\overset{O}{\underset{}{S}}- \longrightarrow -\overset{O}{\underset{O}{S}}-$ Sulphide　Sulphoxide　Sulphone (piperidine) → (lactam) A lactam	Usually involves either the hetero atom or hydroxylation of an α-carbon atom in saturated heterocyclics. Sulphur hetero atoms are usually oxidised to the corresponding sulphoxide or sulphone. Secondary amine hetero groups may be converted to the corresponding lactam
Alkyl halides	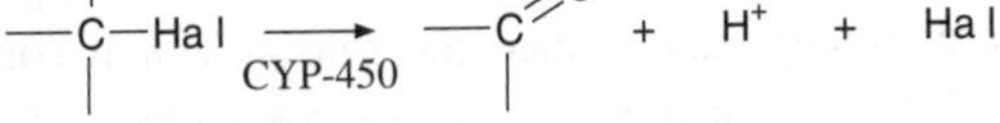 $-\overset{H}{\underset{}{C}}-Hal \xrightarrow{\text{CYP-450}} -C{\overset{O}{\lessgtr}} + H^+ + Hal$	Oxidative dehydrohalogenation is an important route for many alkyl fluorides, chlorides and bromides

Table 12.1 *(Continued)*

Functional group	General reaction	Notes
Aryl halides	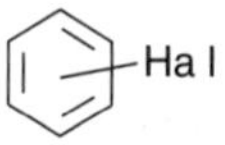	Aryl halides are usually too stable to take part in metabolic reactions
Alcohols	**Primary alcohol** $$RCH_2OH \longrightarrow RCHO \longrightarrow RCOOH$$ **Secondary alcohol** 	Primary and secondary alcohols are normally metabolised by oxidation catalysed by alcohol dehydrogenase. Tertiary alcohols are not usually metabolised by oxidation
Amines	**Primary amines** $$R-NH_2 \rightleftharpoons R-NHOH \rightleftharpoons R-N{=}O \longrightarrow R-NO_2$$ **Secondary amines** 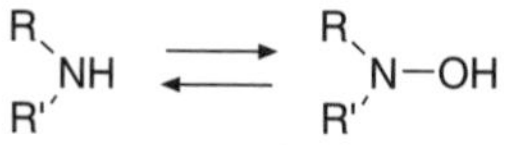 **Tertiary amines** 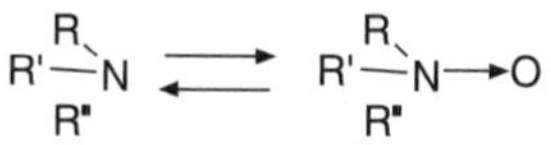	All types of aliphatic amine can be oxidised by CYP-450 and FMO. However, each class of amine yields different types of compound
	$$RCH_2NH_2 \xrightarrow{MAO} RCHO \xrightarrow[\text{oxidation}]{\text{Further}} RCOOH$$	Primary amines attached to a methylene group may also be converted to the corresponding aldehyde by monoamine oxidase (MAO)
	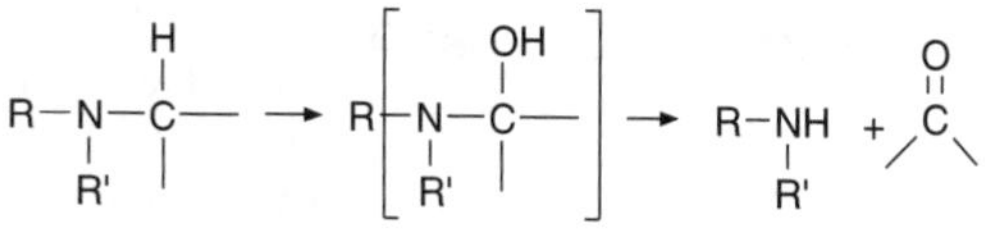 Carbinolamine	All types of amine with α-C–H bonds may undergo α-hydroxylation, catalysed by CYP-450, to an unstable carbinolamine intermediate that spontaneously decomposes to the corresponding carbonyl and nitrogen compounds

Key: R and R′ = or ≠ H

Table 12.1 *(Continued)*

Functional group	General reaction	Notes

Secondary and tertiary amines

$$\underset{R'}{\overset{R}{>}}NH \xrightarrow[\text{CYP-450}]{} RNH_2 + \text{Aldehyde or ketone}$$

$$\underset{R'''}{\overset{R}{\underset{|}{R'-N}}} \xrightarrow[\text{CYP-450}]{} \underset{R'}{\overset{R}{>}}NH + \text{Aldehyde or ketone}$$

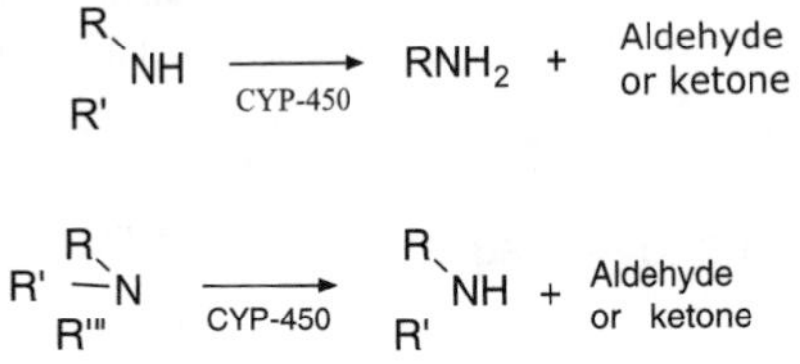

Aliphatic secondary and tertiary amines are metabolised by N-dealkylation to amines, aldehydes and ketones. Tertiary amines dealkylate more readily than secondary amines. Small alkyl groups, such as methyl, ethyl and isopropyl, are rapidly removed as methanal, ethanal and propanone, respectively

Amides including peptide

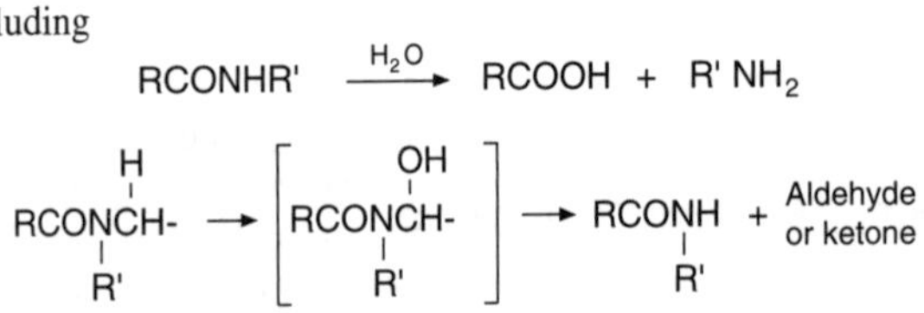

$$RCONHR' \xrightarrow{H_2O} RCOOH + R'NH_2$$

$$\underset{R'}{\overset{H}{\underset{|}{RCONCH-}}} \rightarrow \left[\underset{R'}{\overset{OH}{\underset{|}{RCONCH-}}}\right] \rightarrow \underset{R'}{\overset{}{\underset{|}{RCONH}}} + \text{Aldehyde or ketone}$$

All types of amide are metabolised by hydrolysis Amides may be metabolised by oxidative α-hydroxylation or dealkylation, both of which occur at the α-C–H bond of the amine residue

Aldehydes and ketones

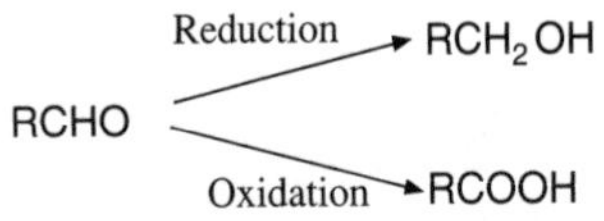

$$RCHO \underset{\text{Oxidation}}{\overset{\text{Reduction}}{<}} \begin{array}{l} RCH_2OH \\ RCOOH \end{array}$$

Aldehydes may be oxidised or reduced. However, ketones are resistant to oxidation and so are normally metabolised by reduction. Reduction of unsymmetrical ketones may give a mixture of isomers

Carboxylic acids

β-Oxidation:

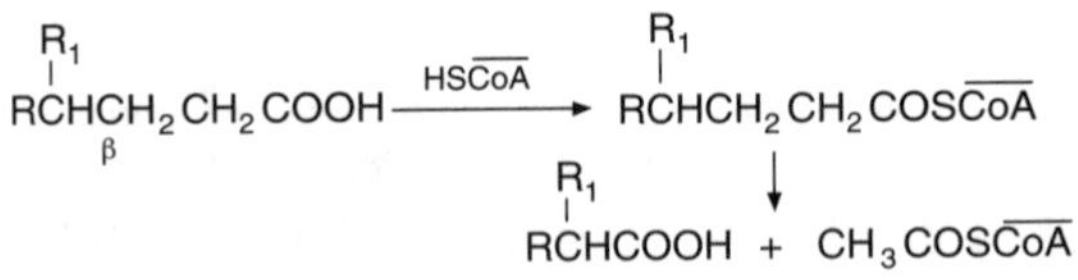

$$\underset{\beta}{\overset{R_1}{\underset{|}{RCHCH_2CH_2COOH}}} \xrightarrow{HS\overline{CoA}} \overset{R_1}{\underset{|}{RCHCH_2CH_2COS\overline{CoA}}}$$

$$\overset{R_1}{\underset{|}{RCHCOOH}} + CH_3COS\overline{CoA}$$

α-Oxidation:

$$\overset{CH_3}{\underset{\alpha}{\underset{|}{RCH_2CH_2CHCH_2COOH}}} \longrightarrow \overset{CH_3}{\underset{|}{RCH_2CH_2CHCOOH}}$$

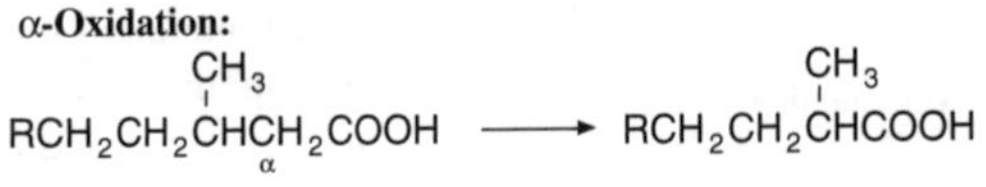

Aliphatic acids are oxidised at their α and β carbon atoms provided that there are methylene groups adjacent to the carboxyl group. β-Oxidation is the most common. It is repeated until a branch in the chain is reached. α-Oxidation can occur when β-oxidation is not possible

Table 12.1 *(Continued)*

Functional group	General reaction	Notes
Epoxides	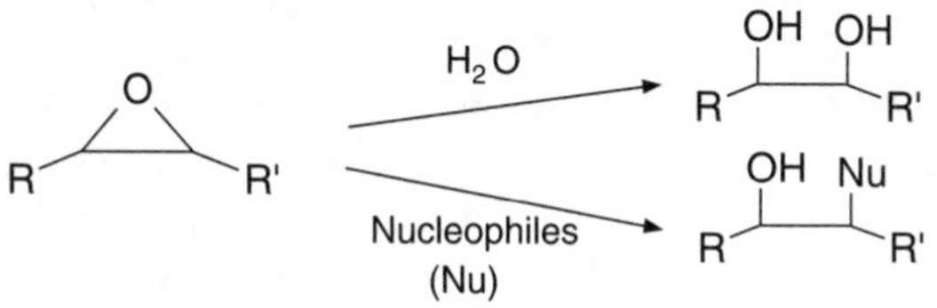 R—(epoxide)—R' $\xrightarrow{H_2O}$ R—CH(OH)—CH(OH)—R'; R—(epoxide)—R' $\xrightarrow{\text{Nucleophiles (Nu)}}$ R—CH(OH)—CH(Nu)—R'	Epoxides are readily hydrated by epoxide hydrases to the corresponding diols. They can also react with the nucleophilic groups of a variety of biological molecules
Ethers	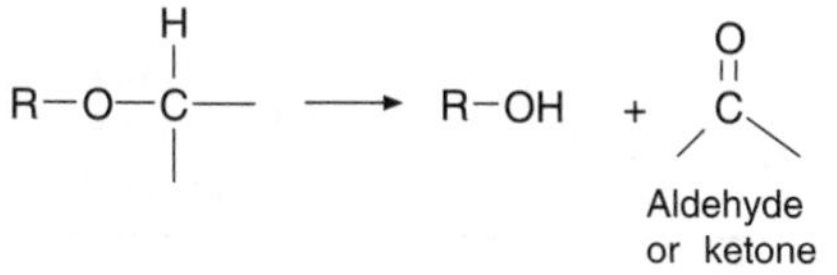 R—O—CH $\longrightarrow$ R—OH + C=O (Aldehyde or ketone)	The main route appears to be oxidative dealkylation. It is catalysed by mixed function oxidases and produces alcohols, phenols, aldehydes and ketones. Small alkyl groups are often removed in preference to larger groups
Esters	RCOOR' $\xrightarrow{H_2O}$ RCOOH + R'OH	Esters are metabolised by hydrolysis. The reaction is catalysed by esterases and other hydrolytic enzymes
Hydrazine	RNHNHR' $\longrightarrow$ RNH_2 + $R'NH_2$;  $\underset{R'}{\overset{R}{\diagdown}}NNH_2$ $\xrightarrow{FMO}$ R—$\underset{R'}{\overset{O\uparrow}{N}}NH_2$	Hydrazines are usually metabolised by reduction to the corresponding amines. However, some *gem*-substituted nitrogen hydrazines are oxidised by FMO enzymes to the corresponding oxide
Nitriles	RCH_2CN $\longrightarrow$ RCHO + CN^-	Aliphatic nitriles are metabolised by oxidation to the corresponding aldehyde with the liberation of highly toxic cyanide ions. The reaction is believed to proceed by an α-hydroxylation mechanism. Aromatic nitriles are usually metabolised by hydroxylation of the aromatic ring
Nitro group	$Ar\,NO_2$ $\longrightarrow$ $Ar\,NH_2$	Nitro groups are usually reduced to the corresponding amine by nitroreductases

Table 12.1 *(Continued)*

Functional group	General reaction	Notes
Sulphides and sulphoxides	$R{-}S{-}R' \underset{\text{Reduction}}{\overset{\text{FMO}}{\rightleftarrows}} \overset{O}{\underset{}{\parallel}} R{-}\overset{O}{\overset{\parallel}{S}}{-}R' \xrightarrow{\text{FMO}} R{-}\overset{O}{\overset{\parallel}{\underset{\parallel}{\underset{O}{S}}}}{-}R'$	Sulphides are oxidised to sulphoxides, which may undergo further oxidation to sulphones
	$R{-}S{-}R' \longrightarrow R{-}SH \; + \; \overset{O}{\underset{/\,\backslash}{\overset{\parallel}{C}}}$	Sulphides are also metabolised by dealkylation
	$R{-}S{-}S{-}R' \longrightarrow RSH \; + \; R'SH$	Disulphides are usually reduced to the corresponding thiols
Thiols	$RSH \xrightarrow{\text{FMO}} R{-}S{-}S{-}R$	Thiols may be oxidised by FMO to the corresponding disulphide

addition, some metabolites accumulate in the body because there is no mechanism for their excretion. These compounds mayor may not have an adverse effect on health.

The existence of this sequence of metabolites increases the possibility of a toxic reaction and so a new drug should be designed to be metabolised and excreted by as short a pathway as possible once it has acted.

12.6   Phase II metabolic routes

Phase II reactions are often known as *conjugation reactions* because they involve the attachment of a group or a molecule to the drug or metabolite. They may occur at any point in the metabolism of a drug or xenobiotic but they are often the final step in the metabolic pathway before excretion. The products formed by these reactions are known as *conjugates*. They are normally water soluble and are usually excreted in the urine and/or bile. The conjugates formed are usually pharmacologically inactive although there are some notable exceptions. For example, it has been suggested that the hepatotoxicity and nephrotoxicity of phenacetin could be due to the formation of the *O*-sulphate esters, which have been shown to bind to microsomal proteins.

Phenacetin (analgesic)   *N*-Hydroxyphenacetin   *O*-Sulphate conjugate

The reactions commonly involved in Phase II conjugation are acylation, sulphate formation and conjugation with amino acids, glucuronic acid glutathione and mercapturic acid (Table 12.2). Methylation is also regarded as a Phase II reaction although it is

Table 12.2 Phase II reactions. These normally produce pharmacologically inert metabolites but a few metabolites, such as *N*-acetylisoniazid and the sulphate conjugates of phenacetin, are toxic

Phase II reaction. Functional group/notes	General reaction	Example
Acylation Primary aromatic amines ($ArNH_2$) Simple sulphonamides ($-SO_2NH_2$) Hydrazines ($-NHNH_2$) Hydrazides ($-CONHNH_2$) Phenols (ArOH)	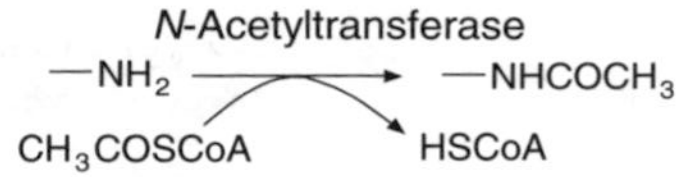N-Acetyltransferase —NH2 —NHCOCH3 CH3COSCoA HSCoA	Sulphanilamide (antibacterial) 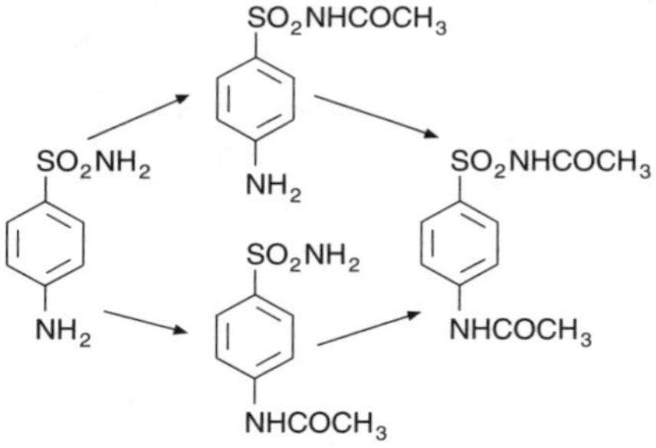SO2NH2 NH2 SO2NHCOCH3 NH2 SO2NH2 NHCOCH3 SO2NHCOCH3 NHCOCH3
Sulphate formation Phenols (ArOH) Alcohols (ROH) Simple sulphonamides ($-SO_2NH_2$) Primary aromatic amines ($ArNH_2$)	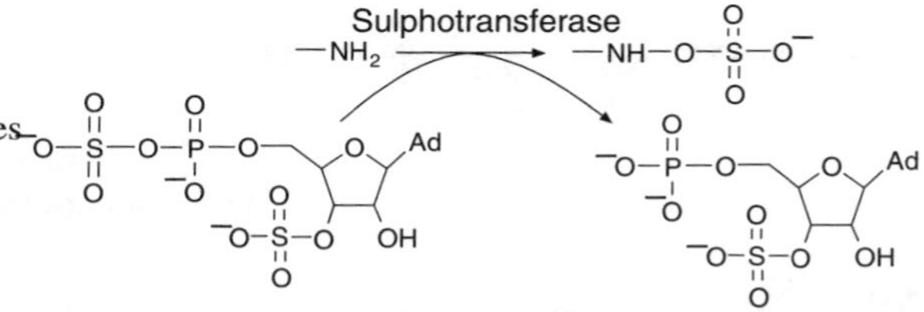Sulphotransferase —NH2 —NH—O—S—O⁻ Ad OH Ad OH 3'–Phosphoadenosine–5'–phosphosulphate (PAPS) 3'–Phosphoadenosine–5'–phosphate (PAP)	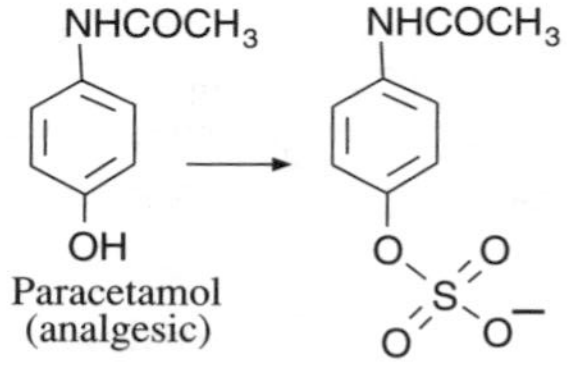NHCOCH3 OH NHCOCH3 Paracetamol (analgesic)
Conjugation with amino acids Acrboxylic acids (-COOH) The main amino acids used to form the conjugates are glycine, glutamine, ornithine (birds), alanine (hamsters and mice), arginine and taurine	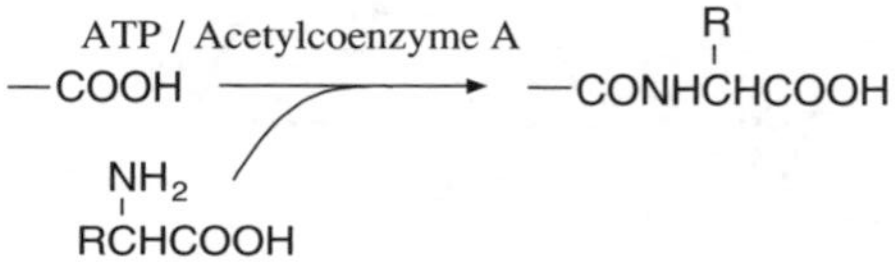ATP / Acetylcoenzyme A —COOH —CONHCHCOOH R NH2 RCHCOOH	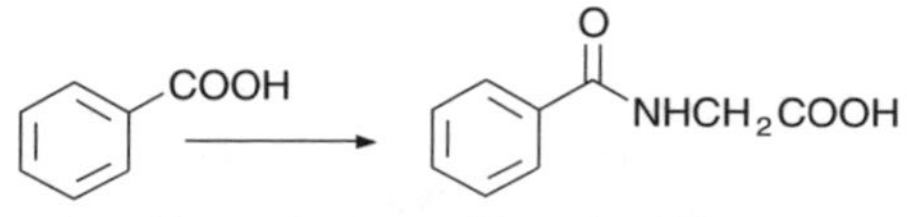COOH NHCH2COOH Benzoic acid (preservative) Hippuric acid, a glycine conjugate

Conjugation with glucuronic acid (Gluc)

Carboxylic acids (RCCH)
Phenols (ArOH)
Alcohols (ROH)
Amines
Thiols (RSH)

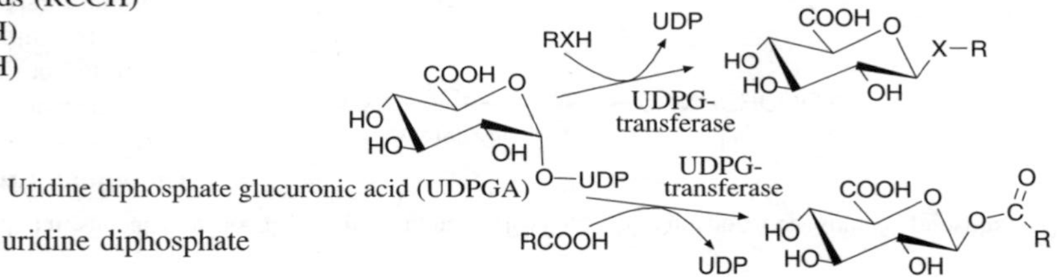

Uridine diphosphate glucuronic acid (UDPGA)

Key: UDP = uridine diphosphate

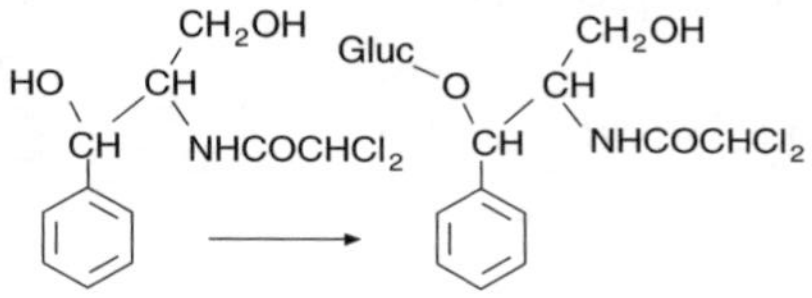

Chloramphenicol
(antibiotic)

RXH → UDP, UDPG-transferase

RCOOH → UDP, UDPG-transferase

Conjugation with glutathione (GSH)

Electrophilic centres caused by:

Halides
Nitro groups
Epoxides
Sulphonates
Organophosphate groups

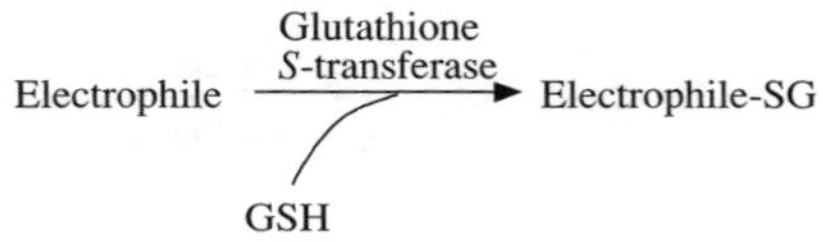

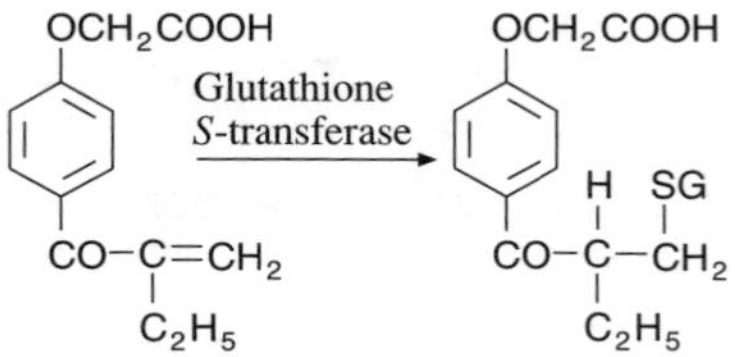

Electrophile + GSH $\xrightarrow{\text{Glutathione } S\text{-transferase}}$ Electrophile-SG

Ethacrynic acid
(diuretic)

Glutathione
S-transferase

Methylation

Phenols (ArOH)
Alcohols (ROH)
Amines
N-Heterocyclics

Methylation detoxifies
a drug but produces
a less polar and so
less easily excreted
metabolite

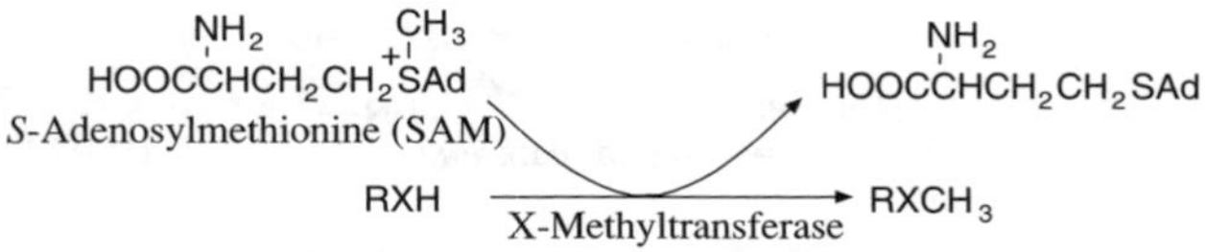

S-Adenosylmethionine (SAM)

RXH $\xrightarrow{\text{X-Methyltransferase}}$ RXCH$_3$

Dimercaprol
heavy metal
poisoning antidote

$\xrightarrow[S\text{-Methyltransferase}]{\text{SAM}}$

normally a minor metabolic route. However, it can be a major route for phenolic hydroxy groups. In all cases, the reaction is usually catalysed by a specific transferase.

12.7 Pharmacokinetics of metabolites

The activity and behaviour of a metabolite will have a direct bearing on the safe use and dose of a drug administered to a patient (see section 12.2). A build-up in the concentration of a metabolite may have a serious effect on a patient. Consequently, when investigating the pharmacokinetics of a drug it is also necessary to obtain pharmacokinetic data concerning the action and elimination of its metabolites. This information is usually obtained in humans by administering the drug and measuring the change in concentration of the appropriate metabolite with time in the plasma. However, as metabolites are produced in the appropriate body compartment, a metabolite may be partly or fully metabolised before it reaches the plasma. In these cases the amount of metabolite found by analysis of plasma samples is only a fraction of the amount of metabolite produced by the body. For simplicity, the discussions in this text assume that *all* the metabolite produced reaches the plasma. Alternatively, the metabolite may be administered separately when independent data concerning its activity and pharmacokinetics are required. However, observations made from metabolite administration can be suspect because its bioavailability is usually different to that when it is produced *in situ* from the drug.

The total administered dose (A) of a drug is excreted partly unchanged and partly metabolised (Fig. 12.6). Since metabolites are produced in the appropriate body compartment, a metabolite may be partly or fully metabolised before it reaches the plasma. In these cases the amount of metabolite found by analysis of plasma samples is only a fraction of the amount of metabolite produced by the body. However, for simplicity, the discussions in **this text will assume** that *all* **the metabolite produced reaches the plasma.**

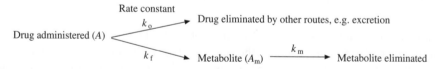

Figure 12.6 A schematic representation of the general elimination routes for a drug in the body

The pharmacokinetics of metabolite formation and elimination can yield useful information concerning the use of the drug. For example, the rate of change of concentration of a *metabolite* (dM/dt) in the plasma is given by:

$$dM/dt = \text{Rate of formation} - \text{Rate of elimination} \qquad (12.1)$$

Since most biological processes exhibit first-order kinetics equation (12.1) becomes:

$$dM/dt = k_f A - k_m A_m \qquad (12.2)$$

where k_f and k_m are the rate constants for the metabolite's formation and elimination processes, respectively. If $k_f > k_m$ there will be an accumulation of the metabolite because clearance of the metabolite is slower than that of the drug responsible for forming the metabolite. This accumulation could pose a problem if the metabolite is pharmacologically active. However, if $k_f < k_m$ the metabolite will not accumulate in the body as the metabolite is eliminated faster than it is formed. However, it is not easy to determine k_f. Consequently, as all the processes involved in drug elimination are normally first order, the overall rate constant k for the rate of elimination of the *drug* (equation 12.3) is used instead of k_f because $k = k_f + k_o$, where k_o is the rate of elimination of the drug by all other routes (Fig. 12.6).

$$\text{Overall rate of drug elimination } \mathrm{d}A/\mathrm{d}t = kA \qquad (12.3)$$

The values of k and k_m can be determined experimentally from log plots of plasma measurements of the drug and metabolite (Fig. 12.7).

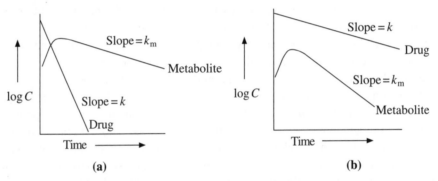

Figure 12.7 Representations of typical log concentration–time plots for a drug and metabolite exhibiting first-order kinetics, showing the general changes when **(a)** $k > k_m$ and **(b)** $k < k_m$

Most metabolic pathways consist of a series of steps. The importance of this series is not the number of steps but whether the pathway has a rate determining step. In other words, is there a metabolite bottleneck where the rate of elimination of a metabolite is far slower than its rate of formation from the drug ($k \gg k_m$)? At such a point the concentration of the metabolite would increase to significant amounts, which could lead to potential clinical problems if the metabolite was pharmaceutically active. Consequently, to avoid problems of this nature a metabolite should be eliminated faster than the drug.

12.8   Drug metabolism and drug design

A knowledge of the metabolic pathway of a drug may be used to design analogues with a different metabolism to that of the lead. This change in metabolism is achieved by modifying the structure of the lead. These structural changes may either make the analogue more stable or increase its ease of metabolism relative to the lead. The structural modifications should be selected so that they not change the nature of the pharmacological activity of the drug. However, it is not possible to accurately predict whether this will be the case and so normally the activity of the analogue may only be found by experiment.

Increasing the metabolic stability and hence the duration of action of a drug is usually achieved by replacing a reactive group by a less reactive one. For example, N-dealkylation can be prevented by replacing an N-methyl group by an N-t-butyl group. Reactive ester groups are replaced by less reactive amide groups. Oxidation of aromatic rings may be reduced by introducing strong electron acceptor substituents such as chloro (–Cl), quaternary amine (—$\overset{+}{N}R_3$), carboxylic acid (—COOH), sulphonate (—SO_3R) and sulphonamide (—SO_2NHR) groups. For example, replacement of the aryl methyl substituent of the antidiabetic tolbutamide by chlorine yielded the antidiabetic chlorpropamide, a compound with a considerably longer half-life.

$$CH_3 \text{—} \underset{}{\langle\text{ }\rangle} \text{—} SO_2NHCONH(CH_2)_3CH_3 \qquad Cl\text{—}\underset{}{\langle\text{ }\rangle}\text{—}SO_2NHCONH(CH_2)_3CH_3$$

Tolbutamide (mean $t_{1/2}$ = 7 hours)   Chlorpropamide (mean $t_{1/2}$ = 35 hours)

However, all these modifications can result in a change of pharmacological activity. For example, the replacement of the ester group in the local anaesthetic procaine by an amide group produced procainamide, which acts as an antiarrhythmic.

$$H_2N\text{—}\underset{}{\langle\text{ }\rangle}\text{—}COOCH_2CH_2N\overset{C_2H_5}{\underset{C_2H_5}{}}\qquad H_2N\text{—}\underset{}{\langle\text{ }\rangle}\text{—}CONHCH_2CH_2N\overset{C_2H_5}{\underset{C_2H_5}{}}$$

Procaine   Procainamide

The ease of metabolism of a drug may be increased by incorporating a metabolically labile group, such as an ester, in the structure of the drug. This type of approach is the basis of prodrug design (see section 12.9). It has also led to the development of so-called *soft drugs*. These are biologically active compounds that are rapidly metabolised by a predictable route to pharmacologically non-toxic compounds. The advantage of this type of drug is that its half-life is so short that the possibility of the patient receiving a fatal overdose is considerably reduced. For example, the neuromuscular blocking agent succinyl choline is almost fully metabolised by hydrolysis in about 10 minutes, which considerably reduces the chances of a fatal overdose. However, this drug can still be fatal as some people do not have the esterase necessary for hydrolysis. A drawback to this approach of incorporating of labile groups into a structure is that it can also change the nature of a drug's biological action.

$$\begin{array}{l} CH_2COOCH_2CH_2N(CH_3)_3 \\ | \\ CH_2COOCH_2CH_2(NCH_3)_3 \end{array} \longrightarrow \begin{array}{l} CH_2COOH \\ | \\ CH_2COOH \end{array} \quad + \quad 2\ HOCH_2CH_2\overset{+}{N}(CH_3)_3$$

Succinylcholine   Succinic acid   Choline

Changing the metabolic pathway of a drug may also be used to develop analogues that do not have the undesirable side effects of the lead compound. For example, the local anaesthetic lignocaine is also used as an antiarrhythmic. In this respect, its undesirable convulsant and emetic side effects are caused by its metabolism in the liver by dealkylation to the mono N-ethyl derivative. The removal of the N-ethyl substituents and their replacement by an α-methyl group gives the antiarrhythmic tocainide. Tocainide cannot be metabolised by the same pathway as lignocaine and does not exhibit convulsant and emetic side effects.

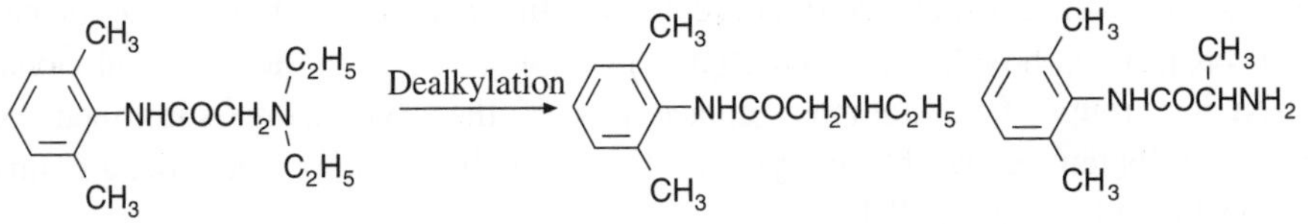

Lignocaine   Tocainide

12.9 Prodrugs

Prodrugs are inactive compounds that yield an active compound in the body. This conversion is frequently carried out by enzyme-controlled metabolic reactions and less frequently by chemical reactions within the body. Prodrugs are used as a way to:

• increase lipid or water solubility;

• improve the taste of a drug to make it more patient compatible;

• alleviate pain when the drug is administered parenterally by injection;

• reduce toxicity;

• increase chemical stability;

• increase biological stability;

• change the length of the time of duration of action;

• deliver the drug to a specific site in the body.

Prodrugs may be broadly classified into two groups, namely, *bioprecursor* and *carrier* prodrugs. They may also be sub-classified according to the nature of their action, for example photoactivated prodrugs (see section 12.9.3).

12.9.1 Bioprecursor prodrugs

Bioprecursor prodrugs are compounds that already contain the embryo of the active species within their structure. They rely on metabolism to produce the active compound. For example, part of the structure of the first prodrug discovered, the antibacterial prontosil, can be converted by metabolic reduction into the active metabolite sulphanilamide (Fig 12.8).

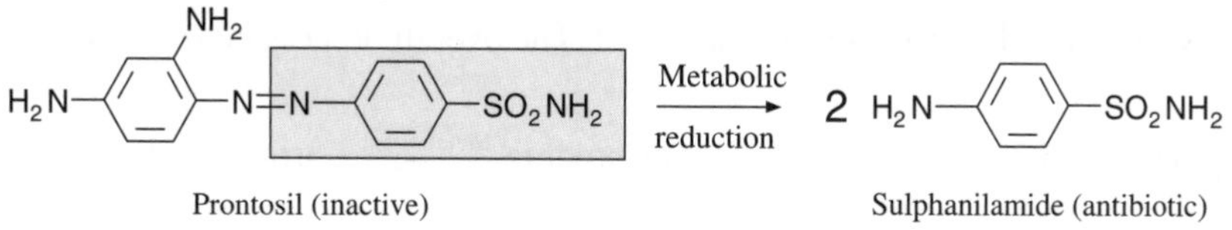

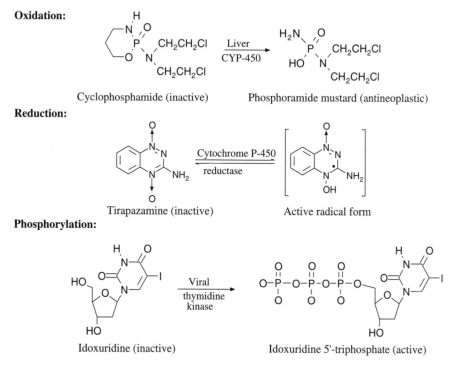

Figure 12.8 Prontosil. The shaded area is the embryonic sulphanilamide

The conversion of a bioprecursor prodrug into its active compound may involve a single step or more commonly a series of steps. A wide variety of reactions are involved in these steps, the most frequent being oxidation, reduction and phosphorylation (Fig. 12.9). Reduction reactions are not as common as oxidations because there are fewer reductive enzymes in the body. Phosphorylation occurs in the activation of a number of antiviral prodrugs (see Table 10.7).

Figure 12.9 Examples of the activation of bioprecursor prodrugs

12.9.2 Carrier prodrugs

Carrier prodrugs differ from bioprecursor prodrugs in that they are formed by combining an active drug with a carrier species to form a compound with the desired chemical and biological characteristics. For example, a lipophilic carrier may be used to improve transport through membranes. The link between carrier and active species must be a group, such as an ester or amide, that can be easily metabolised once absorption has occurred or the drug has been delivered to the desired body compartment. The overall process may be summarised by:

$$\text{Carrier} + \text{Active species} \xrightarrow{\text{Synthesis}} \text{Carrier prodrug} \underset{\text{Metabolism}}{\rightleftharpoons} \text{Carrier} + \text{active species}$$

Carrier prodrugs that consist of the drug linked by a functional group to the carrier are known as *bipartate prodrugs* (Fig. 12.10). *Tripartate prodrugs* are those in which the carrier is linked to the drug by a link consisting of a separate structure. In these systems, the carrier is removed by an enzyme-controlled metabolic process and the linking structure by either an enzyme system or a chemical reaction.

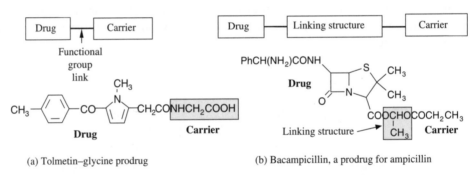

(a) Tolmetin–glycine prodrug   (b) Bacampicillin, a prodrug for ampicillin

Figure 12.10 Examples of **(a)** bipartate and **(b)** tripartate prodrug systems

Ideally, all types of carrier prodrug should meet the following criteria:

• the prodrug should be less toxic than the drug;

• the prodrug should be inactive or significantly less active than the parent drug;

• the rate of formation of the drug from the prodrug should be rapid enough to maintain the drug's concentration within its therapeutic window;

• the metabolites from the carrier should be non-toxic or have a low degree of toxicity;

• the prodrug should have an improved bioavailability if administered orally;

• the prodrug should be site specific.

Carrier prodrugs seldom meet all of these criteria. However, a good example of a carrier prodrug meeting most of these criteria is bacampicillin, one of a number of prodrugs for the antibiotic ampicillin. Only about 40 per cent of an orally administered dose of ampicillin is absorbed. Relatively large quantities of the drug have to be administered in order for it to reach its therapeutic window. This means that significant amounts of the drug remain in the GI tract, damaging the intestinal flora. The prodrug becampicillin is about 98–99 per cent absorbed, which reduces its toxic action in the GI tract. Furthermore, a considerably smaller dose is required in order to maintain the plasma concentration within the drug's therapeutic window. This further reduces the risk of toxic responses. Moreover the metabolism of the prodrug yields carbon dioxide, ethanal and ethanol, all of which are natural metabolites in the body.

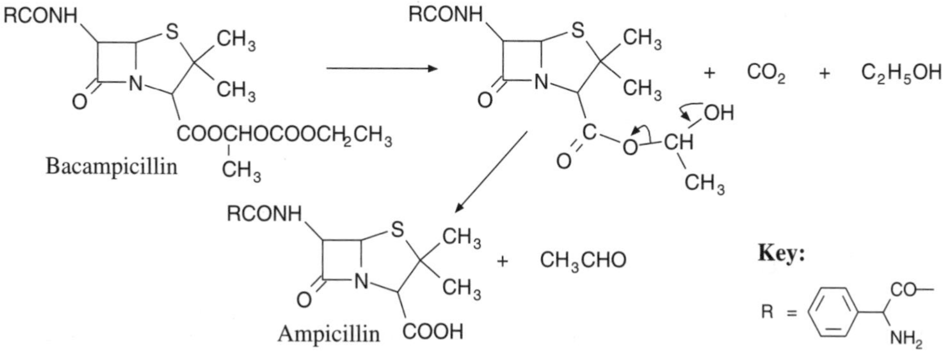

The choice of functional group used as a metabolic link depends both on the functional groups occurring in the drug molecule (Table 12.3) and the need for the prodrug to be metabolised in the appropriate body compartment.

Table 12.3 Examples of the functional groups used to link carriers with drugs

Drug group (D-X)	Type of group	Examples of R groups
Alcohol, phenol (D-OH)	Ester: D-OCOR	Alkyl, phenyl, $-(CH_2)_2 NR_2$, $-(CH_2)_n CONR'R''$, $-(CH_2)_nNHCOR$, $-CH_2OCOR'$
Amines (all types), imides and amides (NH)	Amide: NCOR	Alkyl, phenyl, $-CH_2NHCOAr$, $-CH_2OCOR''$
	Carbamate: NCOR	$-OCHR'OCOR''$, $-OCH_2OPO_2H_3$
	Imine: N=CHR	Aryl
Aldehydes and ketones (C=O)	Acetals: $C(OR)_2$	Alkyl
	Imine: C=NR	Aryl, -OR
Carboxylic acids (D-COOH)	Ester: D-COOR	Alkyl, aryl, $-(CH_2)_n NR'R''$, $-(CH_2)_nCONR'R''$, $-(CH_2)_nNHCOR'R''$, $-CH(R)OCOR$, $-CH(R)OCONR'R''$

The precise nature of the structure of the carrier used to form a carrier prodrug will depend on the intended outcome (see section 12.9.3). For example, valaciclovir, the L-valyl ester of aciclovir, was developed to improve the poor oral absorption of the antiviral aciclovir. It is absorbed using a stereospecific transport protein in the intestine. Once absorbed it is rapidly hydrolysed in the blood stream to aciclovir A unique form of carrier prodrug uses monoclonal antibodies to deliver the drug to its target (see section 10.15.2).

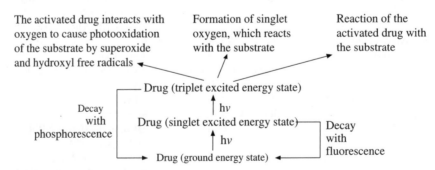

12.9.3 Photoactivated prodrugs

A number of inactive compounds are activated by irradiation with specific wavelengths of visible light or longwave length UV (UV-A). These photosensitizing drugs act by a variety of mechanisms. In general, the visible or UV radiation raises the energy of the drug molecule to an excited state that can either decay back to the ground state or interact by a number of mechanisms with a cellular substrate (Fig. 12.11). This interaction usually results in the destruction of the target cell and a beneficial effect on the patient.

Treatment based on this strategy is known as *photodynamic therapy* (*PDT*). It often requires the use of special lamps, lasers and optical fibres to focus the radiation on particular areas of the body. Although the inactive drug is usually widely distributed in the

The activated drug interacts with oxygen to cause photooxidation of the substrate by superoxide and hydroxyl free radicals

Formation of singlet oxygen, which reacts with the substrate

Reaction of the activated drug with the substrate

Drug (triplet excited energy state)

↑ hν

Decay with phosphorescence

Drug (singlet excited energy state)

↑ hν

Decay with fluorescence

Drug (ground energy state)

Figure 12.11 The activation of a photosensitive drug and its possible modes of action

body it will not be activated in tissues that are not penetrated by the activating light or UV radiation. Consequently, PDT has been limited to easily accessible areas of the body such as the skin, mouth and lungs. For example, in 1997 Dr P Marks, a consultant neurosurgeon at Leeds General Infirmary, used photofrin: a photosensitive drug that preferentially accumulates in tumour tissue to treat advanced brain tumours in a number of patients. Photofrin was injected into the patient and after allowing the drug to accumulate in the tumour (2–3 days) an optical fibre with a laser attached was inserted into the patient's brain through the nose and the drug was activated. Its activation results in the destruction of the tumour cells. However, PDT has to be used with great care since it has been found that the use of these drugs can cause photosensitisation. As a result, the patient cannot tolerate sunlight or other bright lights. In addition skin disfigurement, blistering and other unwanted side effects can also occur.

The photosensitive compounds that have been used with some effect are analogues of psoralen, porphyrin, chlorine and other light-sensitive compounds (Table 12.4). These analogues have often been found to preferentially accumulate in tumour and other target cells, giving a certain degree of specificity to treatments based on the use of these drugs. Drugs are usually administered by topical application, or orally. In the latter case irradiation is only carried out after the drug has had time to reach its intended target in sufficient concentration.

12.9.4 The design of carrier prodrug systems for specific purposes

The nature of the carrier used will largely control a drug's bioavailability. Consequently, the selection of a suitable carrier enables the medicinal chemist to change the biological properties of a drug. Careful selection of a carrier usually improves a drug's performance and in some cases has been used to direct the drug to specific areas.

Improving absorption and transport through membranes

The transport of a drug through a membrane depends largely on its relative solubilities in water and lipids. If the drug is too water soluble it will not enter the membrane but if it is too lipid soluble it will enter but not leave the membrane. Good absorption requires that a drug's hydrophilic–lipophilic nature is in balance. The selection of a suitable carrier can be used to fine-tune this balance and consequently improve absorption of the drug.

Lipophilic carriers are used to increase the lipophilic nature and hence the absorption and passive transport of the drug across membranes. This is usually achieved by combining the carrier with a polar group(s) on the drug (Table 12.5). However, the carrier must be selected so that the new compound is able to act as a prodrug and release the active drug in the body. For example, adrenaline, when used to treat glaucoma, is poorly absorbed through the cornea. However, converting it to the less polar prodrug dipivaloyladrenaline by forming the di-trimethylethanoate derivative masks the polar phenolic hydroxy groups, which makes the molecule more lipophilic and results in a better

Table 12.4 Examples of photosensitive compounds used in treating disease

Drug	Use/Note
8-Methoxypsoralen (8-MOP) (Methoxsalen) 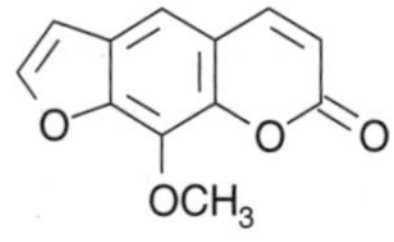 	Administered orally and topically to treat psoriasis and vitiligo (severe skin blemishes). In vitiligo methoxsalen is used to repigment the blemishes. Treatment is often referred to as *PUVA* (psoralenp + UV-A). Unwanted side effects include nausea, erythema, blistering cataract and immune dysfunction Psoralens have also been found to inhibit the proliferation of Ehrlich ascites tumour cells and the infectivity of DNA virus, amongst other activities
Khellin 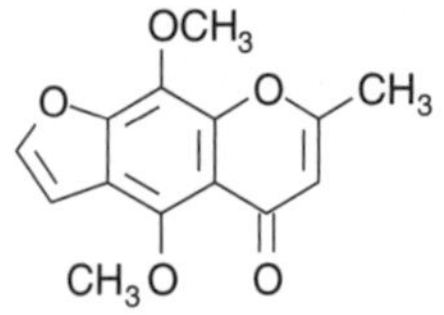 	Used orally to treat vitiligo. Patients are irradiated with UV-A after 2.5 hours. Treatment, which is often referred to as *KUVA*, can take up to 1 year. Unwanted side effects are nausea and dizziness
Photofrin Photofrin is a mixture of oligomers where *n* ranges from zero to eight. The structure given is representative of the mixtures of components	Has been used to treat brain and other tumours. It is injected into the patient and irradiated 24–48 hours later by either a red light source with wavelengths in the 590–640 region or laser light in the red spectral region. The red light source is normally used for tumours embedded in the skin and readily accessible body cavities such as the throat and vagina. Light from laser sources is used for more deep-seated tumours where it is necessary to insert fibre optics into the tumour

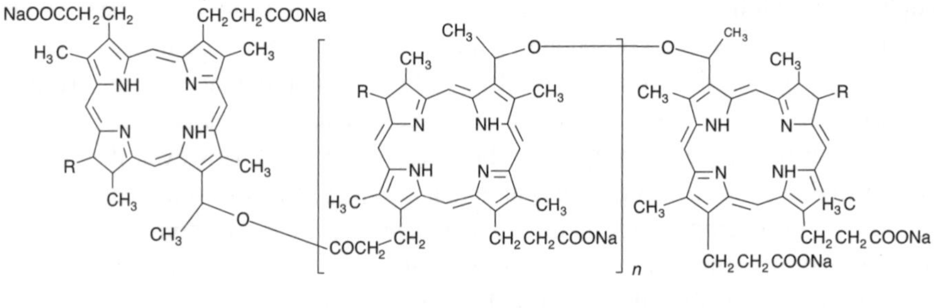

Key:

R = —CH and / or —CH=CH$_2$ (with OH and CH$_3$ substituents)

Table 12.4 (*Continued*)

Drug	Use/Note
Temoporfin (Foscan) 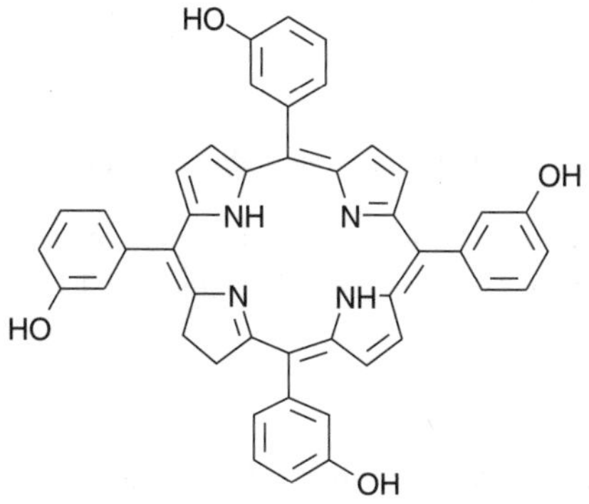	Undergoing clinical trials for the treatment of a number of different cancers

absorption. Adrenaline is liberated by the action of the esterases found in the cornea and aqueous humour.

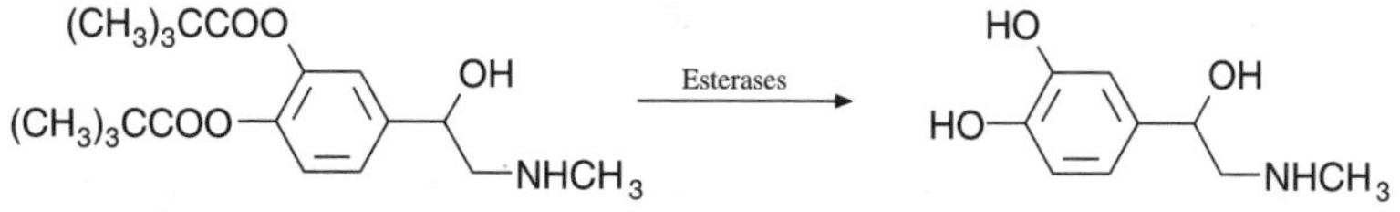

It is difficult to select a lipophilic carrier that will provide the degree of lipophilic character required. If the carrier is too lipophilic the prodrug will remain in the membrane.

The absorption of a drug will also depend on its water solubility. A drug must have a suitable water solubility if it is to be transported through a membrane by passive diffusion (see section 7.3.3). Carriers with water solubilising groups have been used to produce prodrugs with a better water solubility than the active drug. For example, amino acid carriers have been used to prepare water-soluble derivatives of the local anaesthetic benzocaine whilst

Table 12.5 Examples of the reactions used to improve the lipophilic nature of drugs

Functional group	Derivative
Acids	An appropriate ester
Alcohols and phenols	An appropriate ester
Aldehydes and ketones	Acetal
Amines	Quaternary ammonium derivatives, amino acid peptides and imines

the sodium succinate derivative has been used for the glucocorticoid methylprednisolone.

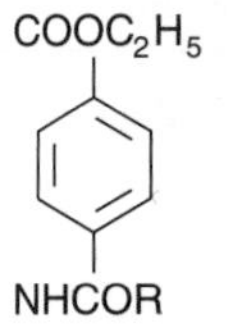

Benzocaine prodrugs R = glycine,
alanine, valine and leucine residues

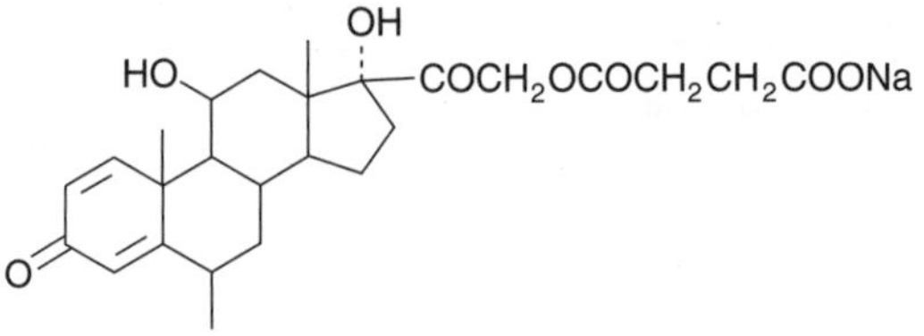

Methylprednisolone sodium succinate

Water solubilising carriers should have either ionisable groups that can form salts or groups that can hydrogen bond with water or both (see sections 2.10 and 2.10.1).

Slow release

Prodrugs may be used to prolong the duration of action by providing a slow release mechanism for the drug. Slow, prolonged release is particularly important for drugs that are used to treat psychoses where the patient requires medication that is effective over a long period of time.

Slow release and subsequent extension of action is often provided by the slow hydrolysis of amide- and ester-linked fatty acid carriers. Hydrolysis of these groups can release the drug over a period of time that can vary from several hours to weeks. For example, the use of glycine as a carrier for the anti-inflammatory tolmetin sodium results in the duration of its peak concentration being increased from about one to nine hours.

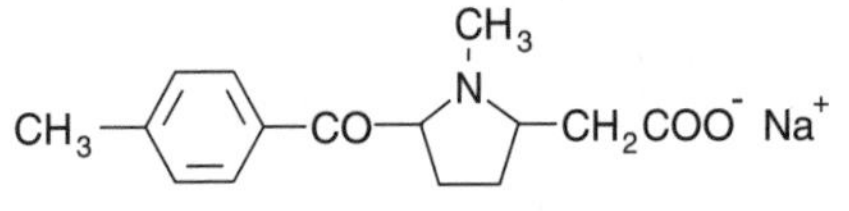

Tolmetin sodium

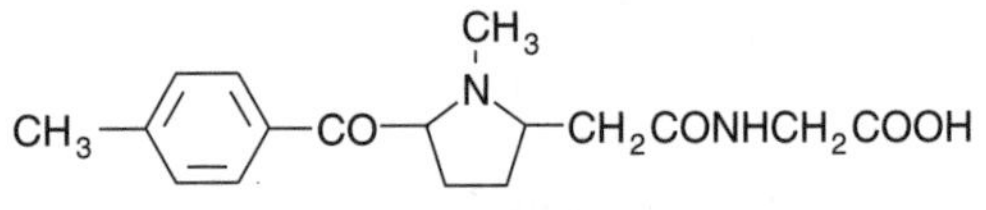

Tolmetin–glycine prodrug

Slow-release carrier prodrugs are also used as the basis of depot preparations that are administered by intramuscular injection. For example, a single dose of the almost insoluble carrier prodrug cycloguanil embonate will slowly release the antimalarial drug cycloguanil at therapeutic levels over a period of several months.

Cycloguanil embonate

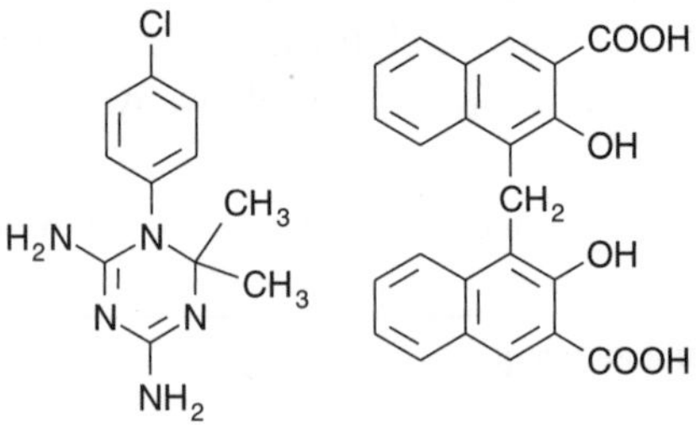

Site specificity

When a drug is absorbed into the body it is not transported to just its site of action but is rapidly distributed through all the available body compartments. This means that one must use a higher concentration than that necessary to achieve a favourable therapeutic result. An undesirable consequence of using higher concentrations is the increased possibility of the unwanted pharmacological effects of the drug becoming significant. Carrier prodrug systems offer a possible solution to this problem. The design of prodrugs is normally based on either exploiting enzyme systems or the tissue pH of the target area to the rest of the body. In theory, it should be possible to design a carrier prodrug that would only release the drug in the vicinity of its site of action. The requirements for such a drug are that once it is released it should remain mainly in the target area and only slowly migrate to other areas. In addition the carrier should be metabolised to non-toxic metabolites. Unfortunately, these requirements have been difficult to achieved.

The site-specific carrier prodrug approach has been successfully used to design drugs capable of crossing the blood–brain barrier. This barrier will only allow the passage of very lipophilic molecules unless there is an active transport mechanism available for the compound. A method developed by Bodor and other workers involved the combination of a hydrophilic drug with a suitable lipophilic carrier that, after crossing the blood–brain barrier, would be rapidly metabolised to the drug and carrier. Once released the hydrophilic drug is unable to recross the blood–brain barrier. The selected carrier must also be metabolised to yield non-toxic metabolites. Carriers based on the dihydropyridine ring system have been found to be particularly useful in this respect. This ring system has been found to have the required lipophilic character for not only crossing the blood–brain barrier but also other membrane barriers. The dihydropyridine system is particularly useful since it is possible to vary the functional groups attached to the dihydropyridine ring so that the carrier can be designed to link to a specific drug. Once the dihydropyridine prodrug has crossed the blood–brain barrier it is easily oxidised by the oxidases found in the brain to the hydrophilic quaternary ammonium salt, which cannot return across the barrier, and relatively non-toxic pyridine derivatives in the vicinity of its site of action (Fig. 12.12).

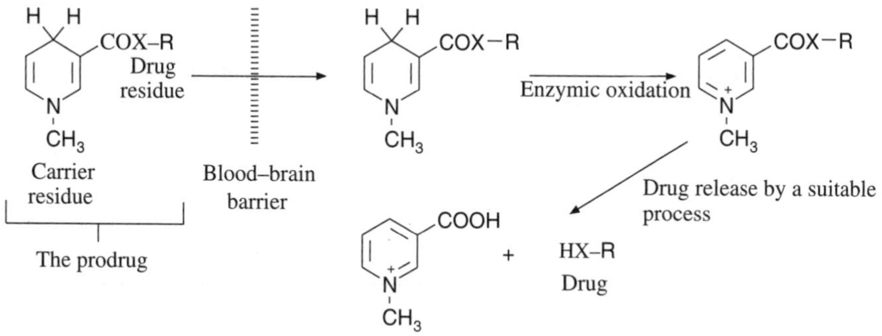

Figure 12.12 The use of dihydropyridine as a carrier to take drugs across the blood–brain barrier

The most commonly used method in prodrug design is to exploit the differences in the nature and concentration of enzymes at the target site to the rest of the body (Table 12.6).

Table 12.6 Examples of site-specific prodrug delivery systems. Carrier residues are shaded

Example	Prodrug	Target	Enzyme that releases the drug
Pilocarpine (antiglaucoma)	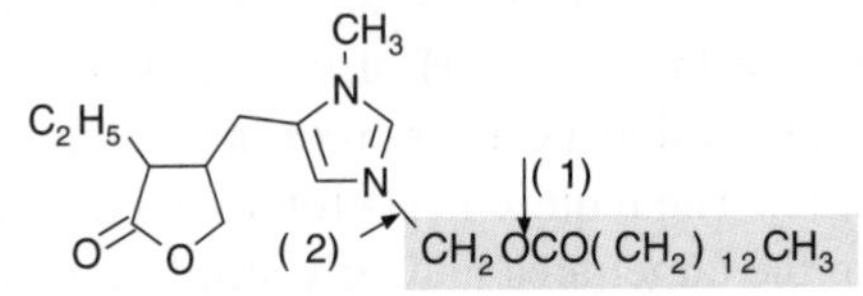	Eye	(1) Esterases followed by (topical application) (2) non-enzymic dealkylation
Adrenaline (antiglaucoma)	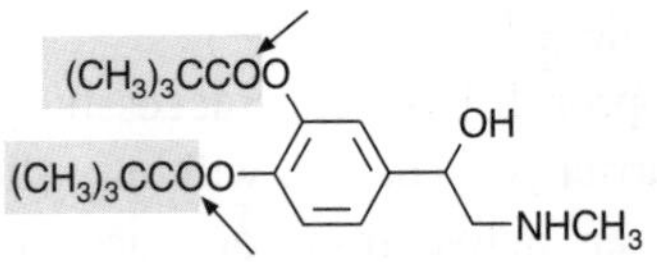	Eye	Esterases (topical application)
Sulphamethoxazole (antibacterial)	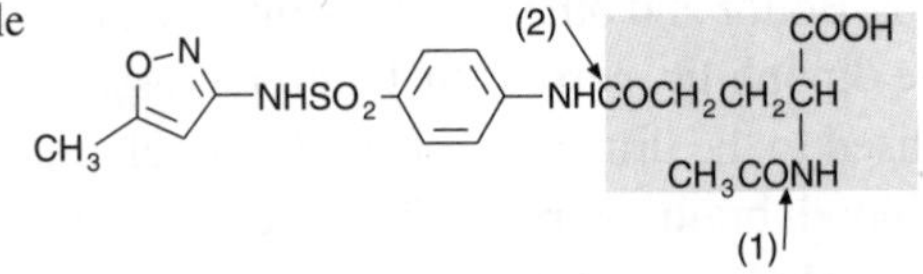	Kidney	(1) N-Acylamino acid deacylase followed by (2) γ-glutamyl transpeptidase
Dopamine (vasodilator)	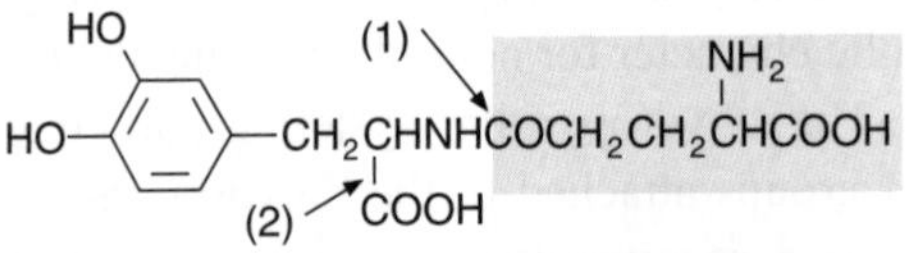	Liver	(1) γ-Glutamyl transpeptidase (2) Dopa decarboxylase
Dexamethasone (anti-inflammatory)	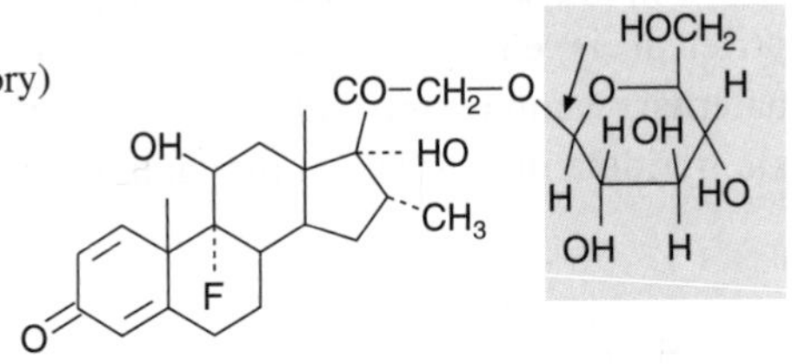	Bowel	β-D-Glucosidase (from colonic microflora)
β-Lactams (antibiotic)	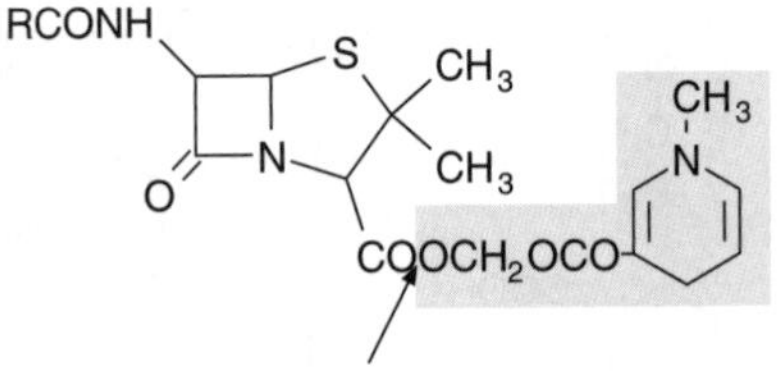	Brain	Esterases

This strategy has been used to design antitumour drugs since tumours contain higher proportions of phosphatases and peptidases than normal tissues. For example, diethylstilbestrol diphosphate (fosfestrol) has been used to deliver the oestrogen agonist diethylstilbestrol to prostatic carcinomas. Unfortunately, this approach has had limited success in producing enzyme activated site specific antitumour drugs.

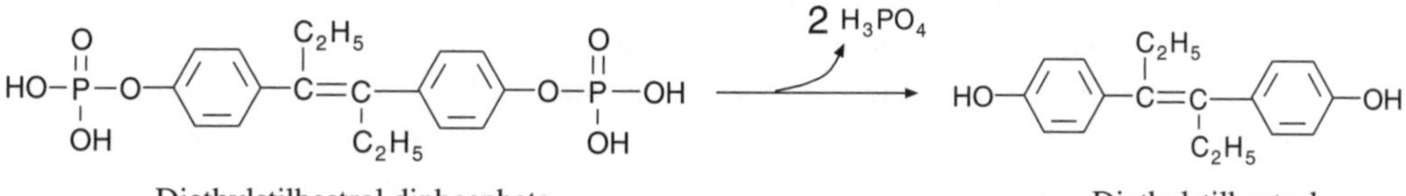

Diethylstilbestrol diphosphate   Diethylstilbestrol

The variation of pH from one compartment of the body to another has also been used as a way of activating prodrugs in specific areas of the body. For example, the antiulcer drug omeprazole inhibits gastric acid secretion by inhibiting gastric $H+$, K^+-ATPase. This enzyme is located in the acid-producing parietal cells. Omeprazole is converted in the acidic compartment of the parietal cell into its active cyclic sulphenamide. This cyclic sulphenamide acts by forming a disulphide bond with the enzyme, which prevents enzyme action.

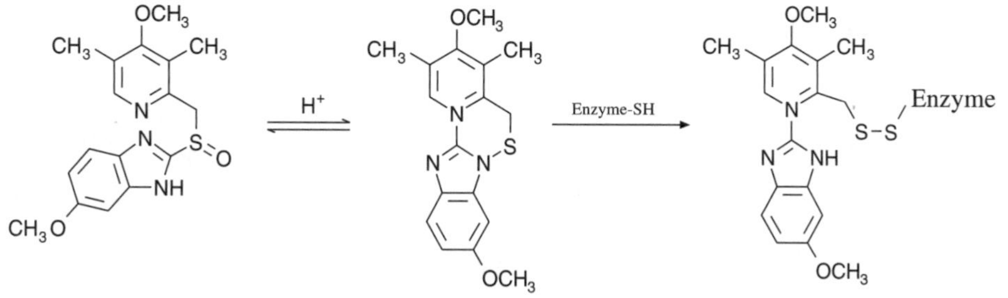

Omeprazole   Active cyclic   Omeprazole–enzyme complex
(inactive)   sulphenamide   (inhibited enzyme)

A unique new method of delivering a drug to its target is to use antibodies as prodrug carriers (see later section on ADEPT)

Minimising side effects

Prodrug formation may be used to minimise toxic side effects. For example, salicylic acid is one of the oldest analgesics known. However, its use can cause gastric irritation and bleeding. The conversion of salicylic acid to its prodrug aspirin by acetylation of the phenolic hydroxy group of salicylic acid improves absorption and also reduces the degree of stomach irritation since aspirin is mainly converted to salicylic acid by esterases after absorption from the GI tract. This reduces the amount of salicylic acid in contact with the gut wall lining.

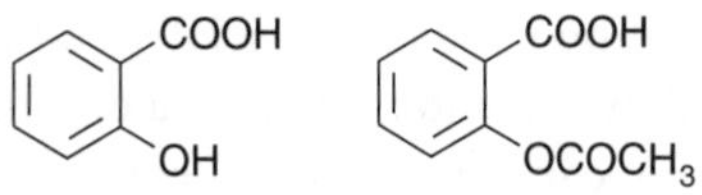

Salicylic acid   Aspirin

Improving drug stability

Leads intended for oral administration may be overlooked because they do not survive first-pass metabolism in sufficient amounts to cause a potentially useful biological response. As most drug metabolism is enzyme catalysed, one approach to increasing stability is to introduce groups into the structure of the lead that sterically hinder the binding of the drug to the enzyme. An alternative approach is to produce prodrugs in which the carrier is attached to the metabolically labile group(s) of the drug in an attempt to prevent the metabolism of the drug until it is released from the carrier at its target site. For example, terbutalin, a β_2-adrenoceptor agonist used to treat asthma, contains two metabolically labile phenol groups that reduce its first-pass bioavailability. Its bioavailability was improved by converting these labile groups to the corresponding N,N-dimethyl carbamate esters. N,N-Disubstituted carbamate esters are usually stable to chemical and enzymic hydrolysis. The resultant prodrug, bambuterol, is converted by a series of metabolic reactions to terbutalin. It is interesting to note that bambuterol inhibits the pseudocholinesterase that catalyses the last step of the metabolism, giving a slow release of terbutalin. A daily dose of bambuterol has a duration of 24 hours.

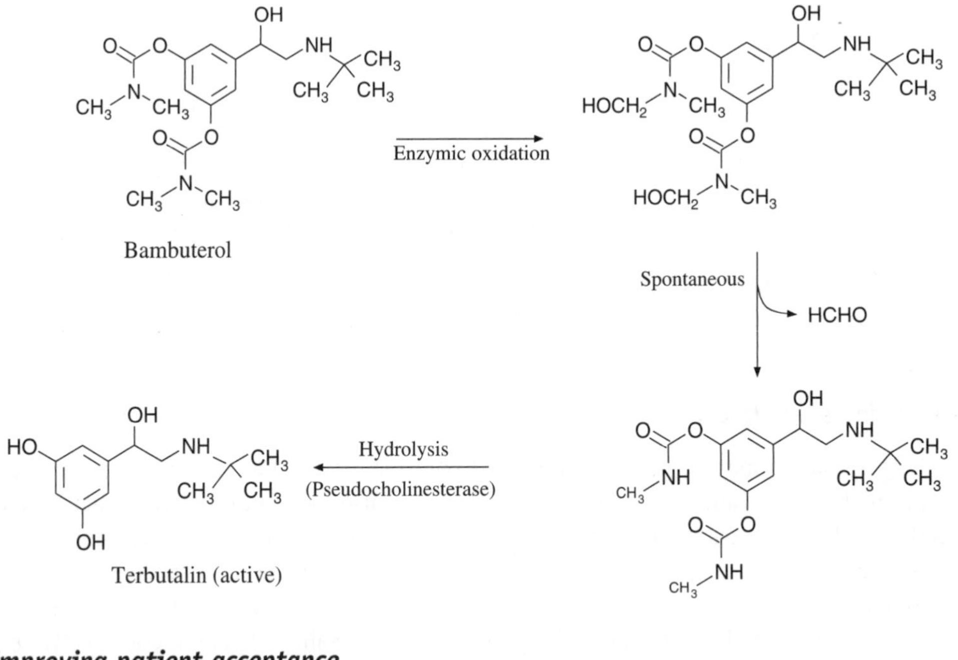

Improving patient acceptance

Odour and taste are important aspects of drug administration. A drug with a poor odour or too bitter a taste will be rejected by patients, especially children. Furthermore, a drug that causes pain when administered by injection can have a detrimental effect on a patient. The formation of a carrier prodrug can sometimes alleviate some of these problems. For example, palmitic acid and other long-chain fatty acids are often used as carriers since they usually form prodrugs with a bland taste. For example, the antibiotic clindamycin has a very bitter taste, which makes it unsuitable for use with children. It also causes considerable pain on injection. However, it was found that the palmitate ester was not bitter and the phosphate caused less pain than the parent drug when injected. In both cases the drug is released by enzyme action. The use of fatty acid esters to improve patient acceptance does, however, reduce the water solubility and increase the lipid solubility of the drug. This could affect the drug's bioavailability.

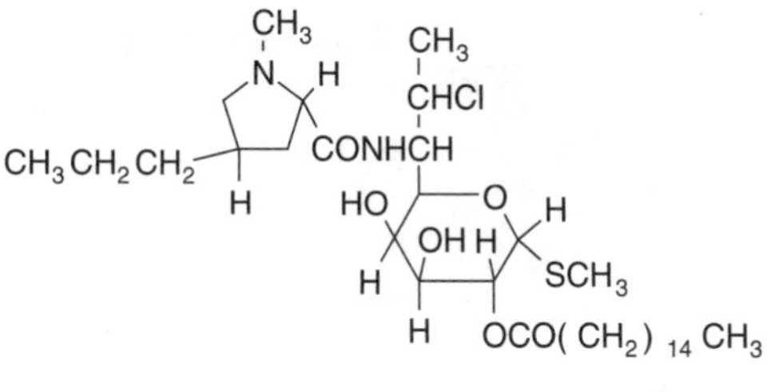

Clindamycin-2-palmitate 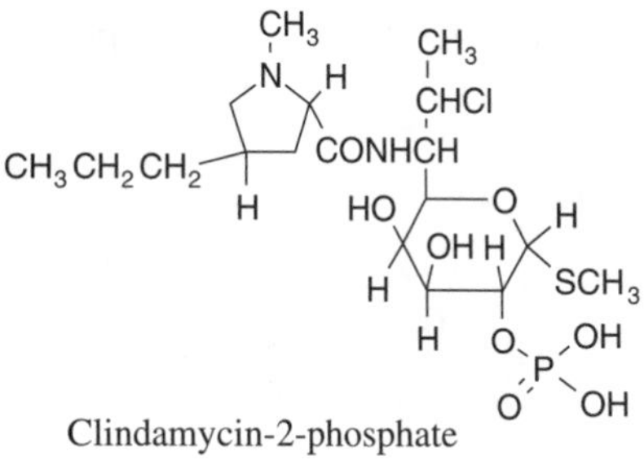   Clindamycin-2-phosphate

Antibody-directed enzyme prodrug therapy (ADEPT)

Antibody-directed enzyme prodrug therapy has been used in an attempt to develop drugs that specifically target cancer cells. The method of approach is based on the observation that many prodrugs are enzyme activated. It uses an antibody–enzyme conjugate to deliver the enzyme to the target. Once a sufficient concentration of the enzyme has reached the tumour, the prodrug is administered. When the prodrug reaches the tumour it is converted by the enzyme carried by the antibody to the active drug. For example, the anticancer agent etoposide is a semisynthetic derivative of podophyllotoxin, a compound isolated from the north American plant *Podophyllum peltatum*. In the ADEPT approach, its phosphorylated derivative is used as the prodrug because it is inactive but can be converted to etoposide by alkaline phosphatase(AP). This enzyme is delivered using an antibody–alkaline phosphatase conjugate whose antibody section recognises specific antigens on the tumour's surface. Subsequent administration of the etoposide phosphate is followed by liberation of etoposide at the surface of the tumour. The etoposide acts by diffusing into the tumour cell and destroying it (Fig. 12.13).

The *ideal* ADEPT approach depends on:

1. Finding an enzyme that does not usually occur in the body but is capable of liberating the drug from the drug–carrier prodrug complex.

2. The enzyme being relatively stable under physiological conditions.

3. Producing a drug–carrier complex that is not metabolised to any extent by the enzyme systems in the body.

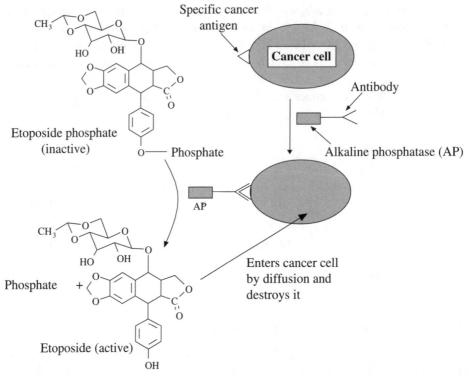

Figure 12.13 An outline of the ADEPT approach illustrated by the etoposide–alkaline phosphatase ADEPT system. Reproduced with permission from G. Walsh, 2004 *Biopharmaceuticals: Biochemistry and Biotechnology*, 2nd Edition, table 10.4, (c) John Wiley and Sons, Ltd

Unfortunately these conditions are difficult to fulfil and to date clinical results for the first generation of antibody-directed prodrugs have been disappointing. One disadvantage of ADEPT is that there may be an immune response to the antibody–enzyme conjugate as the enzyme is a foreign body. However, the risk of this happening may be reduced by the use of humanised antibodies and human enzymes such as alkaline phosphatase and β-glucuronidase. Further disadvantages are the lack of information concerning tumour antibodies and obtaining an enzyme that is sufficiently active to liberate sufficient of the active drug at the target site.

Two similar approaches to ADEPT are also being investigated, namely *antibody-directed abzyme prodrug therapy (ADAPT)* and *gene-directed enzyme prodrug therapy (GDEPT)*. In ADAPT an abzyme is an antibody that can act as a catalyst. This should allow the researchers to develop antibodies that will both attach themselves to the target and catalyse the release of the drug from the prodrug. It should reduce the risk of an immune response and also allow the development of highly specific enzyme systems. The second development, GDEPT, is based on the delivery of a gene, using a suitable vector (see section 10.15), to the tumour cell. The gene encodes the enzyme system that produces an enzyme inside the cancer cell, which in theory will convert the prodrug into its active form only after it has entered the tumour cell. This approach reduces the possibility of an immune response to the enzyme as the enzyme is only produced inside the cell. It also lowers the risk of unwanted side effects. However, it does mean that the prodrug must be able to enter the tumour cell if it is to be effective. One enzyme that has been used in GDEPT studies is viral thymidine kinase, which activates the antiviral drug aciclovir (see section 10.14.4). The activation of this drug by human thymidine kinase is poor and so its active form is largely formed within the virus. Both the ADAPT and GDEPT approaches are still in their initial stages.

12.10 Questions

1 Explain the significance of each of the members of the following pairs of terms: (a) Phase I and Phase II reactions; (b) conjugated structure and conjugation reaction; (c) carrier and bioprecursor prodrugs; and (d) soft drugs and prodrugs.

2 Explain why hippuric acids ($ArCONHCH_2COOH$) are likely to be water soluble.

3 List the main biological factors that could influence drug metabolism. Outline their main effects.

4 Outline the types of pharmacological activities that a metabolite could exhibit.

5 Explain, by means of half electronic equations, how cytochrome P-450 is reduced by cytochrome P-450 reductase when a C–H bond is oxidised by cytochrome P-450 in a metabolic pathway.

6 Outline, by means of general equations, how conjugation with glycine is used to metabolise aromatic acids. Suggest a chemical reason for the product of this process being readily excreted by the kidney.

7 Suggest, by means of chemical equations and/or notes, feasible initial steps for the metabolism of each of the following compounds: (a) pethidine and (b) 4-aminoazobenzene.

8 Explain why glucuronic conjugates are highly water soluble. Suggest reasons for glucuronate conjugation being a major phase II metabolic route.

9 (a) What is the desired objective of the drug metabolism? How is this normally achieved?

   (b) Suggest a series of metabolic reactions that could form a feasible metabolic pathway for *N,N*-diethylaminobenzene.

10 The following scheme represents the hypothetical metabolic pathway of a drug. The figures in parentheses are the rate constants for the appropriate step.

$$\text{Drug} \xrightarrow{(0.04)} \text{A} \xrightarrow{(0.30)} \text{B} \xrightarrow{(4.67)} \text{C} \xrightarrow{(0.004)} \text{D} \xrightarrow{(2.49)} \text{E} \longrightarrow \text{Excreted}$$
$$\text{B} \xdownarrow{(0.04)}$$
$$\text{F} \longrightarrow \text{Excreted}$$

(a) Explain the significance of the rate constants for the metabolism of the drug to stage B.

(b) What is the significance of the rate constants for the metabolism of B to F and C, respectively.

(c) Where is the rate determining step of the series? What is its significance?

11 Why is it necessary in some instances to design drugs with a very rapid rate of metabolism?

12 Design a prodrug that could be used to transport the diethanoate ester of dopamine (A) across the blood–brain barrier. Show by means of notes and equations how this prodrug would function.

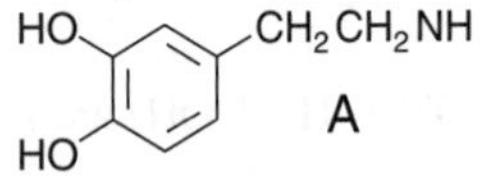

13 Outline the differences between the ADEPT, ADAPT and GDEPT approaches to prodrug design.

13

Complexes and chelating agents

13.1   Introduction

Metallic elements are essential components of many of the processes that are necessary for healthy human and animal life. A number of metallic elements, namely, sodium, potassium, calcium and magnesium, are required in bulk and are sometimes referred to as *minerals* although non-metallic elements such as phosphorus, chlorine (as chloride) and sulphur are also covered by this term. Other essential metallic elements, referred to as *trace elements*, are present in amounts that are less than 0.01% of the average human body mass. Examples of trace elements are: lithium, chromium, manganese, molybdenum, cobalt, nickel, selenium, iodine, tin, and zinc. However, besides trace and mineral elements, the healthy human body normally contains small amounts of other elements, such as silicon, arsenic and boron, which do not appear to be essential for maintaining health although current research may ultimately determine their purpose.

Metal ions and atoms occur in biological systems as either free ions or covalently bound in the structures of complex organic compounds. Free ions fulfil a variety of functions, for example the passage of sodium and potassium ions through the membranes of neurons is responsible for the transmission of an electrical impulse through the neuron (see section 9.4.3) whilst the movement of calcium into a cell can activate the enzyme system that initiates the formation of the messenger molecule nitric oxide (see section 14.4). Covalently bonded metal ions are involved in a variety of biological functions. For example, zinc is the reactive centre of a number of enzymes that initiate the cleavage of bonds whilst iron and copper are the reactive centres of many electron-transfer enzyme systems. Covalently bound metal ions can also act in a structural role, holding the enzyme structure in the correct conformation to activate a distant active site. For example, zinc is believed to act in this capacity in bovine erythrocyte superoxide dismutase (Cu–Zn BESOD), which has a copper–zinc bimetallic

Medicinal Chemistry, Second Edition   Gareth Thomas
© 2007 John Wiley & Sons, Ltd

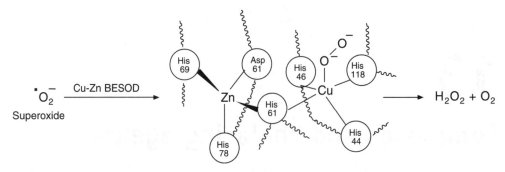

Figure 13.1 A representation of the structure of BESOD and its involvement in the conversion of superoxide to oxygen and hydrogen peroxide

site (Fig. 13.1). The zinc ion holds the structure in the correct conformation for the copper to act as the active centre for the conversion of superoxide to hydrogen peroxide and oxygen.

The concentration of essential metals in human and animal bodies is critical for healthy life. Too low a concentration causes deficiency diseases whilst too high a concentration is toxic. For example, too low a concentration of selenium would result in liver necrosis and white muscle disease but too high a concentration can cause the development of cancers and a disease in cattle known as the blind staggers, characterised by impaired vision and muscle weakness. Consequently, the control of metal ion concentration to prevent or alleviate related pathological conditions is an important aspect of medicine. Too high a concentration can sometimes be reduced by either increasing the rate of excretion by the formation of a suitable complex or removing the source of the problem. Similarly, too low a concentration can be increased by treatment with metal ion supplements. These often taken the form of complexes since metallic ions are more easily absorbed in this form.

Living organisms can also absorb non-essential metals in concentrations that are not beneficial to the welfare of that organism. These metals are absorbed because of pollution of the atmosphere and food chain. For example, lead, mercury, cadmium and other heavy metals may be found in humans eating plants grown on contaminated land. These pollutants can compete for important binding sites and as a result cause disease. Treatment of this so-called *heavy metal poisoning* usually makes use of drugs that act as chelating agents (see section 13.5.1). Metal complexes are also used as anticancer agents (see section 13.5.2), antiarthritics (see section 13.5.3), antimicrobials (see section 13.5.4) and diagnostic aids.

13.2   The shapes and structures of complexes

The term complex is normally used in chemistry to denote a compound whose structure contains one or more metal atoms or ions to which are bonded electrically neutral or charged species, such as chloride, cyanide, water, ammonia and diaminoethane (en), referred to as **ligands**. This use of the term ligand is not to be confused with its use to describe the species binding to a receptor. *Organometallic* compounds are complexes in which the metal is directly bonded to a carbon atom. The shapes of complexes about the metal are normally determined by X-ray crystallography. It is difficult to determine the shapes of biological molecules *in situ* but the advent of specialist nuclear magnetic resonance (NMR) techniques is giving some information in this respect.

Stable complexes are formed when the electronic configuration of the bonded metal corresponds to that of the nearest noble gas in the periodic table. This means that complexes formed by main group metals are stable when the sum of the electrons in their outermost shells and those supplied by the ligands equals eight (the *octet rule*). Similarly, stable transition group metal complexes are formed when the number of electrons in the outermost shell of the metal plus the electrons supplied by the ligand equals eighteen (the *eighteen electron rule*). These rules can be used to predict the number of ligands that could bond to a metal to form a stable complex. However, there are numerous exceptions to these rules.

The structures of complexes are usually explained in terms of molecular orbital (MO) theory. These explanations are usually based on a traditional σ covalent bond being formed between the metal and the ligand plus, in many cases, an extra dative π type of bond formed between the metal and the ligand (see section 13.2.1). The latter means that, in many cases, the single lines used to represent covalent bonds between the metal and the ligand in the structural formulae of complexes do not represent a standard two-electron covalent bond. However, the MO pictures of many complexes are not available as the appropriate calculations are so large that they have not yet been made.

13.2.1 Ligands

Ligands can be classified in a number of ways:

- the number of atoms donating electrons to a metal atom;

- the number of electrons they contribute to a metal atom;

- the type of bond they form with a metal atom.

The number of atoms forming bonds with the metal atom

The number of atoms directly involved in bonding to a metal atom is known as the *coordination number* of the metal atom, the most common being four and six. Ligands may be classified according to their coordination number of donor atoms as uni-, di-, tri-, tetra-dentate, etc. or using the Greek prefixes mono-, bi-, ter-, quadri- dentate, etc. ligands. Both systems are used in the literature.

The geometric arrangements of the coordinated atoms of the ligands bonded to the metal in naturally occurring complexes usually approximate to those shown in Table 13.1. For example, the atoms bonding to a four-coordinate zinc ion are usually arranged in a distorted rather than a regular tetrahedral shape about the zinc ion. It is possible for an element to exhibit more than one geometry in both different complexes and in the same complex (see section 13.2.4).

Table 13.1 The common geometrical arrangements of coordinated atoms (L) about a central metal ion

Coordination number	Common geometric arrangements		Examples of metals that can exhibit this arrangement
2	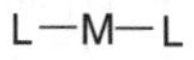 L—M—L Linear		Au(I), Ag(I), Hg(II), Cu(I)
3	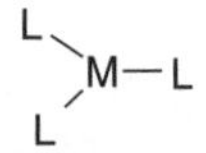 Trigonal planar		Cr(III), Fe(III)
4	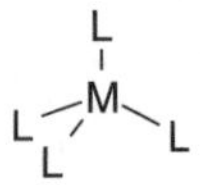 Tetrahedral 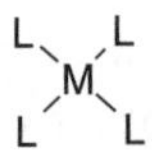	Square planar	**Tetrahedral:** Co(II), Cu(II), Zn(II), Fe(III), Co(IV), Ti(IV), Ni(II) **Square planar:** Cu)II), Pt(II), Ni(II), Cr(II), Mn(III)
5	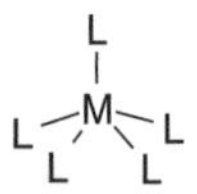 Square pyramidal 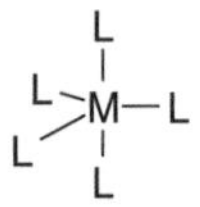	Trigonal bipyramidal	**Square pyramidal:** Mo(IV), Cu(II), Fe(II). **Trigonal bipyramidal:** V(IV), Nb(IV), Ta(V)
6	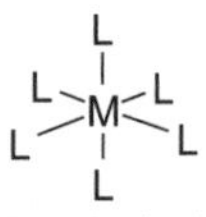 Octahedral		Co(IV), Fe(II), Mg(II), Cr(III)

Multidentate ligands that form ring structures in which the ligand is bonded by more than one atom to a single metal atom are known as *chelating agents*. Chelating agents that bind strongly to metal cations to form stable water-soluble complexes are known as *sequestering agents*. The complexes produced by chelating agents are known as *metal chelates* or *chelation compounds*. The formation of ring systems may impose restrictions on the stereochemistry of the complex. For example, the flexible diethylenetriamine forms rings in which the three bonding atoms and the metal atom do not have to be in the same plane. However, rings will only be formed by the rigid, fully conjugated terpyridine if the bonding atoms and the metal atom are all in the same plane.

The number of electrons donated to the metal atom

Ligands may be classified as one, two, three... eight electron donors to the metal atom. For example, ligands such as methyl, fluorine and hydroxide ions that form a normal covalent bond with the metal are classified as one-electron donors whilst ligands that form dative bonds by the donation of two electrons from the ligand are said to be two-electron donors, and so on (Table 13.2). However, with some ligands, the number of electrons donated will depend on the nature of the complex in which they occur. For example, bromine normally acts as a one-electron donor, but in the form of a ligand bridge it acts as a three electron donor (see section 13.2.2). However, the symbol η^n before the name of a ligand implies that *n atoms* of the ligand are involved in the bonding but not necessarily *n* electrons.

Table 13.2 Examples of the classification of ligands according to the number of electrons donated by the ligand. Electrons are counted as one for unpaired and two for lone pairs and π bonds

Classification	Ligand example	Structure
One-electron donor	Bromine	Br
Two-electron donor	Ammonia	$:NH_3$
Three-electron donor	η^3-Allyl	$H_2C{<}^{CH}_{\;\;}\!\dot{C}H_2$
Four-electron donor	η^4-Conjugated dienes	$\begin{array}{c} CH{-}CH \\ CH_2 \quad\; CH_2 \end{array}$
Five-electron donor	η^5-Cyclopentadienyl (Cp)	
Six-electron donor	η^6-Arene	—R

The type of bond formed with the metal atom

This system classifies ligands as *simple* or *π-acceptor* ligands. Simple ligands, such as methyl, fluorine and ammonia, form classical covalent bonds with the metal atom by sharing an electron with the metal atom or dative bonds by donating a pair of electrons to the metal atom. *π-Acceptor* ligands form a σ bond to the metal by donating electrons in a suitable orbital to the metal atom. The metal atom also donates electrons to an energetically favourable empty π or π* molecular orbital in the ligand, which reduces the electron density of the metal.

Two important biologically active compounds that form complexes whose structures can be explained in this manner are carbon monoxide and nitric oxide (see Chapter 11). In the case of the structure of the metal–carbon monoxide bond of metal carbonyls, the molecular orbital picture of carbon monoxide (Fig. 13.2) shows that carbon has a lone pair that

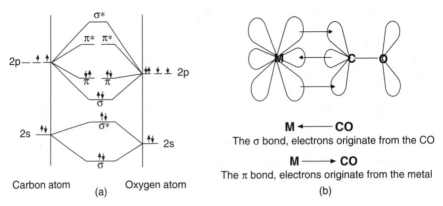

Figure 13.2 **(a)** The molecular orbital picture of carbon monoxide. The 1s orbitals are not shown as they do not contribute to the bonding. **(b)** The structure of the metal–carbonyl bond.

occupies a σ* molecular orbital. This lone pair forms a dative σ bond with a vacant σ molecular orbital in the metal. The structure of carbon monoxide also has two energetically favourable vacant π* molecular orbitals that can interact with filled metal d orbitals to give a significant degree of metal to carbon monoxide π–π* interaction. This type of π bonding is often referred to as *back* or *synergic bonding*. In carbon monoxide, the total degree of back bonding from the metal to the ligand almost balances the electron donation from the ligand to the metal. Consequently, the polarisation of the M–CO bond is low, which accounts for the low dipole moment of M–CO bonds (about 0.5 D). The strength of the π–π* interaction also accounts for the bond's chemical stability.

Nitric oxide has one more electron than carbon dioxide (*see* Fig. 14.1). This electron occupies a 2π* molecular orbital and enables nitric oxide to act as a three-electron donor, unlike carbon monoxide that is a two-electron donor. However, the structure of the M–NO bond is very similar to that of the M–CO bond. Like carbon monoxide, nitric oxide donates its lone pair of σ* electrons to the metal to form a σ bond but, unlike the M–CO bond, the metal atom only contributes three electrons to the M–NO π–π* interaction, the fourth electron being supplied by the nitric oxide. The net result is that the metal π–π* nitric oxide interaction has a full set of four electrons, three originating from the metal and one from the nitric oxide. The high degree of interaction between the metal and the ligand explains the strength of metal–carbon monoxide and metal–nitric oxide bonds. However, the metal–nitric oxide bond appears to be stronger as carbon monoxide is displaced in preference to nitric oxide in complexes containing both NO and CO ligands. The strength of these metal–ligand bonds accounts for the ease and strength with which both nitric oxide and carbon monoxide bind to iron and other metals in biological molecules.

Many ligand–metal bonds are explained in terms of synergic back bonding. However, each case must be considered on its own merits.

13.2.2 Bridging ligands

Ligands can form bridges between metal atoms. For example, in $Mn2(\mu_2 - Br)_2(CO)_8$ bromine acts as a bridge between the manganese atoms by using one of its lone pairs of electrons (dative covalent bond $\rightarrow$) to bond to one manganese atom and its unpaired electron (covalent bond —) to bond to the other manganese atom (Fig. 13.3). The symbol μ in both the names and formula of a complex shows the presence of a ligand bridge in the structure of a complex. Its subscript indicates the number of metal ions linked by that bridge. The nature of the bridge is indicated by the group found in the parentheses containing μ, while the number of bridges is shown by the subscript for the parentheses (Fig.13.3).

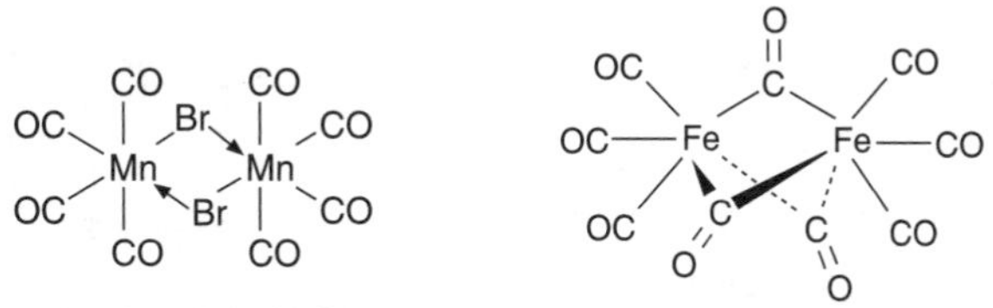

Figure 13.3 Examples of complexes with bridging ligands. In **(a)** manganese atoms are linked by two carbon monoxide bridges. In **(b)** two iron atoms are linked by three carbon monoxide bridges.

13.2.3 Metal–metal bonds

The structures of complexes can contain metal–metal bonds. For example, both cobalt and manganese form carbonyl complexes whose structures are explained by the existence of a metal to metal bond (Fig. 13.4).

Figure 13.4 The metal–metal bonds of **(a)** $Co_2(CO)_8$ and **(b)** $Mn_2(CO)_{10}$ give the metal atoms stable electronic configurations with 18 electrons in the outer shell

13.2.4 Metal clusters

Metal–metal bonding and ligand bridge formation can result in complexes whose structures contain several metal atoms in relatively close proximity. These localised concentrations of metal atoms are commonly referred to as *metal clusters*. Metal clusters are found in some proteins. They are often associated with the biological activity of the molecule and are commonly referred to by their molecular formulae. For example, the enzyme aconitase contains an Fe_3S_4 iron–sulphur cluster (Fig.13.5) with an approximately cubic shape that catalyses the conversion of citrate to aconitate. The Fe_3S_4 is activated by the binding of a fourth Fe^{2+} ion to the vacant corner of the iron cluster. This ion forms a coordination compound with the citrate by covalently bonding to the carboxylate and accepting an electron pair from the hydroxy group. The latter weakens the C–OH bond, allowing the hydroxyl group to act as a leaving group for the formation of the alkene C=C of the aconitate.

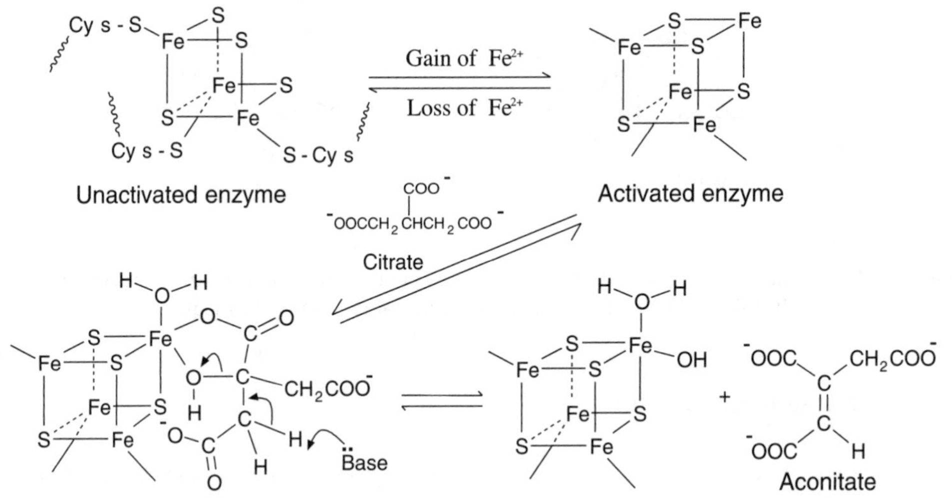

Figure 13.5 The structure of the Fe_3S_4 iron cluster in aconitase and its mode of action. Note that in the interests of clarity the sulphur atoms binding the cysteine to the cluster are not included in the formula used to represent the cluster and the peptide chains are omitted after the first structure

Different types of cluster may be found in the same enzyme, for example, succinate dehydrogenase contains three different iron–sulphur clusters (Fig. 13.6).

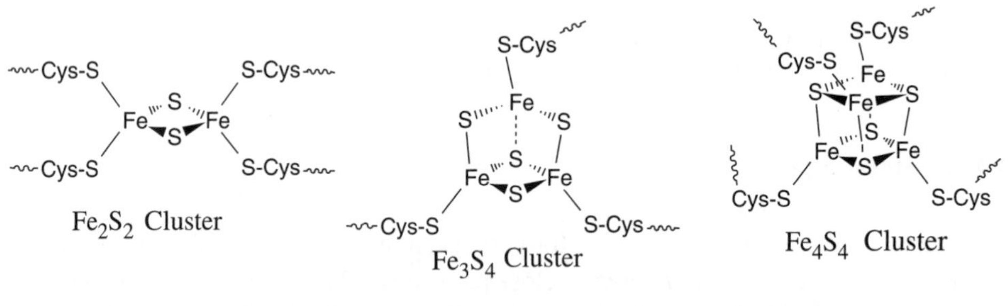

Figure 13.6 The structures of the three iron clusters in succinate dehydrogenase

13.3 Metal–ligand affinities

The stability of complexes plays a major role in their biological and chemical activity. Metals exhibit a preference for particular ligands whilst ligands will preferentially bond to certain metals. In this context a 'new' ligand (L) may displace an 'old' ligand (Y) from a complex:

$$M(Y)_n + L \rightleftharpoons M(Y)_{n-1}L + Y$$

This has important medicinal implications when one considers that most drugs contain groups that can act as ligands and the number of different types of metallic ions the body can contain. Consequently, the affinity of metals for ligands is a major consideration in discussing the reactivity of a complex in the biological context. Attempts to quantify and predict relative metal–ligand affinities in terms of the relative strengths of metal–ligand bonds are based on two different approaches: equilibrium constants and the concept of hard and soft acids and bases.

13.3.1 Affinity and equilibrium constants

This method of affinity assessment assumes that the formation of any complex is a dynamic equilibrium processes. Consequently, it is possible to use the equilibrium constant (K) for the formation of the complex by the direct reaction of the metal and ligand as a measure of its stability and, as a result, its metal–ligand bond strength. Since complexes are mainly formed in aqueous solutions, the majority of quantitative equilibrium constant determinations have been made in aqueous solution. They involve the displacement of hydrated water molecules but, by convention, displaced water molecules are ignored as the concentration of water is so large by comparison to the concentration of the complex that it is effectively the same constant value for all complexes.

The formation of a complex by the direct reaction of a metal ion and a ligand can be regarded as taking place in a series of steps (Fig. 13.7) since it is improbable that all the

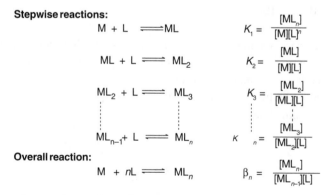

Figure 13.7 The stepwise equations and overall formation stability equilibrium constants for the formation of ML_n

Table 13.3 Examples of the stability constants for a 1:1 ATP–metal complex

Metal ion	log K	Metal ion	log K
Na^+	0.96	K^+	1.00
Ca^{2+}	3.97	Co^{2+}	4.66
Zn^{2+}	4.85	Cu^{2+}	6.13

ligands will bond to the metal at precisely the same time. However, stability formation constants are usually recorded for convenience as log values of the overall stability formation constant (Table 13.3): the larger the value, the more stable the complex and the greater the affinity of the metal for the ligands. The overall stability formation constant (β_n) is related to the stepwise formation constants by the expression:

$$\beta_n = K_1 \cdot K_2 \cdot K_3 \cdot \ldots K_n \tag{13.1}$$

where $K_1, K_2 \ldots K_n$ are the equilibrium constants for step 1, step 2 … step n.

The formation stability constant data accumulated since the 1960s allow a number of *generalisations* to be made with regard to the stabilities of complexes:

1. Complexes formed by a metal and a specific ligand tend to be more stable when the metal is in the +3 as against the +2 oxidation state.

2. Metals of the first transition series in their +2 oxidation state will form complexes whose stability is usually in the order:

$$\text{Hg} > \text{Cu} > \text{Ni} \sim \text{Pb} > \text{Co} \sim \text{Zn} > \text{Fe} > \text{Mn} \quad (Note : \text{Mn} > \text{Mg} > \text{Ca})$$

3. Where ligands contain the same donor atoms, a ligand that can form a chelated ring (L–L) will form more stable complexes than a ligand (L) that does not form a chelated ring, that is, $\beta_{L-L} > \beta_L$.

It is stressed that these statements are generalisations and that there are many exceptions. However, as a general rule, in systems where more than one complex can be formed the complex with the larger formation stability constant will be the most stable and therefore should be formed in the highest yield. However, all the possible compounds will be present in the system at equilibrium.

Equilibrium lies to the right when $\beta_{MY} < \beta_{ML}$

$$MY + L \xrightleftharpoons[\beta_{MY}]{\beta_{ML}} ML + Y$$

Equilibrium lies to the left when $\beta_{MY} > \beta_{ML}$

13.3.2 Hard and soft acids and bases

The hard and soft acid approach to predicting metal–ligand affinities regards the complex formation process as being a Lewis acid–base type of reaction. It classifies the reactants as being either a hard or soft acid or base. The species that donates the electrons is the Lewis base, whilst the species that is able to accept the electrons is the Lewis acid. In most cases the metal is the acid and the ligand is the base.

$$M \quad :L \quad \longrightarrow \quad M\,L$$

The terms hard and soft refer to the availability and mobility of the electrons possessed by the acid or base. Soft species have electrons that are easily removed (relatively mobile) whilst hard species have electrons that are firmly held (not very mobile). Softness is associated with a low charge density while hardness is related to a high charge density. For example, a hard acid will have a high positive charge density and a small size whereas a soft acid would have a low positive charge density and a large size. A large number of species have been classified as hard and soft acids and bases using these definitions but, as expected with all such definitions, a number of borderline cases are also known (Table 13.4).

Table 13.4 Examples of hard and soft acids and bases

	Hard	**Soft**	**Borderline**
Acids	H^+, Li^+, Na^+, K^+, Be^{2+}, Mg^{2+}, Ca^{2+}, Sr^{2+}, Mn^{2+}, Al^{3+}, Cr^{3+}, Co^{3+}, Fe^{3+}	Cu^+, Ag^+, Au^+, Hg^+, Pd^{2+}, Cd^{2+}, Pt^{2+}, Hg^{2+}, Pt^{4+}, Te^{4+}	Fe^{2+}, Co^{2+}, Cu^{2+}, Zn^{2+}, Pb^{2+}, Sn^{2+}, Sb^{2+}, Bi^{3+}, NO^+, SO_2, R_3C^+, $C_6H_5^+$
	Si^{4+}, Ti^{4+}, Zr^{4+}, Sn^{4+}, Hf^{4+}, BF_3, AlH_3, $Al(Me)_3$, SO_3, RSO_2^+, RPO_2^+, HF, HCl	RS^+, I^+, Br^+, I_2, Br_2, O, Cl, Br, I	
Bases	H_2O, OH^-, F^-, Cl^-, RO^-, PO_4^{3-}, ClO_4^-, SO_2^{2-}, CO_3^{2-}, ROH, RO, ROR, $RCOR$, $RCOO^-$, NH_3, RNH_2, NH_2NH_2	R_2S, RSH, RS^-, I^-, SCN^-, CN^-, I^-, R^-, H^-, $S_2O_3^{2-}$, $CH_2{=}CH_2$, C_6H_6, CO, NO	Br^-, NO_3^-, SO_2^{2-}, N_2,$PhNH_2$, pyridine

Imidazole

The concept of hard and soft acids and bases can be used to predict the relative strengths of bonds in complexes. A bond formed between two atoms with the same degree of electron mobility would be stable, having an almost even distribution of electrons. However, a bond formed between atoms with widely differing degrees of electron mobility would be less stable. Consequently, strong bonds are formed between hard acids and hard bases, soft acids and soft bases or borderline acids with borderline bases, whilst hard–soft bonding is generally much weaker. This is borne out in biological systems where Ca^{2+} ions (hard acid) are frequently found coordinated with carboxylate (hard base), Fe^{3+} (hard acid) with either carboxylate or phenoxide (hard base), Cu^{2+} (borderline acid) with the nitrogens of the imidazole ring (borderline base) of histidine residues and cadmium (soft acid) with the sulphydryl groups (soft bases) of proteins. Metallothioneins, a group of small proteins whose structures contain about 30 per cent cysteine, are believed to protect cells by complexing with soft toxic metals such as mercury(II) and lead(II).

In aqueous solution, the stability of complexes formed by soft acids and Lewis bases has been found to be in decreasing order of stability:

$$S \sim C > I > Br > Cl > F$$

Complexes between hard acids and Lewis bases are generally formed between either oxygen or fluorine donor atoms. They are not normally formed by other atoms.

13.3.3 The general medical significance of complex stability

The addition of a xenobiotic to a living organism could affect the balance of metal ions in that organism because the structures of most xenobiotics contain groups that are able to act as ligands. Consequently, the xenobiotic could form complexes that could either remove minerals and trace metals from the system by excretion or prevent essential minerals and trace element metals from carrying out their normal function or initiate a pathological response. These situations are likely to occur if the xenobiotic forms complexes that are more stable than those normally formed between the metal and the naturally occurring ligands in the system.

In view of the large numbers of different ligands found in biological systems it is not usually possible to accurately predict the effect of a xenobiotic on the metal ion balance of the system. Consequently, it is necessary in the development of a potential drug to investigate its effect on the metal ion balance of the body.

13.4 The general roles of metal complexes in biological processes

Naturally occurring metal ion complexes are involved in a wide variety of biological processes including metal storage, transport, detoxification and enzymes as well as in a structural role. This section sets out to give the medicinal chemist a better appreciation of the possible areas of impact of a new drug by presenting a very general snapshot of the types of process in which coordination compounds make a significant contribution.

Metal ions are found as part of the structures of many naturally occurring proteins These proteins are classified as *metalloproteins*. Metalloproteins that act as enzymes are subclassified as *metalloenzymes*. The metal ion in all types of metalloprotein may be bound

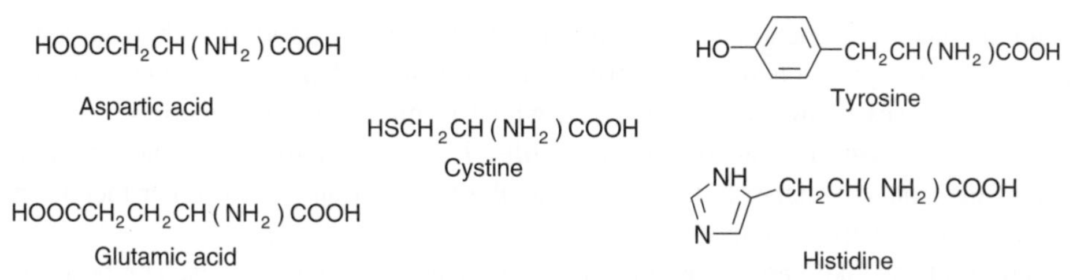

HOOCCH$_2$CH (NH$_2$) COOH

Aspartic acid

HSCH$_2$CH (NH$_2$) COOH

Cystine

HOOCCH$_2$CH$_2$CH (NH$_2$) COOH

Glutamic acid

HO—⟨ ⟩—CH$_2$CH (NH$_2$)COOH

Tyrosine

NH—CH$_2$CH(NH$_2$) COOH

Histidine

Figure 13.8 Amino acids that are frequently found coordinated to metal ions in metalloproteins.

either to an amino acid residue or to a prosthetic group in the protein. In the former case the metal ion is often coordinated to either one or more aspartic acid, cysteine, glutamine, histidine or tyrosine residues (Fig. 13.8). Metal ions are also found as integral parts of the structures of other types of naturally occurring molecules.

The metals found in metalloproteins are often readily substituted by other metals: for example, Ca by Pb, Cd or Sr, Fe by Pu, K by Tl or Cs, Mg by Be or Al and Zn by Cd. These exchanges occur because the resultant complex is more stable than the original naturally occurring complex. It usually results in either the disruption of a biological process or the accumulation of the metal in the body. Both of these processes can lead to pathological conditions.

Metal ions that are not immediately required by the organism are usually stored in the organism in the form of complexes. For example, iron is mainly stored in mammals as ferritin, which is essentially a core of hydrated iron(III) oxide coated with a protein (Fig. 13.9). Ferritin is widely distributed in the organs of mammals, especially the liver, spleen and bone marrow. It also occurs in plants and bacteria. Factors that reduce the iron in a biological system will result in the liberation of replacement iron from ferritin stores. Iron is also stored as haemosiderin, which is considered to be a degradation product of ferritin.

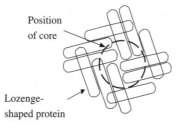

Position
of core

Lozenge-
shaped protein

Figure 13.9 Ferritin. A core of about 8 nm in diameter that is believed to consist of a hexagonal packed array of oxygen ions with Fe(III) ions occupying the octahedral sites. The core is attached to 24 lozenge-shaped proteins by Fe(III) ions and (FeII–O–FeIII) dimers

Little is known about the storage of other metals. However, it is believed that one of the functions of the small, cysteine-rich, multifunctional proteins called metallothioneins is the storage of ions such as copper(I) and zinc(II). In addition these proteins also readily bind toxic ions such as Cd^{2+} and Hg^{2+}.

Naturally occurring metal complexes are involved in the transport of both metal ions and ligands. For example, in aerobic bacteria, iron is transported from its storage sites in the form of complexes known as siderophores, whilst in mammals monomeric glycoproteins (M~80 000) known as transferrins are utilised. Both siderophores and transferrins transport the iron as water-soluble Fe(III) complexes. The transport of other metal ions is also believed to involve complex formation.

Metalloproteins take part in the transport of species other than metals. The best known system is the transport of oxygen, carbon dioxide and other compounds by haemoglobin (Hb). Haemoglobin transports oxygen from the lungs and delivers it to myoglobin in the tissues. The oxygen transport systems of other species also involve metalloproteins.

Metal complexes are involved in the process of detoxification. This is the process by which a living organism converts unwanted species into harmless substances. Natural metal ion detoxification processes usually remove the metal ions from circulation by forming complexes that are stable at physiological pH. Metalloproteins are also involved in the detoxification of unwanted ligands. For example, the detoxification of superoxide by Fe(III)SOD to oxygen and hydrogen peroxide (Fig. 13.1) and the removal of carbon dioxide from tissue by haemoglobin.

Metals with multiple oxidation states often constitute the active sites of the proteins involved in electron transfer reactions. In these cases the process of electron transfer is essentially a redox process in which a change in the oxidation state of a metal in one protein is accompanied by the corresponding change of state in the next adjacent protein. For example, electrons generated by an oxidation process in one protein are transferred to the adjacent protein where they reduce the oxidised state of the metal (Fig. 13.10). These oxidation state changes are often accompanied by a change in the geometry of the metallic site, which necessitates the change taking place via a suitable transition state. Iron clusters (Fig. 13.6) are often involved in the electron transfer processes in mammals.

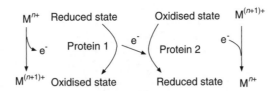

Figure 13.10 A schematic representation of an intramolecular electron transfer process involving metal sites. The two proteins are embedded in close proximity in a membrane

One third of the human body's enzyme systems are metalloenzymes. Their active sites are either coordinated single metal ions or, more commonly, metal clusters. Several active sites with different structures may be located within the same enzyme (see section 13.2.4). Metalloenzymes are classified according to their mode of action (see section 9.2). Each category includes metal clusters with different structures that are capable of bringing about the same substrate changes.

Metals are able to bond with DNA in a number of different ways. Their potential coordination sites are the oxygen and nitrogen atoms of the base residues, the hydroxyl groups of the sugar residues and the oxygen atoms of the phosphate residues. However, the main sites of coordination appear to be those found on the base and phosphate residues. The binding of a metal ion to DNA may stabilise the structure of the DNA and can cause a therapeutic (see cisplatin, section 13.5.2) or toxic response.

All metal ions have a structural role in metalloproteins. Their coordination number determines the configuration of their area of the structure of the protein. For example, a metal with a coordination number of four is likely to have either an approximately tetrahedral or square planar arrangement for its ligands. In addition metal ions influence the conformations of the adjacent peptide chains. Zinc ions are particularly interesting in that they appear to act as both intermolecular and intramolecular bridges in protein structures. Intramolecular Zn^{2+} bridges have been found in the proteins of a number of species. The zinc ion is responsible for the formation of the peptide chain of the protein into a loop of about 12–15 amino acid residues referred to as a zinc finger. The chain is coordinated to the zinc in an approximate tetrahedral configuration through cysteine and histidine residues (Fig. 13.11). This type of zinc finger is classified as a C_2H_2 type in order to distinguish them from the C_x type (where x is usually 4, 5 or 6) in which only cysteine residues are coordinated to the zinc. Zinc fingers can occur either singly or in tandem, with a number of fingers being found in one protein molecule.

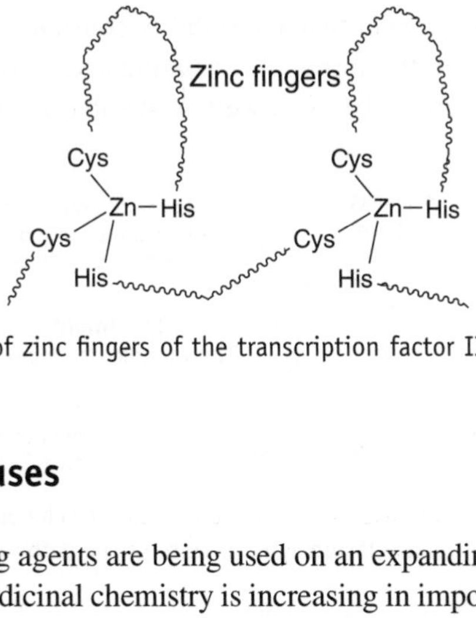

Figure 13.11 The structure of zinc fingers of the transcription factor IIIA of the African clawed toad *Xenopus*

13.5 Therapeutic uses

Complexes and complexing agents are being used on an expanding scale to treat a variety of diseases. This aspect of medicinal chemistry is increasing in importance with advances in the knowledge of the role of metals in the physiological and pathological states of the body.

13.5.1 Metal poisoning

The presence of excessive amounts of a metal in a living organism is responsible for a variety of syndromes. For example, Wilson's disease, symptoms of which are a malfunctioning liver, neurological damage and brown or green rings in the cornea of the eyes, is caused by a copper overload due to a genetically inherited metabolic defect. Excessive amounts of calcium result in calcification of tissue, cataracts and kidney and gall stones.

Metal poisoning occurs when the body's metal management system allows the concentration of the metal to reach toxic levels in sensitive areas of the body. The metal can enter the organism in a number of ways ranging from accidental ingestion, pollution of the food chain, skin absorption and breathed in as atmospheric pollutants. Treatment is based on the use of chelating agents that form stable complexes with the excess metal and are either easily excreted or deposited as harmless solids. However, it is possible that a second line of attack based on identifying the natural detoxification process for the metal will emerge.

The toxic effects of a metal are believed to be due to the formation of a stable complex between the metal and a species that is an essential component of a biological pathway in a living organism. The resulting complex is unable to take part in the pathway and so its efficiency is reduced. Consequently, the organism develops the diseased state associated with toxic levels of the metal. However, these toxic levels can be reduced by the use of chelating agents to form stable complexes that are easily excreted (Fig. 13.12). For this form of therapy to be effective the metal must form more stable complexes with the chelating agent than the naturally occurring ligands it coordinates with in the living organism. In other words, the stability formation constant (β_1) for the formation of the metal–chelate (L'M) must be greater than the stability constant (β_2) for the formation of the metal–natural ligand (LM). Furthermore, the chelating agent must also contain good water solubilising groups (see section 2.10.2) as well as suitable ligands.

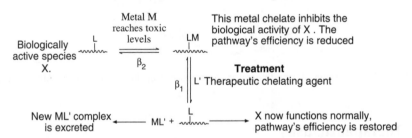

Figure 13.12 A schematic representation of the general action of chelating agents in the treatment of metal poisoning. The addition of the chelating agent L' restores the pathway to full operation

Chelating agents are used instead of monodentate ligands because their complexes tend to be more stable than those formed by monodentate ligands. For example, the stability formation constant log β value for $[Cu(NH_3)_4]^{2+}$ is 11.9 whilst that for the CuEDTA complex is 17.7. In addition, experimental evidence indicates that the stability of complexes usually increases with the number of bonds formed by the chelating agent to the metal and that chelating agents that form five- and six-membered rings will form the most stable complexes. Consequently, a compound that is to be used as a therapeutic chelating agent should have the following characteristics:

- ligand groups that are specific for the metal;

- be a multi-ligand chelating compound;

- form complexes that are more stable than the relevant naturally occurring ligands;

- be easily excreted;

- have an LD_{50} greater than 400 mgkg^{-1}.

Chelating agents commonly used to treat cases of metal poisoning are ethylenediaminotetraacetic acid (EDTA), dimercaprol and penicillamine (Fig. 13.13), which is a degradation product of penicillin. The structure of EDTA contains 'hard' amine and carboxylate groups, which means it readily coordinates with 'hard' metals such as calcium and magnesium. Both of these elements are essential components of living organisms and their depletion would result in the malfunction of a number of biological processes as well as weakening of the bone structure of mammals. Consequently, EDTA is normally used in the form of its $Na_2Ca(EDTA)$ salt in an attempt to reduce the Ca^{2+} depletion. Dimercaprol (British anti-Lewisite, BAL) has the 'soft' sulphydryl groups and so coordinates 'soft' metal ions such as Cd^{2+}, Hg^{2+} and Cu^+. However the discovery of a natural detoxification process in bacteria resistant to mercury poisoning may result in a new genetic method of treatment for mercury poisoning. Penicillamine has both 'hard' and 'soft' ligand groups. It is used in the treatment of Wilson's disease and in chronic cases of lead and mercury poisoning. The

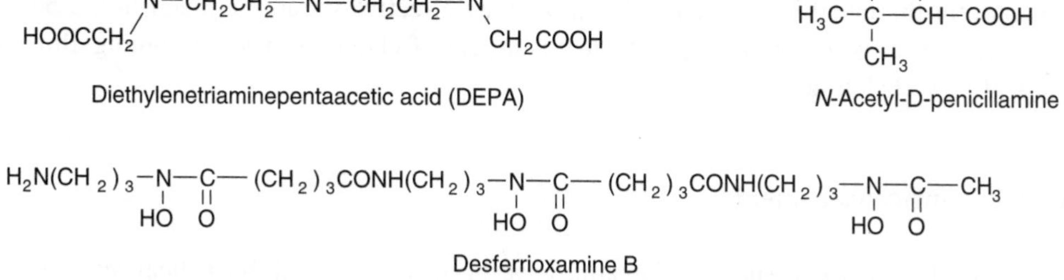

Figure 13.13 Examples of the chelation agents used to treat metal poisoning

siderophore desferrioxamine, which contains hard ligand groups, has been used to treat iron overloads.

A number of compounds with similar types of active structural areas have been developed from the lead compounds described in the previous paragraph. For example, EDTA has lead to the development of polyaminocarboxylic acids such as DEPA (diethylenetriaminepentaacetic acid Fig. 13.13) which has been shown to be effective in cases of plutonium poisoning.

The toxicity and difficulty of administration of BAL has resulted in the use of sodium 2,3-dimercaptosuccinate and unithiol (Fig. 13.14). Complexes formed by these chelating agents are charged in solution. Consequently, these complexes are less likely to cross biological membranes (see section 7.3.3) and be distributed around the body, which reduces the possibility of unwanted side effects. However, these charged complexes will accumulate in the liver and kidneys, which enhances their chances of excretion.

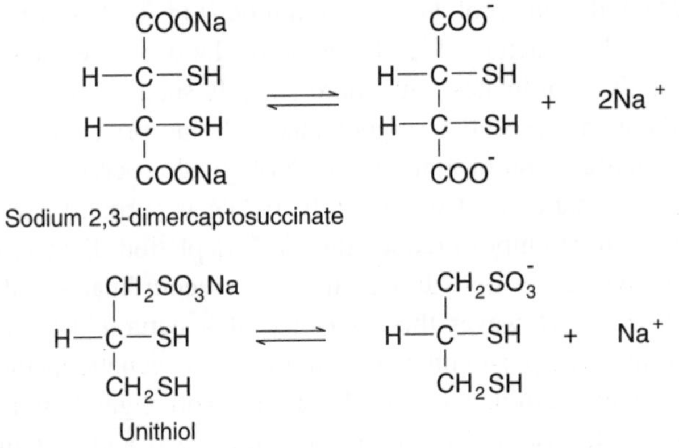

Sodium 2,3-dimercaptosuccinate

Unithiol

Figure 13.14 Sodium 2,3-dimercaptosuccinate and unithiol. The thiol groups have a high affinity for metals

The toxicity of D-penicillamine has resulted in the use of the less toxic *N*-acetyl-D-penicillamine. This drug is very effective in treating methylmercury(II) poisoning and will extract this compound from brain tissue.

Chelating agents can cause worse problems than they are designed to cure. For example, both BAL and penicillamine form complexes with cadmium that are more toxic than the metal that they are being used to remove. Furthermore, some ligands will form complexes that prevent the metal being excreted, that is, they effectively trap the metal in the body. Consequently, it is important to consider this aspect of chelation when designing clinical trials of new chelating agents.

13.5.2 Anticancer agents

Many metal complexes, such as cisplatin, have been found to exhibit anticancer activity (Fig. 13.15). Cisplatin was discovered as a consequence of an investigation into the effects

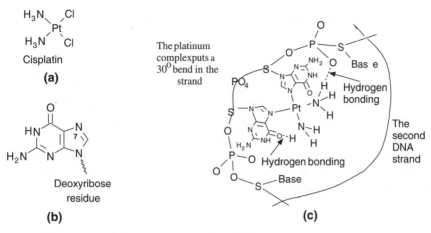

Figure 13.15 **(a)** Cisplatin, **(b)** a guanine residue and **(c)** the proposed structure of the cisplatin–DNA complex. An outline structure is given for guanine in this complex and S represents a sugar residue

of electric fields on the growth of *E. coli* bacteria. Barnett, Rosenberg and collaborators observed that normal cell division was inhibited and cells grew into long filaments. Eventually it was discovered that the cause of this growth was due to *cis*-diaminedichloroplatinum(II) (cisplatin) and *cis*-diamminetetrachloroplatinum(IV) generated *in situ* from the platinum electrodes and ammonium chloride in the solution used in the original study. Cisplatin was a well documented compound and animal testing showed that it was active against testicular, cervical, ovarian, lung and other cancers. Consequently, drug development has focused on this compound and its analogues. However, it is not active against all forms of cancer.

The mechanism by which cisplatin acts is not fully elucidated. In the extracellular fluid cisplatin undergoes little chloride–water interchange because of the fluid's high chloride concentration. Since it is uncharged it can cross cell membranes by passive diffusion (see section 7.3.3). However, once the drug penetrates the cell, the chloride concentration is low enough to allow a significant chloride–water interchange with the formation of species such as $[Pt(NH_3)_2(H_2O)_2]^{2+}$ and $[Pt(NH_3)_2(OH)_2]$. Practical evidence suggests that these species coordinate with either the N-7 atoms of two guanine residues or the N-7 of a guanine residue with an adenine residue to form intrastrand bridges linking two areas of the same strand (Fig. 13.15). This causes a distinct bend in the DNA at the point of platination, which leads to a supression of replication. The mechanism by which the formation of these complexes leads to the death or suppression of reproduction of the cancer cell is not fully understood but has been shown to involve transcription factors known as high mobility group proteins (HMGs). It is not certain how these proteins facilitate the anticancer activity of cisplatin but they appear to bind to the area of the DNA containing the cisplatin adduct and prevent repair of the DNA. Furthermore, as HMGs appear to be expressed mainly in tumour cells their involvement would offer an explanation of why cisplatin is to some degree selective. Transplatin, the geometric isomer of cisplatin, also binds to DNA but less is known about the nature of this binding.

However, it has also been shown to inhibit DNA replication although it has not been used clinically.

Cisplatin suffers from the serious disadvantage that it must be administered by intravenous infusion and it is very toxic. Its use can cause nausea, vomiting, renal dysfunction and leucopenia. In addition some tumours become resistant to the drug and so researchers are actively seeking second-generation platinum(II) and (IV) analogues with less severe side effects (Fig. 13.16a). It is thought that cisplatin resistance may be due to an increase in the rate of repair of the damaged DNA. The first step in operation of this repair system is believed to involve the *cutting out* of the DNA–cisplatin complex by the repair system. The resultant gap between the two sections of the DNA strand is filled by the action of a DNA-polymerase and sealed by a DNA-ligase (Fig. 13.16b).

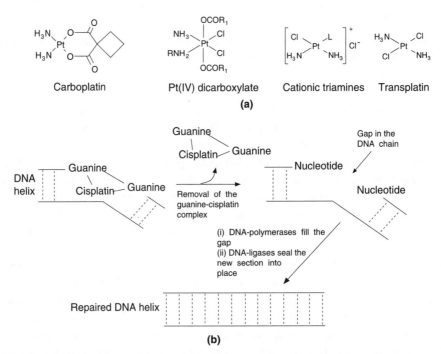

Figure 13.16 **(a)** Examples of platinum complexes that are active against cancer. Carboplatin is less toxic than cisplatin and is now in clinical use. Pt(VI) dicarboxylate analogues are active against a number of ovarian carcinoma cell lines. Cationic triamines where L is pyridine, a para-substituted pyridine, pyrimidine or purine are active against S 180 ascites and L 1210 tumours in mice. **(b)** A schematic representation of the repair system that is believed to cause resistance to the anticancer action of cisplatin

The importance of the discovery of cisplatin has stimulated the investigation of other metal complexes for antitumour activity. A number of compounds of main group (Ga, Ge, Sn and Zn) and transition metals (Ti, V, Cr, Mo, Mn, Fe, Cu and others) have been reported to have antitumour activity (Fig. 13.17), many of which have been modelled on the square planar coordination of cisplatin but a number with octahedral coordination have also been

Vanadocene dichloride

Spirogermanium hydrochloride

Ferrocene salts

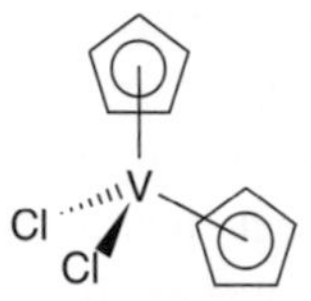

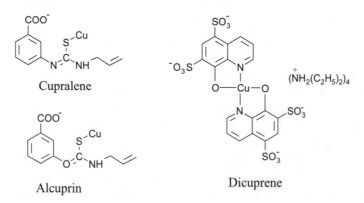

Figure 13.17 Examples of the complexes of other metals that have antitumour activity. Vanadocene dichloride and its titanium, molybdeum, tungsten, niobium and tantalum analogues exhibit antitumour activity. Ferrocene salts can exhibit activity against animal tumours

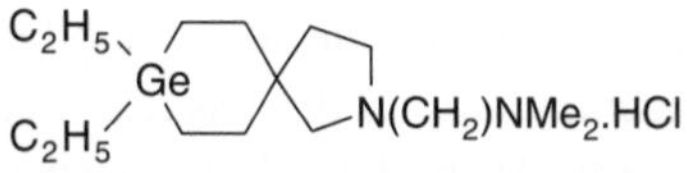

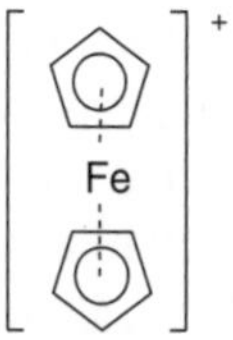

found to be active. However, few compounds have been studied in depth since investigations have been dominated by platinum coordination compounds. Although metal complexes are being used as antitumour agents they can also cause cancers. For example nickel carbonyl (Ni(CO)4) is one of the most carcinogenic compounds known to man. Consequently, this aspect of the pharmacology of metal complexes must be borne in mind when designing clinical trials for complexes.

13.5.3 Antiarthritics

Rheumatoid arthritis affects over 5 per cent of the population of the Western world. Its cause is unknown but it is thought that it results from a failure of the patient's immune system. Practical evidence also suggests that rheumatoid arthritis is related to a *local* imbalance in the concentration of copper. However, the relationship between the copper concentration and rheumatoid arthritis is still not clear.

A number of copper complexes have been found to be active against rheumatoid arthritis (Fig. 13.18). This antiarthritic activity is believed to be due to their anti-inflammatory action. Penicillamine has also been used to treat rheumatoid arthritis even though it

Cupralene

Alcuprin

Dicuprene

Figure 13.18 Examples of copper complexes with antiarthritic activity

increases the rate of excretion of copper by urinary excretion (see section 13.5.1). It is thought that its antiarthritic activity is due to it mobilising the copper in the form of a complex that temporarily accumulates in the tissues, thereby reducing any inflammation.

Since 1940 a number of gold(I) thiolates have been used to treat rheumatoid arthritis (Fig. 13.19), although little is known about their mode of action. However, they are slow to act and it can be several months before any beneficial effects are noticed. Furthermore, with the exception of auranofin, which is administered orally, they are given by a painful intramuscular injection.

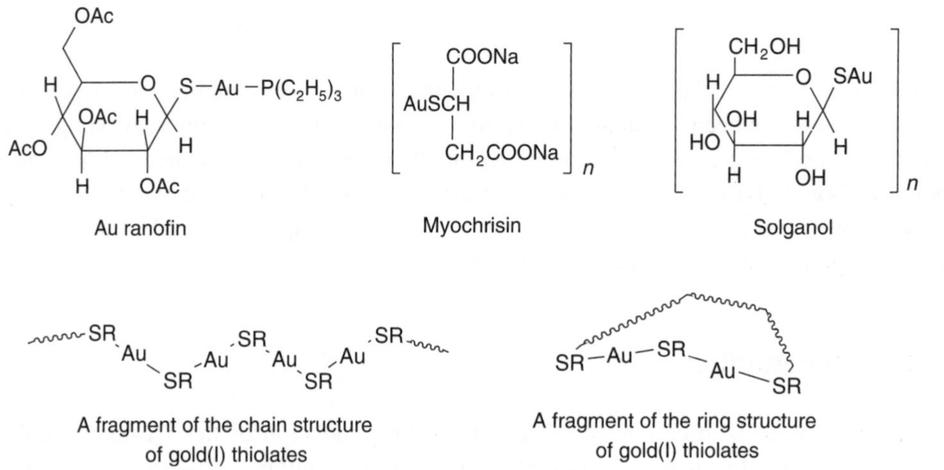

Figure 13.19 Gold(I) compounds used to treat rheumatoid arthritis. Experimental work has indicated that the 1:1 gold thiolate complexes are small chain or ring polymers

13.5.4 Antimicrobal complexes

A large number of metal complexes exhibit antimicrobial activity. The activity of these complexes may be due to either the presence of a toxic metal ion or a biologically active ligand (Fig. 13.20). For example, silver(I) sulphadiazine is used clinically as a topical antifungicide and antibacterial. It appears to depend for its action on the release of a toxic Ag(I) ion rather than the sulphadiazine ligand. Other silver(I) complexes, such as silver(I) imidazole $[Ag(imd)]_n$, are also active against a range of microorganisms. J. A. Urbina *et al.* have shown that a complex of ruthenium(II) with chloroquine shows a significantly increased activity of the parent drug against *Plasmodium falciparum*, a parasite that causes malaria, while a complex of ruthenium(II) with chlortrimazole exhibits up to a 90 per cent increase in the parent drug's inhibition of *Trypanosoma cruzi*, a cause of Chagas' disease. Aluminium, gallium(III) and iron complexes are also active against *Plasmodium falciparum*. For example, $[Ga(madd)]^+$, where madd is 1,12-bis(2-hydroxy-3-methoxy-benzyl)-1,5,8,12-tetraazadodecane, is active against the chloroquine-resistant parasite. The polyoxometalates exhibit antiviral activity. For example, the tungsten–antimony

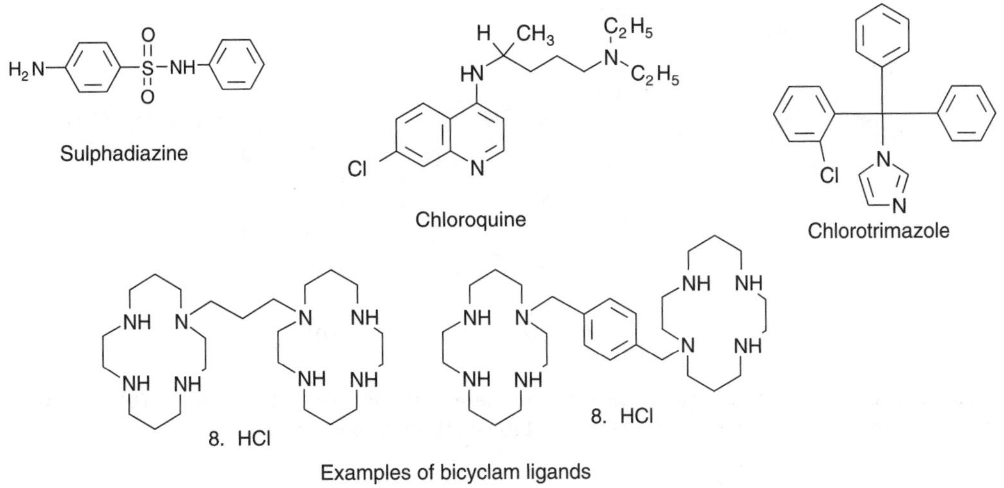

Sulphadiazine

Chloroquine

Chlorotrimazole

8. HCl

8. HCl

Examples of bicyclam ligands

Figure 13.20 Examples of some of the ligands used in biologically active complexes

complex [NaW$_{21}$Sb$_9$O$_{86}$](NH$_4$)$_{17}$(Na) has been shown to be an anti-HIV agent. However, it is too toxic to be of clinical use. A number of other polyoxometalates have also been found to be anti-HIV agents. However, the bioavailability of polyoxometalates is generally poor and needs improvement if a viable drug is to be produced. Bicyclams (Fig. 13.20) are also anti-HIV agents. They have been shown to inhibit the retrovirus replicative cycle. Although numerous metal–complex antimicrobial agents have been discovered, few have been found to be suitable for clinical use.

13.5.5 Photoactivated metal complexes

A number of active metal complexes have had their activity enhanced when irradiated with visible light. For example, irradiation of *trans,cis*-[Pt(OAc)$_2$I$_2$(en)] (Fig. 13.21) with visible light ($\lambda > 375$ nmn) increases the toxicity of this complex to bladder cancer cells by 35 per cent. Furthermore, some inactive metal complexes have been converted into an active form by irradiation (see section 12.9.3). For example, Sessler has shown that both lutetium(III) and gadolinium(II) texaphyrins are antitumour agents These compounds exhibit a selectivity of 10:1 for tumour cells, which is much greater than the 3:1 exhibited by the porphyrins normally used in PDT (see section 12.9.3). Lutetium(III) texaphyrin (Lutex) is inactive until activated by light in the far red region of the visible spectrum. On activation it reacts with oxygen to generate cytotoxic singlet oxygen (see Fig. 12.11). Gadolinium(III) texaphyrin is activated by X-rays since the gadolinium acts as a radiation sensitiser. The presence of the gadolinium atom also makes it possible to use this complex as a marker in magnetic resonance imaging (MRI). The accumulation of gadolinium in the tumour cells is allowing the use of MRI scans to pinpoint the positions of tumours, which allows the doctor to selectively irradiate these areas. Both lutetium(III) and gadolinium(III) texaphyrin are undergoing clinical assessment.

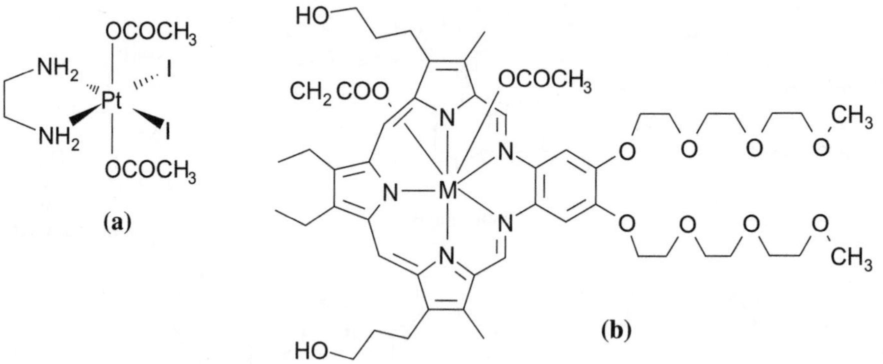

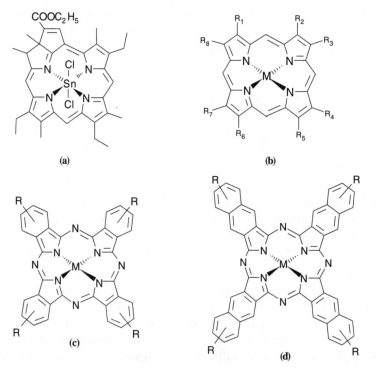

Figure 13.21 Examples of photoactivated metal complexes: **(a)** *trans,cis*-[Pt(OAc)$_2$I$_2$(en)]; **(b)** the general molecular structure of texaphyrins, where M = lutetium(III) or gadolinium(III)

A number of metallophorphyrins are also undergoing evaluation as potential photoactivated drugs. For example, tin(IV) ethyl etiopurpurin preferentially binds to high-density lipoproteins in blood while tin, cobalt and gallium porphyrin complexes (Fig. 13.22) are inhibitors of haem oxygenase. Phthalocyanine and naphthalocyanine

Figure 13.22 Examples of the general structures of photoactivated metal complexes. **(a)** Tin (IV) ethyl etiopurpurin. **(b)** Metal porphyrins. **(c)** Metal phthalocyanines. **(d)** Metal naphthalocyanines

complexes also exhibit photosensitive biological activity (Fig. 13.22). Unfortunately, many photosensitive metal complexes are very hydrophobic and so need specialised delivery systems. For example, the use of liposomes (see section 2.13.1) can result in good tumour accumulation of zinc, gadolinium, indium and tin complexes.

13.6 Drug action and metal chelation

The activities of a number of drugs can be related to their ability to chelate essential metals. For example, the action of 8-hydroxyquinoline (Fig. 13.23) appears to be partly based on its ability to chelate iron. Analogues of 8-hydroxyquinoline that are unable to form chelates exhibit either a much reduced or zero activity. The activity of isoniazid is also attributed to its ability to chelate iron. Furthermore, the mode of action of the tetracyclines is possibly based on their chelation with magnesium. The magnesium is believed to form a bridge binding the tetracycline to the rRNA in the bacteria. This is thought to halt the protein synthesising action of the bacterial ribosomes. Since most drugs contain potential ligand groups, it is possible that many compounds owe part of their pharmacological activity to complex formation.

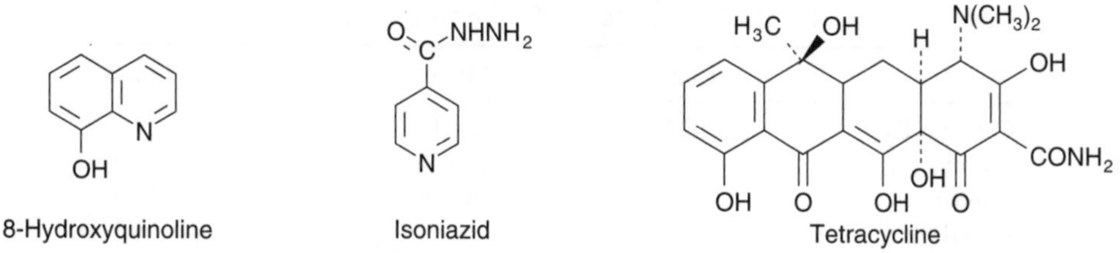

Figure 13.23 Examples of some of the drugs that are believed to owe part of their action to chelation

Xenobiotics, such as drugs, whose structures contain ligands that can form more stable ligands than the endogenous ligand can upset the balance of metal ions in biological systems. This may lead to deficiency diseases.

13.7 Questions

1 Outline, quoting relevant examples, the various uses of metallocomplexes in maintaining health in humans.

2 Draw structural formulae for each of the following ligands. Classify the ligand in terms of its electron donating power or the number of atoms bonding to the metal atom. (a) Cp, (b) EDTA and (c) ammine.

3 Draw the structural formulae of each of the following compounds. Indicate on the formula the atom most likely to coordinate with a metal ion. (a) Cysteine, (b) tyrosine, (c) 8-hydroxyquinoline and (d) dimercaprol.

4 Explain, using suitable examples, the meaning of the terms: (a) synergic bonding, (b) ligand bridge and (c) metal cluster in the context of the structures of complexes.

5 Explain the meaning of the terms hard and soft acids and bases. Predict, using the concept of hard and soft acids and bases, whether it is possible for the following pairs of species to form a stable complex: (a) cysteine and Pt(II), (b) carbon monoxide and tin(IV), (c) magnesium(II) and benzene, (d) copper(I) and sodium ethanethiolate, (e) tin(II) and histidine, (f) cholesterol and iron(III), (g) cadmium and glutamic acid and (h) mercury(II) and acrylic acid. Indicate, in the case of stable complex formation, where the ligand coordinates with the metal ion.

6 Calcium ions are known to initiate blood clotting. Suggest a compound that could be added to blood to prevent blood samples from clotting. Explain how the compound prevents blood clotting.

7 List the general features that a compound should exhibit if it is to be suitable for use as a metal detoxification agent.

8 The stability formation constants of a number of EDTA–metal complexes are given in Table 13.5.

Table 13.5 The log stability constants of some EDTA complexes

Metal	Log β	Metal	Log β
Calcium(II)	10.6	Iron(III)	25.1
Chromium(II)	13.0	Lead(II)	17.0
Copper(II)	18.0	Mercury(II)	21.0
Iron(II)	14.3	Zinc(II)	16.5

(a) What would be the effect of adding 0.01 mmol of EDTA to a solution containing 0.5 mmol of copper(II) and 0.01 mol of calcium(II) ions?

(b) The Cr–EDTA complex is highly toxic. State whether EDTA is suitable to use as a heavy metal antidote for lead when a patient has accidentally ingested a solution containing chromium(II), mercury(II) and lead(II) ions.

(c) Explain the consequences of adding 0.05 mmol of EDTA to a solution containing a mixture of 0.05mmol of iron (II) and 0.05 mmol of iron(III) ions. Why would the addition of excess zinc(II) ions to the solution have no effect on the concentration of iron ions in the mixture.

9 Suggest a series of experiments to demonstrate that chelation with iron(III) may play a part in the action of a potential drug on *Staphylococcus aureus*. Practical details are not required.

10 (a) Outline the mode of action of a photoactivated metal complex?

(b) Suggest a strategy for finding metal complexes that may be used as photoactivated drugs.

14

Nitric oxide

14.1   Introduction

In the late 1980s and early 1990s it was confirmed by several groups of workers that nitric oxide was a chemical messenger released by the endothelium and other tissues in mammals. It was tentatively identified as being the endothelium-derived relaxing factor (EDRF) discovered earlier in the 1990s and has now been linked to a multitude of physiological and pathophysiological states in mammals. For example, it is now known to be involved in the control of blood pressure, neurotransmission and the immune defence system of the body. Excessive production has been linked to atherosclerosis, hypotension, Huntington's disease, Alzheimer's disease and AIDS dementia whilst underproduction has been related to thrombosis, vasospasm and impotence. This diversity of action has resulted in there being a considerable interest in the relationship of nitric oxide to both healthy and diseased states.

Nitric oxide is a colourless paramagnetic gas (boiling point -151.7°C), sparingly soluble in water ($2-3$ mmol dm^{-3}). It is produced in mammals by the enzyme-catalysed interaction of molecular oxygen and arginine (see section 14.4). However, whether it is nitric oxide or a derivative of nitric oxide that is ultimately responsible for the observed physiological response to nitric oxide generation is still under dispute.

14.2   The structure of nitric oxide

Nitric oxide is a linear molecule. It is a relatively stable free radical with one unpaired electron (Fig. 14.1). However, unlike many reactive free radicals, nitric oxide does not

Medicinal Chemistry, Second Edition   Gareth Thomas
© 2007 John Wiley & Sons, Ltd

dimerise appreciably in the gas phase at room temperature and pressure, although it appears to form N_2O_2 in the liquid state.

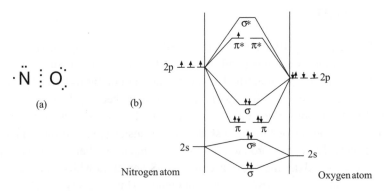

The simple molecular orbital (Mo) picture for nitric oxide shows that the unpaired electron is in an antibonding MO (Fig. 14.1). Consequently, this electron should be readily lost, resulting in the formation of the nitrosonium ion (NO^+). This is in agreement with nitric oxide's low ionisation potential of 9.25 eV compared to a value of 15.56 eV for nitrogen gas.

Figure 14.1 (a) The simple electronic structure of nitric oxide. (b) The MO energy diagram of the nitric oxide. Only the outer shell electrons are shown

14.3 The chemical properties of nitric oxide

It is necessary to understand the chemistry of nitric oxide and related compounds in order to predict and understand their behaviour *in vivo*. The chemistry of nitric oxide is quite varied. It can be oxidised, form salts, act as an electrophile, oxidising agent and free radical and form complexes. It gives rise to a series of salts and complexes with metals. Many of the chemical species identified in laboratory reactions have also been found in biological systems. It is believed that many of these species could be formed in biological systems by reactions similar to those found in the *in vitro* laboratory experiments, although at present few of these reactions have been detected *in vivo*.

In biological systems, nitric oxide often appears to be closely associated with many other simple nitrogen–oxygen containing species such as nitrogen dioxide, nitrogen trioxide, nitrogen tetroxide, nitrite, nitrate and peroxynitrite (Fig. 14.2). Nitrogen dioxide and peroxynitrite are believed to cause tissue damage and so for this reason the relevant chemistry of these substances will also be included in this section.

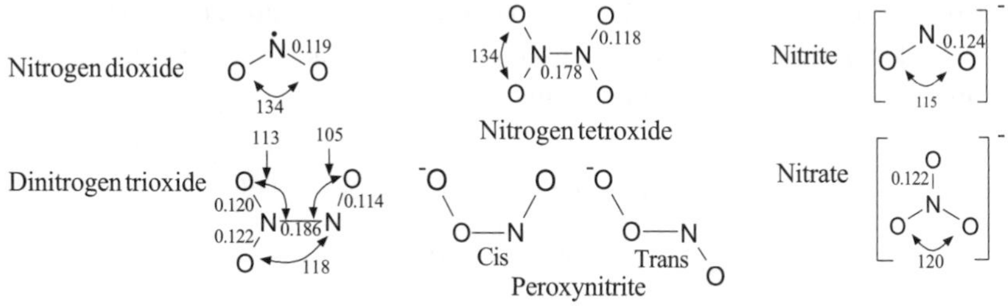

Figure 14.2 The structures of some compounds related to nitric oxide. Bond lengths are in nanometres

14.3.1 Oxidation

Nitric oxide is readily oxidised by oxygen in both the gaseous state and under aqueous aerobic conditions to nitrogen dioxide. The nitrogen dioxide formed in this reaction readily dimerises to nitrogen tetroxide, which reacts with water to form a mixture of nitrite and nitrate ions.

$$2NO + O_2 \longrightarrow 2NO_2$$

$$2NO_2 \rightleftharpoons N_2O_4 \xrightarrow{H_2O} NO_2^- + NO_3^- + 2H^+$$

Practical evidence suggests that this reaction is not a major metabolic route for nitric oxide. However, nitrogen dioxide is a strong oxidising agent, a nitrosylating agent (adds nitroso, NO, groups to a structure) and a nitrating agent (adds nitro, NO_2). Consequently, these reactions could have implications for the action of nitric oxide in biological systems.

Nitric oxide is oxidised by superoxide $^\bullet O_2^-$ to form peroxynitrite. The reaction is very fast and is probably a major route for the metabolism of nitric oxide.

$$^- O{-}O^\bullet \ + ^\bullet NO \longrightarrow ^- OONO$$
$$\text{Superoxide} \qquad \text{Peroxynitrite}$$

Peroxynitrite is a stable anion at alkaline pH ($pK_a = 6.8$ at 37°C). It has been suggested that the stability of the peroxynitrite ion is due to its structure being held in a *cis* conformation by internal forces of attraction. Under acid conditions the *cis* isomer is protonated to the *cis*-peroxynitrous acid, which isomerises to the more stable *trans* isomer.

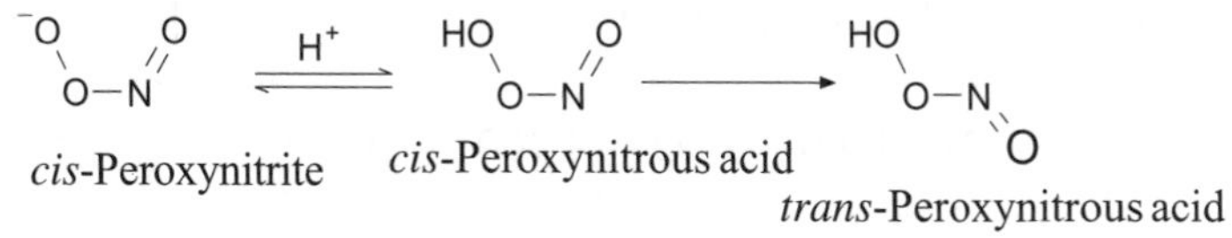

cis-Peroxynitrite   *cis*-Peroxynitrous acid   *trans*-Peroxynitrous acid

At neutral pH peroxynitrite is rapidly protonated to form the unstable peroxynitrous acid, which rapidly decomposes to nitrogen dioxide, hydroxyl radicals in about 20–30 per cent yield and nitrate ions by two separate routes.

$$^-OONO + H^+ \longrightarrow HOONO \longrightarrow NO_2 + \dot{O}H$$
$$NO_3^- \; + \; H^+$$

The stability of peroxynitrite allows it to diffuse considerable distances through biological systems as well as to cross membranes before it reacts. This reactivity is believed to take three main routes:

1. At physiological pH in the presence of hydrogen ions peroxynitrite can result in the formation of an intermediate with hydroxyl free radical-like reactivity.

2. Peroxynitrite can react with metal ions and the metal centres of superoxide dismutase (SOD) to form a nitrating agent with similar reactivity to the nitronium ion ($^+NO_2$). This nitrating agent readily nitrates phenolic residues, such as the tyrosine residues of lysozyme and histone.

3. Peroxynitrite reacts with sulphydryl groups of proteins and other naturally occurring molecules (see section 14.3.4).

Many pathological conditions are thought to result in the tissues simultaneously producing nitric oxide and superoxide. Normally SOD would deactivate the superoxide but kinetic studies show that nitric oxide is produced in large enough quantities and fast enough to prevent this deactivation. Consequently, it has been suggested that in view of its reactivity peroxynitrite may be a major species involved in these conditions.

14.3.2 Salt formation

Both ^+NO and ^-NO ions are known. Their existence may be explained by the low ionisation (9.5 eV) and reduction potentials (0.39 eV) of nitric oxide.

$$\underset{\text{Nitrosonium ion}}{^+NO} + e \; \underset{}{\overset{\text{Oxidation}}{\longleftarrow}} \; NO \; \underset{\text{1e Nitroxylion}}{\overset{\text{Reduction}}{\longrightarrow}} \; ^-NO$$

Nitrosonium salts (^+NO salts) are well known but are readily hydrolysed in water. For example, nitrosonium sulphate rapidly hydrolyses to nitrous and sulphuric acids.

$$NOHSO_4 + H_2O \longrightarrow HNO_2 + H_2SO_4$$

The nitroxyl ion (NO^-) is a less well characterised species. However, it is believed to be formed in the reduction of NO by cuprous (Cu^+) superoxide dismutase (Cu^+SOD).

$$Cu^+SOD + NO \rightleftharpoons NO^- + Cu^{2+}SOD$$

Nitroxyl ions have been shown to react rapidly with molecular oxygen to form peroxynitrite, which would suggest that the nitroxyl ion would have a very short life in oxygenated tissue.

14.3.3 Reaction as an electrophile

Nitric oxide can act as an electrophile because its electronic configuration is one electron short of a stable octet. It readily reacts with thiols, amines and other nucleophiles. For example, nitric oxide reacts in this manner with primary and secondary amines to form adducts known as NONO-ates.

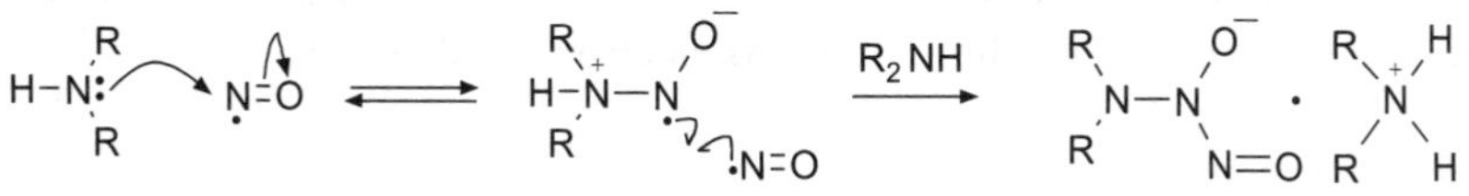

NONO-ate adduct

The NONO-ate adduct is unstable in aqueous solutions, yielding nitric oxide. The rate at which the nitric oxide is produced has been shown to depend on the pH, temperature and structure of the amine. Consequently, NONO-ate adducts may have possible use as drugs to treat cases where the production of endogenous nitric oxide is impaired (see section 14.6.2).

Nitric oxide reacts with proteins containing thiol groups, under physiological conditions, to form *S*-nitrosothiol derivatives. *In vitro* evidence suggests that nitric oxide is transported in the plasma in the form of stable *S*-nitrosothiols, about 80 per cent of which are *S*-nitroso-serum albumins. These *S*-nitrosothiols are believed to act as a depot for nitric oxide maintaining vascular tone. Furthermore, it is thought that *S*-nitrosothiols could be intermediates in the cellular action of nitric oxide (see section 14.4.1)

14.3.4 Reaction as an oxidising agent

Both nitric oxide and nitrogen dioxide have been reported to oxidise thiols under basic conditions.

Nitric oxide :

$$\underset{\text{Thiol}}{RSH} \xrightarrow{\text{Base}} RS^-$$

$$N=O + RS^- \longrightarrow \underset{S\text{-Nitrosophenol}}{RS-\ddot{N}-O^-}$$

$$2RS-\ddot{N}-O^- \xrightarrow{2H^+} \underset{\text{Hyponitrous acid}}{HO-N=N-OH} + \underset{\text{Disulphide}}{RSSR}$$

Nitrogen dioxide :

$$2RSH + NO_2 \longrightarrow RSSR + NO + H_2O$$

Thiol   Disulphide

14.3.5 Complex formation

Complexes containing nitric oxide ligands have been prepared by the direct reaction of nitric oxide with the metal ion of complexes that have unused coordination sites, the displacement of an existing ligand by nitric oxide and the reaction of inorganic (NO_2^-) or organic (RONO) nitrites. The nitric oxide ligand may be bonded to the metal in three distinctly different ways, namely:

1. Complexes in which the MNO bond is linear or almost linear, usually lying between 160° and 180°. In these complexes the nitric oxide donates its odd electron to the metal atom and binds to the metal through a two-electron bond (Fig.14.3a).

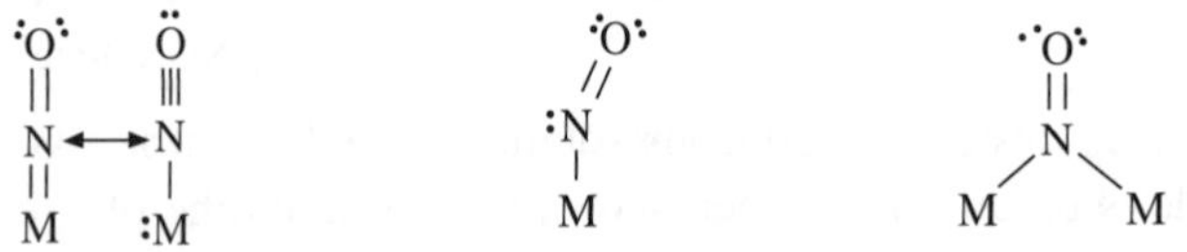

Na$_2$[Fe(CN)$_5$NO]   [Co(NH$_3$)$_5$NO]$^{2+}$ [{Cr(η^5-C$_5$H$_5$)(NO)(μ_2-NO)}$_2$]
[Mn(CO)$_2$(NO)(PPh$_3$)$_2$]   [Rh(Cl)$_2$(NO)(PPh$_3$)$_2$]   [Mn$_3$(η^5-C$_5$H$_5$)$_3$(μ_2-NO)$_3$(μ_3-NO)]
[Co(Cl)$_2$(NO)(PMePh$_2$)$_2$]   [Ir(Cl)$_2$(NO)(PPh$_3$)$_2$]

(a) Linear   (b)Bent   (c) Bridge

Figure 14.3 Representations of the bonding in linear, bent and bridge NO complexes. Examples of linear, bent and bridge complexes are given below each structure. Complexes that contain a mixture of these MNO structures are also known. Bent, linear and bridge NO complexes normally have differences in their IR and NMR spectra that can be used to identify the type of structure

2. Complexes in which the MNO bond angle is bent and lies between 120° and 140°. In these structures the nitric oxide is considered to be a one-electron donor to the metal atom (Fig. 14.3b).

3. Complexes in which the nitric oxide acts as a bridge (Fig.14.3c). More than one bridge may link the metal atoms in these complexes.

14.3.6 Nitric oxide complexes with iron

Iron is widely distributed in mammalian cells, both as free ions and complexed with a wide variety of proteins. Nitric oxide readily forms complexes with both ferrous and ferric ions in the presence of other suitable ligands as well as reacting with the Fe^{II} and the Fe^{III} centres of iron containing naturally occurring molecules. The coordination state of iron in the resulting complexes is often four, the nitric oxide occupying any vacant coordination sites.

Electron paramagnetic resonance spectroscopy showed that nitric oxide reacts with cysteine, histidine and other amino acids in the presence of ferrous ions to form four-coordinate nitric oxide–iron–amino acid complexes (Fig.14.4). Spectroscopy also showed that nitric oxide reacted with proteins in the presence of ferrous ions to form complexes that were associated either with thiol groups or imidazole groups. Proteins whose structures contained a high proportion of thiol groups formed complexes involving the thiol groups in preference to the imidazole groups. Electron spin resonance has also shown that proteins whose structures did not contain iron bound to haem but included thiol groups formed complexes in the presence of free iron where one iron was complexed to two thiol groups and two nitric oxide molecules. These structures are probably similar to that proposed for the nitric oxide–iron–cysteine complex (Fig.14.4a). Complexes of this type are called *dinitrosyl–iron–dithiol* complexes.

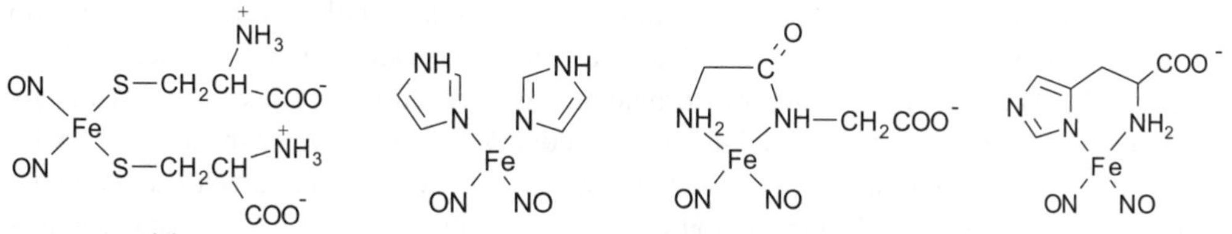

Figure 14.4 Proposed structures for the nitric oxide–iron complexes of (**a**) cysteine, (**b**) imidazole, (**c**) glycylglycine and (**d**) histidine

The interaction of nitric oxide with cellular iron and proteins usually results in a loss of enzyme activity. However, it is not clear in the case of the proteins with thiol groups if the action of nitric oxide is the direct cause of the loss of enzyme activity. This is because the *dinitrosyl–iron–dithiol* protein complexes could also lose their activity by exchanging their original protein–thiol ligands (RSH) with other different protein–thiol ligands (R'SH).

$$(NO)_2Fe(RS)_2 + 2R'SH \longrightarrow (NO)_2Fe(R'S)_2 + 2RSH$$

A further complication is that nitrogen dioxide can also react in the same manner as nitric oxide to form dinitrosyl–iron–dithiols. Consequently, it could be that nitrogen dioxide produced from nitric oxide is the *in vivo* source of dinitrosyl–iron–dithiols, in which case it would be nitrogen dioxide that is responsible for the loss of enzyme activity. The problem is at present unresolved. However, the formation of dinitrosyl–iron–dithiols has been associated with a wide variety of types of tissue damage and it has also been found that cellular iron is a major target of nitric oxide.

Experimental work has also shown that nitric oxide reacts with both the Fe^{II} and the Fe^{III} oxidation states of iron bound to haem in protein molecules. The reaction of nitric oxide with Fe^{II} in haem containing proteins has been shown to form stable ON–Fe^{II}–haem–protein complexes

$$NO + Fe^{II}-Haem\text{-}\text{-}Protein \longrightarrow ON-Fe^{II}\text{-}\text{-}Haem\text{-}\text{-}Protein$$

However, the reaction of nitric oxide with Fe^{III} in haem containing proteins has been shown to yield a nitrosyl–iron complex whose structure is best represented by canonical forms.

$$NO + Fe^{III}-Haem-Protein \longrightarrow Protein-Haem-Fe^{III}-NO \longleftrightarrow Protein-Haem-Fe^{II}-\overset{+}{N}O$$

The electron-deficient nature of the nitrogen explains why these complexes act as electrophiles and react with many nucleophiles, reducing the Fe^{II} to Fe^{III} in the process. For example, Castro and Wade have shown that nitric oxide reacts with metmyoglobin to form a nitric oxide–Fe^{III}–haem complex that nitrosates a wide variety of nucleophiles (Fig. 14.5). The Fe^{II}-haem complex formed in these reactions rapidly reacts with any nitric oxide present to yield an ON–Fe^{II}–haem complex. Similar experimental work with other Fe^{III}–haem containing proteins, such as catalase, peroxidase and human haemoglobin, has indicated that it is the conformation of the protein about the Fe–NO site that controls nitrosation. It has also been suggested on the basis of these investigations that nitric oxide could provide an *in vivo* route for the formation of the highly carcinogenic nitrosamines.

14.3.7 The chemical properties of nitric oxide complexes

Nitric oxide ligands of nitric complexes exhibit a wide variety of different types of reaction that include action as an electrophile or nucleophile, oxidation exchange and displacement reactions, amongst others. The difference in the attachment of the nitric oxide ligand to the metal ion in the structures of linear, bent and bridge nitric oxide complexes accounts for some of the differences in their reactivity. For example, some linear MNO complexes act as *electrophiles* because the nitrogen atom of the nitric oxide has donated three electrons to the metal. This leaves the nitrogen deficient in electrons and so open to attack by nucleophiles such as OH^-, RO^-, RS^- and RNH_2 (see also Fig. 14.5).

$$Protein-Haem-Fe^{III}-No \xrightarrow{RSH} \underset{S\text{-Nitrosothiol}}{RS-N=O} + Protein-Haem-Fe^{II}$$

S-Nitrosothiols slowly decompose, releasing nitric oxide, and so are of potential use as nitric oxide donors. However, the mechanism by which they release their nitric oxide is not

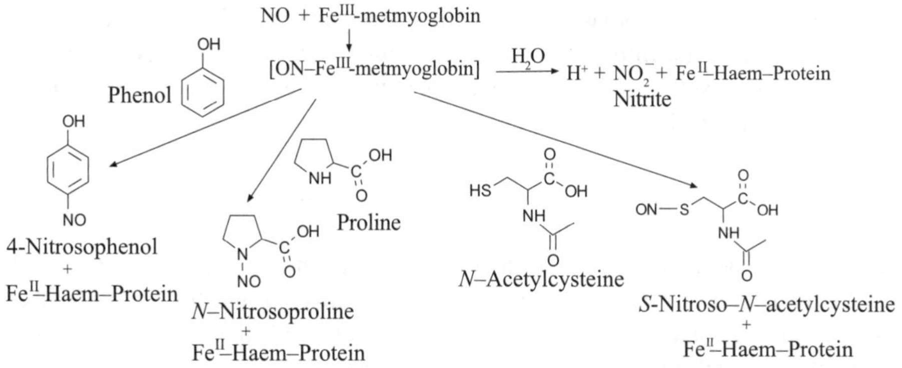

Figure 14.5 Examples of the reactions of nucleophiles with metmyoglobin–nitric oxide mixtures

clear. Furthermore, there is evidence to suggest that S-nitrosothiols are one of the agents that act on soluble guanylyl cyclase. The nitric oxide ligands of both bent and bridge nitric oxide complexes act as *nucleophiles*, reacting with H^+ and other electrophiles. This is because the nitrogen of the nitric oxide has only donated one electron to the metal atom and so it has a lone pair of electrons that can react with electrophiles. It has also been reported that the nitric oxide ligand in some nitric oxide complexes is *oxidised* to nitrogen dioxide.

$$L_nM(NO) \xrightarrow{\frac{1}{2}O_2} L_nM(NO_2)$$

These nitrogen dioxide ligands have been shown to be involved in oxygen transfer reactions to alkenes, disulphides and other organic species, which means that nitric oxide could react through this route in biological systems.

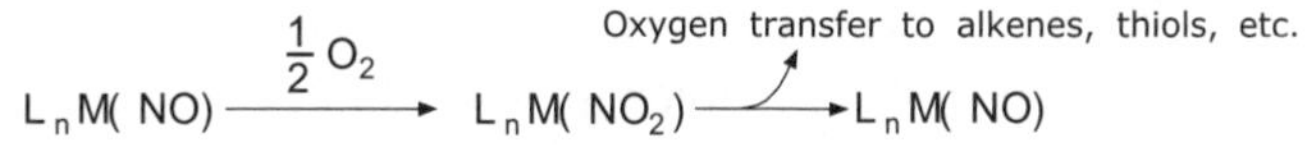

Nitric oxide ligands also undergo *exchange reactions* in which the nitric oxide is exchanged with a ligand in another complex and *displacement reactions* in which nitric oxide displaces other ligands from their complexes. The latter type of reaction is believed to be responsible for activation of the enzyme guanylyl cyclase by nitric oxide in cells. The binding of nitric oxide to the iron atom of the haem nucleus of this enzyme releases a histidine residue. It has been suggested that this histidine residue acts as either a catalyst or a nucleophile, which increases the activity of the guanylyl cyclase. Other ligands that bind to the iron of haem do not liberate a histidine residue or activate guanylyl cyclase. Furthermore, removal of the nitric oxide deactivates the enzyme.

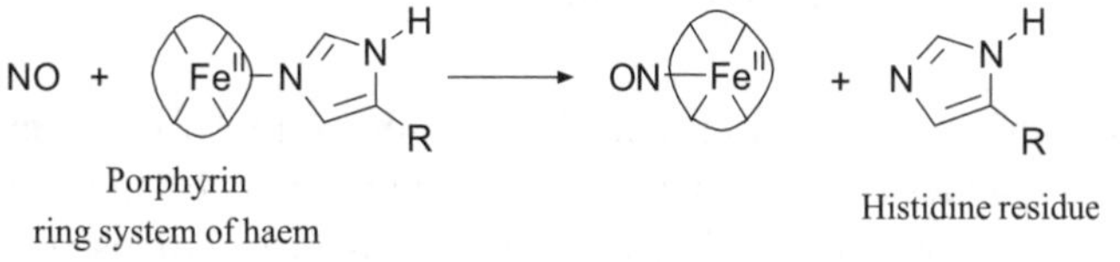

Nitric oxide has been reported to react with some metal complexes to form a range of products including cyclic nitrogen–oxygen compounds, nitrous oxide (N_2O) and nitrate. For example, nitric oxide rapidly reacts with oxymyoglobin (O_2-My-Fe^{II}) to form nitrate and metmyoglobin (My-Fe^{III}).

$$O_2 - My - Fe^{II} \xrightarrow{NO} My - Fe^{III} + NO_3$$

Furthermore, oxyhaemoglobin has been found to react in a similar fashion to produce methaemoglobin in a reaction that is thought to be a major route for the metabolism of nitric oxide in red blood cells. Both of these reaction are also believed to involve the initial formation of peroxynitrite, which decomposes to nitrate

$$Fe^{II}-Hb - O_2 + NO \longrightarrow Fe^{III}-Hb + ONOO^-$$
$$ONOO^- \longrightarrow NO_3^-$$

14.3.8 The chemistry of related compounds

It is still not certain that the physiological and pathological effects attributed to nitric oxide are directly due to that species. Many of the simple nitrogen compounds that can be formed from nitric oxide also react with the same compounds as nitric oxide. Consequently, the biochemistry of nitric oxide cannot be considered in isolation. The chemistry of compounds where nitric oxide may be a biological precursor must also be considered.

Nitrogen dioxide (NO₂)

Nitrogen dioxide is a free radical (structure Fig. 14.2) and an oxidising agent ($E^{\theta'} + 0.99$ V). It has been reported that it initiates the auto-oxidation of unsaturated fatty acids in lipids. It is known to attack pulmonary lipids leading to membrane damage.

At low concentrations nitrogen dioxide has been reported to react with thiols to form nitric oxide and thiol free radicals. The reaction is thought to proceed via a nitrosothiol intermediate although the reaction has not been observed in biological systems.

$$2RSH + NO_2 \longrightarrow RS\bullet + \underset{\text{Nitrosothiol}}{RS-NO} + H_2O$$

$$RS-NO \longrightarrow RS\bullet + \bullet NO$$

Nitrogen dioxide can also act as a nitrating agent. For example, nitrogen dioxide readily nitrates tyrosine residues in proteins to form nitrotyrosine and tyrosine biphenyl derivatives, the reaction rate increasing with pH. It has been proposed that this reaction occurs through a hydrogen abstraction, which would also account for the formation of tyrosine biphenyl derivatives.

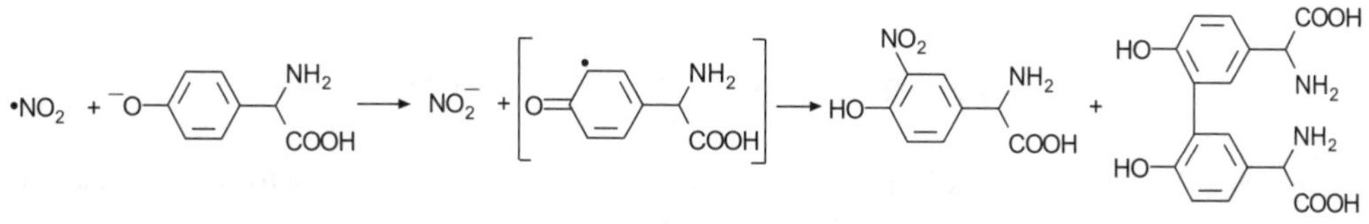

Nitration of tyrosine residues of proteins is known to change protein function. Consequently, nitration could be a major pathological route for tissue injury, especially as extensive nitration has been found in a number of pathological conditions.

Dinitrogen trioxide (N_2O_3)

Dinitrogen trioxide (structure Fig. 14.2) is not stable, decomposing to nitric oxide and nitrogen dioxide at room temperature and above.

$$N_2O_3 \longrightarrow NO + NO_2$$

Dinitrogen trioxide is effectively the anhydride of nitrous acid, forming an unstable blue solution in water and non-polar solvents at room temperature.

$$N_2O_3 + H_2O \rightleftharpoons \underset{\text{Nitrous acid}}{2\,HNO_2}$$

Dinitrogen trioxide is a powerful oxidising agent and a strong nitrosating agent. For example, it has been shown to rapidly oxidise oxymyoglobin (O_2-My-Fe^{II}) to metmyoglobin (My-Fe^{III}) and oxyhaemoglobin (O_2-Hb-Fe^{II}) to methaemoglobin (Hb-Fe^{III}).

$$O_2\text{-My-Fe}^{II} \xrightarrow{N_2O_3} \text{My-Fe}^{III} + NO_3^-$$

$$O_2\text{-Hb-Fe}^{II} \xrightarrow{N_2O_3} \text{Hb-Fe}^{III} + NO_3^-$$

Dinitrogen trioxide in the form of nitrous acid also nitrosates amines and thiols.

Nitrite

Nitrite (structure Fig. 14.2) is ingested in food where it is used as a preservative, and is produced in mammals as a metabolite of nitric oxide, nitrate and other substances. For example, nitrate is reduced to nitrite in the oral cavity by bacteria and in the stomach by intestinal flora. Nitrite is also formed indirectly from nitric oxide (see section 14.3.1).

Nitrite is mildly toxic (tolerance limit $\sim$100 mg kg^{-1} per day), autocatalysing the oxidation of oxyhaemoglobin to methaemoglobin, nitrogen dioxide and hydrogen peroxide.

Methaemoglobin is formed in sufficient quantity to cause methaemoglobinaemia. Hydrogen peroxide is a very strong oxidising agent and is highly toxic when produced *in vivo*. It also forms weak complexes with methaemoglobin.

$$Fe^{II}\text{-Hb-}O_2 \xrightarrow{NO_2^-} Fe^{III}\text{-Hb} + NO_2 + H_2O_2$$

An increase in the concentration of nitrate and nitrite ions in the urine is frequently used as an indication of the involvement of nitric oxide in a biological system.

Peroxynitrite

Peroxynitrite (structure Fig.14.2) is formed *in vivo* by the reaction of nitric oxide and superoxide (see section 14.3.1). Superoxide is normally scavenged by superoxide dismutase (SOD) but nitric oxide reacts so rapidly with superoxide that it outcompetes SOD to form peroxynitrite. Peroxynitrite is a strong oxidising agent. It reacts with SOD and other metalloproteins to form the nitronium ion ($^+NO_2$), which is a powerful nitrating agent that readily nitrates the tyrosine residues found in protein. This nitration has been shown to result in a loss of the protein's activity.

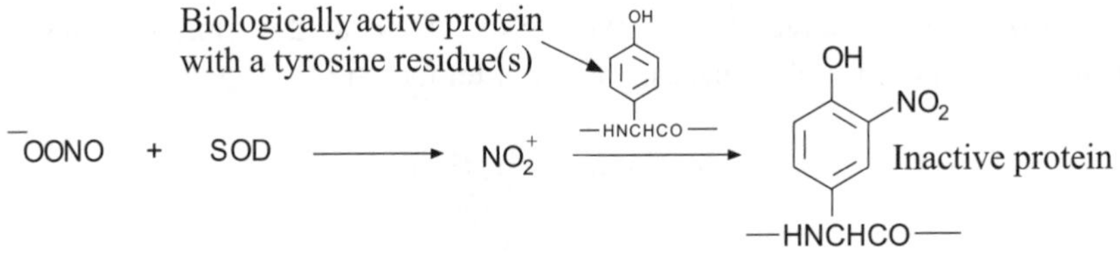

It has been proposed that nitration initiated by peroxynitrite is a factor in amyotrophic lateral sclerosis (ALS). It is suggested that SOD mutants have a reduced superoxide scavenging effect, which results in overproduction of peroxynitrite. The excess peroxynitrite nitrates tyrosine residues, which prevents the normal signal transduction by growth factors that support the survival of motor neurons. This would lead to a gradual loss of motor neurons and their associated activity, which is the principal feature of ALS.

14.4 The cellular production and role of nitric oxide

Nitric oxide is produced *in vivo* by the catalytic oxidation of L-arginine by a family of enzymes known as nitric oxide synthases (NOS). The reaction requires nicotinamide adenosine diphosphate as a cofactor (NADPH) and produces nitric oxide and L-citrulline (Fig. 14.6) in a 1:1 molar ratio. As a result, the concentration of citrulline is often used as an estimate of the *in vivo* concentration of nitric oxide.

Nitric oxide synthases have been broadly classified as constitutive NOS (cNOS) and inducible NOS (iNOS) enzymes. The cNOS enzymes are present in endothelial and neural

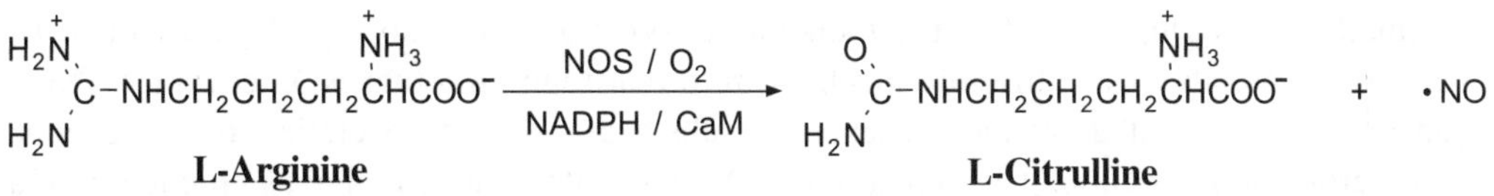

Figure 14.6 A schematic representation of the formation of nitric oxide

tissue and are usually referred to as eNOS and nNOS enzymes, respectively (Table 14.1). These enzymes are not identical but have similar properties. They appear to be present at an approximately constant level in the host cell but only produce nitric oxide when activated by the Ca^{2+} ion binding protein calmodulin (CAM). Conversely, iNOS enzymes are not present in the cell but are produced in response to stimulants of host and bacterial origin. Activation of cNOS results in the production of a short burst of nitric oxide at a low concentration whilst activation of iNOS results in the continuous production of nitric oxide at a high concentration.

Table 14.1 Some characteristic properties of cNOS and iNOS

	cNOS		iNOS
	nNOS	eNOS	iNOS
Cellular location	Cytosolic (aqueous medium)	Particulate (membrane bound)	Cytosolic (aqueous medium)
Ca^{2+} dependent	Yes	Yes	No

It has been shown that the endothelial and neural tissue production of nitric oxide is initiated by agonists such as acetylcholine, ADP, bradykinin and glutamate. The binding of these agonists to appropriate receptors causes an increase in cellular Ca^{2+} ions. These Ca^{2+} ions bind to and activate CaM (see section 9.4.2), which in turn activates the cNOS present in the cell to produce nitric oxide. It has also been observed that an increase in the shear stress in blood flow on the endothelium due to exercise can also stimulate the synthesis of nitric oxide.

Experimental work has shown that the first step in the formation of nitric oxide is the synthesis of N^{ω}-hydroxy-L-arginine as an enzyme-bound intermediate by a two-electron oxidation involving molecular oxygen, NADPH and CaM. This intermediate is converted to citrulline with the liberation of nitric oxide by an overall three-electron oxidation that also involves molecular oxygen, NADPH and CaM. The concentration of cellular L-arginine is maintained by the recycling of the L-citrulline to L-arginine. However, it has been shown that a low local concentration of cellular L-arginine results in impairment of endothelium-dependent relaxation, which is alleviated by an infusion of L-arginine.

iNOS is found in a wide variety of cells such as mast cells, macrophages, Kupffer cells and neutrophiles. It is also found in endothelial cells and vascular smooth muscle. Unlike cNOS it is not calcium dependent. However, it has been found that in macrophages iNOS calmodulin is tightly bound to the inducible enzyme and so probably plays a part in its action but possibly by a different mechanism to that found in other NOS enzymes. iNOS enzymes are activated by the presence of substances such as bacterial toxins, gamma interferon and interleukin-1β. Activation of iNOS results in the continuous production of a high concentration of nitric oxide.

$$
\begin{array}{c}
\underset{H_2N}{\overset{+}{H_2N}} \\
\backslash\backslash \\
C-NHCH_2CH_2CH_2CHCOO^- \quad \xrightarrow[\text{NADPH / CaM}]{\text{NOS / O}_2} \\
/ \quad \overset{|}{NH_3} \\
H_2N \quad \text{L-Arginine}
\end{array}
\qquad
\begin{array}{c}
HON \quad \overset{+}{NH_3} \\
\backslash\backslash \quad | \\
C-NHCH_2CH_2CH_2CHCOO^- \\
/ \\
H_2N \quad N^\omega\text{-Hydroxy-L-arginine}
\end{array}
$$

L-Arginine → NOS / O₂, NADPH / CaM → Nω-Hydroxy-L-arginine

NOS / O₂
NADPH / CaM

$$
\underset{H_2N}{\overset{O}{\backslash\backslash}}C-NHCH_2CH_2CH_2\overset{\overset{+}{NH_3}}{\underset{}{C}}HCOO^- \quad + \quad \cdot NO
$$

L-Citrulline is recycled   L-Citrulline

These general forms of NOS reflect the two distinct general modes of action of NO. With cNOS the enzyme produces bursts of NO that transmit a message to the target cells without damaging those cells. With iNOS the enzyme is responsible for the continuous production of NO in sufficient concentration to damage and kill cells that may or may not be of benefit to the organism. For example, activated immune cells produce amounts of NO that are lethal to harmful target cells such as those found in cancers and invasive parasites but overproduction of nitric oxide has been linked to the death of pancreatic β cells in insulin-dependent diabetes mellitus.

14.4.1 General mode of action

Nitric oxide has a short biological life, undergoing rapid metabolism to nitrate, nitrite and other species (see sections 14.3.1 and 14.3.6). This short life means that nitric oxide is not able to diffuse any distance through the system before there is a significant decrease in its concentration. As a result, its targets must be close to its source and activated by the concentration of nitric oxide in their vicinity. For example, release of nitric oxide from the endothelium by the cNOS route is now known to cause a local relaxation of the underlying smooth muscle surrounding blood vessels, which results in a reduction of blood pressure.

The prevailing school of thought is that nitric oxide synthesised through the cNOS route is believed to act by binding to the iron in the haem unit (see section 14.3.6) that constitutes the active site of soluble guanylyl cyclase (GC). This alters the conformation of the enzyme, which activates it to act as a catalyst. However, some workers believe that the nitric oxide is converted to an *S*-nitrosothiol and it is this compound that nitrosates the soluble GC. Activation of the soluble GC enzyme, by either of these routes, results in the conversion of guanosine triphosphate (GTP) to cyclic guanosine monophosphate (cGMP) in the target cell (Fig. 14.7). cGMP is known to interact with various proteins and in doing so changes their biological activity. For example, increase in cGMP concentration has been shown to inhibit Na^+ channels of the kidney and to decrease the $[Ca^{2+}]$ in smooth muscle

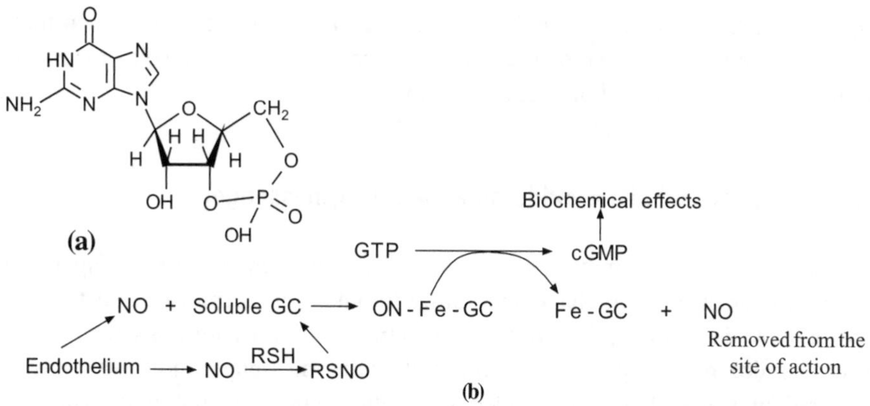

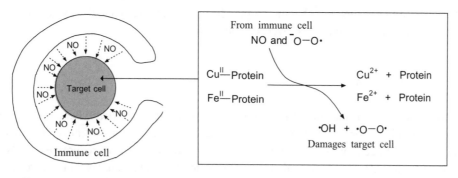

Figure 14.7 describing elements:

(a)

Biochemical effects

$$GTP \longrightarrow cGMP$$

$$NO + \text{Soluble GC} \longrightarrow ON\text{-}Fe\text{-}GC \qquad Fe\text{-}GC + NO$$

Removed from the
site of action

$$\text{Endothelium} \longrightarrow NO \xrightarrow{RSH} RSNO$$

(b)

Figure 14.7 **(a)** Cyclic guanosine monophosphate (cGMP). **(b)** An outline of the formation of cyclic guanosine monophosphate (cGMP) from guanosine triphosphate (GTP) by soluble guanylyl cyclase (Soluble GC)

and platelets. Dissociation of the nitric oxide from the active ON–Fe^{II}–haem–enzyme complex deactivates the guanylyl cyclase.

The nitric oxide released by the cNOS route that does not bind to a haem target area may take part in nitrosation reaction or react with thiols (see section 14.3.7) to form nitrosothiols, which decompose to release nitric oxide. It has been suggested that nitrosothiols, such as nitrosocysteine, may act as a depot for nitric oxide, thereby prolonging its action. Furthermore, it has also been suggested that nitric oxide binds to the thiol groups of mammalian albumin and as such is transported in the plasma.

Nitric oxide generated by the immune system through the iNOS pathway acts as a killer molecule. The high concentration of nitric oxide (nanomols) produced by this process causes lethal oxidative injuries to the target cells, such as cancer and parasite cells. Little is known about this process but it is now thought that the nitric oxide reacts in conjunction with superoxide, which is also produced by activated immune system cells. The immune cells increase their surface area and fold around their target cells or microorganisms. Once in position they release nitric oxide, which attacks the copper- and iron-complexed proteins in the target cell, liberating copper and iron ions from these proteins (Fig. 14.8). This is

Figure 14.8 A schematic representation of the action of nitric oxide as a killer molecule

accompanied by the formation of hydroxyl free radicals and molecular oxygen that cause massive oxidative injury to the target cell. It is thought that this may also be the mechanism of cellular injury in some pathological conditions.

14.4.2 Suitability of nitric oxide as a chemical messenger

Nitric oxide is a unique chemical messenger. All the other chemical messengers transmit information by means of their shape and ability to bind to a receptor (see sections 8.2 and 8.6). Nitric oxide does not depend on its shape to transmit information. Its action appears to be due to its redox reactivity. Furthermore, unlike other messenger molecules, nitric oxide is not stored *in situ* and released under a specific stimulation but it is synthesised as required. In other words, unlike classical messengers it is synthesised on demand and then diffuses to its target. Verma *et al.* have suggested that carbon monoxide may also be a messenger of this type.

Nitric oxide is ideally suited as a locally acting chemical messenger in spite of its toxic nature because:

- it is soluble in both water (about 2 mmol dm^{-3} at 20°C) and lipids;

- its small size enables it to cross cell membranes as easily as oxygen and carbon dioxide;

- it diffuses faster in water than oxygen and carbon dioxide;

- it is a short-lived species in biological systems, reacting with oxygen and other biological molecules;

- its short life in biological systems results in a decrease in nitric oxide concentration as distance from its source increases, which results in its "message" being localised;

- of its reactivity and the ease of formation of ON–Fe^{II}–haem complexes;

- the reversibility of ON–Fe^{II}–haem complex formation enables guanylyl cyclase to switch off after the nitric oxide is removed by the system.

14.4.3 Metabolism

The main metabolic route for nitric oxide is diffusion into the blood where it reacts with oxyhaemoglobin to form methaemoglobin, nitrate and small amounts of nitrite (see section 14.3.7). The methaemoglobin is converted back to haemoglobin by reductases whilst the nitrate is transferred to the serum and excreted in the urine. Small amounts of nitrate are thought to be reduced by bacterial action in the intestinal tract and be exhaled as ammonia and nitrogen.

A minor route for the metabolism of nitric oxide is oxidation to nitrogen dioxide (see section 14.3.1). A more important process is the reaction with superoxide to form peroxynitrite, which has been implicated in some pathological conditions (see section 14.3.1).

The higher concentration of nitric oxide produced by iNOS activity leads to a very much higher concentration of nitrate in the urine. However, this is not a reliable indicator of the involvement of endogenous NO formation in the action of the immune system in man because nitrate is also produced from external sources. For example, it is produced from nitric oxide inhaled from the atmosphere, tobacco smoke and from nitrite used as a preservative in food.

14.5 The role of nitric oxide in physiological and pathophysiological states

Nitric oxide is both a chemical messenger and a cytotoxic agent, the former being initiated by cNOS and the latter by iNOS. In its chemical messenger mode it acts as an initiator for biological processes that are essential for a healthy organism. However, in its cytotoxic mode it can be both beneficial and harmful. Nitric oxide forms part of the immune system but it is also believed to be the toxic agent that initiates tissue damage involved in some pathological conditions. This section sets out to survey the role of nitric oxide in some of the many normal physiological and pathophysiological situations in which it is found.

14.5.1 The role of nitric oxide in the cardiovascular system

The cardiovascular system is the network of blood vessels, including the heart, through which blood flows to all parts of the body. In 1980 it was shown by Furchgott and Zawadzki that removal of the endothelium prevented the relaxation effect of acetylcholine on blood vessels. This led to the discovery that stimulation of the endothelial cells resulted in the release of a substance that Furchgott called endothelium-derived relaxing factor (EDRF). This was followed by the discovery that many vasoactive substances were found to release EDRF from endothelial cells. The biological action and chemical properties of EDRF were found to resemble those of nitric oxide and so in 1987 it was proposed independently by Furchgott and Ignarro that EDRF was nitric oxide. Later work by Moncada *et al.* indicated that nitric oxide was responsible for the biological activity of EDRF. As a result, a number of workers have suggested that EDRF is a S-nitrosothiol produced by the action of nitric oxide on a suitable thiol. However, in spite of the controversy about EDRF it has become apparent that nitric oxide plays an important role in both the healthy and the unhealthy states of the cardiovascular system. Evidence suggests that the generation of too little nitric oxide can result in hypertension (high blood pressure), angina and impotence whilst the synthesis of too high a concentration of nitric oxide is thought to be related to circulatory shock, inflammation and strokes.

Nitric oxide generated in the endothelium targets the underlying smooth muscle cells, causing these cells to relax. This dilates (widens) the blood vessel, allowing an increased flow rate that reduces blood pressure. It has also been suggested that nitric oxide reduces the adhesion and aggregation of platelets and leucocytes, which would also help to increase the flow of blood through the vessel. Therefore, it appears that the role of nitric oxide produced by the cNOS route in a healthy endothelium is to control and maintain a healthy blood flow. If this is so, damage to the endothelium that results in endothelium dysfunction could be responsible for some cardiovascular diseases.

Endothelium dysfunction has been shown either to be caused by a genetic defect or be acquired as a result of a poor diet, smoking or sedentary life style. It inhibits the relaxation of the endothelium, which it is believed could result in an increase in blood pressure and further endothelium damage. Endothelium dysfunction is thought to be due to an increase in superoxide production in the endothelium. This superoxide reacts with any available nitric oxide to form peroxynitrite (see section 14.3.1), which reduces the amount of nitric oxide diffusing to the smooth muscle cells. This makes it less likely for them to relax and as a result the blood vessels are less likely to dilate to reduce hypertension. In addition, the formation of the highly toxic peroxynitrite ion could cause further issue damage. Furthermore, the loss of nitric oxide could allow endothelin, the endogenous vasoconstrictor produced by the endothelium, to counterbalance the vasodilating effect of nitric oxide, to constrict the blood vessel and increase blood pressure.

The formation of a large concentration of nitric oxide leads to circulatory shock (hypotension or reduced blood pressure) due to excessive vasodilation. Experimental evidence suggests that these large doses of nitric oxide are produced by the iNOS route and are also associated with the lowering of blood pressure that occurs with endotoxin, haemorrhagic and septic shock. Septic shock is also accompanied by inflammation. Consequently, it is possible that iNOS inhibitors could be used to treat these conditions.

14.5.2 The role of nitric oxide in the nervous system

A cNOS enzyme has been isolated from rat brain. It has been found to be very similar to endothelial cNOS in that it is calmodulin dependent (see section 14.4). The enzyme is mainly located in the neurons of the granule cell layer of the cerebellum. Little cNOS was found in the remainder of the brain's neurons. *In vitro* experimental evidence has indicated that this brain cNOS is activated by glutamate stimulation and that the nitric oxide produced appears to target Purkinje cells. Purkinje cells have a high concentration of guanylyl cyclase (see section 14.4.1) and stimulation of nitric oxide production by glutamate increases the concentration of cGMP (Fig. 14.7) in these cells. NOS has also been found throughout the CNS and the peripheral nervous system. It exists exclusively in neurons but, unlike its occurrence in the brain, no organised pattern of distribution has been observed and only a small percentage of the neurons contain NOS.

The function of nitric oxide in the brain is not clear. Evidence suggests that it could act as a neurotransmitter, with its action initiated by glutamate. It has been suggested that the glutamate is released from the presynaptic nerve terminal by exocytosis triggered by a nerve signal. It diffuses across the synaptic gap to the postsynaptic nerve terminal where it interacts with *N*-methyl-D-aspartate receptors (NMDA receptors). These receptors are coupled to calcium channels and their activation allows calcium to flow into the postsynaptic nerve terminal. This calcium combines with calmodulin to activate cNOS, which converts L-arginine to citrulline and nitric oxide (Fig. 14.9). It is proposed that the nitric oxide formed either diffuses to nearby target cells or may react with specific thiols in the neuron to form *S*-nitrosothiols, which are stored in vesicles until released by a voltage-dependent mechanism. One target of the nitric oxide, irrespective of whether it is synthesised directly or released from an *S*-nitrosothiol, appears to be the presynaptic nerve where it stimulates the conversion of GTP to cGMP. The purpose of this conversion is not known.

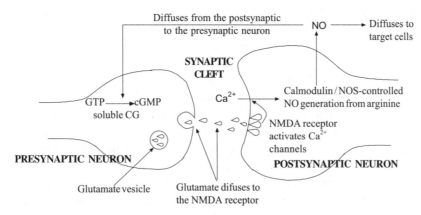

Figure 14.9   A schematic outline of the mechanism for the action of nitric oxide as a neurotransmitter. The cGMP produced in the presynaptic neuron by the action of NO initiates phosphorylation and other biochemical processes

The sustained release of high concentrations of nitric oxide has been linked to strokes as well as possibly causing other forms of cell damage and death. In strokes, cells lose their ability to exclude Ca^{2+} ions. This means that the calmodulin in the cell will remain active (see section 9.4.2). Since cNOS is Ca^{2+} ion dependent the presence of this Ca^{2+}-activated calmodulin will switch on cNOS, resulting in the production of large amounts of nitric oxide. It has been suggested that this nitric oxide combines with superoxide to form peroxynitrite and it is this powerful oxidising agent that destroys cell membranes either by direct reaction with components of the membrane or by producing hydroxy and other free radicals that attack cell membranes.

Experimental evidence also suggests that nitric oxide can act as a neuromediator and that the controlled release of nitric oxide in low concentration is part of the normal function of the brain. It has been suggested that it could be a mediator for blood flow and neuronal activity and have an influence on the underlying mechanisms of long-term memory and depression. In addition, it has also been suggested that one role of the nitric oxide produced in the brain could be to protect some neurons against oxidative damage.

Nitric oxide acts as a transmitter in the peripheral nervous system of the urogenital and gastrointestinal tracts. It appears to play a major part in gastric dilation and maintaining the compartmentalisation of the gastrointestinal tract. Its absence has been linked to infantile pyloric stenosis and male impotence (see section 14.5.4).

14.5.3   Nitric oxide and diabetes

Destruction of pancreatic β cells is known to occur in insulin-dependent diabetes mellitus (IDDM or type I diabetes) The loss of 80 per cent or more of these cells results in insulin deficiency, which reduces the body's ability to control blood glucose. Cell death is believed to be caused by a variety of endogenous agents and also the autoimmune system. The mechanism by which these agents act is not known and is the subject of much controversy. However, it was thought that nitric oxide may be involved both in normal cell operation and IDDM.

In vitro experimental evidence has indicated that the overproduction of nitric oxide may destroy pancreatic β cells during the development of IDDM. It has been proposed by Corbett and McDaniel that IL-1 released from macrophages binds to specific receptors on the pancreatic β cells. This activates tyrosine kinase, which, through a series of intermediaries, activates the expression of iNOS. The high concentration of nitric oxide produced by this enzyme inhibits the activity of other essential enzymes by interaction with their iron–sulphur centres. This brings about cell dysfunction and ultimately cell distruction. Although this proposal has yet to be proved, specific iNOS inhibitors could be of value in preventing the development of IDDM.

14.5.4   Nitric oxide and impotence

NOS has been located in the adventitial layer of the penile arteries. Release of nitric oxide has been shown to cause a dose-dependent rapid relaxation of the human corpus cavernosum with a subsequent penile erection. NOS inhibitors have been shown to prevent this relaxation whilst nitric oxide sources mimics the NOS effect. Consequently, it appears that nitric oxide is an important mediator in penile erection. This erection is believed to be due to the action of cyclic guanosine monophosphate (cGMP) formed from guanosine triphosphate (GTP) by the action of guanylyl cyclase activated by nitric oxide (see section 14.4.1). Biological activity is terminated by the cGMP being converted to 5′-guanosine monophosphate (5′-GMP) by the action of phosphodiesterase type 5 (PDE 5). Consequently, it is possible that impotence could be treated by either injections of nitric oxide donors into the corpus cavernosum or inhibition of PDE 5. The latter approach led to the discovery of sildenafil (see section 14.6.2). Sildenafil (Viagra) inhibits the action of PDE 5, which prevents the deactivation of cGMP and as a result the termination of the physiological response. The drug is effectively compensating for low nitric oxide production by allowing cGMP to accumulate.

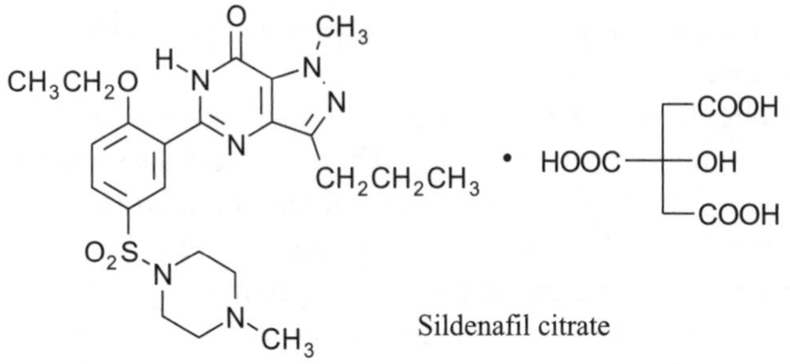

Sildenafil citrate

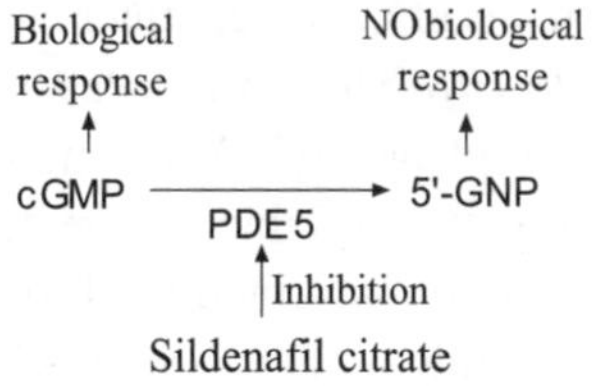

It is likely that impotence in chronic diabetic men is due to a failure of nitric oxide synthesis.

14.5.5 Nitric oxide and the immune system

The immune system is an organism's natural defensive system against pathogens (microorganisms and viruses). In mammals immunity is due to the different members of a group of white cells known as lymphocytes. These cells are produced in the bone marrow and effectively police the body by being able to move through the intercellular spaces as well as the blood stream. They operate (immune response) in two general ways:

1. cellular immunity: the defensive mechanism is triggered by the presence of foreign antigens but without the immediate production of antibodies. It is mediated by T lymphocytes or T cells.

2. humoral immunity: B lymphocytes or B cells produce antibodies in response to a foreign antigen such as a foreign macromolecule, carbohydrate, nucleic acid or protein. These antibodies trigger a defensive mechanism, which destroys the invading foreign antigen.

The cellular immune response is triggered when a macrophage envelopes and partly decomposes a foreign antigen. The decomposition products are displayed on the surface of the macrophage bound to surface proteins known as major histocompatibility complex (MHC) proteins. These structures are recognised by cytotoxic T cells, which causes the macrophage to release interleukin-1, which stimulates the production of large numbers of cytotoxic T cells. These killer T cells bind specifically to the foreign antigen, releasing perforin, a protein that lyses the target cell.

Humoral immunity is triggered when a foreign antigen binds to an immunoglobulin displayed on the surface of the B cell. The cell responds by engulfing and partly decomposing the foreign antigen. These partial decomposition products are displayed on its surface in the form of complexes with Class II MHC protein found on the surface of B cells. This stimulates mature helper T cells to bind to the B cell, causing that cell to divide and produce large numbers of specialist plasma cells. These cells secrete antibodies that are specific for the foreign antigen. The antibodies bind to the foreign antigen and effectively label that antigen for destruction either by ingestion by phagocytes (phagocytosis) or activation of the complement system.

In vitro studies in 1987 by Hibbs, Vavrin and Taintor showed that cytotoxic-activated macrophages (CAMs) required L-arginine for their activity. These workers also observed that the L-arginine was metabolised to L-citrulline, nitrite and nitrate. Furthermore, NOS inhibitors prevented the formation of L-citrulline, nitrite and nitrate, as well as preventing the cytotoxic action of CAM. Electron paramagnetic resonance spectroscopy has shown that the action of CAMs is accompanied by the formation of nitrosyl–iron complexes. Furthermore, superoxide, which readily reacts with nitric oxide (see section 14.3.1), inhibits the cytotoxic action of CAMs. These observations indicated that nitric oxide is involved in the operation of the immune system. It is now believed that one cytotoxic mechanism employed by macrophages involves engulfing the target cell and flooding it with nitric oxide (see section 14.4.1).

14.6 Therapeutic possibilities

Nitric oxide is involved in both physiological and pathophysiological conditions. The pathophysiological effects can be conveniently divided into two types, those due to an excess of nitric oxide and those due to a lack of nitric oxide. Consequently, the main approaches to drug design are based on producing compounds that prevent the overproduction of nitric oxide or act as a source of nitric oxide. Gene manipulation is also being investigated as a means of controlling nitric oxide production. This section outlines only some of the approaches being followed in these areas.

14.6.1 Compounds that reduce nitric oxide generation

A wide range of pathophysiological states have been associated with the overproduction of nitric oxide. For example, overproduction of nitric oxide by iNOS has been associated with osteoporosis, inflammation, rheumatoid arthritis and morphine dependence while over-production of nNOS is associated with Alzheimer's disease, strokes and schizophrenia. Consequently, these NOS enzymes offer a potential target for drugs to treat these conditions. However, as nitric oxide production is essential for human health (see section 14.5) any drugs must selectively target the relevant NOS enzyme.

Reduction of nitric oxide production can be achieved by inhibiting the action of NOS or its activating processes, such as calcium ingress. Based on a knowledge of the process for the production of nitric oxide (Fig. 14.6) an obvious line of investigation was to develope NOS

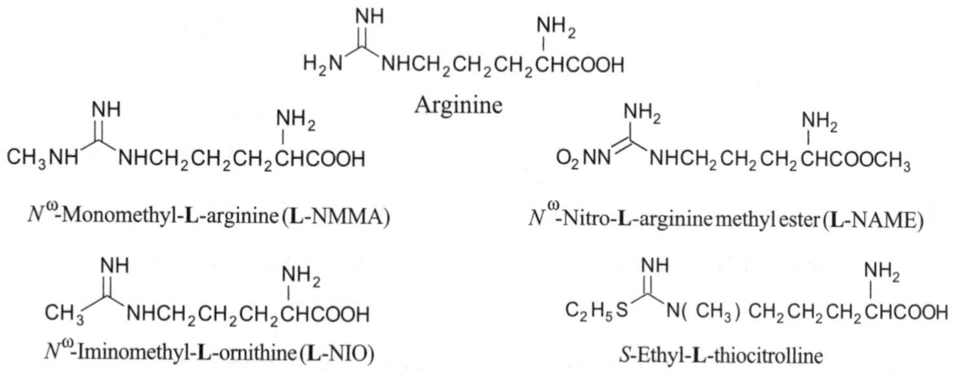

Figure 14.10　Examples of arginine analogues used to block NOS activity

inhibitors by synthesising analogues of L-arginine. A number of these analogues have been found to inhibit the formation of nitric oxide by acting as NOS blocking agents (Fig. 14.10). These inhibitors have been extensively used to investigate the action of nitric oxide. N^ω-Monomethyl-L-arginine (L-NMMA) has been found to increase blood pressure in man and other species. However, its selectivity is low. N-Iminomethyl-L-ornithine (L-NIO) is an irreversible inhibitor of NOS in activated macrophages. N^ω-Nitro-L-arginine methyl ester (L-NAME) has been reported to exhibit a selectivity of 300:1 for nNOS as against iNOS, while S-ethyl-L-thiocitrolline has been reported to exhibit a 50:1 preference for nNOS against eNOS. However, none of these compounds have been accepted for general clinical use.

Several simple guanidino compounds (Fig.14.11) have also been found to inhibit nitric oxide synthesis. It is believed that these compounds inhibit nitric oxide synthesis by preventing the second stage of the oxidation of arginine (see section 14.5).

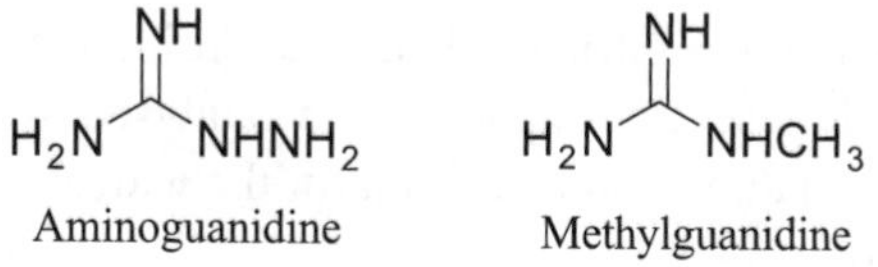

Figure 14.11　Examples of guanidine derivatives used as NOS inhibitors

Aminoguanidine has been shown to be a selective inhibitor for iNOS in animal models. It has a minimal effect on the cNOS that is required to maintain blood pressure. The selectivity of aminoguanidine is believed to be due to the presence of the hydrazine residue since replacement of this moiety by a methyl group, which has similar overall shape and size, resulted in the loss of selectivity and a considerable loss of activity.

Many other classes of compound have been examined as sources of selective NOS inhibitors. For example, 2-iminoazaheterocycles and imidazole analogues (Fig.14.12) show a preference for iNOS against nNOS. Dipeptides containing an N^ω-nitroarginine residue

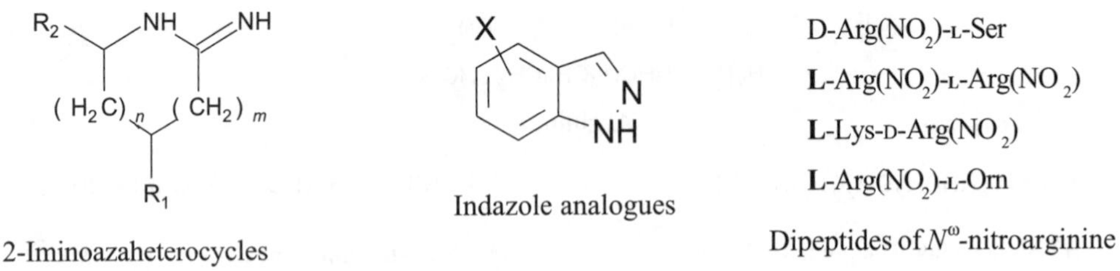

Figure 14.12 Examples of structures whose analogues inhibit NOS enzymes

have also shown a range of different selectivies towards NOS isoenzymes. For example, those dipeptides listed in Figure 14.12 exhibit a greater selectivity for nNOS than either eNOS or iNOS.

14.6.2 Compounds that supply nitric oxide

Sodium nitroprusside and organic nitrates and nitrites (Fig. 14.13) have been used for over 100 years to treat angina. Glycerol trinitrate has also been used to relieve impotence.

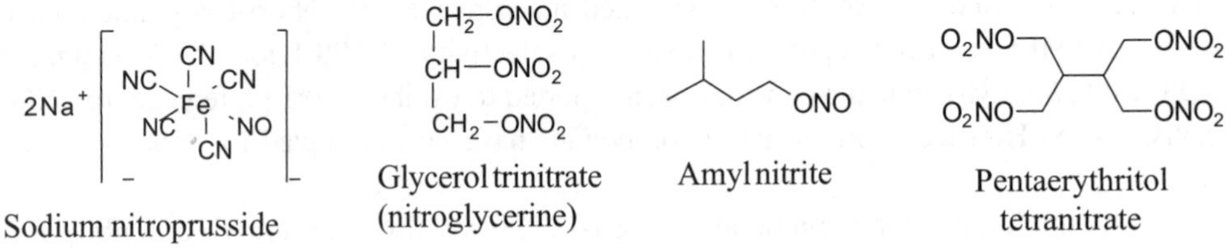

Figure 14.13 Examples of compounds used to treat cardiovascular diseases

It has been demonstrated that sodium nitroprusside and organic nitrates and nitrites act by forming either nitric oxide or a nitric oxide adduct during their metabolism (Fig.14.14). The metabolic pathways of these drugs appear to be catalysed by enzymes that are specific for each drug. This specific nature would account for the wide diversity of pharmacological action of each of these drugs.

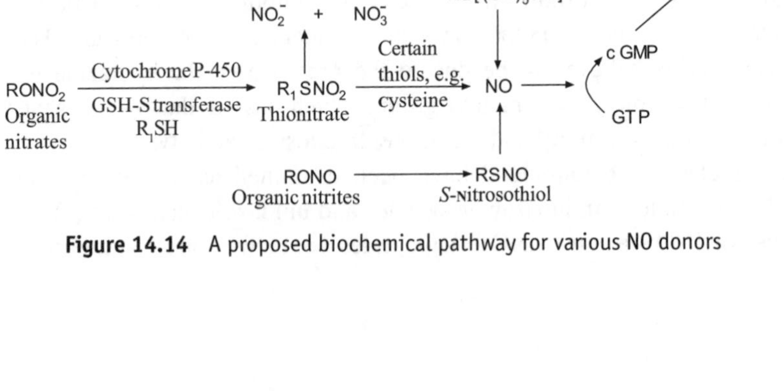

Figure 14.14 A proposed biochemical pathway for various NO donors

A knowledge of the chemistry and biochemical pathway of nitric oxide has resulted in several groups of compounds being investigated as leads to new drugs. For example, the suggestion that EDRF is an *S*-nitrosothiol has resulted in the investigation of a number of these compounds as potential drugs. *S*-Nitrosocaptopril, *S*-nitroso-*N*-acetylcysteine and *S*-nitroso-*N*-acetylpenicillamine have all been shown to have vasodilator properties in animals and may have some use in humans.

NONO-ates (see section 14.3.3) are also being investigated as potential sources of drugs. These compounds are prepared by the direct action of nitric oxide on a nucleophile.

$$2NO \ + \ Nucleophile(X^-) \longrightarrow \overset{\overset{\displaystyle O^-}{|}}{X\text{–}N\text{–}N}{=}O$$
$$NONO\text{-}ate$$

They are stable solids that spontaneously decompose in water. The rate of decomposition depends on the temperature, pH and the nature of the nucleophilic residue X. Since the rate of release of nitric oxide depends on the nature of X, NONO-ates could be useful as slow-release drugs. Furthermore, the spontaneous nature of the generation of nitric oxide means that the release of nitric oxide *in vivo* would depend on the chemical nature of the NONO-ate rather than the intervention of another biological process such as a redox system.

Sydnomines are a group of compounds used to treat angina. Molsidnomine is metabolised in the liver to 3-morpholino-sydnomine (SIN-1), which spontaneously releases nitric oxide under aerobic conditions. SIN-1 does not produce nitric oxide under anaerobic conditions, which suggests that the intervention of a redox system is necessary for nitric oxide release.

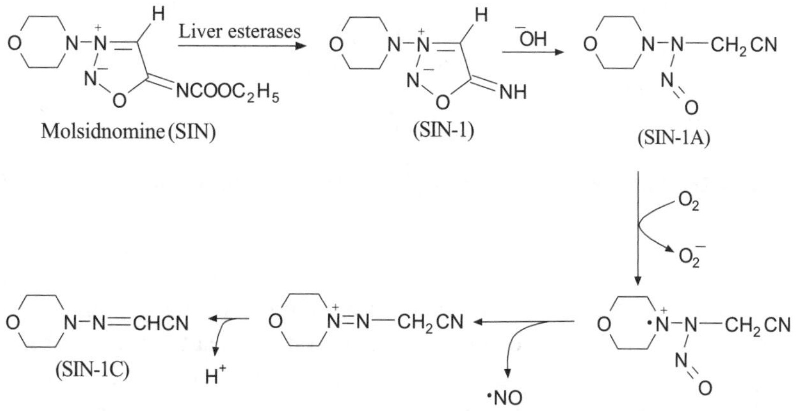

Sildenafil, a nitric oxide releasing agent

Sildenafil (Viagra) and a number of similar nitric oxide releasing compounds are used to treat erectile dysfunction in men (see section 14.5.4). Viagra, like many drugs in common use today, was discovered by accident. In 1985 S. Campbell and D. Roberts at Pfizer UK in Sandwich, Kent started a programme to develop a compound that would lower blood pressure by enhancing the activity of atrial natriuretic peptide. This peptide was a vasodilator known to act by causing the enzyme guanylyl cyclase (see section 14.4.1) to increase its production of cyclic guanosine monophosphate (cGMP, Fig. 14.15), which ultimately results in vasodilation. However, cGMP is metabolised by the enzyme phosphodiesterase type 5 (PDE 5), which terminates its action. Consequently, it was decided at Pfizer to find an inhibitor of PDE 5. After a literature search, zaprinast (Fig.14.15), which had a similar electronic distribution to cGMP, was selected as a lead since it was a weak inhibitor of PDE 5. Initial SAR investigations centred on the heterocyclic ring system of zaprinast. The compounds synthesised were tested for PDE 5 inhibition. The most promising compound was a pyrazolopyrimidinone with about ten times the potency of zaprinast. At this point the benzene ring became the focus of the SAR investigation. Crucially a sulphonamide group was introduced in an attempt to mimic the cyclic phosphate of cGMP and a butyl group replaced the 3-methyl in the pyrazolopyrimidinone ring system. This approach ultimately led to the synthesis of sildenafil in 1989, a PDE 5 inhibitor that was a hundred times stronger than zaprinast, as well as being highly specific in its action.

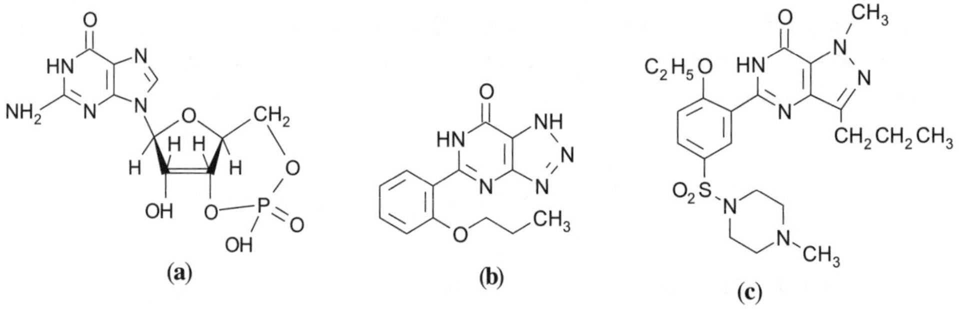

Figure 14.15   (a) Cyclic guanosine monophosphate (cGMP). (b) Zaprinast. (c) Sildenafil

By 1989 the original objective of the Pfizer programme had changed and sildenafil was used in trials to assess its use in treating coronary heart disease. Results were disappointingly but a side effect was revealed when the men in a trial in Merythr Tydfil showed a reluctance to hand in unused tablets at the end of the trial. Careful questioning revealed that the drug was causing an increase in erectile function. Further clinical investigations showed that it was of use in treating erectile dysfunction. Since the discovery of sildenafil a number of other compounds that inhibit PDE 5 have been discovered (Fig. 14.16).

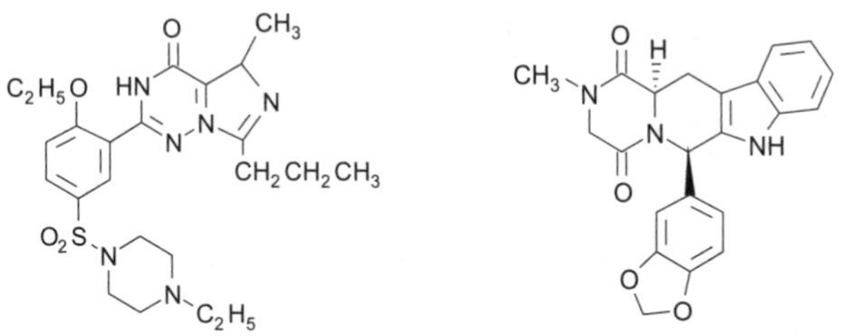

Vardenafil   Tadalafil

Figure 14.16 Examples of PDE 5 inhibitors

14.6.3 The genetic approach

The cloning of a gene can enable researchers to understand the nature of the control of the expression of that gene (see section 10.15.1). This information is the starting point for the development of compounds that can either block or enhance the action of a specific gene. Each of the different isoforms of NOS is produced by a different gene. Consequently, control of the relevant gene would influence the production of the relevant NOS isoform and the subsequent generation of nitric oxide produced by that isoform. This should enable medicinal chemists to design specific NOS inhibitors and stimulants. A number of NOS enzymes have been cloned from a variety of sources (Table 14.2) but no compounds have yet been developed for clinical use.

Table 14.2 Sources of cloned NOS

Type of NOS (Human)	Source (Human)	Type of NOS	Source (Other)
Neuronal cNOS	Human cerebellum	Neuronal cNOS	Rat cerebellum
Endothelial cNOS	Human endotheial cells	Endothelial cNOS	Bovine endotheial cells
iNOS	Human hepatocytes	iNOS	Murine macrophages
		iNOS	Rat vascular smooth muscle

14.7 Questions

1 Name each of the following compounds: (a) N_2O_3; (b) ONOO; (c) $ONSCH_2CH(NH_2)COOH$ and (d) $(C_2H_2)_2NNONO$.

2 Predict how nitric oxide might be expected to react with each of the following: (a) oxygen; (b) nitrogen dioxide; (c) diethyl amine; (d) ethanethiol in the presence of ferrous ions; and (e) a mixture of metmyoglobin and glycine.

3 Explain the meaning of the terms nitrosation and nitration. Illustrate the answer by reference to the *in vivo* nitrosation and nitration of the tyrosine residues of proteins.

4   Outline the biological significance of the reaction of nitric oxide with intercellular iron.

5   Describe the fundamental differences in the action of iNOS and cNOS with respect to nitric oxide production.

6   What general types of biological action are associated with iNOS and cNOS.

7   Describe, by means of equations and notes, how nitric oxide is generated in the endothelium. Give details of any intermediates and cofactors that are involved in the process.

8   What enzyme does nitric oxide activate in endothelial smooth muscle cells. Show, by means of a chemical equation(s), the chemical process catalysed by this enzyme. Give one physiological result of this process.

9   Suggest three general chemical approaches that could be used to deal with the pathological effects of nitric oxide. Illustrate the answer by reference to classes of compounds that are either used as drugs or have a potential drug use.

10   List the evidence that suggests that nitric oxide is involved in insulin-dependent diabetes.

15

An introduction to drug and analogue synthesis

15.1 Introduction

This chapter is intended to introduce some of the strategies used in the design of synthetic pathways. These pathways may be broadly classified as either *partial* or *full* synthetic routes. Partial synthetic routes are a combination of traditional organic synthesis and other methods. However, these routes tend to be more concerned with the large-scale production of proven drugs rather than the synthesis of leads and analogues.

Partial synthetic routes are often used in both drug and analogue synthesis and drug manufacturing processes when the required compound has a number of chiral centres in its structure. They use biochemical and other methods to produce the initial starting materials and traditional organic synthetic methods to convert these compounds to the target structure. For example, fermentation (a microorganism source) is used to produce benzylpenicillin, which is used as the starting point for the manufacture of a number of other penicillins (see section 15.3.3), pig insulin (an animal source) is used as the starting material for the production of human insulin, and diosgenin, extracted from a number of *Dioscorea* species (a plant source), is the starting point for the synthesis of a number of steroids. In contrast, full synthetic pathways start with readily available compounds, both synthetic and naturally occurring, but only utilise the standard methods of organic synthesis to produce the desired product (see sections 15.2 and 15.3).

Medicinal Chemistry, Second Edition   Gareth Thomas
© 2007 John Wiley & Sons, Ltd

15.2 Some general considerations

15.2.1 Starting materials

The choice of starting materials is important in any synthetic route. Common sense dictates that they should be chosen on the basis of what will give the best chance of reaching the desired product. Furthermore, in all cases the starting materials should be cheap and readily available. However this is not always possible when carrying out the initial synthesis of a specific analogue. The job of the process chemist is to convert these expensive analogue syntheses into more viable manufacturing methods.

15.2.2 Practical considerations

The chemical reactions selected for the proposed synthetic pathway will obviously depend on the structure of the target compound. However, a number of general considerations need to be borne in mind when selecting these reactions:

1. The yields of reactions should be high. This is particularly important when the synthetic pathway involves a large number of steps.

2. The products should be relatively easy to isolate, purify and identify.

3. Reactions should be stereospecific as it is often difficult and expensive to separate enantiomers. However, the exclusive use of stereospecific reactions in a synthetic pathway is a condition that is often difficult to satisfy.

4. The reactions used in the research stage of the synthesis should be adaptable to large-scale production methods. The reactions used by research workers frequently use expensive exotic reagents and it is the job of pharmaceutical development chemists to find simpler cost-effective alternatives.

15.2.3 The overall design

All approaches are based on a knowledge of the chemistry of functional groups and their associated carbon skeletons. The design may result in either a *linear synthesis*, where one step in the pathway is immediately followed by another:

$$A \longrightarrow B \longrightarrow C \longrightarrow D \longrightarrow \text{etc. to the target molecule}$$

or a *convergent synthesis*, where two or more sections of the molecule are synthesised separately before being combined to form the target structure (see section 15.4.2). In both cases, the *disconnection* approach (see section 15.4.1) may be used to design the pathway and identify suitable starting materials. Alternative design strategies that can also be usefully employed to design a synthesis are:

- finding compounds with similar structures to the target molecule and modifying their synthetic routes, if known, to produce the target compound;

- modifying natural products whose structures contain the main part of the target structure (see section 15.5).

An important aspect of all medicinal chemistry synthetic pathway design is divergency. Ideally, the chosen route should be such that it is relatively easy to modify the structure of the lead compound either directly or during the course of its synthesis. This is an economic way of producing a greater range of analogues for testing and hence increasing the chance of discovering an active compound. Initially these modifications would normally take the form of changing the nature of side chains or introducing new substituents in previously unsubstituted positions. The synthetic pathway for the preparation of the lead compound should include stages where it is possible to introduce these new side chains and substituents. For example, the presence of an amino group in a structure opens out the possibility of introducing different side chains by N-acylation (Fig. 15.1).

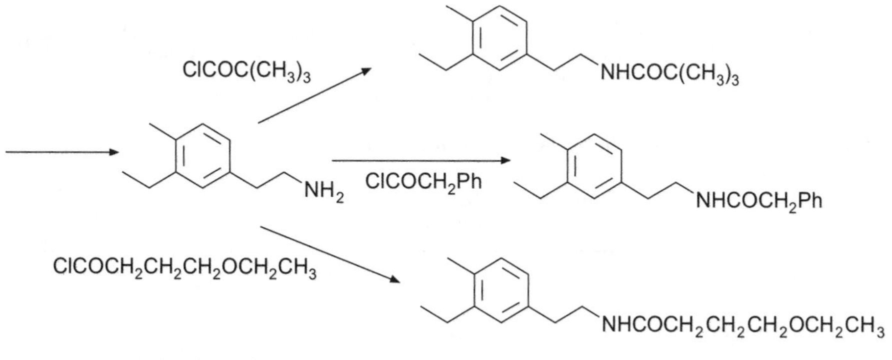

Figure 15.1 A stage in a hypothetical drug design pathway, illustrating some of the possibilities provided by the presence of an amino group in the structure of an intermediate. The products of the reactions illustrated could be the final products of the design pathway or they could be intermediates for the next stage(s) in the synthetic pathway

15.2.4 The use of protecting groups

The design of synthetic pathways often requires a reaction to be carried out at one centre in a molecule, the primary process, whilst preventing a second centre from either interfering with the primary process or undergoing a similar unwanted reaction. This objective may be achieved by careful choice of reagents and reaction conditions. However, an alternative is to combine the second centre with a so-called *protecting group* to form a structure that cannot react under the prevailing reaction conditions. A protecting group must be easy to attach to the relevant functional group, form a stable structure that is not affected by the reaction conditions and reagents being used to carry out the primary process and should be easily removed once it is no longer required (Table 15.1). However, in some circumstances, protecting groups may not be removed but converted into another structure as part of the synthesis.

Table 15.1 Examples of protecting groups. The conditions used will vary and so only the principal reagents are shown

Functional Group	Protecting Group	Removal
Alcohols/phenols	**Benzyl ether group** -OH $\xrightarrow[\text{Base}]{\text{PhCH}_2\text{Cl}}$ -O-CH$_2$Ph A benzyl ether	Hydrogenolysis -O-CH$_2$Ph $\longrightarrow$ -OH (catalytic hydrogenation)
Alcohols	**Triphenylmethyl (trityl) ether group** -OH $\xrightarrow[\text{Base}]{\text{Ph}_3\text{CCl}}$ -O-CPh$_3$ A triphenylmethyl ether	-O-CPh$_3$ $\xrightarrow{\text{CH}_3\text{COOH}}$ -OH Acid conditions
Carboxylic acids	***t*-Butyl ester group** $\xrightarrow[\substack{\text{Isobutene} \\ \text{H}_2\text{SO}_4}]{\text{CH}_3{-}\text{C}{=}\text{CH}_2,\ \text{CH}_3}$ -COOH $\longrightarrow$ -COO-C(CH$_3$)$_3$ A *t*-butyl ester	-COO-C(CH$_3$)$_3$ $\xrightarrow{\text{Dry acid conditions}}$ -COOH
Amines	**Ethanamide (acetamide) amide groups** -NH$_2$ $\xrightarrow{\text{CH}_3\text{COOCOCH}_3}$ -NHCOCH$_3$ An ethanamide	-NHCOCH$_3$ $\xrightarrow[\text{Hydrolysis}]{\text{H}^+/\text{H}_2\text{O}}$ -NH$_2$

15.3   Asymmetry in syntheses

The presence of an asymmetric centre or centres in a target structure means that its synthesis requires either the use of *non-stereoselective reactions* and the separation of the resulting stereoisomers or the use of *stereoselective reactions* that mainly produce one of the possible enantiomers. This section introduces some of the general methods used to incorporate stereospecific centres into a target molecule. However, for a more comprehensive discussion the reader is referred to *Stereoselective Synthesis* by R. S. Atkinson, published by John Wiley and Sons (1995) and *Selectivity in Organic Synthesis* by R. S. Ward, published by John Wiley and Sons (1999).

15.3.1 The use of non-stereoselective reactions to produce stereospecific centres

Non-stereoselective reactions produce either a mixture of diastereoisomers or a racemic modification. Diastereoisomers exhibit different physical properties. Consequently, techniques utilising these differences may be used to separate the isomers. The most common methods of separation are fractional crystallisation and appropriate forms of chromatography.

The separation (*resolution*) of a racemic modification into its constituent enantiomers is normally achieved by converting the enantiomers in the racemate into a pair of diastereoisomers by reaction with a pure enantiomer (Fig. 15.2). Enantiomers of acids are used for racemates of bases whilst enantiomers of bases are used for racemates of acids (Table 15.2). Neutral compounds may sometimes be resolved by conversion to an acidic or basic derivative that is suitable for diastereoisomer formation. The diastereoisomers are separated using methods based on the differences in their physical properties and the pure enantiomers are regenerated from the corresponding diastereoisomers by suitable reactions. For example, () octan-2-ol, a neutral compound, can be resolved into its separate enantiomers by conversion to the corresponding racemic hydrogen phthalate followed by treatment with (−)brucine (Fig. 15.3, page 537).

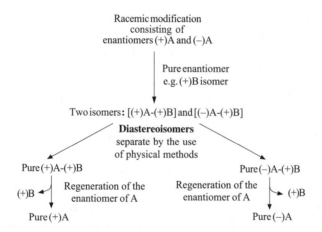

Figure 15.2 A schematic representation of the use of diastereoisomers in the resolution of racemic modifications

The incorporation of the resolution of a racemic modification into a synthetic pathway considerably reduces the overall yield of the synthesis because the maximum theoretical yield of an enantiomer is 50 per cent.

15.3.2 The use of stereoselective reactions to produce stereogenetic centres

Stereoselective reactions are those that result in the selective production of one of the stereoisomers of the product. The extent of the selectivity may be recorded as the

Table 15.2 Examples of the pure enantiomers used to resolve racemic modifications by forming diastereoisomers. In all regeneration processes there is a danger of the racemic modification being reformed by racemisation

Functional group	Enantiomers used (Resolving agents)	Diastereoisomers	Regeneration
Carboxylic and other acids	A suitable base, e.g. (−) Brucine (−) Strychnine (−) Morphine	Salts	Treatment with a suitable acid, e.g. HCl
Amines and other bases	A suitable acid, e.g. (+) Tartaric acid (−) Malic acid (+) Camphorsulphonic acid	Salts	Treatment with a suitable base, e.g. NaOH
Alcohols	A suitable acid (see above)	Esters	Acid or base hydrolysis

enantiomeric excess (e.e.) when the reaction produces a mixture of enantiomers and as the *diastereoisomeric excess* (d.e.) when it produces a mixture of diastereoisomers. Both of these parameters are defined as the difference between the yields of the isomers expressed as a percentage of the total yield of the reaction (equation 15.1)

$$\text{e.e. or d.e.} = \frac{(\text{Yield of the major product} - \text{Yield of the minor product}) \times 100}{\text{Yield of the major product} + \text{Yield of the minor product}} \quad (15.1)$$

$$= \% \text{ Major stereoisomer} - \% \text{ Minor stereoisomer} \quad (15.2)$$

The values of e.e. and d.e. are obtained by measuring the yields of the individual stereoisomers. An e.e. or d.e. value of zero means that the stereoisomers are produced in equal amounts. In the case of enantiomeric mixtures the product is likely to be in the form of a racemic modification. Conversely, an e.e. or d.e. value of 100 per cent indicates that only one product is formed. This rarely occurs in practice and most reactions yield a mixture of isomers.

The principal factors that appear to influence the stereochemistry of a reaction are:

- the shape of the substrate about the reaction centre;

- the nature of the reagent;

- the mechanism of the reaction;

- the catalyst used;

- the relative activation energies of the pathways used to produce the isomers.

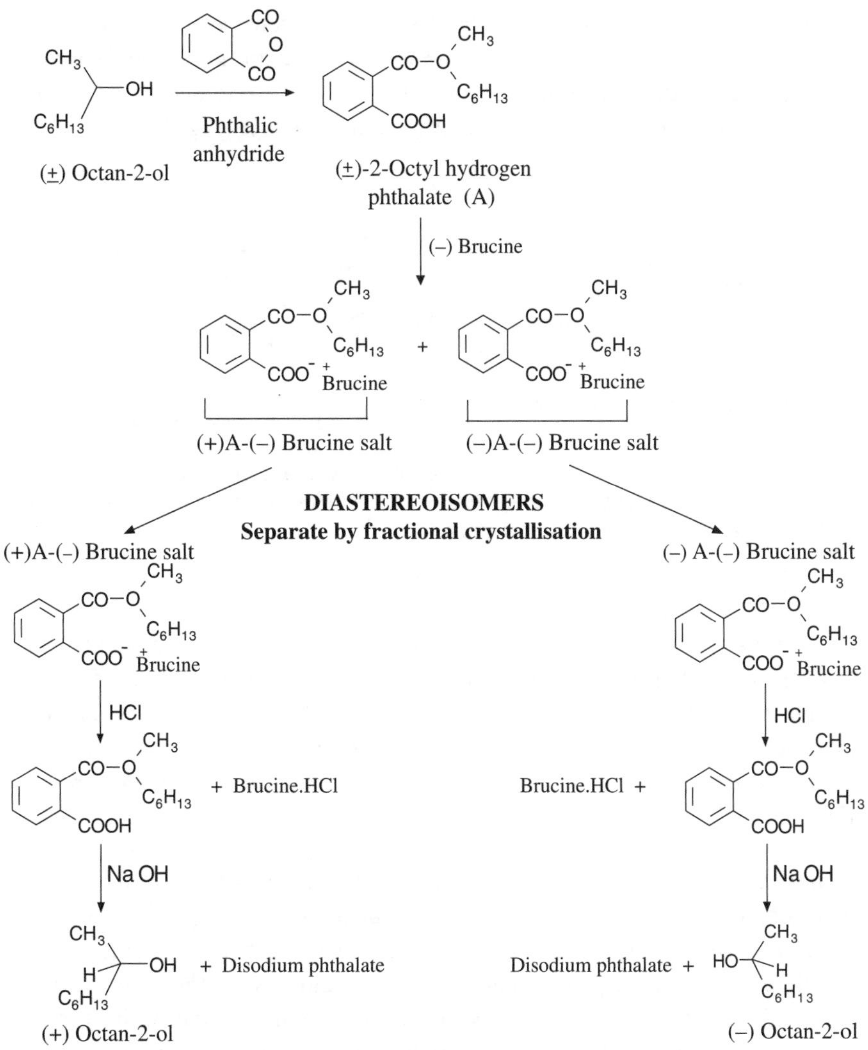

Figure 15.3 The reaction sequence used to resolve a racemic mixture of octan-2-ol

These factors are interrelated and should not be considered in isolation when assessing the stereochemical potential of a reaction. However, it is more convenient to consider each of these factors separately in order to illustrate their influence on the stereoselectivity of a reaction.

The shape of the substrate molecule A reaction will yield a mixture of enantiomers when there is an equal possibility of the reagent approaching the reactive centre of the substrate from opposite directions (see Figs 15.5 and 15.6). However, the same type of reaction will be stereospecific if steric hindrance reduces the chances of the reagent from attacking from more than one direction. For example, Davies *et al.* have developed a synthesis of the antihypertensive drug *S,S*-captopril (see section 9.12.2), which involves the introduction of the side chain chiral centre by alkylation of an enolate. This reaction is stereoselective because the iron–phosphorus–benzene ring complex of the substrate only allows the unhindered approach of the *t*-butylthiomethylbromide from one side of the molecule. Consequently, reaction occurs mainly from that side and produces the *S* configuration product (Fig. 15.4).

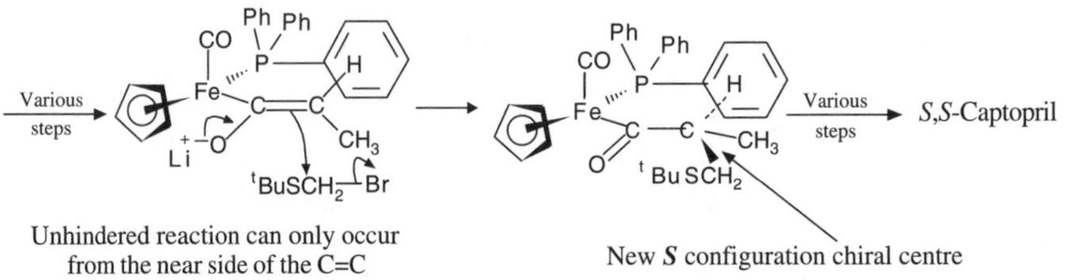

Figure 15.4 Stereoselective alkylation of an enolate in the synthesis of captopril (see section 9.12.2). *Key:* The heavier straight lines are bonds in the plain of the paper while the thin straight lines are bonds behind the plain of the paper. ^{t}Bu = a tertiary-butyl group

The nature of the reagent The nature of the reagent may affect the stereochemistry of the product of a reaction. Different reagents undergoing the same general type of reaction with the same reaction centre of a substrate may yield products that have different types of stereochemistry. For example, hydroxylation of the C=C bond of *E*-but-2-ene by osmium tetroxide yields a racemate (Fig. 15.5a) but bromination of this compound produces the *meso*-dibromide (Fig. 15.5b).

The mechanism The mechanism by which a reaction proceeds could influence the stereochemistry of the product(s). For example, in theory nucleophilic substitution of a chiral alkyl halide by an S_N1 mechanism should result in the formation of a racemate because the nucleophile can attack the planar intermediate from either side (Fig.15.6a). However, the time taken for the halide ion to diffuse away from the carbonium ion means that for this period of time the attack of the nucleophile is restricted to one side of the intermediate. Consequently, reactions proceeding by an S_N1 mechanism normally yield a product that consists of a mixture of an enantiomer with the opposite configuration to the substrate and the racemic modification. However, nucleophilic substitution by an S_N2 mechanism at a chiral alkyl halide centre will produce one enantiomer because the attack of the nucleophile can only take place from the side opposite to the halogen atom (Fig. 15.6b). It also causes an inversion of configuration.

The catalyst used The action of enzymes catalysing reactions that produce asymmetric centres is usually stereospecific. Consequently, a number of enzymes have been used to bring about a number of stereoselective transformations (see section 15.3.3). Furthermore, a number of non-enzyme catalysts have also been developed to catalyse the formation of chiral centres (see section 15.3.3).

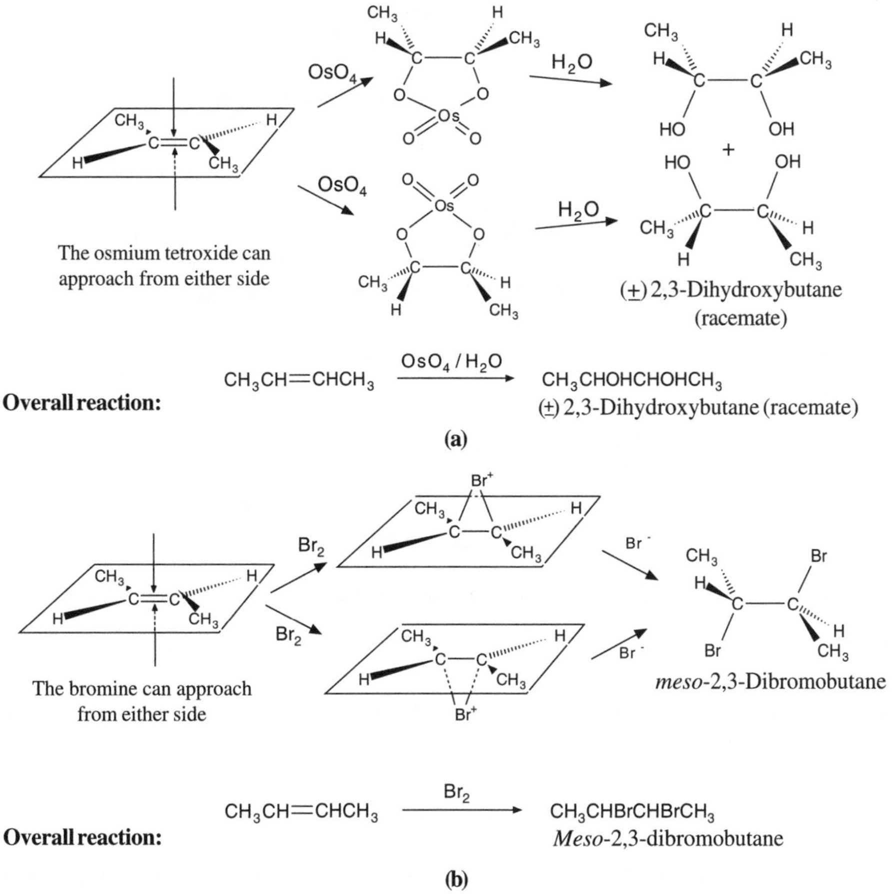

Figure 15.5 Examples of the effect of the nature of a reagent on the stereochemistry of a reaction. In both examples the reagent has an equal chance of attacking the CC from either side

The activation energy of the process The usefulness of a reaction in stereoselective synthesis will also depend on the activation energy of a process. Consider, for example, a reaction that produces a mixture of two stereoisomers. The activation energies for the formation of these stereoisomers will be the same if the stereoisomers are enantiomeric but *may be different* if the stereoisomers are diastereoisomeric (Fig. 15.7). The relative proportions of the diastereoisomers produced in the reaction mixture will depend on the relative values of the activation energies of the pathways producing the stereoisomers. The greater this difference, the greater the chance that the reaction will be diastereoselective with respect to the product formed via the lowest activation energy pathway. This is because reactants will find it easier to acquire the energy necessary to overcome the lower activation energy barrier (ΔG_1) than the higher (ΔG_2) and so there is a greater chance of the reaction proceeding by the lower energy pathway. Furthermore, it can be shown that the

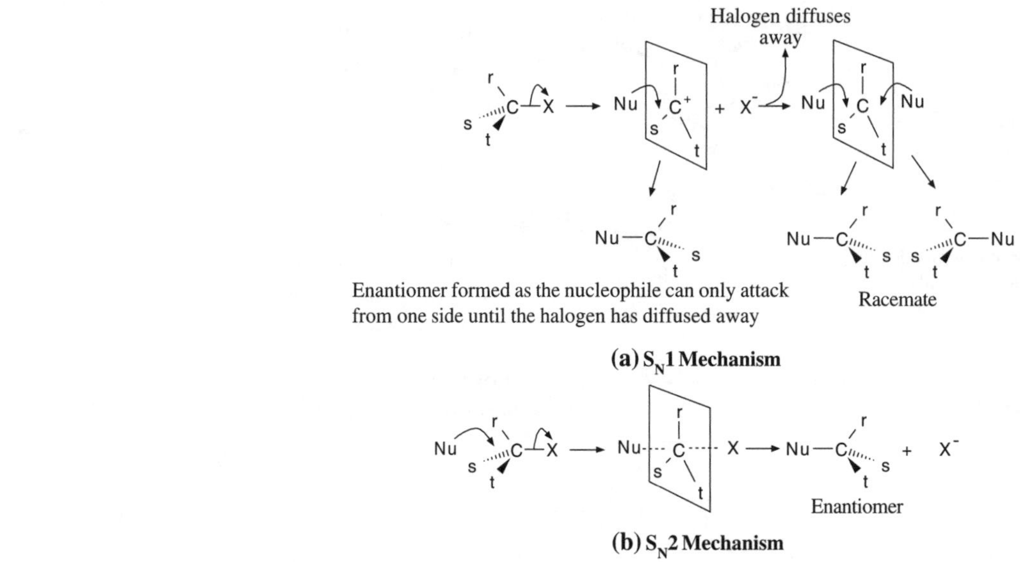

Halogen diffuses away
Enantiomer formed as the nucleophile can only attack from one side until the halogen has diffused away
Racemate
(a) S_N1 Mechanism
Enantiomer
(b) S_N2 Mechanism

Figure 15.6 The stereochemistry of nucleophilic substitutions at a chiral alkyl halide centre

relative rates of the two reactions as expressed in terms of their rate constants are given by the equation:

$$k_1/k_2 = e^{[-\Delta G_1 + \Delta G_2]/RT} \tag{15.3}$$

where k_1 is the rate constant for product 1, k_2 is the rate constant for product 2, R is the ideal gas constant and T is the temperature in K. Equation 15.3 shows that decreasing the temperature increases the ratio of k_1 to k_2, which results in an increase in the yield of product 1. In other words, lowering the temperature of a stereoselective reaction tends to make it more stereoselective.

Energy of the system
Reaction ordinate
Product 2
Product 1

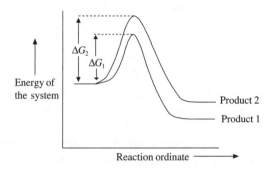

Figure 15.7 A schematic representation of the energy pathways of a reaction that produces two diastereoisomers by processes that have different activation energies

15.3.3 General methods of asymmetric synthesis

There is no set method for designing an asymmetric synthesis. Each synthesis must be treated on its merits and in all cases success will depend on the skill and ingenuity of the research worker. The range and scope of the reactions used in asymmetric synthesis are extremely large and consequently they are difficult to classify. In this text they are discussed under the broad headings of reactions that either require a catalyst or those that do not require a catalyst for their stereoselectivity. However, it is emphasised that this and the subdivisions used are a simplification and many reactions can fall into more than one category. Furthermore, it should be realised that several different general approaches may be used in the design of a synthetic pathway.

Stereoselective methods that depend on the use of a catalyst

Methods using enzymes as catalysts These methods can, in theory, use all types of substrate and reagents as the enzyme control will give the process stereoselectivity. These methods are economical in its use of chiral material but suffer from the disadvantage that they can require large quantities of the enzyme to produce significant quantities of the drug.

Enzyme-catalysed processes may be *single* or *interrelated multistep* processes. The latter usually uses enzymes or microorganisms to produce asymmetric starting materials from basic raw materials. For example, the production of the various semisynthetic penicillins uses benzylpenicillin, phenoxymethylpenicillin and cephalosporin C as starting materials because these can be produced from naturally occurring raw materials by fermentation. This section is only concerned with examples of the use of enzymes or microorganisms to catalyse a single transformation.

A wide variety of enzyme-controlled stereoselective transformations are known. These transformations include oxidations, reductions, reductive aminations, addition of ammonia, transaminations and hydrations (Fig.15.8). The microorganisms used as enzyme sources in these transformations have either been produced in bulk after isolation from natural sources or produced from existing microorganisms by genetic engineering (see section 10.15). The configuration of the new asymmetric centre produced by a particular enzyme or organism will depend on the structure of the substrate. However, substrates whose reactive centres have similar structures will often produce asymmetric centres with the same configuration.

The enzyme preparations used in transformations may take the form of a solid isolated from its natural source. These preparations are referred to as *cell-free* enzymes and do not usually include the cofactors and coenzymes essential for enzyme action (see section 9.1). Consequently, the use of cell-free preparations often requires the addition of the appropriate cofactors and coenzymes. The use of complete microorganisms means that the necessary cofactors and coenzymes are already present but in order for the microorganism to be effective the substrate must be able to penetrate the cell envelope (see section 7.3).

Enzymes normally act in aqueous media, usually at room temperature and about pH 7 (see section 9.7). However, many of the substrates of interest to medicinal chemists are insoluble in water. Changing the solvent can have an effect on the structure of the active site of an enzyme and, as a consequence, its activity. Consequently, a cell-free enzyme preparation

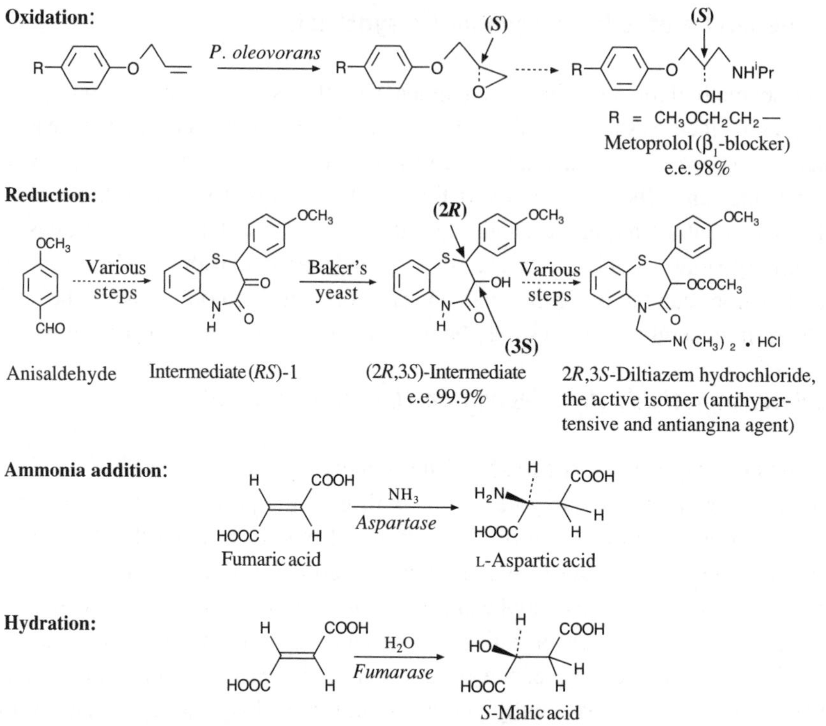
Oxidation:
P. oleovorans
(S)
(S)
NHiPr
OH
R = CH3OCH2CH2—
Metoprolol (β1-blocker)
e.e. 98%
Reduction:
Various steps
Baker's yeast
Various steps
(2R)
(3S)
Anisaldehyde
Intermediate (RS)-1
(2R,3S)-Intermediate
e.e. 99.9%
2R,3S-Diltiazem hydrochloride, the active isomer (antihyper-tensive and antiangina agent)
Ammonia addition:
NH3
Aspartase
Fumaric acid
L-Aspartic acid
Hydration:
H2O
Fumarase
S-Malic acid

Figure 15.8 Examples of enzyme-controlled transformations

must be capable of acting in non-aqueous solvents. Lipases, a group of enzymes that catalyse ester hydrolysis and acetyl transfer reactions, satisfy this requirement in that they are active in cyclohexane and toluene, amongst other hydrocarbon solvents. They have the advantages of being commercially available, many are inexpensive and they do not require a cofactor.

Methods using non-enzyme catalysts A number of stereospecific non-enzyme catalysts have been developed that convert achiral substrates into chiral products. These catalysts are usually either complex organic (Fig. 15.9) or organometallic compounds (Fig. 15.10). The

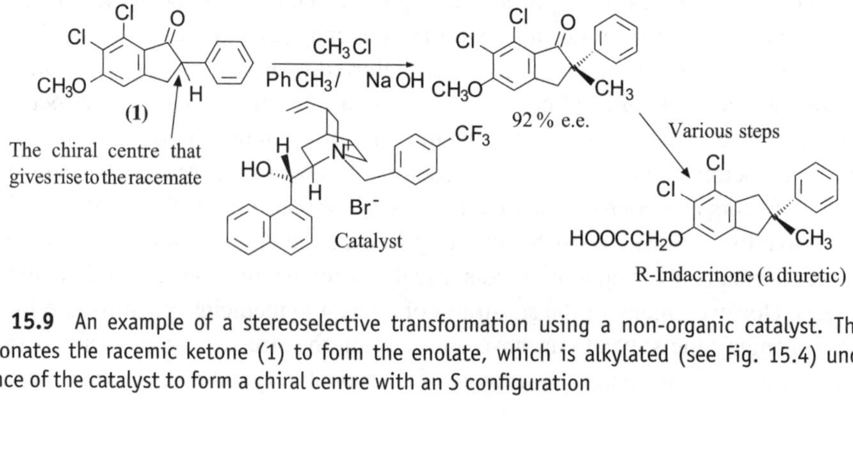
CH3Cl
PhCH3/ NaOH
(1)
The chiral centre that gives rise to the racemate
92 % e.e.
Various steps
Catalyst
R-Indacrinone (a diuretic)

Figure 15.9 An example of a stereoselective transformation using a non-organic catalyst. The base deprotonates the racemic ketone (1) to form the enolate, which is alkylated (see Fig. 15.4) under the influence of the catalyst to form a chiral centre with an *S* configuration

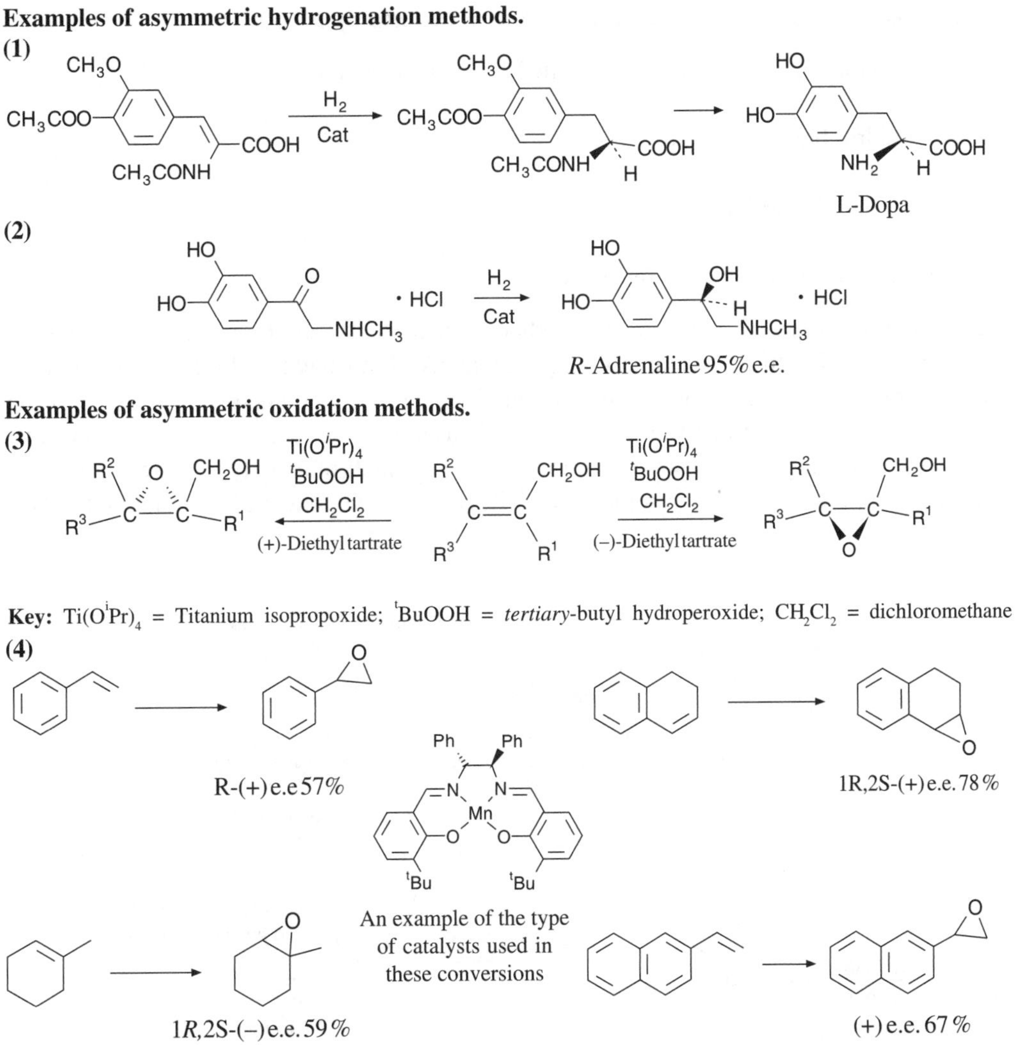
Examples of asymmetric hydrogenation methods.
(1)
H2 Cat
L-Dopa
(2)
H2 Cat
R-Adrenaline 95% e.e.
Examples of asymmetric oxidation methods.
(3)
Ti(OiPr)4 tBuOOH CH2Cl2 (+)-Diethyl tartrate
Ti(OiPr)4 tBuOOH CH2Cl2 (–)-Diethyl tartrate
Key: Ti(OiPr)4 = Titanium isopropoxide; tBuOOH = tertiary-butyl hydroperoxide; CH2Cl2 = dichloromethane
(4)
R-(+) e.e 57%
1R,2S-(+) e.e. 78 %
An example of the type of catalysts used in these conversions
1R,2S-(–) e.e. 59 %
(+) e.e. 67 %

Figure 15.10 Examples of the use of non-enzyme catalysts in stereoselective synthesis. (1) The stereogenetic step in the process Monsanto used to produce L-dopa. The catalyst is a rhodium complex with chiral phosphine ligands. (2) The synthesis of *R*-adrenaline using a rhodium–iron-based catalyst. (3) The Sharpless–Katsuki epoxidation. The stereochemistry of the product depends on which enantiomer of diethyl tartrate is used in the preparation. (4) Examples of epoxidations using manganese complexes of chiral Schiff bases, modified from W. Zhang, J. L. Loebach, S. R. Wilson and E. N. Jacobsen, *J. Am. Chem. Soc.*, **112**, 2801 (1990)

organometallic catalysts are usually optically active complexes whose structures usually contain one or more chiral ligands. An exception is the Sharpless–Katsuki epoxidation, which uses a mixture of an achiral titanium complex and an enantiomer of diethyl tartrate. The selection or development of a catalyst, reagent and reaction conditions for a transformation are normally made by considering similar stereoselective reactions in the literature.

Stereoselective methods that do not require the use of a catalyst

These general approaches can be classified for convenience as:

- using chiral building blocks;

- using a chiral auxiliary;

- using achiral substrates and reagents.

Using chiral building blocks   These methods depend on the use of enantiomerically pure building blocks with the required configurations. A chiral building block is treated with either chiral or achiral reagents to introduce the desired asymmetric centre(s) into the product. The stereochemistry of the substrate is used to make the reaction stereoselective (see Fig. 15.12 stage 1 and also section 15.3.2). The products of these types of reaction range from a mixture of diastereoisomers (Fig. 15.11), which may be separated into their

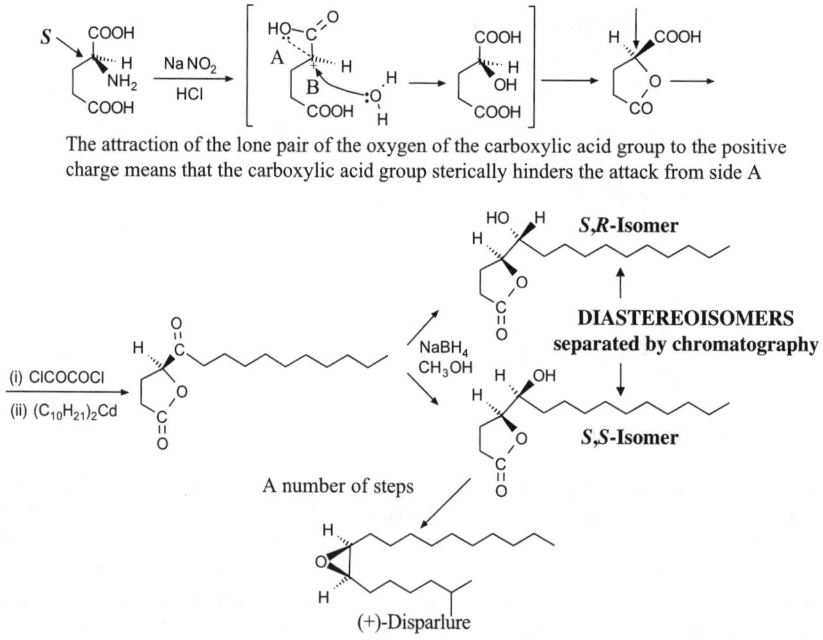

The attraction of the lone pair of the oxygen of the carboxylic acid group to the positive charge means that the carboxylic acid group sterically hinders the attack from side A

Figure 15.11   A scheme for the synthesis of (+)-disparlure, a pheromone produced by the female gypsy moth. This method starts with S-(+)-glutamic acid, which contains one of the required asymmetric centres. In the first stage the presence of the adjacent carboxylic acid group prevents nucleophilic attack of water occurring from side A (behind the paper) of the molecule. It only allows the nucleophile to attack from side B (in front of the paper) and so the configuration of the glutamic acid residue is retained in the product. The second asymmetric centre is introduced by the non-selective reduction of a ketone intermediate to a mixture of hydroxyacetone diastereoisomers that are separated by chromatography and the S,S-isomer is converted by a series of seven steps into (+)-disparlure.

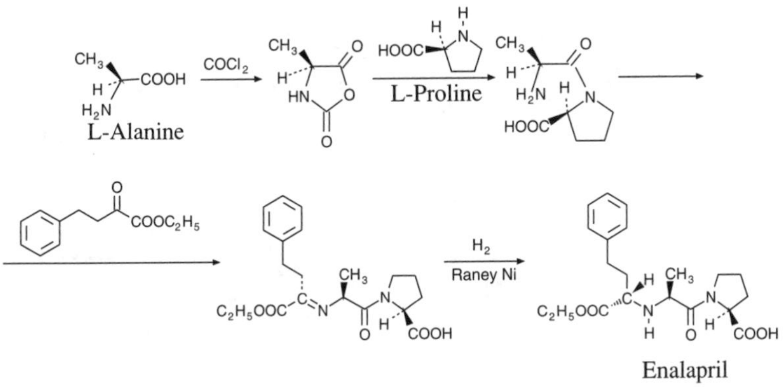

Figure 15.12 A synthetic route for the preparation of the ACE inhibitor enalapril (see section 9.12.2). The configurations of L-alanine and L-proline (the reagent for stage 2) are retained in the final product.The reduction of the intermediate is stereoselective, giving the S,S,S-isomer in 87% yield

constituents (see Fig. 15.3 and section 15.3.1), to a single enantiomer (Fig. 15.12). In all cases, the reactions used in further stages of the synthesis should not affect the configurations of the chiral centres of the building blocks. However, in some instances reactions that cause an inversion of configuration may be used (Fig. 15.13). The main sources of enantiomerically pure substrates and reagents are naturally occurring compounds, such as amino acids, amino alcohols, hydroxyacids, alkaloids, terpenes and carbohydrates. These materials are usually cheap and available in bulk.

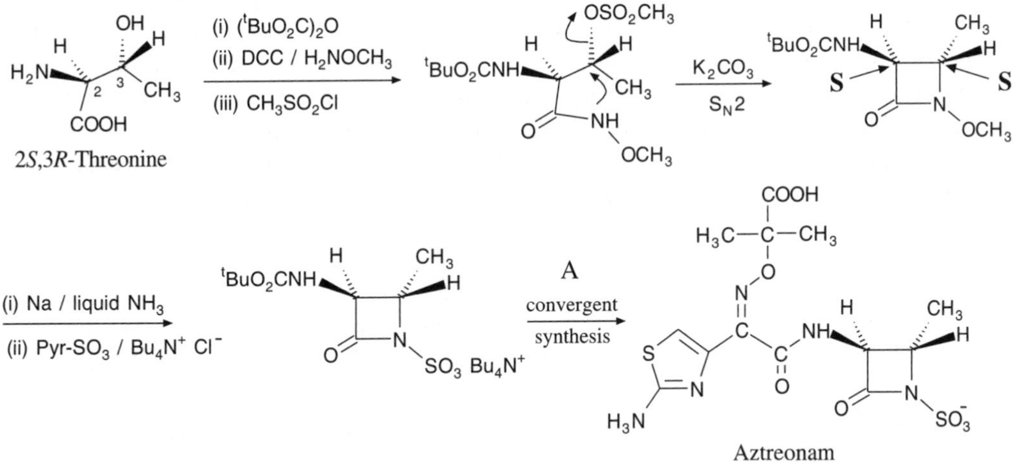

Figure 15.13 A scheme for the synthesis of the antibacterial aztreonam. The synthesis starts with the naturally occurring 2S,3R-threonine. The amino and hydroxy groups are protected before the acid is converted into the corresponding hydroxamate. An internal S_N2 closes the ring and also inverts the configuration of carbon 3 (see Fig. 15.6a). The N-methoxy group is removed by reduction and the resulting β-lactam is sulphonated. The sulphonated β-lactam is converted to the free amine, which is coupled to the rest of the molecule (a convergent synthesis)

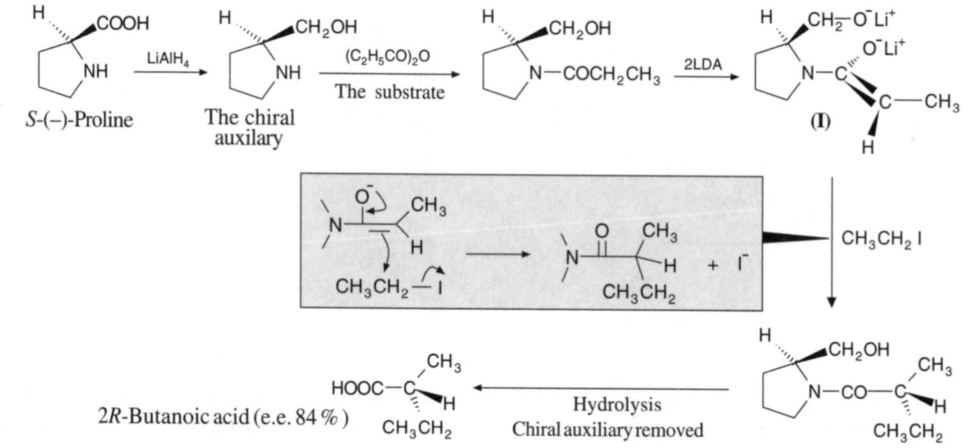

Figure 15.14 The synthesis of 2*R*-methylbutanoic acid, illustrating the use of a chiral auxiliary. The chiral auxiliary is 2*S*-hydroxymethyltetrahydropyrrole, which is readily prepared from the naturally occurring amino acid proline. The chiral auxiliary is reacted with propanoic acid anhydride to form the corresponding amide. Treatment of the amide with lithium diisopropylamide (LDA) forms the corresponding enolate (I). The reaction almost exclusively forms the *Z*-isomer of the enolate in which the OLi units are well separated and possibly have the configuration shown. The approach of the ethyl iodide is sterically hindered from the top (by the OLi units or H's) and so alkylation from the lower side of the molecule is preferred. Electrophilic addition to the appropriate enolate is a widely used method for producing the enantiomers of α alkyl-substituted carboxylic acids

Using a chiral auxiliary This method is based on a three-step process. The achiral substrate is combined with a chiral reagent known as a *chiral auxiliary* to form a chiral intermediate. Treatment of this intermediate with a suitable reagent produces the new asymmetric centre. The chiral auxiliary causes, by steric or other means (see section 15.3.2), the reaction to favour the production of one of the possible stereoisomers in preference to the others. Completion of the reaction is followed by removal of the chiral auxiliary, which may be recovered and recycled, thereby cutting down development costs (Fig. 15.14).

$$S \xrightarrow{(A^*)} S\text{-}A^* \xrightarrow{\text{Stereospecific reaction}} P^*\text{-}A^* \longrightarrow P^* + A^*$$

S: Achiral substrate; (A*): Chiral auxiliary, A pure enantiomer; P*-A*: Intermediate of the product (P*) and the chiral auxiliary; P* + A*: Product formation and regeneration of the chiral auxiliary

An advantage of this approach is that where the reaction used to produce the new asymmetric centre has a poor stereoselectivity, the two products of the reaction will be diastereoisomers as they contain two different asymmetric centres. These diastereoisomers may be separated by crystallisation or chromatography (see section 15.3.1) and the unwanted isomer discarded.

Using achiral substrates and reagents A wide variety of achiral substrates and reagents can give rise to asymmetric centres. However, the usefulness of these reactions in stereoselective synthesis will depend on their degree of stereoselectivity (see section 15.3.2). For example, electrophilic addition of hydrogen chloride to butene gives rise to a racemic mixture of the *R* and *S* isomers of 2-chlorobutane because Cl⁻ addition has an equal chance of occurring from either side of the C=C bond (Fig. 15.15).

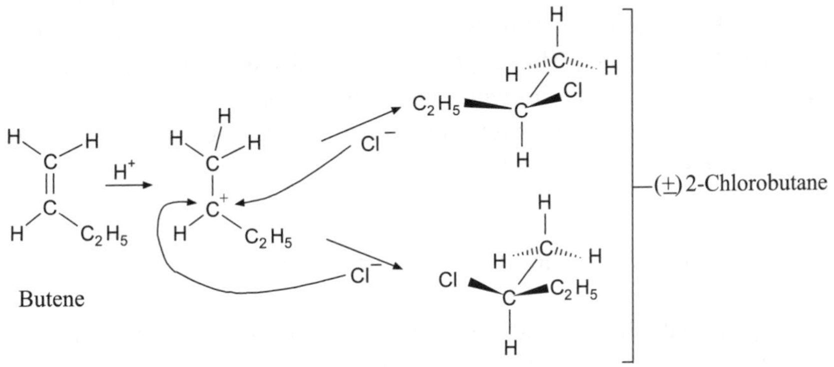

Figure 15.15 The addition of hydrogen chloride to butene

15.3.4 Methods of assessing the purity of stereoisomers

An understanding of the relationship of a stereoisomer of a chiral drug candidate requires an accurate assessment of the purity of that stereoisomer. A wide range of methods for determining this purity are available. They include, amongst others:

1. *Optical rotation.* The specific rotation of the enantiomer is compared with a literature value for the enantiomer.

2. *High-pressure liquid chromatography (HPLC).* This is carried out using a chiral stationary phase (CSP), which forms diastereomeric complexes with the enantiomers being separated. These complexes move at different speeds through the column.

3. *Gas chromatography (GC).* This is carried out using a chiral stationary phase (CSP) that also forms diastereomeric complexes that move at different speeds through the column. The method suffers from the disadvantage that it requires vaporisation, without decomposition, of the sample.

4. *Electrophoresis.* Capilary electrophoresis (CE) is carried out in coiled silica capillaries up to a metre long. A so-called *chiral selector* is added to the electrolyte to form diastereoisomers that move at different speeds through the capilliary.

Ideally the purity should be assessed by several different methods since different analytical procedures can give different results.

15.4   Designing organic syntheses

The synthetic pathway for a drug or analogue must start with readily available materials and convert them by a series of inexpensive reactions into the target compound. There are no obvious routes as each compound will present a different challenge. The usual approach is to work back from the target structure in a series of steps until cheap commercially available materials are found. This approach is formalised by a method developed by S. Warren that is known as the *disconnection approach* or *retrosynthetic analysis*. In all cases the final pathway should contain a minimum of stages in order to keep costs to a minimum and overall yields to a maximum.

15.4.1   An introduction to the disconnection approach

This approach starts with the target structure and then works *backwards* by artificially cutting the target structure into sections known as *synthons*. Each of these backward steps is represented by a double shafted arrow ($\Longrightarrow$) whilst ⌇⌇⌇⌇ is drawn through the disconnected bond of the target structure. Each of the possible synthons is converted on paper into a real compound known as a *reagent* whose structure is similar to that of the synthon. All the possible disconnection routes must be considered. The disconnection selected for a step in the pathway is the one that gives rise to the best reagents for a reconnection reaction. This analysis is repeated with the reagents of each disconnection step until readily available starting materials are obtained. The selection of the reagents and the reactions for their reconnection may require extensive literature searches (Table 15.3, page 549).

In the disconnection approach, bonds are usually disconnected by either homolytic or heterolytic fission (Fig. 15.16). However, some bonds may be disconnected by a reverse pericyclic mechanism (see Table 15.4, pages 550-551, Diels–Alder).

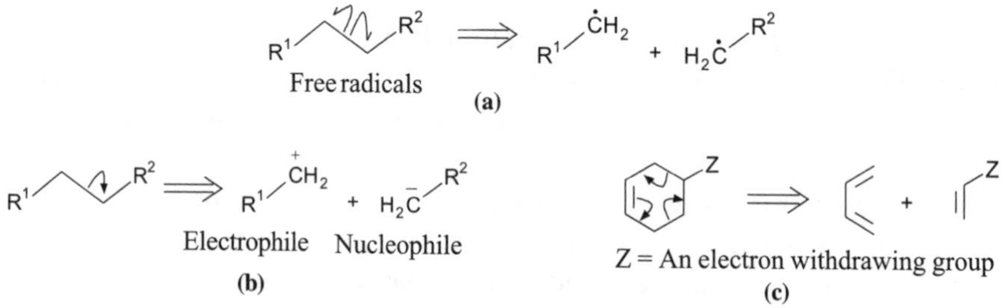

Figure 15.16   (a) Homolytic, (b) heterolytic and (c) pericyclic bond disconnections

Free radical disconnections are usually disregarded because it is difficult to predict the outcome of reconnection reactions that proceed through a free radical mechanism as these reactions tend to produce mixtures. Heterolytic disconnections result in the formation of electrophilic and nucleophilic species. The most useful heterolytic disconnections are those that give rise to a stable species or occur by a feasible disconnection mechanism, such as hydrolysis, because these disconnections are more likely to have a corresponding reconnection

Table 15.3 Examples of books and data bases that catalogue chemical reactions

Title	Author	Classification Used
Books:		
Synthetic Organic Chemistry	Wagner and Zook	Lists reactions according to the functional group being produced
Organic Functional Group Preparations	Sandler and Karo	Lists reactions according to the functional group being produced
Reagents for Organic Synthesis	Fieser and Fieser	Lists reagents and their uses in alphabetical order. (Includes suppliers)
Carbanions in Synthesis	Ayres	Lists carbanion transformations
Oxidations in Organic Chemistry	Hudlicky	Lists and discusses transformations that can be brought about by oxidation
Reduction in Organic Chemistry	Hudlicky	Lists and discusses transformations that can be brought about by reduction
Data bases:		
CASREACT	The Chemical Abstracts Research Service	Information from 1985. Covers single and multistep reactions. Includes CAS Registry numbers of reactants, products, reagents, catalysts and solvents
ISI Reaction Centre	Institute for Scientific Information	Data from 1840. Classified according to reaction type. Includes bioassays
Crossfire	Beilstein Information Service	Three main types of data, structural properties and reactions, including preparations and chemical literature references

reaction. Synthons are converted to a real reagent by converting them to a structurally similar compound that has the relevant electrophilic or nucleophilic centre. For example, a carbanion synthon with the structure RCH_2^- could correspond to a Grignard reagent RCH_2MgBr. Similarly, an electrophilic synthon $R\overset{+}{C}O$ could correspond to an acid halide $RCOCl$ or ester $RCOOR'$. Where disconnection does not produce an obvious electrophilic or nucleophilic synthon, all possible structurally related reagents should be considered.

Disconnections may be made by disconnecting functional groups or by the carbon skeleton. It is normal to first attempt to disconnect the sections that are held together by functional groups such as esters, amides and acetals as it is usually easier to find reconnection reactions for these functional groups. Consider, for example, the synthesis of the local anaesthetic benzocaine. The most appropriate disconnections are the ester and amine groups. At this point it is a *matter of experience* as to which disassembly route is followed. The normal approach is to pick the synthons that give rise to reagents that can most easily be reformed into the product. Consequently, in this case the ester disconnection would appear to be the most profitable pathway as ester formation is relatively easy but it is not possible to directly introduce a nucleophilic amino group into a benzene ring.

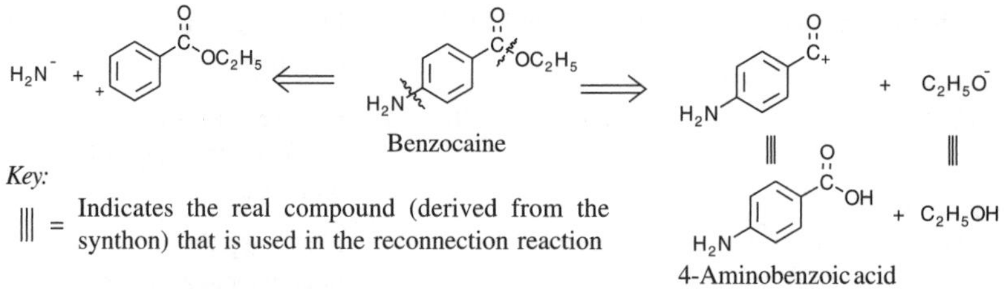

Benzocaine

Key:

||| = Indicates the real compound (derived from the synthon) that is used in the reconnection reaction

4-Aminobenzoic acid

Table 15.4 Examples of the most useful disconnection and reconnection systems. The compounds used for the reconnection reactions are those normally associated with the synthons for the disconnection. These reactions are not the only reactions that could be used for reconnection. Source: S. Warren, *Organic Synthesis, the Disconnection Approach*, John Wiley and Sons, Ltd, 1982

Disconnections	Examples of reconnection reactions

Functional group disconnections:

Ether:

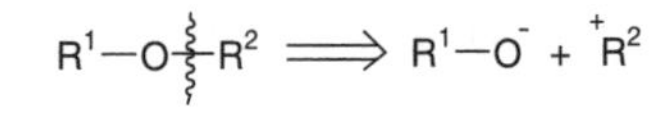

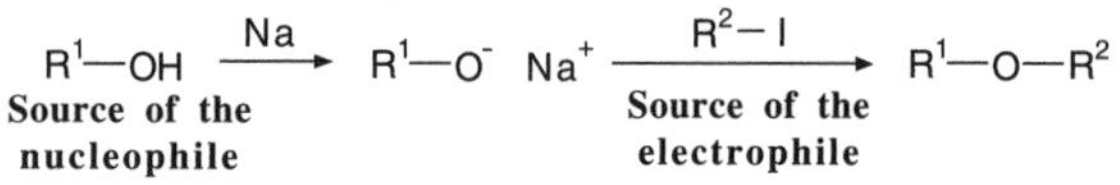

$R^1 - O \xrightarrow{} R^2 \implies R^1 - O^- + {}^+R^2$

$R^1 - OH \xrightarrow{Na} R^1 - O^- \ Na^+ \xrightarrow{R^2 - I} R^1 - O - R^2$

Source of the nucleophile   Source of the electrophile

For methyl ethers use dimethyl sulphate instead of R^2I

Amide:

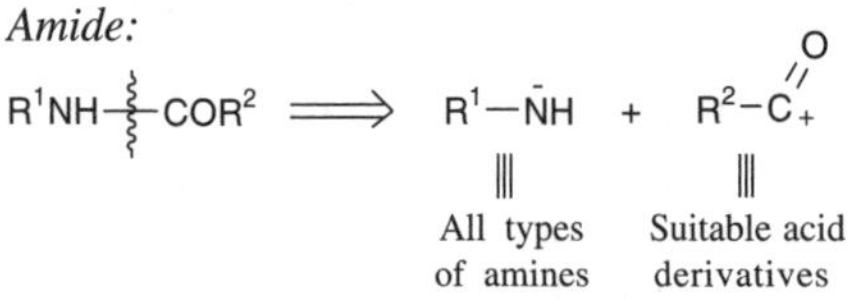

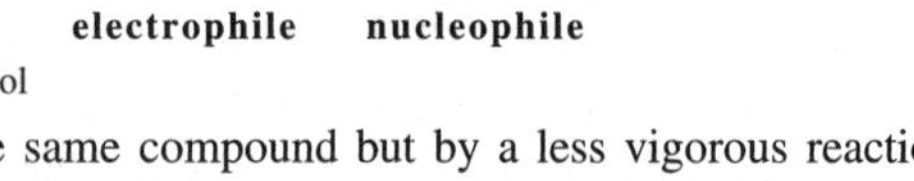

$R^1 NH - COR^2 \implies R^1 - \bar{N}H + R^2 - \overset{O}{\overset{\parallel}{C}}{}^+$

||| All types of amines ||| Suitable acid derivatives

$R^1 NH_2 + R^2 COCl \longrightarrow R^1 NH - COR^2$

Source of the nucleophile   Source of the electrophile

Note: The acid anhydride will give the same compound but by a less vigorous reaction

Ester:

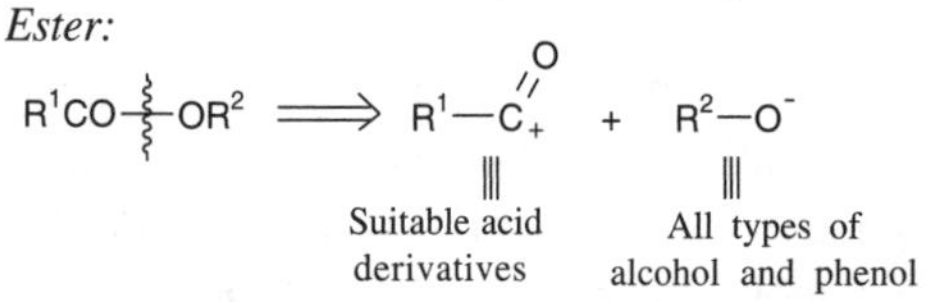

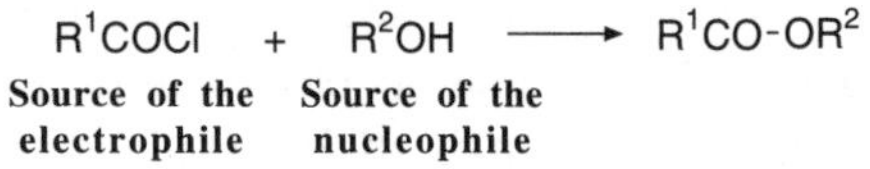

$R^1 CO - OR^2 \implies R^1 - \overset{O}{\overset{\parallel}{C}}{}^+ + R^2 - O^-$

||| Suitable acid derivatives ||| All types of alcohol and phenol

$R^1 COCl + R^2 OH \longrightarrow R^1 CO - OR^2$

Source of the electrophile   Source of the nucleophile

Note: The acid anhydride will give the same compound but by a less vigorous reaction

Lactone:

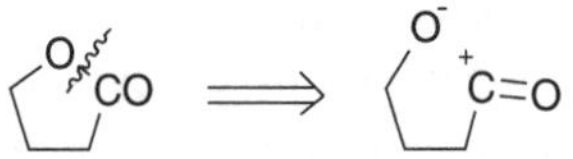

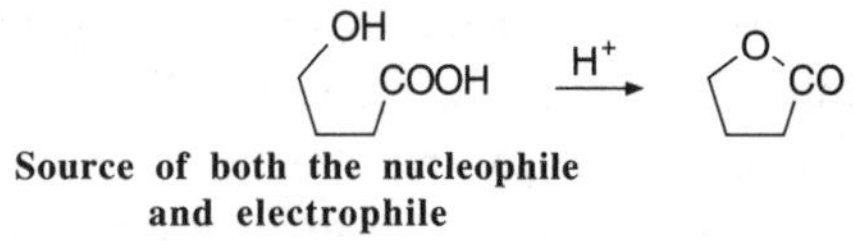

Source of both the nucleophile and electrophile

Acetal

Source of the electrophile   Source of the nucleophile

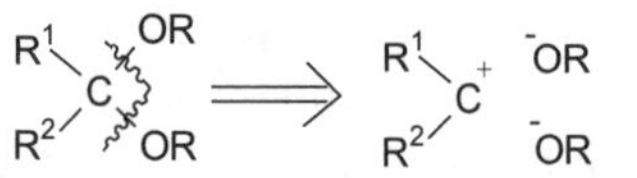

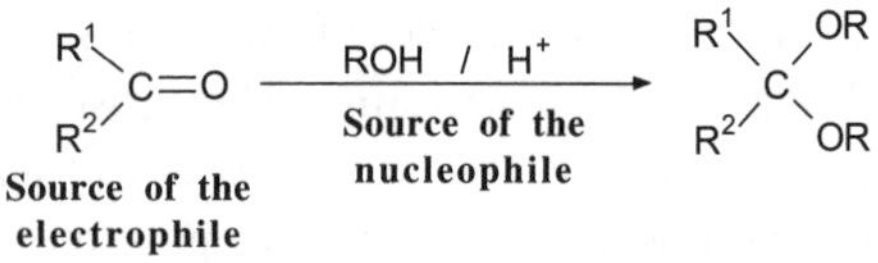

Table 15.4 *(Continued)*

Disconnections	Examples of reconnection reactions

Carbon–carbon disconnections:

Carbanion–electrophile disconnections: Disconnection should normally occur adjacent to an electron withdrawing group. In each case, one of the compounds derived from the synthon should be able to form a carbanion whilst the structure of the other should contain an electrophilic centre. Reconnection is by means of a suitable carbanion substitution or condensation reaction

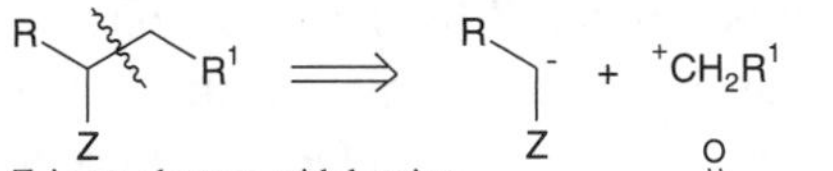

Z is an electron withdrawing group, e.g. NO_2, COOR, SOR, SO_2R, CN

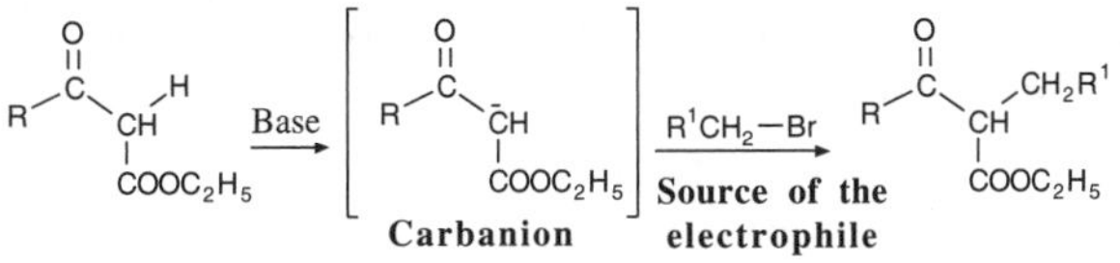

Base → Carbanion → R^1CH_2—Br Source of the electrophile

Friedel–Crafts disconnections:

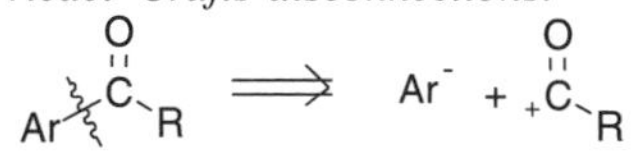

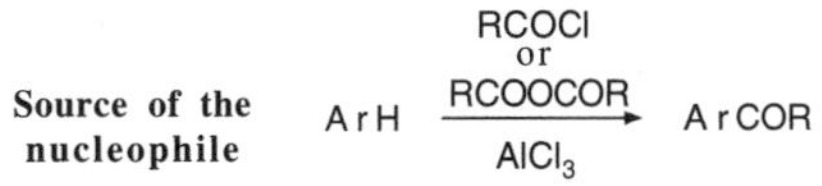

Source of the nucleophile

$ArH \xrightarrow[AlCl_3]{\begin{array}{c}RCOCl\\or\\RCOOCOR\end{array}} ArCOR$

Ar = an aromatic system

The position of substitution will depend on the nature of the substituents on the aromatic (Ar) ring system

Adjacent to a ketone:

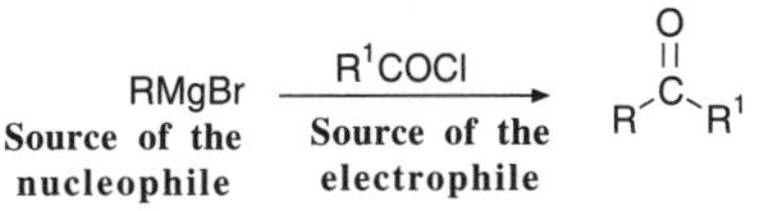

RMgBr Source of the nucleophile

$\xrightarrow{R^1COCl}$ Source of the electrophile

Adjacent to an alcohol:

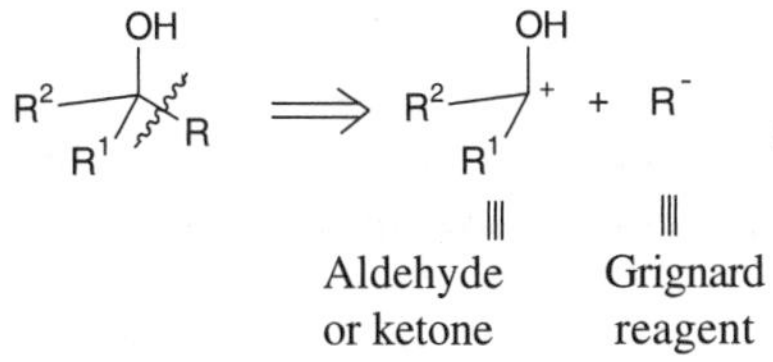

‖ Aldehyde or ketone

‖ Grignard reagent

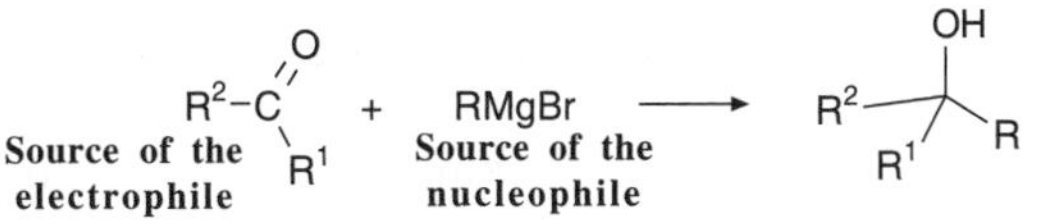

Source of the electrophile + RMgBr Source of the nucleophile →

Carboxylic acids:

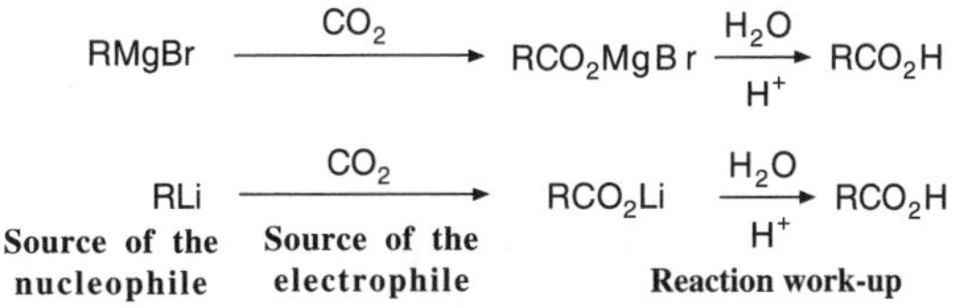

$R\text{—}COOH \implies R^- + {}^+CO_2H$

$RMgBr \xrightarrow{CO_2} RCO_2MgBr \xrightarrow[H^+]{H_2O} RCO_2H$

$RLi \xrightarrow{CO_2} RCO_2Li \xrightarrow[H^+]{H_2O} RCO_2H$

Source of the nucleophile   Source of the electrophile   Reaction work-up

Diels–Alder: Disconnection is by means of a reversed pericyclic mechanism

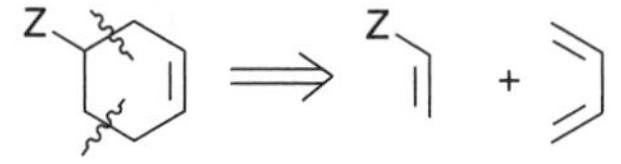

Z is an electron withdrawing group

Epoxides:

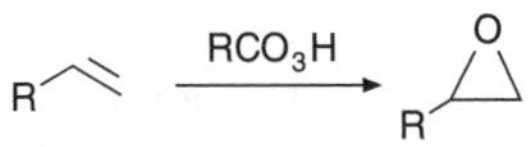

$\xrightarrow{RCO_3H}$

Ethanol is a readily available starting material but 4-aminobenzoic acid is not. Therefore, the next step is to consider the disconnection of the amino and carboxylic acid groups of 4-aminobenzoic acid. However, there are no simple inexpensive reactions for the reconnection of these groups. Consequently, the next step has to be a *functional group interconversion* (FGI). Disconnection arrows are usually used for FGIs but as no synthons are involved it is customary to use real structures in the relationships. FGIs are found by searching the literature for suitable reactions. This search, in the case of the current example, reveals that aromatic carboxylic acids may be produced by the oxidation of an aromatic methyl group whilst an aromatic amine may be produced by reduction of the corresponding nitro group. Since amino groups are sensitive to oxidation the disconnection via FGI follows the order:

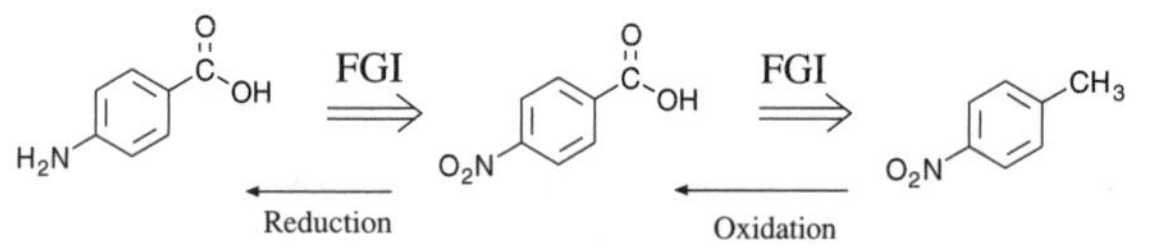

The final stage is to consider the disconnection of both the methyl and nitro groups.

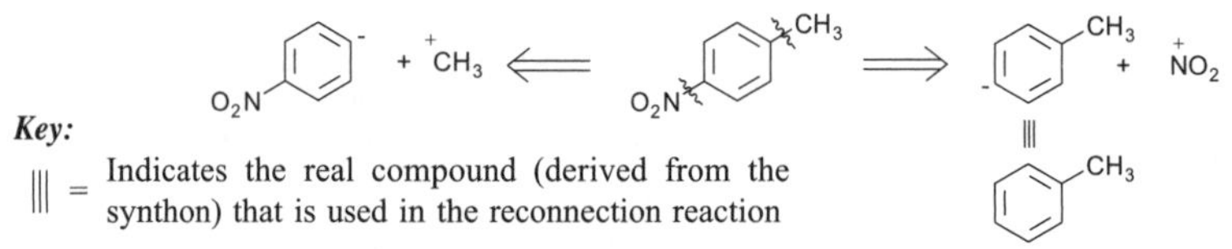

Key:

‖ = Indicates the real compound (derived from the synthon) that is used in the reconnection reaction

Toluene is a readily available compound so the best disconnection is the nitro group. This is also supported by the fact that it is easy to form 4-nitrotoluene by nitration of toluene. Consequently, the complete synthesis is:

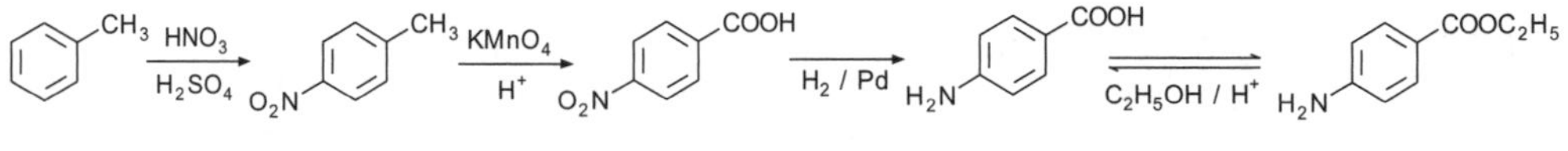

Carbon skeletons are usually disassembled at a point adjacent or near to a functional group or at a branch point in the structure. Consider, for example, the synthesis of 4-phenyl-1,4-butyrolactone (Fig. 15.17). The scheme for the disconnection starts with the lactone. It is followed by an FGI of the alcohol to the ketone. Although there are other possibilities, these are the most obvious logical disassemblies for these stages. However, there are several possible disconnections at this point. Routes 1, 2 and 3 give synthons that do not give rise to readily available compounds. Further disconnections would be needed. However, route 4 gives access to synthons corresponding to readily available inexpensive starting materials that can undergo reconnection by a carbanion nucleophilic substitution. Consequently, this route, which is the shortest, becomes the preferred pathway for the

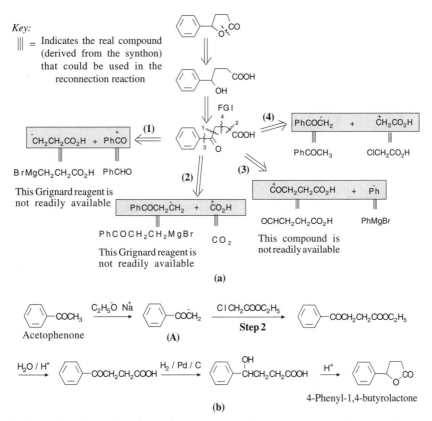

Figure 15.17 (a) The disconnection scheme for the synthesis of 4-phenyl-1,4-butyrolactone. The shaded structures are synthons. (b) The complete synthesis is based on the disconnection flow chart. It is necessary to protect the acid at step 2 in the form of its ethyl ester in order to prevent it reacting with the carbanion A

synthesis. The design of a synthesis using the disconnection approach must take into account:

- the order of disconnection could influence the ease and direction of subsequent reactions;

- the need to protect a reactive group in a compound by the use of a suitable protecting agent (see sections 15.2.4 and Table 15.1);

- the need to incorporate the appropriate chiral centres into the structure (see section 15.3.3).

A wide range of disconnections linked to suitable reconnections are known (Table 15.4). However, a number of functional groups are usually best introduced by FGI (Table 15.5). Once again it cannot be overemphasised that their use will depend on the experience of the designer, which only improves with usage and time.

Table 15.5 Examples of reactions used to reverse FGIs. R can be both an aromatic and an aliphatic residue but Ar is only an aromatic residue

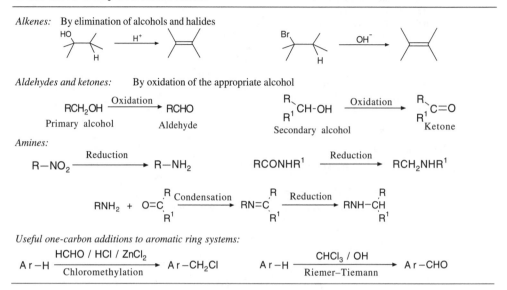

15.4.2 Convergent synthesis

The yields of the final product in a linear multistep synthesis (Fig. 15.18a) may be small even though the yields of the individual steps are large. For example, a ten-step synthesis in which each of the steps has a yield of 90 per cent will only have an overall yield of the target molecule of about 35 per cent. A strategy known as *convergent synthesis* may be

A ——→ B ——→ C ——→ D ——→ E ——→ F ——→ G ——→ H ——→ I ——→ J ——→ Target structure
Overall yield of the target structure 35 %
(a)

A ——→ B ——→ C ——→ D ——→ E
F ——→ G ——→ H ——→ I ——→ J
→ Target structure
Overall yield of the target structure 53 %
(b)

A ——→ B ——→ C
D ——→ E ——→ F
G ——→ H ——→ I
→ J
→ Target structure
Overall yield of the target structure 66 %
(c)

Figure 15.18 A schematic representation of (**a**) linear and (**b,c**) convergent syntheses. The overall yields are based on a 90 per cent yield for each step

used to improve this overall yield. In convergent syntheses, sections of the target molecule are prepared and joined together to form the target molecule (Fig. 15. 18b). A two-path convergent synthesis would have the effect of increasing the overall yield of the same ten-step synthesis to about 53 per cent and also reducing the number of steps. It is not necessary to restrict convergence to two paths and the overall yield could be further improved by the use of several convergent pathways (Fig. 15.18c).

The process of disconnection may be applied to target molecules by initially disconnecting large sections of the structure that could be suitable for individual linear synthesis. Consider, for example, the synthesis of the antihistamine N,N-dimethyl-2-[1-(4-chlorophenyl)-1-methyl-1-phenylmethoxy]ethylamine ({I} Fig. 15.19). Disconnection of the ether group will divide the target molecule into two reasonably sized synthons. These synthons correspond to the nucleophilic and electrophilic precursors of the drug, namely the tertiary alcohol and the substituted alkyl halide. Since these compounds may be readily combined to form the drug it is now possible to design a convergent synthesis in which each of these compounds is prepared separately before being combined to form the antihistamine (Fig.15.19). First consider the synthesis of the tertiary alcohol. As tertiary alcohols are easily prepared using Grignard reagents, the obvious disconnections for

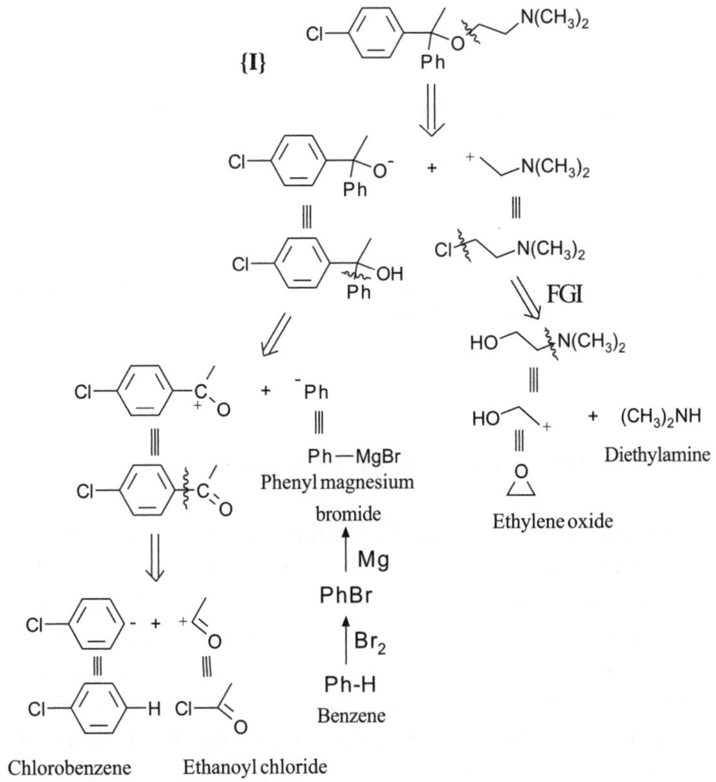

Figure 15.19 The disconnection scheme for the synthesis of the antihistamine N,N-dimethyl-2-[1-(4-chlorophenyl)-1-methyl-1-phenylmethoxy]ethylamine

simplifying the structure are the aromatic ring systems. The phenyl group is the better disconnection as it gives rise to the synthons of 4-chloroacetophenone and benzene. The latter can be reconnected to the 4-chloroacetophenone by a Grignard reaction using phenyl magnesium bromide. However, disconnection of the 4-chlorophenyl would have given the synthon of 1,4-dichlorobenzene, which would require the mono Grignard derivative for the reconnection step. Synthesis of this derivative would not be simple. The final step is the disconnection of the ketone from the 4-chloroacetophenone, which yields the readily available chlorobenzene and ethanoyl chloride.

The disconnection of the substituted alkyl halide in the second arm of the convergent synthesis starts with an FGI as halides are usually produced by either direct action of a halogen or substitution of an alcohol. Alcohols are produced from epoxides by nucleophilic substitution and so the disconnection of the secondary amino group leads to ethylene oxide and diethylamine as the readily available starting compound for a convergent synthesis (Fig. 15.20).

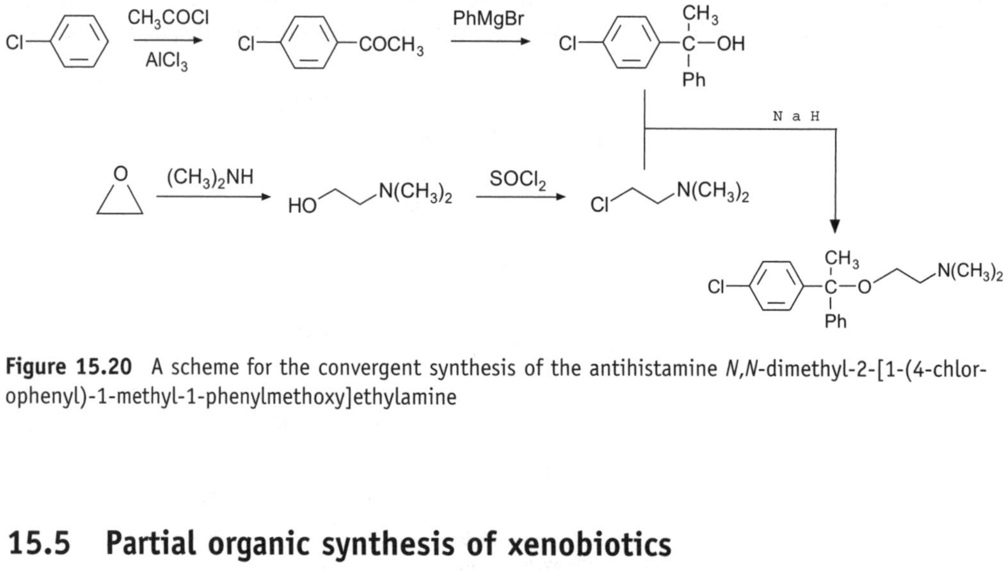

Figure 15.20 A scheme for the convergent synthesis of the antihistamine *N*,*N*-dimethyl-2-[1-(4-chlorophenyl)-1-methyl-1-phenylmethoxy]ethylamine

15.5   Partial organic synthesis of xenobiotics

Partial synthetic pathways use biochemical and other methods to produce the initial starting materials and traditional organic synthesis to convert these compounds to the target structure. The principal methods are based on starting with compounds produced by microbiological transformations, the use of enzymes or extracted from other natural sources. These methods are used to produce starting materials because it usually cuts down the cost of production and produces compounds whose structures have chiral centres with the required configurations. For example, the total synthesis of steroidal drugs is not feasible because of the many chiral centres found in their structures. Consequently, partial synthesis is the normal approach to producing new analogues and manufacturing steroidal drugs. For example, the starting material for the production of progesterone is diosgenin obtained from a number of *Dioscorea* species (a plant source). Diosgenin may be converted to pregnenolone acetate by a series of steps (Fig. 15.21). This compound serves as the starting material for the synthesis of a number of steroidal drugs, including progesterone.

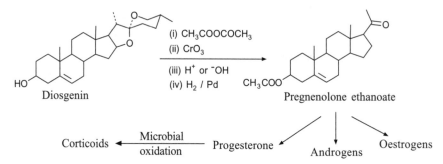

Figure 15.21 An outline of the synthesis of progesterone from diosgenin

The disconnection approach may also be used to design the steps in the partial synthesis of the target molecules from these types of starting material compounds.

15.6 Questions

1 Explain the meaning of the terms (a) linear, (b) convergent and (c) partial synthetic pathways.

2 Outline the practical considerations that need to be taken into account when selecting reactions for use in a synthetic pathway.

3 What are the requirements for a good protecting group?

4 Describe the factors that appear to influence the stereochemistry of a chemical reaction.

5 Describe the use of catalysts in asymmetric synthesis.

6 What is a chiral auxiliary? Suggest a feasible stereospecific synthesis for 2R-methylhexanoic acid starting from S-(−)-proline and propanoic anhydride.

7 Explain the meaning of the term 'synthon'. Draw the best synthons and their corresponding real compounds for each of the following compounds:

(a)   (b)   (c)   (d)

8 What is the significance of the initials FGI in the disconnection approach to designing synthetic pathways? Suggest the best disconnection sequences for the synthesis of each of the following compounds:

(a)

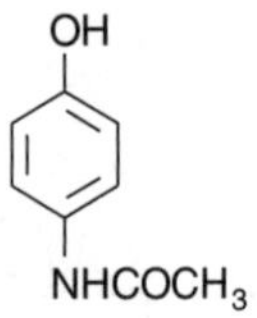

OH

NHCOCH$_3$

(b)

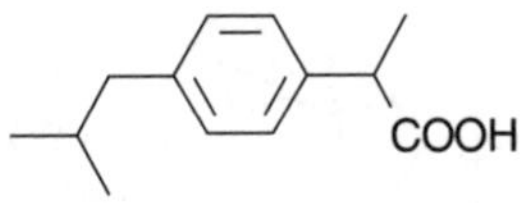

COOH

16

Drug development and production

16.1 Introduction

Development is the conversion of a biologically active compound into a safe marketable product. It is a multi-disciplinary process that requires the collaboration of teams of workers from many different disciplines. Its success is dependent on their skills and judgement. This chapter will *outline* the work carried out by these teams in the main areas of the development process. The activities in many of these areas are interdependent, which means that they should take place consecutively or at the same time. Consequently, as speed is of the essence in all development work, these activities will require careful planning and coordination.

The development process normally takes between seven and ten years from initiation to marketing the drug. Furthermore, only one in 400–1000 drug candidates considered for development ever reach the market. Consequently, development is very expensive, the average cost of successfully developing a drug now being estimated to be about 450 million pounds. As a result, the high-risk nature of development means that it is necessary to plan a comprehensive strategy to reduce both the pharmaceutical and financial risks. The first step in this strategy is to define product pharmaceutical and financial targets and assess whether the drug will reach these goals. Consequently, it is essential to assess whether the new drug will be able to compete with existing competitors and what advantages it may have over those competitors. This assessment should be repeated at appropriate points in the development to ensure that the development is still viable. The management of this and other aspects of the development is often made by monitoring its *critical path* (Fig. 16.1). This consists of the activities that determine the time taken for the drug to become registered by the appropriate government body. It does not include all the activities that are necessary to develop and evaluate a drug. However, monitoring the critical path does provide a way of making certain that the project remains on schedule.

Medicinal Chemistry, Second Edition   Gareth Thomas
© 2007 John Wiley & Sons, Ltd

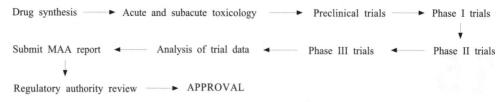

Figure 16.1 An example of the activities that are likely to form the critical path for the development of a drug. MAA is the *marketing authorisation application* that a company has to submit in Britain in order to produce and market a drug

The development of a drug consists of converting the original research discovery into a cost-effective manufacturing process. This is coupled with preclinical and clinical testing (see section 16.3) of the drug to ensure that it is safe to use. These two aspects of the development are often carried out in parallel because the results of the testing may lead to changes being made to the structure of the drug or even its rejection. For example, preclinical testing of the drug may reveal unwanted features in its toxicology while clinical testing may show that the drug has unwanted side effects. In these types of situation, it may be possible to modify the structure of the drug to remove or minimise these unwanted features early in the drug development process, thus reducing both the chance of rejection of the drug and its cost. Alternatively, if these features cannot be changed and they are serious problems, the development can be stopped in order to avoid unnecessary costs.

16.2   Chemical development

The reactions used by research workers frequently use expensive reagents and only produce sufficient quantities of the drug for very limited biological testing. In addition, the research routes often involve techniques such as chromatography that cannot be adapted to large-scale production methods. Consequently, the *first* step in the chemical synthesis development is to find safe, cost-effective alternatives. These alternative reactions should start from cheap readily available materials, consist of more economical reactions that have high yields and be capable of being safely carried out on a large laboratory or manufacturing scale (see section 16.2.4). They should yield sufficient quantities of the lead in an acceptable purity for the initial activity and toxicology tests. Initially several large-scale synthetic routes will be investigated in order to determine the optimum route. Convergent routes are usually preferred over linear syntheses as they give better overall yields (see section 15.4.2). However, in the early stages of development, speed of production of sufficient amounts of the lead for comprehensive testing is often more important than devising a chemically efficient synthesis. This is because the results of these initial tests will enable the company managers to decide whether the compound is worth further development.

The progression of the drug through the development process will lead to an increased demand for larger quantities of the drug. This demand is normally met by conversion of the most promising laboratory synthesis to pilot plant scale. Conversion of a synthesis to pilot plant scale, which is effectively a mini-manufacturing process, may require changes in the synthesis to allow for the use of large-scale equipment (see section 16.2.1). These changes, waste disposal and the hazards associated with carrying out the synthesis using larger amounts of the reagents should be evaluated before the conversion is made (see section 16.2.2). The pilot plant should produce sufficient quantities of the drug to carry out the comprehensive tests that are required by the regulating authority. Once a licence to manufacture the drug has been granted, the pilot plant synthesis is converted to the full manufacturing process. The main considerations in this conversion are chemical engineering issues, environmental impact, waste disposal, safety quality control and cost.

16.2.1 Chemical engineering issues

Initially, the chemical engineers will have to decide whether the reactions can be safely carried out in the existing plant or if it is necessary to construct a new plant. Since the latter could be very expensive they should also consider whether it is possible to modify the synthesis so that they could utilise existing equipment. The standard large-scale reaction vessel is a stainless-steel container. It is usually equipped with a stirring paddle, a heating jacket for controlling the reaction temperature, openings that allow solids and liquids to be placed in or removed from the vessel and provision for either distilling or refluxing liquids (Fig. 16.2). Reaction vessels are usually connected by suitable pipework to other pieces of specialised plant, such as filtration, crystallisation and drying equipment. A variety of methods are used to transfer the product from the reaction vessel to these additional pieces of equipment. They include the use of pumps, nitrogen gas under pressure, screw and conveyor systems. Separations are usually carried out using the reaction vessel as the separator and running the lower liquid phase out of the bottom of the reaction vessel.

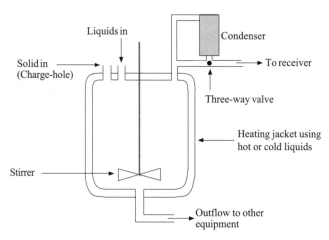

Figure 16.2 A diagrammatic representation of a typical reaction vessel. The three-way valve enables the condenser to be used to either reflux or distil the liquids in the reaction vessel

Control of the reaction and, if required, subsequent operations is achieved by the use of appropriate sensor systems and small armoured glass windows known as sight-glasses, which enable the plant operators to view the contents of the vessel and/or pipework.

The reactions selected for pilot and manufacturing processes must take into account the limitations of the available plant, for example the reaction vessel and the ancillary equipment required for isolation and purification of the product. Reaction vessels usually have an operating range of between about -15 to about $+140°C$. Consequently, as both heating and cooling are expensive, the reactions selected should give a satisfactory yield as near to room temperature as possible and certainly within the temperature control limits of the vessel. Furthermore, large-scale reactions require longer processing times because they take longer to reach their working temperature, produce a satisfactory yield and cool down than laboratory preparations. This means that the reagents, products, solvents and catalysts used must have an adequate degree of stability under the prevailing operating conditions.

16.2.2   Chemical plant: health and safety considerations

Health and safety is of paramount importance when selecting the reactions for either the pilot or manufacturing processes. A full chemical hazard assessment should be carried out before commencing the reaction on a large scale. This should cover the chemistry of the process, the safety of the plant operatives, waste disposal and its impact on the environment. The chemical and biological hazards associated with all the chemicals and solvents used in the reactions should be documented, together with protocols for dealing with accidents, leaks and spillages. Even in the best regulated chemical plants there will be some leakage of compounds into the atmosphere, such as solvent vapour and chemical dust, and so monitoring equipment must be appropriate for the substances being used. Reactions that involve handling highly toxic substances should not be used unless there is no alternative. In this case a stringent protocol governing their use must be drawn up and enforced. One important consideration for all health and safety protocols is the nature of the thermochemistry of the reactions. The heat generated by a large-scale reaction may exceed the cooling capacity of the existing equipment with disastrous results. Consequently, if they are not known, the heats of reaction for each of the stages used in the process must be accurately determined and taken into account when assessing the safety of the plant and its operatives.

Waste disposal and its environmental impact are becoming increasingly important and must now be planned for as part of the development process. In the past solid wastes were buried in special landfill sites but now, because of potential water pollution, it is regarded as undesirable unless alternative methods such as incineration are not suitable. Aqueous liquids may be discharged into the sewage system, rivers and the sea. However, this requires removal of high concentrations of impurities, especially metals such as copper, zinc, mercury and cadmium and organic halogen compounds such as chloroform, which are highly toxic to many forms of aquatic life. Solvents are usually recovered by distillation but this is not always a practical proposition for some solvents, such as dimethyl sulphoxide and

N,N-dimethylformamide. Gases can often be prevented from entering the atmosphere by reaction with suitable absorbents in absorbent towers. However, gases like nitric oxide that cannot be disposed of in this manner require special treatment, which can be very expensive. In all cases the disposal of highly toxic waste is to be avoided as it can require expensive special treatment.

An idea of the impact of a process on the environment may be determined from its effluent load factor (ELF). This is defined as:

$$\text{ELF} = \frac{\text{Mass of all ingredients} - \text{Mass of the product}}{\text{Mass of the product}} \qquad (16.1)$$

The ELF is the amount of waste produced by the process per unit mass of the product. It can refer to individual stages in the synthesis or the whole process. In the ideal situation where all the ingredients are converted into the product, the ELF value will be zero. This does not happen in synthetic processes. In pharmaceutical processes the ELF is usually in the order of 100. The process design chemists will aim to minimise the ELF, in order to reduce expense, by selection of the reactions and the operation of the plant.

16.2.3 Synthesis quality control

The efficiency of drug production will depend on being able to identify and assess the chemical purity of the drug and also that of the intermediate compounds involved at each step in the synthesis. This means that it is normally necessary to devise appropriate chemical quantitative and qualitative protocols for all of the compounds involved in the pilot and manufacturing routes. Physical methods such as HPLC and GC are often used for these purposes. It is also necessary to be able to identify impurities and any closely related structures or stereoisomers as this will be required by the licencing authority. Consequently, some thought should be given to how all these analyses may be made when designing the production route.

In some synthetic routes, an intermediate product may be used in the next stage of the synthesis without it being isolated and purified. This procedure is known as *telescoping*. It has the advantage of avoiding handling very toxic intermediates. It also makes it easier to deal with non-crystalline and oily products. For telescoping to be effective, the initial reaction has to produce a relatively pure product. Consequently, when this technique is used an analytical procedure must be available to ensure that the purity of the product is high enough for the next reaction to be carried out in an efficient manner.

16.2.4 A case study

The release of thromboxane A_2 (Fig. 16.3a) in the body can cause a number of toxic reactions, including bronchoconstriction, blood vessel constriction and platelet

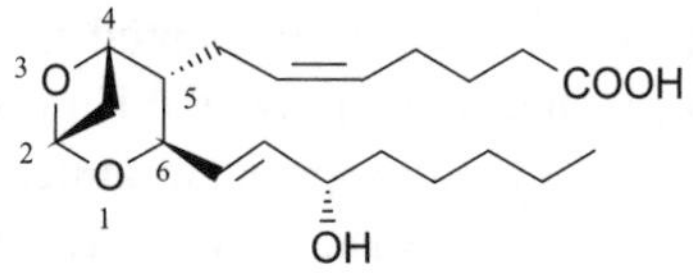

(a)

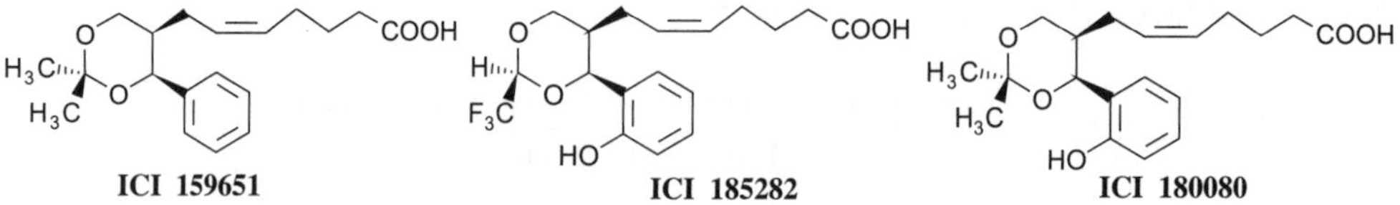

ICI 159651   ICI 185282   ICI 180080

(b)

Figure 16.3 **(a)** Thromboxane A$_2$ (TXA$_2$). *Note:* The locants given are those of the dioxan ring only and not those for the complete molecule. They are used for reference purposes only (see text). **(b)** Examples of some of the thromboxane antagonists developed by ICI

aggregation. This activity led ICI to develop a number of compounds that can act as antagonists for thromboxane (Fig.16.3b). It was found that significant antagonist activity is obtained when:

- at position 2 there is a substituent *cis* to the other substituents of the dioxanring;

- the phenolic and alkenoic acid side chains at positions 5 and 6, respectively, have a *cis* orientation;

- the C=C of the alkenoic acid side chain has a *cis* configuration.

These active analogues were initially synthesised in gram quantities by routes that allowed numerous analogues to be produced for the preliminary biological evaluation (Fig. 16.4).

The potential manufacturing route for the active thromboxane antagonists synthesised by the initial research route outlined in Figure 16.4 must include a way of controlling the stereochemistry of the product, which will also give a single enantiomer. In addition it must also include alternatives to: chromatographic separation, since this technique is not practical on a large scale; ozonolysis, which is very expensive and needs specialised equipment; and sodium ethanethiolate, which would need special containment since it is a volatile liquid with a vile odour.

The initial manufacturing route (Fig. 16.5) used the formation of a suitably substituted 1,4-butyrolactone by a Perkin reaction to control the stereochemistry of the substituents of the dioxan ring. In these reactions any phenolic hydroxyl groups were protected by forming their methyl ethers. The investigators found that the 3,4-substituents of these lactone rings should be *trans* to each other in order to produce a final product in which these substituents had the required *cis* orientation in the dioxan ring. The reduction of the lactone to the lactol enabled the team to use a Wittig reaction to introduce the *cis*-alkenoic acid side chain into

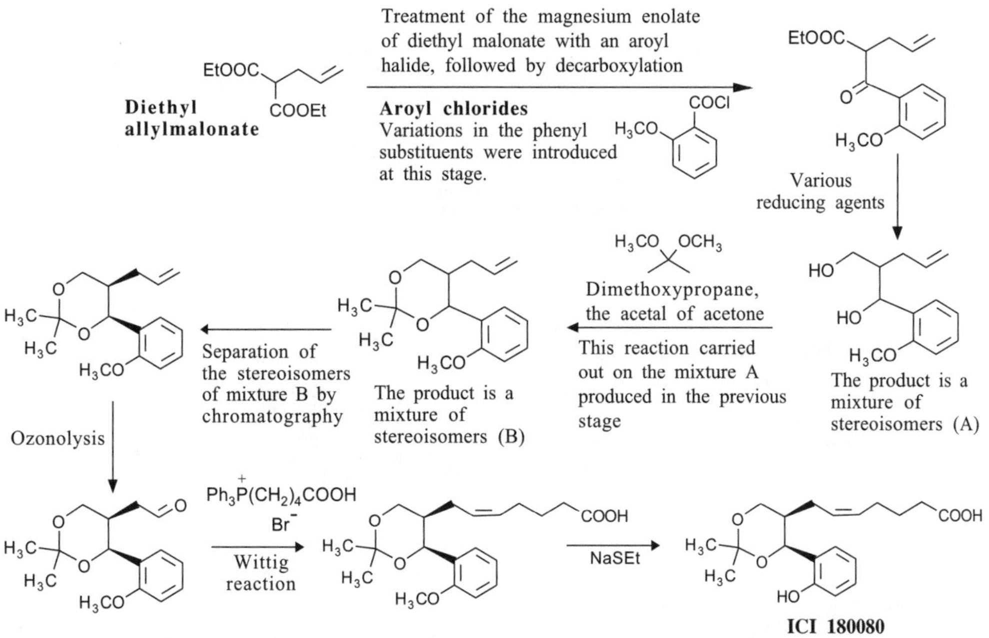

Figure 16.4 An outline of the initial research route for the synthesis of ICI 180080 and other analogues. The intermediates and products were produced as racemates. Geometric stereoisomers were synthesised as shown in the reaction scheme

the structure. Although this reaction produces a mixture of isomers it was relatively easy to isolate a relatively pure *cis* product. The use of the Wittig reaction also avoids the use of ozonolysis and the expense of specialised plant. Cyclisation of the product from the Wittig reaction was achieved by a variety of routes, depending on the nature and stability of the final product, and was used to form the dioxan ring. Finally, where appropriate, the use of sodium ethanethiolate to remove the methyl protecting groups of any phenolic hydroxy groups was avoided by the use of lithium diphenylphosphide. Demethylation was simply carried out by adding the methyl ether to freshly prepared lithium diphenylphosphide, which made this reaction highly suitable for large-scale production methods. It was reported that, using the route outlined in Figure 16.5, products containing less than 1 per cent of stereoisomers were produced on an industrial scale.

16.3 Pharmacological and toxicological testing

Extensive pharmacological and toxicological testing must be carried out on any new drug before it is marketed. These tests are carried out in two stages, namely, *preclinical* and *clinical trials* (Fig. 16.1). They take several years and assess the risks involved with the use of the new drug. In addition, they may provide vital information concerning pharmacokinetic properties of the drug, which can be used in other areas of the

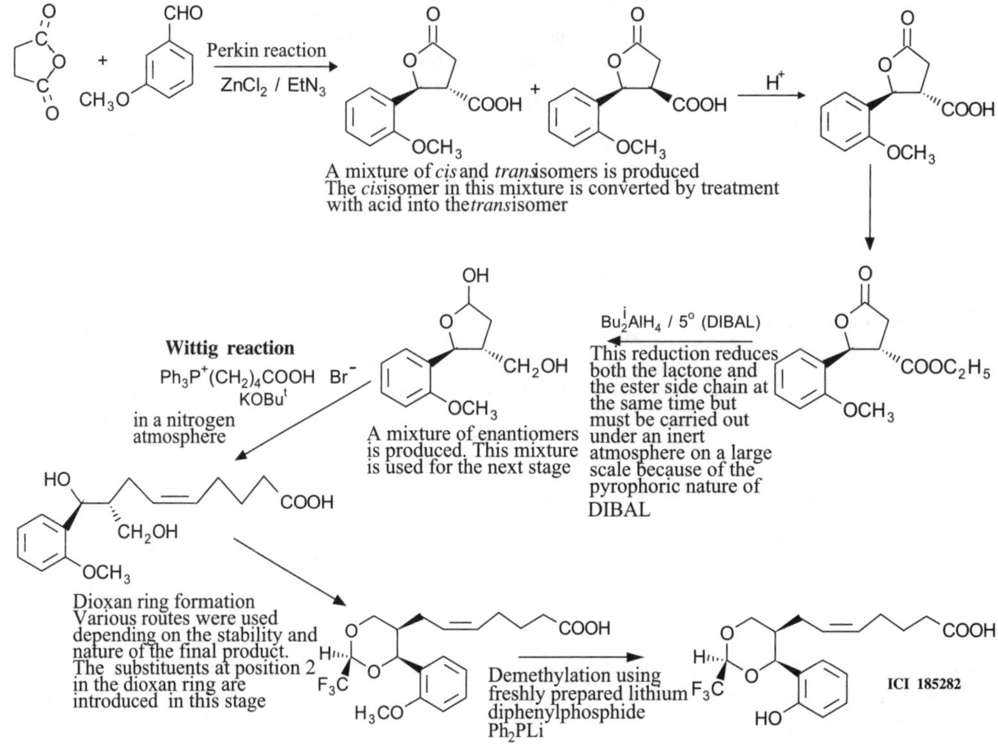

Figure 16.5 An outline of the manufacturing route developed for the production of ICI 185282 and its analogues. Both succinic anhydride and the appropriate aromatic aldehydes are readily available

development. Details of a trial are given in the MAA for the drug, which is submitted to the appropriate government body (see section 16.8). This body will either approve of the trials programme or specify modifications. To develop and market the drug the producer must comply with all the terms of the MAA, which has replaced the older drug licence. It is essentially a very wide ranging scientific assessment of the safety, efficacy and quality of the new product. The extent of the required tests is such that preclinical screening and clinical testing of a drug often form the bulk of the development costs for that drug.

Preclinical trials are essentially toxicity and other biological tests carried out by microbiologists on bacteria and by pharmacologists on tissue samples, animals and sometimes organ cultures to determine whether it is safe to test the drug on humans. The animal tests investigate the effect of the drug on various body systems, such as the reproductive, respiratory, nervous and cardiovascular systems. They are carried out under both *in vivo* (in the living organism) and *in vitro* (in an artificial environment) conditions. These preliminary tests also provide other information concerning the drug's pharmacokinetic properties and its interaction with other drugs and over-the-counter medicines. If necessary, any interactions that enhance or reduce the drug's activity should be investigated further during the clinical trials and the results noted in the product labelling and literature. It should be noted that the results of toxicity tests may differ depending on the route used to synthesise the drug. This is because each route could result in different impurities being present in the final product. Consequently, the manufacturing route must be specified in the MAA.

The preclinical tests should help decide whether it is safe to give the drug to humans and also the toxic dose for humans. The latter enables the investigators to set a dose level to start the Phase I clinical trials. These will include dose-ranging pharmacokinetic studies and bioavailability via chosen administration routes. However, relating animal tests to humans is difficult and the results are only acceptable if the dose–organ toxicity findings include a substantial safety margin.

Once the drug has passed the preclinical trials it undergoes clinical trials in humans. These trials can raise legal and ethical problems and so must be approved by the appropriate legal and ethical committees before the trials are conducted. In most countries this approval requires the issuing of a certificate or licence by the appropriate Medicines Control Agency (see section 16.8).

Clinical trials are chronologically classified as *Phase I, Phase II, Phase III* and *Phase IV trials.* In order to accurately assess the results of a clinical trial, the results must be compared with the normal situation and so in the trials conducted on healthy humans 50 per cent of the subjects are normally given an inactive substance in a form that cannot be distinguished from the test substance. This inactive dosage form is known as a *placebo.* Furthermore, the results of a trial must be reliable and not subject to influence by either the person conducting the trial or the recipient of the drug. Consequently, it is now common practice to carry out a *double-blind* procedure where both the administrator of the drug and the recipient are unaware whether they are dealing with the drug itself or a placebo. In addition, subjects are randomly chosen to receive either the placebo or the drug.

Trials conducted on healthy subjects do not demonstrate the beneficial action of the new drug. It is necessary to carry out double-blind trials on unhealthy patients to assess its efficacy. However, the use of a placebo with patients who are ill raises moral and ethical considerations. Placebos may still be used if the withdrawal of therapy causes no lasting harm to patients. If this is not possible, the effect of the new drug is compared to that of an established drug used to treat the medical condition. This reference drug should be carefully selected. It should not be chosen so as to give the new drug an inflated degree of potency that could be used to give the manufacturer an unfair commercial advantage and the patient an inaccurate idea of the medicine's effectiveness. A third alternative is to use *crossover* trials. Halfway through the trial the patients receiving the drug are switched to either the placebo or the reference drug and the patients receiving the placebo or reference drug are given the new drug. This is ethically more acceptable as both groups have been exposed to the benefits of the new drug.

The first clinical trials (*Phase I* trials) are usually conducted on small groups of healthy volunteers that do not include children and the elderly. Before clinical trials may be carried out on humans in Britain a clinical trials authorisation must be obtained from the

Medicines and Health Regulatory Agency (MHRA). Researchers are also required to design and conduct trials according to the principles of the international ethical and scientific standards for designing, reporting and conducting clinical trials when using human subjects. The volunteers undergo an exhaustive medical examination before the tests and are strictly monitored at all times during the trial. The trials are conducted in either in-house clinics or specialist outside facilities. The objective of these trials on healthy humans is to ascertain the behaviour of the new drug in the human body. They also yield information on the dosage form, absorption, distribution, bioavailability, elimination and side effects of the new drug. The relation of these side effects to specific metabolites allows medicinal chemists to eliminate the side effects by designing new analogues that do not give rise to that metabolite (see section 12.8). In addition, the trials give information concerning the level of the drug in the blood after intravenous and oral dosing, the rate of excretion in the urine and via the bowel and the effect of gender on these parameters. At all times in the trials the function of the kidney and other organs in the body is monitored for adverse reactions. The dose administered to the volunteers is initially a small fraction of that administered by the same route to animals.

Once the safety of the drug has been assessed in healthy volunteers the testing programme moves to *Phase II* trials. However, before these trials can be carried out in Britain the company must obtain a clinical trials certificate, which is issued by the regulating authority. Phase II trials are conducted on small numbers of patients with the condition that the drug has been designed to treat. They assess the drug's effectiveness in treating the condition and also help to establish a dose level and dosage regimen for the drug. In Britain, Phase II trials may only be carried out after a local ethics committee has evaluated and approved the trials programme (see section 16.8). The success of this phase leads to *Phase III* trials where the new product is tried out on large numbers of patients. *Phase III* trials are carried out using both placebos and comparison standards. They are particularly useful for obtaining safety and efficacy data in order to satisfy the product licencing authorities. In both Phase II and III trials a few subjects will exhibit *adverse drug reactions* (ADRs). These are described as responses that are either unwanted or harmful, which occur at the doses used for therapy. They exclude therapy failure. These ADRs are noted and added to the drug's data sheet. However, unless a high percentage of subjects exhibit the same ADR they do not usually result in the drug being withdrawn from use.

When the new drug has been released onto the market the performance of the drug is monitored using very large numbers of patients both in hospital and general practice. This monitoring is often referred to as the *Phase IV* trials. It provides more information about the drug's safety and efficacy. In addition trials are conducted on specific aspects of the drug's use with smaller specialist groups, for example its kinetics in the elderly, infants, neonates and ethnic groups.

The interpretation of the results of all trials requires the close collaboration of clinicians and statisticians. Reliable results are only obtained if at least the minimum number of patients for statistical viability are involved in the preliminary trials. It is often difficult to measure precisely the parameter chosen for assessment. Consequently, results are usually quoted in terms of a *probability coefficient*: the lower the value of this coefficient, the more accurate the results. However, very reliable results will only be obtained from clinical trials if large groups of patients are tested. This is seldom feasible. Consequently, manufacturers and licencing authorities usually settle for the best statistical compromise. Since some adverse effects do not manifest themselves for years it is necessary to constantly monitor the drug (Phase IV trials) after it has been released for general use. However, no matter how carefully the testing and trials of a drug have been carried out, it is still possible that a few individuals will experience a serious adverse drug response even after that drug has entered general clinical use. This is because variations in the biochemistry of individuals allow a drug to hit an alternative target to that originally intended in the human body. Consequently, the public need to be informed that there is always a slight risk when undergoing any drug therapy, although the drug companies have tried to minimise that risk.

Several useful 'spin offs' from drug testing can occur, for example when a drug's *in vivo* activity is greater than its *in vitro* activity. This indicates that a metabolite is more active than the original drug and consequently allows the possible development of a more effective drug that can produce the same outcome but using a lower dose. In other situations a side effect discovered during trials may prove to be more useful than the original drug, for example sildenafil (see section 14.6.2).

16.4 Drug metabolism and pharmacokinetics

Drug metabolism and pharmacokinetics (DMPK) studies are used to show how the concentration of the drug and its metabolites varies with the administered dose of the drug and the time from administration. They are normally carried out using suitable animal species and in humans in Phase I trials. The information obtained from animal studies is used to determine safe dose levels for use in the Phase I clinical trials in humans. However, the accuracy of the data obtained from animal tests is limited since it is obtained by extrapolation (see section 11.7). In addition, it is necessary to determine the dose that just saturates the absorption and elimination processes so that the toxicological and pharmacological events may be correctly interpreted. A more effective interpretation of pharmacological and toxicological data may usually be made if the ADME of the drug and its metabolites are well defined (see section 11.4). Tissue distribution data are usually obtained using single-dose studies but repeated-dose studies should be undertaken when:

- the tissue $t_{1/2}$ of the drug or metabolite is much larger than its plasma $t_{1/2}$ value;

- the C_{ss} of the drug or its metabolites is found to be very much higher than that predicted from single-dose studies;

- the drug is targeted at a specific site;

- particular types of tissues show unexpected lesions.

16.5   Formulation development

The form in which a drug is administered to patients is known as its *dosage form*. Dosage forms can be subdivided according to their physical nature into liquid, semisolid and solid formulations. Liquid formulations include solutions, syrups, suspensions and emulsions. Creams, ointments and gels are normally regarded as semisolid formulations, whereas tablets, capsules, suppositories, pessaries and transdermal patches are classified as solid formulations. However, all these dosage forms consist of the drug and ingredients known as *excipients*. Excipients have a number of functions, such as fillers (bulk providing agents), lubricants, binders, preservatives and antioxidants. A change in the nature of the excipient can significantly affect the release of the drug from the dosage form. Consequently, manufacturers must carry out bioavailability and any other tests specified by the licencing authority if they make changes to the dosage form before marketing the new dosage form.

The type of dosage form required will depend on the nature of the target and the stage in the drug development. Since many promising drug candidates fail at the preclinical and Phase I stages a simple dosage form, such as a oral solution, is often used for the preclinical and early Phase I clinical trials This is in order to keep costs to a minimum at these high-risk stages of drug development. However, the manufacturer must use the dosage form of the drug that he proposes to use in the later clinical trials.

The types of dosage form used must satisfy criteria such as stability and pattern of drug release. Stability studies are used to determine whether the dosage form has an adequate potency after an appropriate period of time, usually 2–3 years. This will determine its shelf life and recommended storage conditions. Drug release is directly influenced by the excipients and any slow-release mechanisms employed. In both of these examples suitable chemical and biological experiments must be designed to obtain or check the relevant data. The results of these experiments may lead to improvements in the design of the dosage form. They are usually carried out in parallel with the clinical trials.

16.6   Production and quality control

The manufacture of the new drug must be carried out under the conditions laid down in the marketing authorisation (MA) (see section 16.8). Since it is not usually practical for manufacturers to dedicate a plant to the production of one particular drug it is essential that the equipment used is cleaned and tested for adulterants before use. Many pharmaceutical manufacturers estimated that production line equipment is only used to produce the product for about 10 per cent of its time. For most of the remaining time it is being stripped down, cleaned and reassembled.

The quality control of drugs and medicines during and after production is essential for their safe use. It was only achieved when accurate analytical methods were developed in the mid-nineteenth century. This led to the publication of national pharmacopoeias and other documents that specified the extent and the nature of the identification tests and quantitative assessments required to ensure that the product reaching the public is fit for purpose. These documents now cover the production, storage and application of pharmaceutical products. They are the subject of constant review but unfortunately this does not completely prevent the occurrence of product-related problems. However, the continual updating of these documents does reduce the possibility of similar problems occurring in other products. It is gratifying to note that since the thalidomide disaster very few drugs have been removed from the market on safety grounds. The development of reliable analytical methods for the trials, production quality control and identification, limit and assay procedures for inclusion into the relevant pharmacopoeias is normally carried out in parallel with the critical path development stages. These analytical methods must be described in detail in the product licence application.

16.7 Patent protection

The high cost of drug development and production makes it essential for a company to maximise its returns from a new drug. This can only be achieved by preventing rivals from unrestricted copying of a new product. Patents are used to prevent rival companies from manufacturing and marketing a product without the permission of the originator of the product. However, many companies do market other manufacturers' products under licence from the original patentor.

Patents have been used, in one form or another, as a means of industrial protection from the early fourteenth century to the present day. Originally, they were intended to encourage the development of new industries and products by granting the developer or producer the monopoly either to use specific industrial equipment or produce specific goods for a limited period. This monopoly, enforced by the appropriate government office, enabled an innovator to obtain a just reward for his efforts. In most countries the awarding of a patent prevents third parties from manufacturing and selling the product without the consent of the innovator. However, patents do encourage and protect the development of new ideas by the publication of new knowledge.

Originally each country issued its own patent laws but by the mid-eighteenth century it was recognised that patent rights should extend beyond national boundaries. The first international agreement was the *Paris Convention* of 1883. This has been revised on numerous occasions, the current treaties in operation being the Europeans Union's European Patent Convention (EPC) of 1978 and the Patent Cooperation Treaty (PCT) signed in Washington in 1970. The former is only open to European countries and administered by the European Patent Office (EPO). The latter is open to all countries of the world and is administered through the national patent offices of the country subscribing to the treaty.

The protection offered by a patent is of paramount importance to a company at the research, development and production stages of drug production. It is essential that a patent is filed as soon as new compounds have been made and shown to have interesting properties, otherwise a rival company working in the same field might pre-empt the patent, which means that a large amount of expensive research work would be unproductive with regard to company profits. In this respect it is particularly important that the patent fully covers the relevant field and does not give a rival manufacturer an exploitable loophole.

The time required for the development of a drug from discovery to production can take at least 7–15 years. In many countries, patents normally run for 20 years from the date of application. Consequently, the time available for a manufacturer to recoup the cost of development and show a profit is rather limited, which accounts for the high cost of some new drugs. Furthermore, it also means that some compounds are never developed because the patent-protected production time available to recoup the cost of development is too short.

A number of strategies are used by companies to increase the life of a patent. Consider, for example, the case of a drug that has been marketed by a company for a number of years as a racemate. The enantiomers that make up this racemate could have different biological properties (see section 2.3 and Table 1.1). However, if this racemate contains enantiomers that have similar activities but different potencies it is possible for the company to separate and market the most potent enantiomer as a new drug and, as such, take out a new patent. This strategy is called *chiral switching*. Examples of drugs that have undergone chiral switching are given in Figure 16.6. Chiral switching is only allowed if the company can show that the 'new drug' is better than the existing one. However, since the early 1980s most drugs have been marketed as either achiral or pure enantiomers.

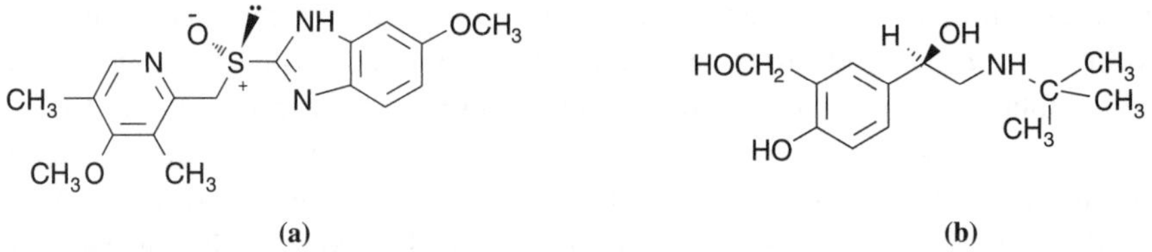

Figure 16.6 Examples of drugs that have undergone chiral switching. **(a)** Omeprazole, an antiulcer agent. This was originally produced as a racemate. The sulphur is a chiral centre because it is tetrahedral and its lone pair makes it asymmetric. It is now marketed as its more potent *S*-enantiomer, esomeprazole. **(b)** Salbutamol, a drug used to treat asthma. Originally marketed as its racemate, the *R*-isomer is 68 times more active and the drug is now marketed as levalbuterol

16.8   Regulation

The release of a new drug onto the market must be approved by the regulating authority for that country. For example, in Britain this is the Medicines Control Agency (MCA), in the European Union it is the European Medicines Evaluation Agency (EMEA) and in the USA it is the US Food and Drug Administration (FDA). These bodies, which are essentially consumer protection agencies, issue a so-called *product licence* or *marketing authorisation* (MA) when they are satisfied as to the method of production, efficacy, safety and quality of the product. To obtain a product licence the pharmaceutical company is required to submit a comprehensive dossier that contains a statement of what the drug should achieve, the relevant pharmacological, formulation and toxicological data, full details of all aspects of the production processes and dosage forms. In the UK and Europe this submission is known as the *marketing authorisation application (MAA)* and its equivalent in the USA is the *new drug application (NDA)*. The issuing of a product licence gives the pharmaceutical company or the person ordering its production the right, sometimes subject to conditions, to produce and sell the new product in the issuing country. Licences may be revoked if the producer does not strictly keep to the conditions laid down in the licence. Any changes, at a later date, to the production process, dosage forms and indications (usage) must also be approved and it is the responsibility of the company to carry out any additional tests that are required.

16.9 Questions

1 (a) What is the critical path in drug development? (b) List the main stages in the critical path of the development of a drug.

2 Explain the meaning of the terms: (a) dosage form, (b) ELF, (c) telescoping in drug production, (d) excipient and (e) double-blind trial.

3 Outline the chemical factors that need to be considered when scaling up a research synthesis to pilot plant scale.

4 Outline a possible production scale route for the preparation of the thromboxane analogue A.

(A)

5 Explain the differences between: (a) preclinical and clinical trials; and (b) Phase I and Phase II trials.

6 What is a patent? Why is it necessary to patent drugs?

Selected further reading

General chemistry

G. Thomas, *Chemistry for Pharmacy and the Life Sciences including Pharmacology and Biomedical Science*, Prentice Hall, 1996.

F. A. Cotton and G. Wilkinson, *Basic Inorganic Chemistry*, Fifth Edition, John Wiley and Sons, 1988.

J. G. Morris, *A Biologist's Physical Chemistry*, Second Edition, Edward Arnold, 1974.

Synthetic chemistry

S. Warren, *Organic Synthesis, the Disconnection Approach*, John Wiley and Sons, 1982.

R. S. Aitken, *Stereoselective Synthesis*, John Wiley and Sons, 1995.

Biochemistry

D. Voet, J. G. Voet and C. W. Pratt, *Fundamentals of Biochemistry*, John Wiley and Sons, 1999.

R. H. Garrett and C. M. Grisham, *Biochemistry*, Saunders College Publishing, Harcourt Brace Publishers, 1995.

Pharmacology

H. P. Rang, M. M. Dale and J. M. Ritter, *Pharmacology*, Third Edition, Churchill Livingstone, 1995.

A. Galbraith, S. Bullock, E. Manias, B. Hunt and A. Richards, *Fundamentals of Pharmacology*, Addison Wesley Longman Limited, 1997.

Inorganic medicinal chemistry

S. J. Lippard and J. M. Berg, *Principles of Bioinorganic Chemistry*, University Science Books, 1994.

J. Lancaster (Editor), *Nitric Oxide, Principles and Actions*, Academic Press, 1996.

Medicinal chemistry

J. H. Block and J. M. Beale (Editors), *Wilson and Gisvold's Textbook of Organic Medicinal and Pharmaceutical Chemistry*, Eleventh Edition, Lippincott-Raven, 2004.

H. J. Smith (Editor), *Smith and Williams' Introduction to the Principles of Drug Design and Action*, Third Edition, Harwood Academic Publishers, 1998.

F. D. King (Editor), *Medicinal Chemistry, Principles and Practice*, Second Edition, The Royal Society of Chemistry, 2002.

C. G. Wermuth (Editor), *The Practice of Medicinal Chemistry*, Second Edition, Academic Press, 2003.

M. E. Wolf (Editor), *Burgers Medicinal Chemistry*, Fifth Edition, John Wiley and Sons, 1997.

G. L. Patrick, *An Introduction to Medicinal Chemistry*, Third Edition, Oxford University Press, 2005.

Combinatorial chemistry

S. R. Wilson and A. W. Czarnick (Editors), *Combinatorial Chemistry, Synthesis and Application*, John Wiley and Sons, 1997.

N. K. Terrett, *Combinatorial Chemistry*, Oxford University Press, 1998.

K. Burgess, *Solid Phase Organic Synthesis*, Wiley-Interscience, 2000.

Historical

W. Sneader, *Drug Development from Laboratory to Clinic*, John Wiley and Sons, 1986.

W. Sneader, *Drug Discovery, a History*, John Wiley and Sons, 2005.

Pharmacokinetics

M. Rowland and T. N. Tozer, *Clinical Pharmacokinetics*, Third Edition, Williams and Wilkins, 1995.

Drugs from natural sources

P. M. Dewick, *Medicinal Natural Products, a Biosynthetic Approach,* Second Edition, John Wiley and Sons, 2003.

G. Walsh, *Biopharmaceuticals, Biochemistry and Biotechnology*, Second Edition, John Wiley and Sons, 2003.

W. C. Evans (Editor), *Trease and Evans Pharmacognosy*, Fifteenth Edition, Elsevier, 2002.

G. Evans (Editor), *A Handbook of Bioanalysis and Drug metabolism*, CRC Press, 2004.

J. P. Devlin (Editor), *High Throughput Screening, The Discovery of Bioactive Substances*, Marcel Decker Inc., 1997.

Computational chemistry

G. H. Grant and W. G. Richards, *Computational Chemistry*, Oxford University Press, 1995.

A. R. Leach, *Molecular Modelling, Principles and Applications*, Second Edition, Prentice Hall, 2001.

M. T. D. Cronin and D. J. Livingstone, *Predicting Chemical Toxicity and Fate*, CRC Press, 2004.

Answers to questions

Numerical answers may be slightly different from the ones you obtain due to differences in calculating methods and equipment. However, a correct answer is one that approximates to the given answer. Answers that require the writing of notes are answered by either an outline of the points that should be included in the answer or a reference to the appropriate section(s) of the text. Where questions have more than one correct answer only one answer is given.

Chapter 1

1 (a) The *ortho* ethyl groups will sterically hinder the ester group. This will slow the rate of hydrolysis of the compound, which is likely to either increase its duration of action or inactivate it by preventing the compound from binding to its target site.

   (b) The trimethylammonium group is permanently positively charged. This permanent charge will reduce the ease with which the molecule passes through biological membranes. In particular it would probably prevent the compound passing the blood – brain barrier.

   (c) The replacement of an ester group by an amide group can reduce the rate of metabolism of the compound by hydrolysis, which may prolong the action of a drug. It may also change the biological activity of the compound.

2 The nature of the pathological target. Its site of action, nature of desired action, stability, ease of absorption and distribution, metabolism, dosage form and regimen.

Medicinal Chemistry, Second Edition   Gareth Thomas
© 2007 John Wiley & Sons, Ltd

3 (a) See Section 1.3. (b) See Section 1.6. (c) See Section 1.6. (d) See Section 1.6. (e) See Section 1.8.4. (f) See Section 1.3. (g) See Section 1.6.

4 For definitions see section 1.7. The factors affecting the pharmacokinetic phase are absorption, distribution, metabolism and elimination. The factor affecting the pharmacodynamic phase is the stereoelectronic structure of the compound.

5 The reduction in pH reduces the negative charge of the albumin and so increases its electrophilic character. Therefore, as amphetamine molecules are nucleophilic in nature their binding should improve with the decrease in pH. Part of this binding will involve salt formation between the amphetamine and the albumin. Amphetamine is more likely to form salts in which it acts as the positive ion as the electrophilic nature of the albumin increases.

6 See 1.4.3.

7 Replace the ester (lactone) with a less easily hydrolysed amide group. Introduce bulky ethyl or propyl groups on either side of the lactone group to reduce the ease of hydrolysis. Use an enteric coating.

8 See 1.3.

9 See 1.4.1. (a) Yes. This structure only disagrees with one rule, namely the value of logP is too high. (b) No, This structure does not comply with two rules, namely: too many hydrogen bond donors and too high a molecular mass. (c) Yes. This structure only disagrees with one rule, namely the value of logP is too high.

10 See 1.4.

Chapter 2

1 See Section 2.2.

2 See Section 2.3.

3 (a) See Section 2.3.1, (b) see Section 2.3.2 and (c) see Section 2.3.3.

4 See Section 2.6.

5 (a) See Section 2.5. (b) See Section 2.5.

6 Convert mmHg to atmospheres and use Henry's Law. Solubility is 0.95 mg per 100 g of water.

7 (a) High polar group to carbon atom ratio (see Section 2.7).

(b) The presence of polar groups that can hydrogen bond to water molecules (see section 2.9) and ionise in water (see Section 2.7).

8 (a) Form salts that would improve water solubility but would break down to yield the drug in the biological system (see Section 2.8).

(b) Introduce water solubilising groups into a part of the structure that is not the pharmacophore of the drug (see Section 2.9).

(c) Formulate as a suitable dosage form (see Sections 2.10 and 2.13).

9 For general and specific methods see Section 2.9.4:

(a) Any method for carboxylic acid, phosphate and sulphonic acid groups.

(b) Any method for basic groups.

(c) Any method for polyhydroxy and ether residues.

10 (a) The degree of ionisation of codeine in the stomach is 99.99 per cent. Most of the codeine is in the form of the corresponding ions and so the drug will not be readily absorbed in the stomach.

(b) The degree of ionisation of codeine in the intestine is 98.44 per cent. This is slightly less than the degree of ionisation in the stomach and so the absorption of codeine will be slightly better in the intestine than the stomach. However, a considerable concentration of the codeine is in the form of the corresponding ions and so the drug will still not be readily absorbed from the intestine.

11 See Section 2.13.

12 (a) Olive oil/water, (b) *n*-octanol/water, (c) chloroform/water.

13 See Section 2.13.

Chapter 3

1 Structure–activity relationships. These are the general relationships obtained from a study of the changes in activity with changes in the structure of a lead. These changes are used to find or predict the structure with the optimum activity. Example: see Section 3.6.

2 (a) See Section 3.4.2. (b) CF_3. It is approximately the same size as a chlorine atom.

3 (a) See Section 3.5 and Table 3.2. (b) See Section 3.1.

4 (a) See Section 3.4.5. (b) See Section 3.4.1. (c) See Section 3.4.6.

5 See Section 3.7.

6 Lipophilicity: see Section 3.7.2 (P, π and D).
Shape: see Section 3.7.4 (E_s and MR).
Electronic effects: see Section 3.7.3 (σ).

7 (a) See Section 3.7.4, Hansch analysis.

(b) (i) n = the number of compounds used to derive the equation; s = the standard deviation for the equation; r = the regression constant and the nearer its value to 1, the better the fit of the data to the Hansch equation. Equations are normally said to have an acceptable degree of accuracy if r is greater than 0.9.

(ii) See Section 3.7.4, Hansch analysis, and in particular equation (3.20).

(iii) A more polar substituent would have a negative π value (see Table 3.5), which could reduce the activity of the compound.

8 (a) See Section 3.7.4, Craig plots.

(b) Use substituents in the upper right-hand corner of Figure 3.16.

9 (a) Compounds 1–6 show the effect of alkyl substituent groups on the activity. The use of these groups will increase the lipohilic nature of the drug. However, the results show that with the exception of compound 6 the increase in lipophilic character increases the activity of the analogue.

(b) Compounds 7–23 show the effect of basic ring system substituent groups on activity. Significant increases in activity occur in analogues with a two-carbon chain and a pyrrolidine residue. The pyrrolidine residue will increase the polar nature of the analogue. Furthermore it appears that unsubstituted and 3-halophenyl-substituted pyrrolidine rings result in analogues with the highest activity.

10 See the discussion of equation (3.23). Equation (1): the high coefficient for E_s indicates that steric factors play a part in deciding the activity of the lead. Furthermore, the value of r is high enough, even though it is below 0.9, to suggest that this is not the only factor. In Equation (2) the small value of the coefficient of σ and the low value of r show that the activity is not due to electronic factors alone, that is, more than one parameter is required in the QSAR equation. In Equation (3) the similar values of the coefficients of both E_s and s indicate that both steric and electronic factors contribute to the lead's activity. However the r value of 0.927 indicates that suitable parameters were used to produce the equation.

Chapter 4

1 See Fig. 4.1b, Fig. 4.3a and Fig. 4.3c.

2 (a) See Section 4.2. (b) See equation (4.4) in Section 4.2.

(c) $E_{\text{coulombic}} = \dfrac{Q_{c_1}Q_{c_2}}{Dr_{c_{1-2}}} + 3\dfrac{(Q_{c_1}Q_{\text{CH}})}{Dr_{c_1-\text{H}}} + 3\dfrac{(Q_{c_2}Q_{\text{CH}})}{Dr_{c_2-\text{H}}}$

where Q represents the point charge on the specified atoms and r the distance between the specified atoms.

(d) *Advantages:* less computing time, very large molecules may be modelled and can be used to give information on the binding of ligands to the target site. *Disadvantages:* accuracy of the structure depends on selecting the correct force field and parameter values. Structures are normally determined at zero Kelvin so they will have different conformations at room and body temperatures.

3 See Section 4.2.1 and Fig. 4.5. Link the fragments methanoic acid and methane to form the acetyl side chain. Link this side chain and methanoic acid to benzene. Check that the hybridisation of the atom is correct and energy minimise.

4 See Section 4.2.1 and in particular Fig. 4.6.

5 (a) See Section 4.2. (b) See Section 4.5.

6 (a) It is based on the concept that all material particles exhibit wave-like properties. This means that the mathematics of wave mechanics can be used to describe and predict these properties.

(b) Schrodinger equation: $H\Psi = E\Psi$.

(c) *Advantages:* useful for calculating the values of physical properties of structure, the electron distribution in a structure and the most likely points at which a molecule will react with electrophiles and nucleophiles. *Disadvantages:* can only be used for structures that contain several hundred atoms. Requires a great deal of computing time.

7 See Section 4.7.

8 Flexible molecules are better able to adjust to the binding site.

Compound B, as the ethane residue joining the benzene rings can form different conformers. The ethene residue joining the benzene rings has a ridged structure.

9 (a), (b) and (c): see Section 4.9.

Chapter 5

1 See Section 5.1 and 5.1.1.

2 • The objectives of the synthesis: do they require the formation of a library of separate compounds or mixtures?
   • The size of the library.
   • Solid or solution phase?
   • If solid phase is selected, use parallel synthesis or Furka's mix and split?
   • The nature of the building blocks and their ease of availability.
   • The suitability of the reactions used in the sequence.
   • The method of identification of the structures of final products.
   • The nature of the screening tests and procedure.

3 See Section 5.2, in particular Fig. 5.5.

4 See Section 5.2.2.

5 Adapt the scheme given in Fig. 5.4 using the appropriate R groups.

6 See Sections 5.3.1, 5.3.2 and 5.3.3.

7 See Section 5.5.

8 See Section 5.6.

9 (a) See Sections 5.6.1 and 5.6.2.

   (b) See Section 5.6.3.

   (c) Low: a run of inactive compounds. High: a run of active compounds; too low a criterion for a hit; assay too general, resulting in it detecting active compounds with different types of activity.

Chapter 6

1 See Section 6.1.

2 See Section 6.1 and Fig. 6.1.

3 (a) Acid could cause hydrolysis and denaturation.

   (b) Extract with a very dilute aqueous solution of NaOH, filter the aqueous solution from the plant debris and neutralise the filtrate with dilute HCl to precipitate the water-insoluble acids.

   (c) Supercritical fluid extraction. Carbon dioxide.

   (d) See Section 6.6.3. Remove chlorophyll and inorganic salts from the extract.

4 See Section 6.6.1.

5 Clean up the extract by removing the chlorophyll and sodium chloride (see sections 6.6.3 and 6.7.2). Either freeze dry the solution to obtain the proteins or precipitate them from solution using large quantities of ammonium sulphate.

6 See Section 6.6.2.

Chapter 7

1 Consult index to find the appropriate pages.

2 See Section 7.2. The integral proteins that have transmembrane spans normally have their C-terminals (COOH) in the intracellular fluid and their N-terminals in the extracellular fluid. These groups ionise in these fluids to form anionic (negative) and cationic (positive) ions. These ions are responsible for the charges on the surfaces of the membrane.

3 See Section 7.2.1 and Fig. 7.3.

4 See Section 7.2.5.

5 See Fig. 7.7 and Section 7.2.5 for the essential features.

6 (a) See Section 7.4.1.

   (b) See Section 7.3.6 and 7.3.7.

   (c) See Section 7.1.

   (d) See Section 7.3.4 and 7.3.5.

7 The degrees of ionisation are calculated from the Henderson–Hasselbalch equation for bases (see equation 2.5). The higher the degree of ionisation, the greater the amount of the drug existing in the form of its cation. Since charged compounds are usually less easily absorbed than the electrically molecules, the activity of the drug will decrease with an increase in the degree of ionisation. Consequently, using this as the basis for the prediction the activities of the drugs will be in the order: benzocaine (0.000126:1) > cocaine (0.016:1) > procaine (5.62:1), where the figures in parentheses are the ratios of ionised to unionised form of the drug.

8 See Section 7.3.5. Example: levodopa.

9 See Section 7.4.2 for definitions and mode of action. When the concentrations of the ions being transferred by the ionophore are the same on both sides of the membrane.

10 See Section 7.4.2. β-Lactam antibiotics have to penetrate the cell walls of the bacteria in order to inhibit the cross-linking of the cell wall during its regeneration. The thicker cell envelope of Gram-negative bacteria makes it more difficult for the drug to reach its site of action. Not all the prion channels in this envelope will transfer β-lactam drugs. β-Lactamases in the periplasmic space will also hydrolyse the drug.

11 See Section 7.4.3 and Fig. 7.40.

12 (a) and (b): see section 7.4.2 and Fig. 7.29. Base the mehanism on that given for penicillin.

13 (a) The carboxylic acid group makes the analogue more polar than B and so it is likely to be less easily absorbed than B.

(b) The amino group also makes the analogue more polar than B and so it is likely to be less easily absorbed than B.

(c) The ester group makes the molecule less polar than B and so it is likely to be more easily absorbed than B.

Chapter 8

1 (a) See Section 8.1, (b) see Section 1.5.7, (c) see Section 8.1, (d) see Section 8.1, (e) see Sections 8.3 and 8.4, (f) see Section 8.1.

2 This is a Type 2 muscarinic cholinergic receptor for acetylcholine. It is a member of the Type 2 superfamily.

3 A schematic representation of the possible bonds involved in the binding of the drug to the receptor. Hydrophobic bonding will occur between the carbon chains. This is not shown on the diagram.

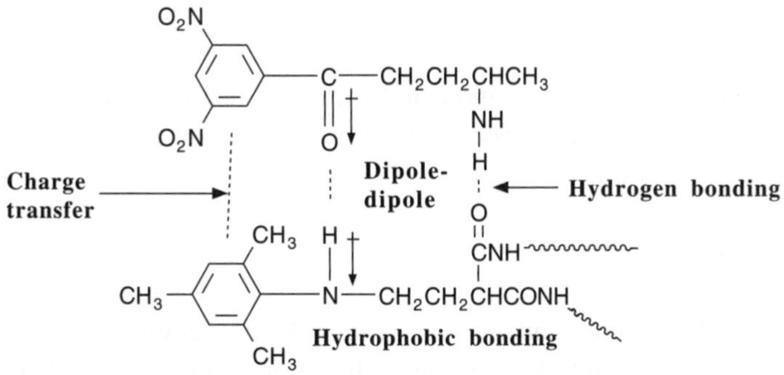

4 See Section 8.4.

5 See Section 8.5.

6 See Section 8.4.2.

7 (a) See Fig. 8.12. (b) DAG activates PKC, which controls a number of cellular processes. IP_3 initiates the rapid release of Ca^{2+} ions. These ions act as secondary messengers initiating a range of cellular responses.

8 Competitive antagonists compete for the same receptor as the agonist. However, competitive antagonists do necessarily bind to the same site as the agonist. They may bind to an alternative site close to the agonist's receptor site. The binding of competitive antagonists is reversible. Consequently, the effect of a competitive antagonist may be reversed by increasing the concentration of the agonist. The increased concentration of the agonist displaces the antagonist from the receptor and restores the cellular response to the agonist Non-competitive antagonists also bind to or near to the same receptor site as the agonist. In this case the binding is not reversible and so increasing the concentration of the agonist does not restore the cellular response to the agonist.

9 See Section 8.6.2.

10 See Section 8.7.1.

Chapter 9

1 (a) See Section 9.1,(b) See Sections 9.1 and 9.4.3, (c) see Section 9.3, (d) see Sections 9.4.2 and 9.4.3, (e) see Section 9.8.2.

2 See Section 9.2.

3 See Section 9.4.2.

4 See Section 9.6.

5 The Lineweaver–Burk plots will indicate the general nature of the enzyme's action and hence its general mechanism. See Section 9.9.1. Assume a single substrate reaction and reversible inhibition.

6 See Sections 9.9.1. and 9.9.2.

7 See Section 9.11. Compound A is a reversible non-competitive inhibitor with a Lineweaver–Burk plot pattern the same as that shown in Figure 9.20b. Compound B is a reversible uncompetitive inhibitor with a Lineweaver–Burk plot pattern the same as that shown in Figure 9.21. Irreversible inhibitors are more suitable as they cannot be reversed by a build-up of the substrate that occurs when the enzyme is inhibited. However, they do tend to have more side effects.

8 See Section 9.9.2, suicide inhibitors. (a) A must bind to the enzyme's active site and so must contain a group that is converted to a group that can react with the enzyme's active site. (b) See Fig. 9.25 for mechanism. Improvement of action: insert an electron acceptor group α to the C–Cl bond. This increases the electrophilic nature of the carbon, making it more susceptible to nucleophilic attack by a group from the active site of the enzyme.

9 (a) Lactate (lactic acid).

   (b) Prepare analogues based on the structure of lactate in order to ensure that they will bind to the active site. For example, incorporate an electrophilic group into the structure of lactate by replacing the OH by a reactive halide group (eg Cl, Br or I). Reason: the active site will have nucleophilic groups that can react with the iodide via a nucleophilic substitution to form a covalent bond with the lactate analogue, thus permanently blocking the active site.

10 See Section 9.10. The groups normally found at the active site of an enzyme are nucleophilic in nature. Pyrrole has an electron-deficient ring and so to increase binding an electron acceptor substituent should be used, whilst to decrease binding an electron donating substituent should be used.

11 See Section 9.13.

12 (a) See Section 9.12.3. Either a hydroxy heptanoic acid side chain or a seven-carbon side chain including a lactone ring and at least one hydroxy group. An unsaturated acyclic ring system.

(b) Increase and decrease the length of the hydroxy side chain with CH_2 units. Low increases in the numbers of CH_2 groups would require no extra polar groups but larger increases would require additional polar groups to to reduce CNS absorption and to balance the lipophilicity of the analogue. Use different rings systems of approximately the same size and shape. Include aromatic rings. Design structures that are readily synthesised by known methods.

Chapter 10

1 (a) See Section 10.1. (b) See Section 10.4. (c) See Sections 10.7 and 10.8.

2 They are flat ring structures (fully conjugated structures). The positions of the appropriate functional groups when considering bond lengths and angles are complementary. The distance apart of these complementary functional groups is short enough to allow hydrogen bonding.

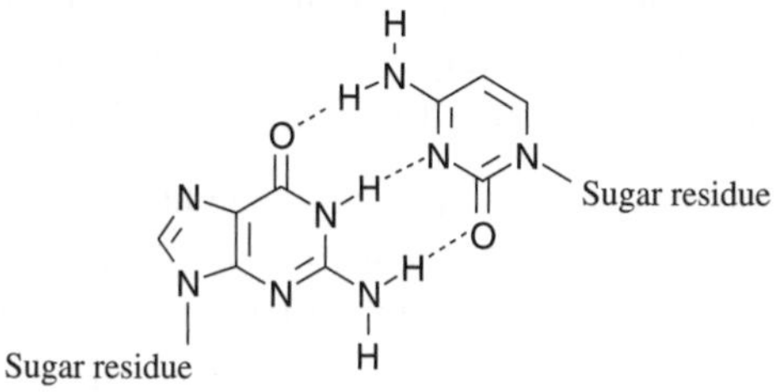

3 (a) TTAGGCATCG, (b) AAUGGCUACG.

4 Replication and acting as a source of genetic information for the synthesis of all the proteins in the body.

5 See Sections 10.1, 10.2 and 10.6. The main differences are:
(a) DNA molecules usually have a very large RMM value compared to RNA molecules.
(b) The structure of RNA contains the sugar residue ribose whilst that of DNA contains the sugar deoxyribose.
(c) RNA molecules consist of a single strand of nucleotides whilst DNA molecules consist of two nucleotide strands in the form of a supercoiled double helix.

6 See Section 10.7.

7 Thymine does not occur in any RNA molecules.

8 The P site is the site where the growing peptide is attached to the ribosome. The A site is the site where the next amino acid–tRNA complex binds to the ribosome prior to the incorporation of the amino acid residue into the growing peptide. The E site is where the empty tRNA molecule exits from the ribosome.

9 The first four codons are not involved in protein synthesis. Protein synthesis starts with AUG and stops with UAA. The peptide coded by the codons between the stop and start signals is:

Met-Pro-Arg-Gly-Gly-Try

10 See Sections 10.11.1. and 10.11.2.

11 See Table 10.4 and section 10.12.2 for evidence. Changes to the substitution pattern of the aromatic ring and the side chain do not lead to analogues with an increased activity.

12 Prepare a series of analogues based on the structure of tetracycline and test their effectiveness against suitable bacterial cultures. It is not easy to design a suitable sequence of analogues and it will vary in detail from worker to worker. However, in this instance a list should include the following general considerations:

(a) A series of compounds with one less ring than tetracycline but retaining the same substituent groups on the relevant rings. These substituent groups should also have the same stereochemistry as the original tetracycline molecule.

(b) The stereochemistry of the groups on tetracycline should be changed.

(c) Minor changes should be made to the positions of functional groups and ring types.

(d) Substituent groups should be changed for functional groups of similar size, shape and electron configuration (see section 3.5).

(e) Substituents should be replaced by either larger or smaller substituents.
     Computer modelling is a useful technique for deciding which substituents should be used (see Chapter 4). Analogues that are more active than tetracycline should be investigated in preference to any other analogues.

13 See Fig. 10.38 and Section 10.13.4.

14 See Section 10.14.4, Nucleic acid synthesis inhibitors. The tetrahydrofuran residue in the structure of the drug does not have a 3'-hydroxy group, which prevents it from forming a bond with a phosphate residue to continue the nucleic acid chain.

15 Proflavine acts by intercalation (see section 10.13.3). The amino groups lock the drug in place by forming ionic bonds with the negative charges of the oxygens of the phosphate groups.

16 See Section 10.13.5.

17 See Section 10.13.1. Examine the chemistry of the biological pathway of the process. Decide which steps offer the best chance of intervention. Synthesise compounds with similar structures to the endogenous compounds being used in the pathway. Use fluorine and thiol groups to replace existing amino and methyl groups.

18 See Section 10.14.2.

19 See Section 10.14.4.

20 See Section 10.15.1.

21 See Section 10.15.2, Gene therapy.

22 The planar anthracene nucleus can fit into the DNA helix (via a major groove). The analogues would be based on this flat structure and groups that are likely to covalently bond to the purine bases and pyrimidine bases. This type of bonding is usually permanent and would therefore prevent replication and transcription of the DNA, leading to cell death.

23 See Section 10.15.2, Monoclonal antibodies.

24 (a) The drug should be more cytotoxic than the normal cytotoxic drugs.

(b) The monoclonal antibody is selective for an antigen that is mainly found on the target cells.

(c) The monoclonal antibody is capable of cell penetration by receptor-mediated endocytosis.

(d) The monoclonal antibody is able to release the drug once the conjugate has reached the target.

Chapter 11

1 (a) See Section 11.1, Fig. 11.2. (b) See Section 11.1.2. (c) See Section 11.4. (d) See Section 11.4.1, Clearance and its significance. (e) See Section 11.5.

2 See Section 11.2.

3 Absolute bioavailability and half-life as a measure of the rate of elimination. These parameters would give an indication of the relative effectiveness of each of the compounds. Absolute bioavailability would indicate the compound with the best absorption characteristics whilst half-life would show which compound was the most stable *in situ* and so have the best chance of being therapeutically effective. They would also indicate which compound would be effective when using a minimum dose (the lower the dose, the lower the chances of unwanted side effects).

4 Frequent small doses, but cyclosporin will be given more frequently than digoxin.

5 Plot a graph of $\log C$ against t. The slope is equal to $k_{el}/2.303$. (a) 1.84 hr^{-1}, (b) 4.12 dm^3, (c) 7.58 dm^3 hr^{-1}. The assumption made is that the elimination exhibits first-order kinetics.

6 At one minute if the clearance rate is 5 cm^3 min^{-1}, 5 cm^3 will be clear of the drug, that is, 5/50 of the drug will have been removed leaving 45 mg of the drug in the compartment. In the next minute another 5/50th of the remaining amount of the drug will be removed as the clearance rate is constant but this will be removed from the 45 mg leaving 40.5 mg, and so on. The figures corresponding to the times are:

Time lapse (minutes):	1	2	3	4	5	6	7	8	9	10
Drug remaining (mg):	45	40.5	36.5	32.8	29.5	26.5	23.8	21.4	19.3	17.4

A logarithmic plot using logarithms to *base 10* of the concentration against time is a straight line with a slope of 0.4584. Assuming a first-order elimination process, the value of k_{el} (See Fig. 5.10) calculated from this slope is:

$$2.303 \times 0.4584 = 1.056$$

and the value of $t_{1/2}$ calculated from k_{el} using equation (5.8) is 0.0656 minutes.

7 Absolute bioavailability is defined by equation (11.40). Calculate the AUC for the 30 mg IV dose by substituting equation (11.28) in equation (11.18). This gives:

$$\text{AUC} = \frac{\text{IV dose}}{V_d k_{el}} = \frac{30}{4.12 \times 1.84} = 3.96$$

Substitute this value in equation (5.39) to give the absolute bioavailability:

$$\text{Absolute bioavailabaility} = \frac{5.01/50}{3.96/30} = 0.76$$

This value indicates that the IV bolus administration gives a significantly better bioavailability than oral administration.

8 The parameters that could be used to compare the biological activities of the analogues are:

(a) Half-life (calculate using equation 5.8), which would give a measure of the duration of the action. The longer the half-life, the longer the time the drug is available in the body.

(b) Absolute bioavailability (calculate using equation 5.39). The bigger the absolute bioavailability, the greater the chance of a favourable biological action.

Analogue:	A	B	C	D
Half-life (minutes):	5	25	15	40
Absolute bioavailability:	1.036	1.526	0.812	1.175

The best analogue is D because it has the longest half-life and a reasonable bioavailability.

9 $C_{ss} = k_0/Cl_P$ and $Cl_P = V_d k_{el}$.
Calculate k_{el} from the value of $t_{1/2}$ (use equation 11.8).

$t_{1/2} = 0.2772\,h^{-1}$.

Convert V_d to cm^3 and calculate the value of Cl_P. Substitute Cl_P in the expression for C_{ss}. Answer: rate of infusion $= 4.35\,\mu g\,cm^3\,h^{-1}$.

Chapter 12

1 (a) See Section 12.1. (b) See Section 12.6. (c) See Sections 12.9.1 and 12.9.2. (d) See Sections 12.8 and 12.9.

2 A chemical species that can interact with water molecules will usually be more water soluble than one that does not interact with water molecules. The formation of hippuric acid increases the interaction of the molecule with water molecules and so is a more water-soluble conjugate. The carboxylic acid group will ionise in water to form an anion that could form ion-dipole bonds with water molecules. In addition the amide link can form hydrogen bonds with water molecules.

3 See Sections 12.1.2 and 12.1.5.

4 See Section 12.2.

5 See Fig. 12 4.

6 See Section 12.6.3. See the answer to question 3. The pH of the solution is correct for kidney excretion.

7 (a)

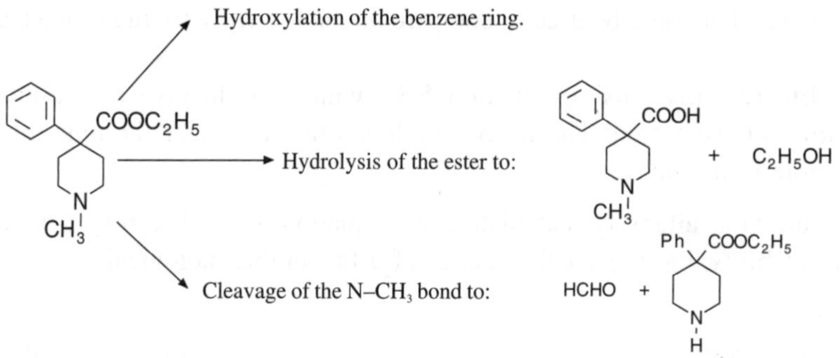

Hydroxylation of the benzene ring.

Hydrolysis of the ester to:

Cleavage of the N–CH₃ bond to:

(b)   Hydroxylation of the benzene ring.

Acylation

Reductive cleavage

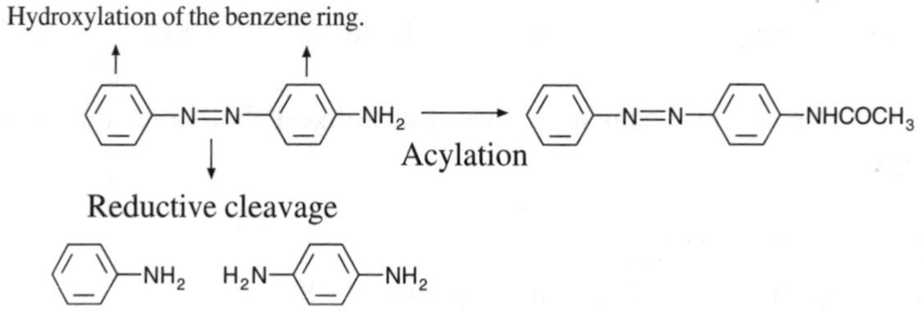

8 A chemical species that can interact with water molecules will usually be more water soluble than one that does not interact with water molecules. There is a considerable interaction between glucuronic conjugates and water molecules because of the ionisable acid group and the numerous hydroxy groups that can hydrogen bond with water. There is a good supply of glucuronic acid in the body, which is capable of reacting with many different functional groups on xenobiotics.

9 (a) To inactive the drug. Metabolism to inert metabolites that are sufficiently water soluble to be readily excreted via the kidney.

(b) The α-carbon of an ethyl group *tert*-amine is hydroxylated and cleaved to form ethanal and the *N*-ethylaminobenzene. Ethanal could be excreted via the lungs or be metabolised further to ethanoic acid. The *N*-ethylaminobenzene could be metabolically oxidised to the corresponding *N*-hydroxy compound or dealkylated to aminobenzene and ethanal.

10 (a) A is metabolised faster than the drug so it does not accumulate in the body.

(b) B to C is the main metabolic route since this is a very much faster process than B to F.

(c) C to D. C will accumulate in the body and if C is pharmacologically active this could pose potential clinical problems for a patient.

11 To avoid fatal overdoses.

12 See Section 12.9.4 (Site specificity) and Fig 12.12. Use an *N*-methyldihydropyridine derivative (B) as carrier. This carrier would require a substituent group that can bond to the dopamine diethanoate. The best group for this purpose is probably a carboxylic acid group since amides are slowly hydrolysed. This means that the prodrug has a good chance of reaching the blood–brain barrier in sufficient quantity to be effective. Once the prodrug has entered the brain it is easily oxidised to the quaternary salt, which, because of its charge, cannot return across the blood–brain barrier.

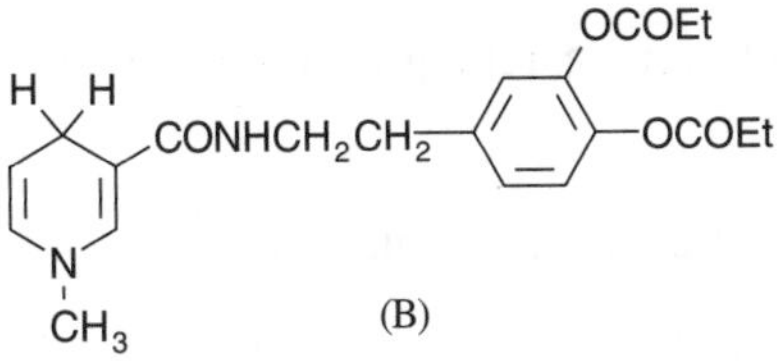

13 See Section 12.9.4, ADEPT.

Chapter 13

1 This is an essay-style question. A suggested approach is to consider the action of metallo complexes from two major points of view, namely: their role in maintaining normal biological functions and their use in drug therapy to treat certain pathological conditions. Examples (see section 13.1): active sites of enzymes, maintaining enzyme structure, treatment of heavy metal poisoning, cancer, rheumatoid arthritis.

2 (a) Cyclopentadienyl, structure in Table 13.2, pentadentate, five-electron donor.

(b) Ethylenediaminetetraacetic acid, structure in Fig. 13.13, hexadentate, 12-electron donor.

(c) Ammonia, NH_3, monodentate, two-electron donor.

3 (a) See Fig. 13.8. The S of the thiol is the most likely but the N of the amino and the O of the carboxylic acid groups could also coordinate.

(b) See Fig. 13.8. The N of the amino and the O of the carboxylic acid groups.

(c) See Fig. 13.23. The ring nitrogen and the O of the phenolic hydroxy group.

(d) See Fig. 13.13. The S of the thiol groups.

4 (a) See Section 13.2.1, The type of bond formed with the metal atom. (b) See Section 13.2.2. (c) See Section 13.2.4.

5 See Section 13.3.2 for definitions.

    (a) Yes (Hard to Hard). S of Cys to metal. (b) No. (c) No. (d) Yes (Soft to Soft), S of thiolate to metal. (e) Yes (Borderline to Borderline), $=N-$ of the imidazole. (f) Yes, OH of cholesterol to metal. (g) No. (h) Yes (Soft to Soft), both alkene and COOH.

6 EDTA. The EDTA removes Ca^{2+} ions from the blood by forming a complex. This inhibits the start of blood clotting.

7 The general requirements for designing a compound for use in metal detoxification are:

    (a) It must be able to act as a multidentate ligand.

    (b) The ligands groups in the structure should be specific for the metal.

    (c) Preferably it should form water-soluble complexes that can be easily excreted.

    (d) The compound should be able to form five- or six-membered rings.

    (e) It should form complexes that are charged in solution in order to prevent side effects.

8 (a) Cu^{2+} forms a more stable complex with EDTA than Ca^{2+} ions. Consequently, the concentration of Cu^{2+} ions will be reduced more than the concentration of Ca^{2+} ions.

    (b) Yes. It will chelate the lead and mercury in preference to the chromium since the $\log K$ values of the EDTA complexes of these metals are higher than that for the EDTA–chromium complex. However, some EDTA–chromium complexes will be formed since all complexes are formed by dynamic equilibria so the patient must be carefully monitored during treatment.

    (c) The concentration of Fe^{3+} ions will be reduced more than the concentration of Fe^{2+} ions since Fe^{3+} ions form a more stable complex with EDTA than Fe^{2+} ions. The addition of an excess of zinc ions has no effect because iron forms a more stable complex than zinc.

9 Three types of experiment must be carried out:

    (a) Determine the growth of the *Staphylococcus aureus* on treatment with the potential drug but in the absence of iron. This should show a normal growth pattern if the presence of iron is essential to the action of the drug.

    (b) Determine the growth of the *S. aureus* in the presence of iron but the absence of the drug. This would show a normal growth rate if the *S. aureus* is not affected by the iron alone.

(c) Determine the growth of the *S. aureus* in the presence of both the drug and iron(III). The growth should be significantly reduced as both of the active factors are present.

10 (a) See Section 13.5.5. (b) Look for an organic structure with suitable ligands that has an absorption in the visible or UV regions. Combine these ligands with a compatible metal using the theory of hard and soft acids and bases to find suitable metals.

Chapter 14

1 (a) Dinitrogen trioxide. (b) Peroxynitrite ion. (c) *S*-Nitrosocysteine. (d) DiethylaminoNONO-ate.

2 It forms:
(a) nitrogen dioxide, (b) dinitrogentrioxide, (c) diethylaminoNONO-ate,

$$
\text{(d)} \qquad
\begin{array}{c}
CH_3CH_2S \\
\\
CH_3CH_2S
\end{array}
\begin{array}{c}
\diagdown \quad \diagup \; NO \\
Fe \\
\diagup \quad \diagdown \; NO
\end{array}
\qquad \text{and (e)} \quad ONNHCH_2COOH.
$$

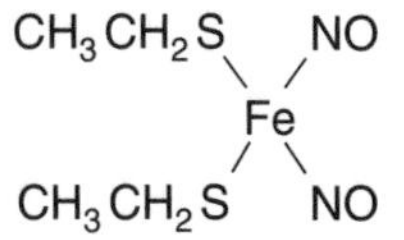

3 Nitrosation: the bonding of a nitroso group (-NO) to a structure (see section 14.3.6). Nitration: the bonding of a nitroso group (-NO) to a structure (see section 14.3.8, Peroxynitrite).

4 Haem iron: main metabolic pathway, protein-bound iron; prevents the normal enzyme function and so is responsible for some pathological conditions.

5 iNOS is not Ca dependent and activation results in the continuous production of NO in a high concentration. cNOS is Ca dependent and produces small amounts of NO in short bursts.

6 iNOS is cytotoxic either in the physiological sense as part of the immune system or in the pathological sense in that it damages tissue. cNOS initiates messenger NO for activating essential biological processes.

7 See Section 14.4.

8 Soluble guanylyl cyclase. GTP cGMP.
Relaxes smooth muscle leading to vasodilation and a reduction in blood pressure.

9 (a) For cases where there is an excess of nitric oxide: drugs that inhibit its production. See Section 14.6.1 for examples.

   (b) For cases where there is a deficiency of nitric oxide: drugs that release nitric oxide. See Section 14.6.2 for examples.

   (c) Genetic approach to control both types of problem. See Section 14.6.3 for examples.

10 See Section 14.5.3.

Chapter 15

1 (a) and (b) See Section 15.2.3. (c) See Section 15.1.

2 See Section 15.2.

3 See Section 15.2.4.

4 See Section 15.3.2.

5 See Section 15.3.3.

6 See Fig. 15.14. Use propyl iodide instead of ethyl iodide in Fig. 15.14.

7 See Section 15.4.1.

(a)

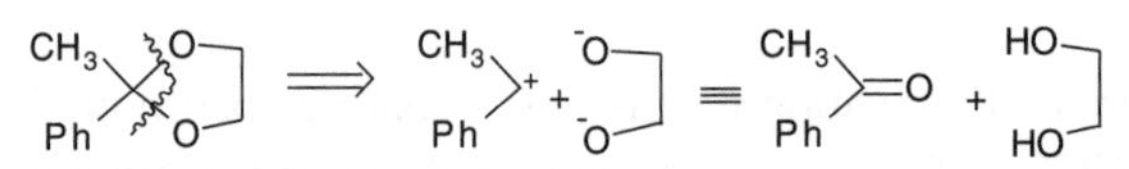

(b)

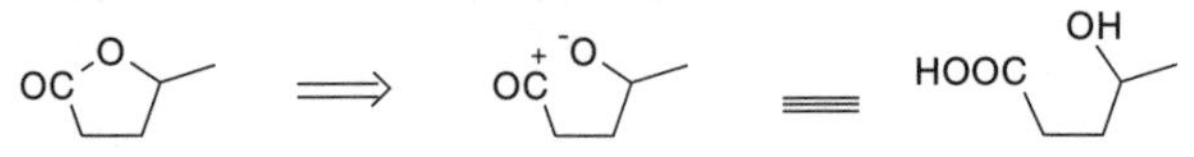

(c)

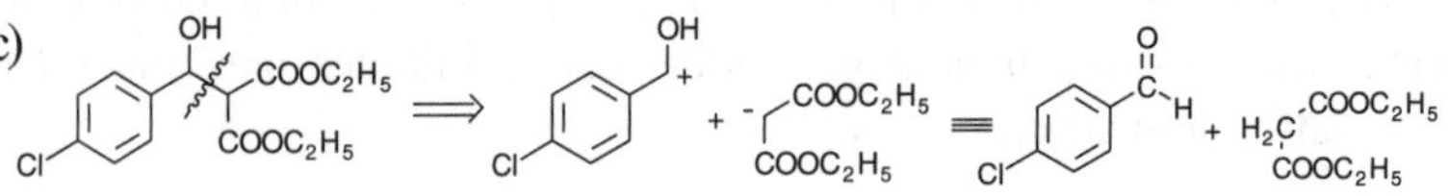

(d) A pericyclic disconnection.

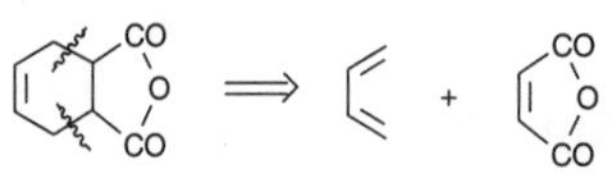

8 See Section 15.4.1.

(a)

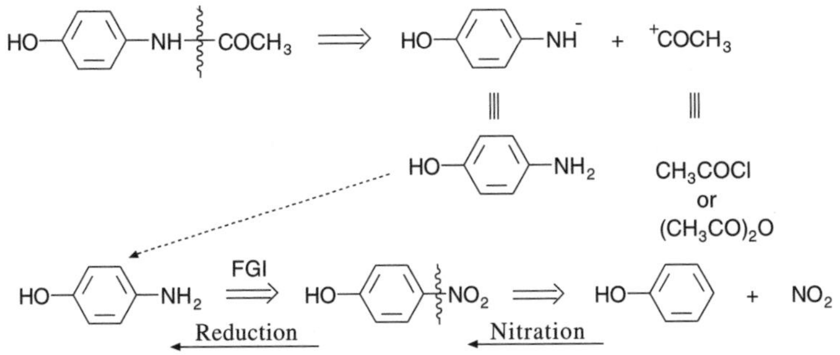

(b)

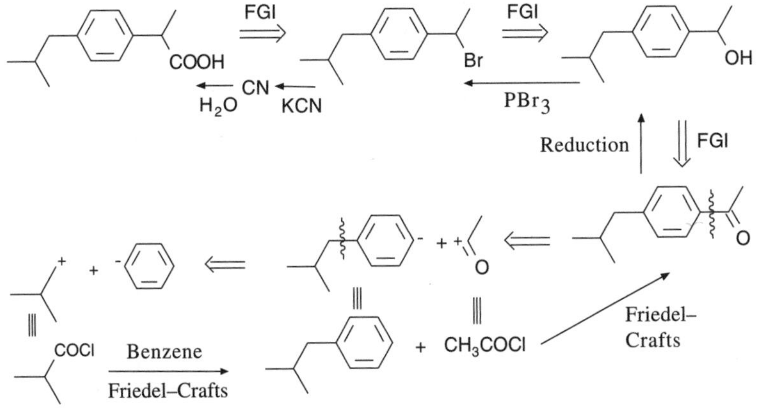

Chapter 16

1 (a) See Section 16.1. (b) List given in Fig. 16.1.

2 (a) See Section 16.5. (b) See Section 16.2.2 and equation (16.1) for a definition. (c) See Section 16.2.3. (d) See Section 16.5. (e) See Section 16.3.

3 See Sections 16.2 and 16.2.4.

4 Use the route given in Fig. 16.5. Use dimethoxypropane to form the dioxan ring (see corresponding stage in Fig. 16.4).

5 (a) See Section 16.3. (b) See Section 16.3.

6 See Section 16.7.

Index